Optimal Water Management and Sustainability in Irrigated Agriculture

Optimal Water Management and Sustainability in Irrigated Agriculture

Editors

Pantazis E. Georgiou
Dimitrios K. Karpouzos

Basel • Beijing • Wuhan • Barcelona • Belgrade • Novi Sad • Cluj • Manchester

Editors

Pantazis E. Georgiou
Aristotle University
of Thessaloniki
Thessaloniki
Greece

Dimitrios K. Karpouzos
Aristotle University
of Thessaloniki
Thessaloniki
Greece

Editorial Office
MDPI AG
Grosspeteranlage 5
4052 Basel, Switzerland

This is a reprint of articles from the Special Issue published online in the open access journal *Agronomy* (ISSN 2073-4395) (available at: https://www.mdpi.com/journal/agronomy/special_issues/optimal_water_management).

For citation purposes, cite each article independently as indicated on the article page online and as indicated below:

Lastname, A.A.; Lastname, B.B. Article Title. *Journal Name* **Year**, *Volume Number*, Page Range.

ISBN 978-3-7258-1519-7 (Hbk)
ISBN 978-3-7258-1520-3 (PDF)
doi.org/10.3390/books978-3-7258-1520-3

Contents

About the Editors

Pantazis E. Georgiou

Dr Pantazis E. Georgiou is a Professor of Irrigation Water Management and Irrigation at the School of Agriculture, Department of Hydraulics, Soil Science and Agricultural Engineering, Aristotle University of Thessaloniki—AUTH (Greece). He received his B.Sc. in Agriculture from AUTH, M.Sc. in Agricultural Hydraulics and Land Reclamation from AUTH, and Ph.D. in Agricultural Engineering and Water Resources from the same university. His primary research interests include irrigation, crop water requirements, irrigation scheduling, precision irrigation, the crop yield response to water, the design and operation of irrigation reservoirs, water resource management, climate change, and droughts. He has participated in several projects regarding the above subjects. He has published a lot of articles in scientific journals and congress proceedings. He has served, among others, as an editorial board member and guest editor of national and international scientific journals, as well as serving as the former President of the Hellenic Association of Agricultural Engineers.

Dimitrios K. Karpouzos

Dr Dimitrios K. Karpouzos is a Professor at the Department of Hydraulics, Soil Science and Agricultural Engineering at School of Agriculture of Aristotle University of Thessaloniki (AUTH). He holds a Diploma and PhD in Civil Water Resources Engineering. His main research interests lie in the field of water resources and irrigation system management and optimization, hydroinformatics, simulation and optimization models in surface and groundwater hydrology, climate change, and drought analysis. He has participated in several national and EU projects, and has published a lot of articles in scientific journals and congress proceedings. He has served, among others, as an editorial board member and guest editor of national and international scientific journals, the former President of the Hellenic Hydrotechnical Union, and is a current board member of the UNESCO Cat. II Center for Integrated and Multidisciplinary Water Resources Management.

Article

Uncertainty Analysis of HYDRUS-1D Model to Simulate Soil Salinity Dynamics under Saline Irrigation Water Conditions Using Markov Chain Monte Carlo Algorithm

Farzam Moghbel [1,2,3], Abolfazl Mosaedi [1,*], Jonathan Aguilar [2,3,*], Bijan Ghahraman [1], Hossein Ansari [1] and Maria C. Gonçalves [4]

1 Department of Water Science and Engineering, Faculty of Agriculture, Ferdowsi University of Mashhad, Mashhad 9177948978, Iran
2 Southwest Research-Extension Center, Kansas State University, 4500 E. Mary St., Garden City, KS 67846, USA
3 Biological and Agricultural Engineering Department, Kansas State University, 1016 Seaton Hall 920 N, Martin Luther King Jr. Drive, Manhattan, KS 66506, USA
4 Instituto Nacional de Investigação Agrária e Veterinária (INIAV), Av. República, 2780-157 Oeiras, Portugal
* Correspondence: mosaedi@um.ac.ir (A.M.); jaguilar@ksu.edu (J.A.); Tel.: +1-(620)-275-9164 (J.A.)

Abstract: Utilizing degraded quality waters such as saline water as irrigation water with proper management methods such as leaching application is a potential answer to water scarcity in agricultural systems. Leaching application requires understanding the relationship between the amount of irrigation water and its quality with the dynamic of salts in the soil. The HYDRUS-1D model can simulate the dynamic of soil salinity under saline water irrigation conditions. However, these simulations are subject to uncertainty. A study was conducted to assess the uncertainty of the HYDRUS-1D model parameters and outputs to simulate the dynamic of salts under saline water irrigation conditions using the Markov Chain Monte Carlo (MCMC) based Metropolis-Hastings algorithm in the R-Studio environment. Results indicated a low level of uncertainty in parameters related to the advection term (water movement simulation) and water stress reduction function for root water uptake in the solute transport process. However, a higher level of uncertainty was detected for dispersivity and diffusivity parameters, possibly because of the study's scale or some error in initial or boundary conditions. The model output (predictive) uncertainty showed a high uncertainty in dry periods compared to wet periods (under irrigation or rainfall). The uncertainty in model parameters was the primary source of total uncertainty in model predictions. The implementation of the Metropolis-Hastings algorithm for the HYDRUS-1D was able to conveniently estimate the residual water content (θ_r) value for the water simulation processes. The model's performance in simulating soil water content and soil water electrical conductivity (ECsw) was good when tested with the 50% quantile of the posterior distribution of the parameters. Uncertainty assessment in this study revealed the effectiveness of the Metropolis-Hastings algorithm in exploring uncertainty aspects of the HYDRUS-1D model for reproducing soil salinity dynamics under saline water irrigation at a field scale.

Keywords: Bayesian; HYDRUS-1D; irrigation; leaching; MCMC; Metropolis-Hastings; prior distribution; posterior distribution; salinity; uncertainty

Citation: Moghbel, F.; Mosaedi, A.; Aguilar, J.; Ghahraman, B.; Ansari, H.; Gonçalves, M.C. Uncertainty Analysis of HYDRUS-1D Model to Simulate Soil Salinity Dynamics under Saline Irrigation Water Conditions Using Markov Chain Monte Carlo Algorithm. *Agronomy* 2022, *12*, 2793. https://doi.org/10.3390/agronomy12112793

Academic Editors: Pantazis Georgiou and Dimitris Karpouzos

Received: 13 September 2022
Accepted: 7 November 2022
Published: 9 November 2022

Publisher's Note: MDPI stays neutral with regard to jurisdictional claims in published maps and institutional affiliations.

1. Introduction

The severity of water shortage for irrigated agriculture is becoming a global catastrophe. On the other hand, the sustainability of agricultural systems and water resources has been extremely challenging due to the rising competition in water demands of the agricultural, industrial, and municipal sectors [1]. Irrigation with saline waters could be a proper alternative to freshwater to sustain the agricultural industry in arid and semi-arid regions facing different water scarcity levels [2]. These waters are mainly obtained from agricultural drainage water, municipal wastewater, and low-quality groundwater. It has been indicated in several studies that when saline water is used for irrigation purposes, special focus should be considered on controlling salinity accumulation in the crops' root zone to achieve long-term productivity in the region [3–6]. The unsuitable management of saline water irrigation can potentially restrict crop water and nutrient uptake due to the inducing salinity build-up in the soil [7]. The salinity distribution and its levels depend on interactions of irrigation, rainfall, evapotranspiration, and the drainage condition of the agricultural fields [8]. Thus, implementing an effective method to alleviate concentrated salinity in the crops' root zone is crucial. Leaching excessive salts downward from crops' root zone through the application of more water than the crops' water requirement during the growing season has been an effective method for salinity control. Leaching application is a critical factor in managing soil-soluble salts brought by saline water irrigation [9]. Therefore, balanced management between applying extra water to control salinity and conserving groundwater resources due to additional water withdrawal is expected. The ratio of water that passes the root zone to the amount of irrigation water is called the leaching fraction (LF) [9]. The minimum leaching fraction of irrigation water with particular quality that needs to be applied over the growing season to keep soil salinity below the crops' salinity thresholds is called the leaching requirement (LR) [10]. To determine the correct value of LR, steady-state and transient-state are two existing approaches and introduced methods in the literature. In steady-state analysis, specific assumptions have been made that include the continuous downward flowing of irrigation water at a constant rate, constant evapotranspiration rate during the growing season, and constant soil soluble salt concentration at any point [8,11]. Comparing field observations proves that none of these assumptions are realistic [8,11]. Thus, using transient-state methods is preferable to compute suitable LR values when the source of irrigation water is saline. One of the well-distinguished transient numerical models is the HYDRUS-1D, model which has the capability of simulating water flow and transport of solutes and ions in unsaturated conditions of soils [12]. The HYDRUS-1D along with its solute transport and root water uptake modules have been proven to be a reliable model for investigating management scenarios regarding long-term and short-term effects of using marginal quality waters as irrigation water on soil salinity and water content [13,14]. Gonçalves et al., 2006 have shown that the HYDRUS-1D model successfully simulated soil water content, overall salinity, and soluble ions over and out of the irrigation season. The model was able to reasonably describe the contribution of rainfall in leaching soluble salts deposited by irrigation during the season [15]. Noshadi et al., 2020 used soil column lysimeters to calibrate and validate the HYDRUS-1D model under different controlled groundwater depths. They revealed that statistical indices of normalized root mean square error (NRMSE) and degree of agreement (d) values were 9.6% and 0.64 for simulating soil water content, and 6.2% and 0.98 for simulating soil salinity, respectively [16]. It has been shown by Phogat et al., 2010 that the HYDRUS-1D predictions of water percolation and soil electrical conductivity of sandy loam soil in mini lysimeters study were not statistically different from the measurements [17]. Salma et al., 2019 have investigated the reliability of the HYDRUS-1D model in terms of investigating the effects of different irrigation regimes, including full and deficit irrigation, on root water uptake and root zone salinity. Their results indicated good accuracy of the model by achieving the root mean square error (RMSE) values of 0.008 m^3 m^{-3} and 0.28 dS/m for simulating soil water content and soil water electrical conductivity (ECsw), respectively [18]. Liu et al., 2019 calibrated and validated the HYDRUS-1D model to investigate the soil salt dynamic of the winter wheat–summer maize root zone irrigated with brackish water in a semi-arid region. The results have shown the model's good performance regarding soil water content (SWC) and ECsw simulations. The RMSEs for simulating SWC and Ecsw were in the range of 0.017–0.04 cm^3 cm^{-3} and 0.059–0.069 dS/m, respectively. In addition, the coefficient of determination (R^2) values were from 0.78 to 0.92 for SWC simulations and from 0.67 to 0.89 for Ecsw simulations [19]. Helalia et al., 2021 have simulated water use and soil salinity of drip-irrigated almond and pistachios root zones by the HYDRUS-1D model under different irrigation water salinities for multiple locations in San Joaquin Valley (SJV), California. Their results have shown good accuracy of the model as NRMSE values between simulated and measured volumetric water content varied from 0.033 to 0.28. Moreover, the coefficient of correlation (R) values varied from 0.4 to 0.72 for ECsw, showing erratic behavior of the model in different locations due to multiple factors such as the non-uniformity of root and water redistributions under drip irrigation [20]. A challenge to obtaining acceptable outputs of the model is the calibration of its parameters, which is usually time and money consuming. Inverse modeling has been known as a convenient technique for calibrating HYDRUS-1D based on observational data [21–24]. Nevertheless, estimating model parameters are still associated with uncertainties that are reflected in model outputs. To achieve valid results from the model, it is necessary to quantify this type of uncertainty [25]. To date, very few studies have explored the uncertainty analysis of HYDRUS-1D parameters, particularly for simulating the salinity dynamic of the soil under saline irrigation conditions [26–28]. This study, by pursuing the Bayesian theorem, investigates the conditional probability distribution of the HYDRUS-1D parameters, grounded on the measurements. The Bayesian theorem consists of three main terms: (a) the likelihood function that incorporates the model outputs and measurements to identify the errors, (b) prior probability that covers all of the uncertainty without considering measurements (initial information), and (c) the posterior probability that combines the previous knowledge (prior) and new information (likelihood) to gain uncertainty of the model [29,30]. Various methods have been implemented to determine the uncertainty analysis of the models based on Bayesian inferences. Among these methods, those based on Markov Chain Monte Carlo (MCMC) approach have gained significant popularity in studies, particularly in hydrology [31–36]. The MCMC has been built based on formal Bayesian inference concepts to estimate the models' parameters and quantification of uncertainty. The MCMC creates a Markov chain whose target distribution is posterior distribution. The Markov chain illustrates the sampling design for simulating Monte Carlo [34,37]. Kunnath-Poovakka et al., 2021 have investigated the parameters' uncertainty of a hydrological model known as the Australian Water Resource Assessment Landscape (AWRA-L) model using an MCMC method and incorporating remotely sensed evapotranspiration and soil moisture data [38]. Yang et al., 2020 have taken advantage of the MCMC approach to develop a method for drought risk assessment under uncertainty [39]. Xu et al., 2018 adopted an MCMC-based algorithm to estimate the parameters and uncertainty of two-parameter non-stationary Lognormal distribution to analyze flood frequency in a river basin [40]. So far, most of the attention has been on implementing these Bayesian theorem-based methods for hydrological problems and few studies have been accomplished to use these approaches for addressing problems in irrigation science. The main aim of this study is to couple a Bayesian MCMC algorithm known as Metropolis-Hastings (M-H) with the numerical model of HYDRUS-1D to quantify the uncertainty of the parameters of water flow, solute transport, and root water uptake modules of the model for simulating soil salinity dynamics over the growing season and out of the growing season in under field conditions.

2. Materials and Methods

2.1. Site Description and Experimental Data

The field experiments were conducted in Alvalade do Sado, Alentejo, Portugal [15]. Soil monoliths were built for the aims of the study. The monoliths had 1.2 m^2 × 1.00 m deep dimensions, which were isolated by plastic to restrict the lateral seepage of water and solute. The soil surface was exposed to atmospheric conditions, and the bottom had free drainage conditions. The vegetation cover was annual spontaneous plants such as gramineous, leguminosae, and compositae. To measure soil water content and pore ware salinity, TDR probes using waveguides from the Trase System (Soil Moisture Equipment Corp., Goleta, CA, USA) and ceramic suction cups were installed at 10, 30, 50, and 70 cm, respectively. The electrical conductivity (EC) of extracted water, using ceramic suction cups, was analyzed as soil water salinity (ECsw). The experiments were conducted for two growing years (505 days), starting on 15 May 2003, to 30 September 2004. The first irrigation season was started on 29 May 2003 and ended on 20 August 2003. The irrigation events of the next growing season were initiated on 23 June 2004 and continued until 20 August 2004. The monoliths were exposed to atmospheric conditions for the days out of irrigation seasons. It should be mentioned that no rainfall occurred from 21 August 2003 until 1 November 2003, and just one rainfall occurred from 21 August 2004 until 30 September 2004. The crop's daily evapotranspiration and precipitations during the experiments are presented in Figure 1, and the irrigation seasons are specified with orange bars.

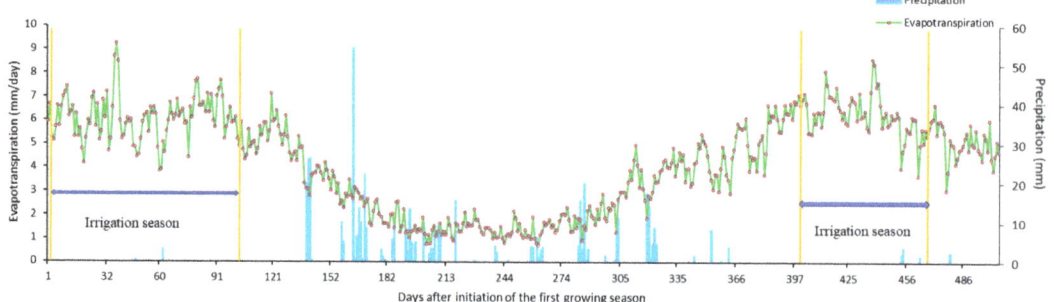

Figure 1. The daily crop evapotranspiration and the precipitations during the experiments starting on 29 May 2003 until 30 September 2004. Orange bars are indicators of irrigation seasons.

In addition, Table 1 shows the average values of weather variables for the study period. The electrical conductivity of irrigation water (ECiw) was 1.6 dS/m (Table 2). Irrigation scheduling followed a fixed 10 mm irrigation depth for each irrigation event during the growing season. The soil profile was refilled whenever the depletion of total soil water reached 10 mm based on soil water content readings. Therefore, variable irrigation intervals were pursued in this study. The monoliths were irrigated using a surface irrigation method similar to the basin irrigation system [15]. The amount of seasonal irrigation depth was 500 mm for each growing season. The total amount of rainfall was 323.8 mm, and 170.1 mm for our experimental period in 2003 and 2004.

Table 1. The average values of the weather parameters for study period in 2003 and 2004 near Alvalade do Sado, Alentejo, Portugal.

Parameter	2003	2004
Tmax (°C)	27.66	26.06
Tmin (°C)	12.31	10.36
Taverage (°C)	19.36	17.58
Relative Humidity$_{average}$%	67.97	70.05
Solar radiation (kj/m^2)	17,789.04	18,675.11
Wind speed (m/s)	2.31	2.20
Precipitation (mm)	323.8	170.1

Table 2. Chemical characteristics of the irrigation water [15].

EC dS/m	SAR (mmol$_{(c)}$L^{-1})$^{-0.5}$	Ions mmol$_{(c)}$L^{-1}				USSL Classification
		Ca^{2+}	Mg^{2+}	Na$^+$	Cl$^-$	
1.6	3.0	3.16	6.32	6.52	16	C3-S1

EC = Electrical Conductivity, SAR = Sodium Adsorption Ratio, USSL = United State Salinity Laboratory.

The soil texture was silty loam from the soil surface to 85 cm depth and loam from 85 cm to 170 cm soil depth. The physical and chemical characteristics of the soil obtained through laboratory analysis including are presented in Table 3.

Table 3. The physical and chemical characteristics of the soil [15].

Parameter	Soil Depth		
	0–48 cm	**48–85 cm**	**85–170 cm**
Coarse sand (%)	6.2	5.1	6.1
Fine sand (%)	53.2	46.8	42.8
Silt (%)	29.6	29.2	28.2
Clay (%)	11	18.9	22.9
Texture	Silty-Loam	Silty-Loam	Loam
Bulk density (g·cm^{-3})	1.49	1.51	1.61
pH	5.94	6.58	6.74
CEC, mmol$_{(c)}$·kg^{-1}	69.96	87.66	92.30

CEC = Cations exchange capacity.

In addition, the initial chemical conditions and the volumetric soil water content of the soil profile of the monoliths are presented in Table 4. As it has been shown, the average soil water salinity (ECsw), soil water content (θv), and sodium adsorption ratio (SAR) were 0.6 dS/m, 0.276 cm^3 cm^{-3}, and 1.92 (mmol$_{(c)}$L^{-1})$^{0.5}$, respectively.

Table 4. The initial conditions of the soil profile of the monoliths [15].

Parameters	Soil Depths			
	0–10 cm	**10–30 cm**	**30–50 cm**	**50–100 cm**
θv (cm^3 cm^{-3})	0.228	0.275	0.296	0.305
ECsw (dS·m^{-1})	0.182	0.32	0.55	1.35
Soluble cations (meq·L^{-1})				
Na$^+$	1.24	2.16	1.54	3.16
Ca^{2+}	0.38	0.58	2.12	4.48
Mg^{2+}	0.34	0.62	0.6	4.28
SAR (meq·L^{-1})$^{0.5}$	2.07	2.79	1.32	1.51

θv = volumetric soil water content, ECsw = electrical conductivity of soil water, SAR = sodium adsorption ratio.

The meteorological data were obtained from the weather station located 10 m from the experimental field and the Penman-Monteith method, Allen et al., 1998 was used to calculate the daily evapotranspiration [41]. To conduct ET partitioning for evaporation and transpiration, the average leaf area index (3.5 m^2 m^{-2}) and its corresponding soil cover factor (SCF) were used.

More details concerning experimental setup and study conditions can be found in Gonçalves et al., 2006.

2.2. HYDRUS-1D Model Simulations

The simulations were accomplished for the first 100 cm of the soil profile where the monoliths were built to prevent lateral water flow.

2.2.1. Water Flow and Root Water Uptake Modeling

The HYDRUS-1D model simulates water flow in the soil under saturated and unsaturated conditions by numerically solving the Richards equation (Equation (1)) using the Galerkin finite element method:

$$\frac{\partial \theta}{\partial t} = \frac{\partial}{\partial z}\left[K(h)\left(\frac{\partial h}{\partial z}\right) - K(h)\right] - S(z, t) \tag{1}$$

where θ is volumetric water content ($L^3 L^{-3}$), h is suction head (L), t is time (T), z is vertical coordinates (L), K(h) is soil unsaturated hydraulic conductivity ($L T^{-1}$), and S(z,t) is sink term known as root water uptake ($L^3 L^{-3} T^{-1}$).

The relationship between h and θ is described by the Van Genuchten-Mualem function (Equation (2)) as follows [42]:

$$\theta(h) = \frac{\theta_s - \theta_r}{1 + (-\alpha(h)^n)^m} + \theta_r \quad (2)$$

where θ_s is the saturated soil volumetric water content ($L^3 L^{-3}$), θ_r is the residual soil volumetric water content ($L^3 L^{-3}$), and α, n, and m are empirical parameters (m = 1 − 1/n).

The unsaturated hydraulic conductivity (Equation (3)) is described as a function of h [42]:

$$K(h) = K_s S_e^l \left[1 - \left(1 - S_e^{\frac{1}{m}} \right)^m \right]^2 \quad (3)$$

where K_s is the saturated hydraulic conductivity ($L T^{-1}$), l is the empirical parameter, and Se is the effective saturation (Se = $(\theta - \theta_r)/(\theta_s - \theta_r)$).

The root water uptake or the sink term in Richards' equation (Equation (1)). It represents the volume of water removed by plants' roots (Equation (4)) from the unit of soil per time unit [43]:

$$S(z) = \beta(z)\alpha(h, \pi)T_p \quad (4)$$

where S(z) is the sink term in Richard's equation ($L^3 L^{-3} T^{-1}$), $\beta(z)$ is the normalized root density distribution (L^{-1}), Tp is the potential transpiration rate ($L^3 L^{-2} T^{-1}$), $\alpha(h,\pi)$ is the dimensionless root water and salinity stresses response function.

The normalized linear root density distribution function (Equation (5)) based on root depth is as follows:

$$\beta(z) = \begin{cases} -0.0067z + 1, \ 0 < z < 30 \text{ cm} \\ -0.0051z + 0.359, \ 0 < z < 70 \text{ cm} \\ 0, \ z > 70 \text{ cm} \end{cases} \quad (5)$$

where z is root depth (L). Based on the observations in this study, 80% of root density was distributed in the first 30 cm of soil root zone, and the remaining 20% root density was extended to 70 cm soil depth [15].

The combined water and salinity stress response function is as follows [43,44]:

$$\alpha(h, \pi) = \frac{1}{1 + \left(\frac{h}{h_{50}} \right)^{P1}} [1 + b(\pi - a)] \quad (6)$$

where h is the soil water pressure head (L), b is the slope of water uptake reduction as a function of the average root zone salinity (-), and a is the threshold of plant root water uptake to average root zone salinity (dS/m), π is root zone average salinity (dS/m), and P1 and h_{50} (L) are empirical crops, soil, and climate-based parameters. The h_{50} is known as the water pressure head, which reduces the root water uptake by half [44].

The partitioning of evaporation and potential transpiration can be computed by application of Beers' law (Equation (7)) to calculate potential evapotranspiration as follows [45]:

$$T_p = ET_p \left(1 - e^{-k*LAI} \right) \quad (7)$$

where ET_p is potential evapotranspiration ($L T^{-1}$), and k is the dimensionless coefficient of radiation attenuation which was considered equal to 0.4 [14,46,47].

2.2.2. Solute Transport Modeling

The advection-dispersion equation was solved numerically in HYDRUS-1D as follows [48]:

$$\frac{\partial \theta c}{\partial t} = \frac{\partial}{\partial z}\left(\theta De \frac{\partial c}{\partial z}\right) - \frac{\partial(q_w c)}{\partial z} \tag{8}$$

where θ is the volumetric soil water content, c is the solute concentration (M L^{-3}), q_w is the soil water flux (L^3 L^{-2} T^{-1}), z is the soil depth (L), and De is the effective dispersion coefficient [L^2 T^{-1}].

The effective dispersion coefficient is the combination of diffusion and dispersion coefficients [48]:

$$D_e = D_l^s + D_{lh} \tag{9}$$

where D_l^s is the effective diffusion coefficient [L^2 T^{-1}] and D_{lh} is the coefficient of hydrodynamic dispersion [L^2 T^{-1}].

$$D_{lh} = \lambda v \tag{10}$$

where λ is a proportionality constant called dispersity [L] and v is the average pore water velocity [L T^{-1}]. The v is obtained from the results of water flow flux [L^3 L^{-2} T^{-1}].

$$D_l^s = \frac{\theta^{\frac{7}{3}}}{\theta_s^2} D_l^w \tag{11}$$

where θ is the volumetric soil water content, θ_s is the volumetric saturated soil water content, and D_l^w is the diffusion coefficient in free water.

$$l = \frac{\theta^{\frac{7}{3}}}{\theta_s^2} \tag{12}$$

where l is known as the tortuosity factor (l).

2.3. The HYDRUS-1D Model Setup

The simulations of soil salinity dynamics have been conducted by the HYDRUS-1D model for soil profile 100 cm in the monoliths. As the soil texture was relatively uniform (Table 3) for the observational points at 10, 30, 50, and 70 cm soil depths, the soil material and layer were considered equal to 1. The simulations were conducted on a daily basis for 505 days (from 15 May 2003 to 30 September 2004). To perform the water flow simulation and describe the unsaturated soil hydraulic properties, the Van Genuchten-Mualem (Equations (1) and (2)) hydraulic model was chosen. The initial values of the water flow parameters for calibration and uncertainty purposes were determined by the default values in the HYDRUS-1D library for the silt loam soil. The atmospheric boundary condition with the surface layer was chosen to represent the study condition at the soil surface. The atmospheric boundary condition was determined using time series irrigation, precipitation, evaporation, and transpiration fluxes. The free drainage boundary condition was used for the bottom of the soil profile. The soil adsorption was negligible based on the field observations [15]. Thus, the adsorption was eliminated by selecting the linear equilibrium adsorption model and assigning the zero value to the distribution coefficient [48]. The upper and bottom boundary conditions for the Advection-Dispersion equation (Equation (8)) were concentration flux and zero concentration, respectively. The root water uptake simulations were carried out by the S-Shape [49] model, and simultaneously the slope and threshold salinity stress reduction function was taken into account in a multiplicative approach (Equation (6)). Moreover, the initial condition was determined by the soil water content and ECsw measurements at 10, 30, 50, and 70 cm as presented in Table 4.

To benefit from two years of observational data, the uncertainty analysis was undertaken for the first year of the study: 15 May 2003 to 22 June 2004, which included the growing season and out of growing season period. The outputs of the M-H algorithm as calibration results were validated for simulating ECsw and soil water content in long-term aspect, with continues observational data from 15 May 2003 until 30 September 2004. The simulations were carried out for the period of the beginning of the first growing season until 45 days after end of the second growing season in the validation process.

2.4. Uncertainty Analysis and Metropolis-Hastings

The term uncertainty has been defined as the degree of confidence in the decision-making process based on the target outputs from the models [25]. The Bayesian inference is convenient in terms of predicting the models' parameters (posterior distribution) by operating the uncertainty analysis grounded on previous knowledge regarding the model parameters (prior distribution), observational data, and likelihood function. The Bayesian concept or inference is demonstrated as follows:

$$P(\theta|Y) = \frac{L(\theta|Y)P(\theta)}{\int L(\theta|Y)P(\theta)d\theta} \tag{13}$$

where θ is the parameter, Y is the observational data, $P(\theta \mid Y)$ is the posterior distribution, $L(\theta \mid Y)$ is the likelihood function, and $P(\theta)$ is the prior distribution. The likelihood is computed from the probability distribution of residuals between observations (Y). The residuals are frequently assumed uncorrelated, normally distributed, and independent [50]. Thus, the outcoming likelihood function is as follows:

$$L\left(\theta|Y, \sigma^2\right) = \prod_{i=1}^{n} \frac{1}{\sqrt{2\pi\sigma_i^2}} \exp\left(-\frac{1}{2}\left(\frac{Y_i(\theta) - Y_i}{\sigma_i}\right)^2\right) \tag{14}$$

where σ is the standard deviation of model errors, Y_i is the observational data, and $Y_i(\theta)$ is the model output. To gain more numerical stability and simplicity, the logarithmic form of the likelihood function was used in this study:

$$L(\theta|Y, \sigma^2) = -\frac{n}{2}\log(2\pi) - \sum_{i=1}^{n}\log(\sigma_i) - \frac{1}{2}\sum_{i=1}^{n}\left(\frac{Y_i(\theta) - Y_i}{\sigma_i}\right)^2 \tag{15}$$

To conduct an uncertainty analysis of the HYDRUS-1D model for simulating solute transport under saline irrigation conditions, we used the posterior distribution of the following parameters:

Water flow parameters = $[\theta_r, \theta_s, \alpha, K_s, n, l]$;
Solute transport parameters = $[\lambda, D_l^w]$;
Salinity Stress for root water uptake = $[a, b]$;
Water Stress for root water uptake = $[h_{50}, P1]$.

The uncertainty analysis was performed in the Bayesian framework using the Metropolis-Hastings (M-H) algorithm for the Markov Chain Monte Carlo method combined with Gibbs sampling. To execute the M-H algorithm, the initial step is identifying the proposal distribution to generate new parametric candidates. In this study, we assumed the priors to be uniformly distributed [51] over space, to assign the same probability to all possible values of parameters. Hence, uniform distribution was considered as the proposal distribution to generate the samples using the Monte Carlo approach. Therefore, the Metropolis-Hastings algorithm was applied as follows [52]:

1. Determine the length of the Markov Chain, T
2. Draw an arbitrary initial candidate (θ_i)
3. Calculate the proposed initial candidate density: $P(\theta_i \mid Y) = L(\theta_i \mid Y)*P(\theta_i)$
4. For n = 2, ... , T do:
5. Generate a candidate (θ_t) from proposal distribution $(q(\theta))$.
6. Calculate the target density: $P(\theta_t \mid Y) = L(\theta_t \mid Y)*P(\theta_t)$

7. Calculate the Random Walk Metropolis acceptance criteria:

$$\alpha = \min\left[1, \frac{L(\theta_t|Y)P(\theta_t)}{L(\theta_{t-1})*P(\theta_{t-1})}\right] \tag{16}$$

8. Generate a random sample from the uniform distribution as U(0,1);
9. If $U \le \alpha$ then $\theta_n = \theta_t$;
10. Else $\theta_n = \theta_{t-1}$;
11. End of the loop.

To achieve the goals of the study, the M-H algorithm was implemented in R-Studio V 1.4.1717 (Posit Corp, Boston, MA, USA) environment for the HYDRUS-1D model. To apply M-H, the model was run 100,000 times. The sets of parameters that resulted in the non-convergence of the model were discarded for any kind of analysis. To combine the Gibbs sampling algorithm, the parameter sampling was carried out one at a time and the rest of the parameters were treated like constant values. Then, metropolis acceptance criteria were evaluated for the sampled parameter at the time.

Prior information (Table 5) about the parameters was obtained from the HYDRUS-1D library, Maas and Hoffman, 1977, Grieve et al., 2012, Skaggs et al., 2006, and measurements were undertaken by Gonçalves et al., 2006 [15,44,53,54].

Table 5. The prior distribution of the parameters used in HYDRUS-1D.

Parameters	Units	Min	Max	Mean	SD	CV
θ_r	-	0.05	0.08	0.065	0.009	0.133
θ_s	-	0.3	0.5	0.400	0.058	0.144
α	1/cm	0.001	0.2	0.101	0.057	0.572
K_s	cm/days	18	100	59.000	23.671	0.401
n	-	1	3	2.000	0.577	0.289
l	-	0.1	1	0.550	0.260	0.472
λ	cm^2/day	5	30	17.500	7.217	0.412
$D_l{}^w$	cm^2/day	1	2	1.500	0.289	0.192
a	dS/m	8	14	11.000	1.732	0.157
b	%	2	7	4.500	1.443	0.321
h_{50}	cm	−5000	−800	−2900	1212.436	−0.418
P1	-	1.5	3	2.25	0.433	0.1924

SD = standard deviation, CV = Coefficient of variation.

In addition, in this research, the relative sensitivity of the parameters was analyzed for the observational data by comparing the coefficient of variation (CV) of parameters' posterior distributions generated by the M-H algorithm. Those parameters with higher values of CV were more sensitive to the soil salinity dynamic under saline irrigation water.

To determine the HYDRUS-1D model output (predictive) uncertainty, the model was run for 1000 parameter sets obtained from 95 confidence intervals (CI) of the parameter's posterior distribution. To have an overall predictive uncertainty analysis for each soil depth, the p-factor (percent of observation data covered by 95 CI band) and r-factor (average width of 95 CI band divided by the standard deviation of observations) were computed [55,56]. To calculate the r-factor, the following formula was used:

$$r - \text{fatctor} = \frac{\frac{1}{K}\left(\sum_{i=1}^{K}(q_U - q_L)\right)}{\sigma} \tag{17}$$

where K is the total number of observations, q_U is the model outputs (salinity) for 97.5% quantile, q_L is the model outputs for 2.5% quantiles, and σ is the standard deviation of observation errors.

For the final step in uncertainty assessments, it seemed interesting to analyze the performance of the model in terms of simulating soil water content (SWC) that accounts for model outputs in terms of soil water movement and soil water uptake under saline water irrigation. The 50% quantiles of the parameters' posterior distributions were used for this purpose. The root mean square error (RMSE), normalized root mean square error (NRMSE), and coefficient of agreement (d):

$$RMSE = \sqrt{\frac{1}{n} \sum_{i=1}^{n} (M_i - S_i)^2} \tag{18}$$

$$NRMSE = \frac{RMSE}{M} \tag{19}$$

$$d = 1 - \left(\sum_{i=1}^{n}(S_i - M_i)^2\right)/\left(\sum_{i=1}^{n}\left(|S_i - \overline{M}| + |M_i - \overline{M}|\right)^2\right) \tag{20}$$

where is M_i, S_i, and $\overline{M}$ are observational values, simulated values, and the average value of measurements. Values of d close to 1 indicate a good performance model. The RMSE and NRMSE close to zero indicate good matching between simulated values and observations. The NRMSE ranges of <10%, 10–20%, 20–30%, and >30% evaluates the model performance as excellent, good, fair, and poor calibration [57].

3. Results

The 100,000 simulation runs of the HYDRUS-1D model after discarding the parameter sets resulted in 5000 iterations of each parameter for the MCMC algorithm. The uncertainty analysis results are presented in this section that assess the probability of the model parameters and outputs for simulating the salinity (ECsw) distribution during and out of the growing season.

3.1. Parameters Posterior Distribution

The posterior distributions of the model parameters represent the existing uncertainty in the model parameters after combining the prior information and observational data through the Bayes theorem. Therefore, comparing parameters priors and posteriors could be beneficial (Tables 5 and 6). As it was mentioned in the previous section, the priors were obtained mainly from the HYDRUS-1D model database and other generic values in the literature. However, the posteriors are specific values for the parameters for a specific field, the vegetation cover, and the region's prevailing climate. The average value of acceptance rate (AR), which is a fraction of accepted parameter samples to proposed ones, was 0.38. Thus, the AR value was in the acceptance range.

Table 6. Statistical indices of the model posterior distribution obtained from the M-H algorithm.

Parameters	Min	Max	Mean	SD	CV
θ_r	0.065	0.079	0.076	0.003	0.040
θ_s	0.435	0.499	0.490	0.013	0.026
α	0.005	0.02	0.010	0.003	0.303
K_s	19	99.127	86.080	11.578	0.135
n	1.295	2.75	1.394	0.178	0.128
l	0.293	0.894	0.540	0.087	0.161
λ	5.011	29.99	17.150	7.270	0.424
D_l^w	1	2	1.501	0.292	0.195
a	8	13.99	11.030	1.763	0.160
b	2.001	6.99	4.516	1.441	0.319
h_{50}	−802.359	−800	−800.755	0.636	−0.001
P1	2.776	3	2.929	0.035872	0.012247

SD = standard deviation, CV = Coefficient of variation.

The mean values of the parameters' posteriors were noticeably different from their corresponding prior values. The range of the parameters shown by min and max values was reduced, and a significant reduction in the standard deviation (SD) and coefficient of variation (CV) was observed as well. As it is shown, for instance, the prior range of α was from 0.001 to 0.2, however, its posterior range was from 0.005 to 0.02, which was more limited than its prior values. The prior mean value of α was 0.101, which was changed to 0.01 for its posterior value. In addition, the SD value of α was changed from 0.057 to 0.003. Very similar to α considerable changes were found for other parameters for water flow simulation. However, the parameters that describe diffusion and hydrodynamic dispersion in the advection-dispersion equation ($D_l{}^w$ and λ) and threshold and slope of salinity stress water uptake reduction (a and b) did not noticeably change from priors. Figures 1–3 indicate the histogram of the posterior distribution of parameters for water flow, solute transport, and root water uptake simulations. It should be mentioned that the x-axes of the histograms were considered to be equal to priors.

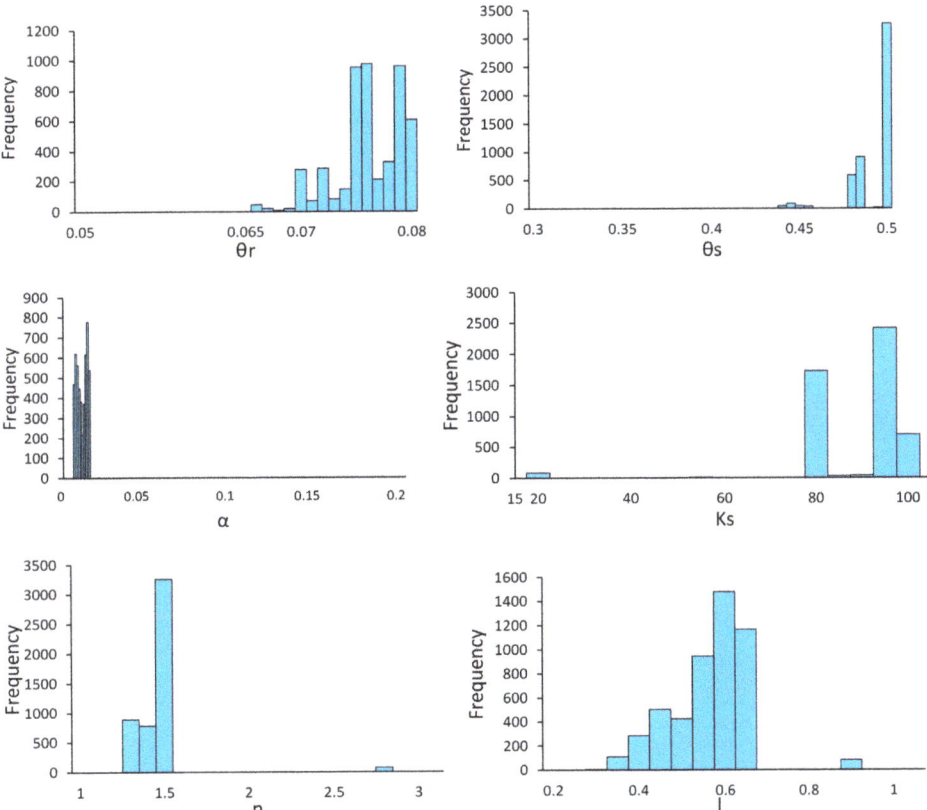

Figure 2. Posterior distribution of soil water flow parameters: θ_r = residual soil water content, θ_s = saturated soil water content, K_s = is saturated hydraulic conductivity, l = tortuosity factor, α = empirical parameter, and n = empirical parameter. The x-axes of the plots are fixed to prior distribution (range) of the parameters.

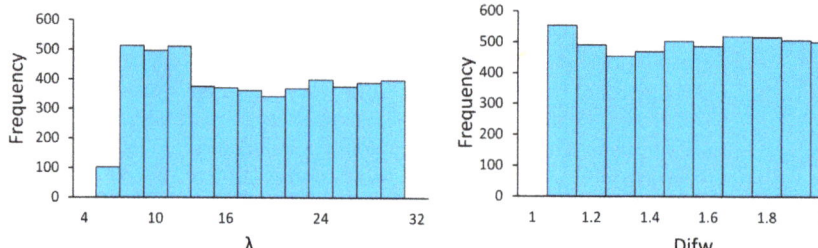

Figure 3. Posterior distribution of solute transport parameters: Difw = diffusion coefficient and λ = dispersivity. The *x*-axises of the plots are fixed to prior distribution (range) of the parameters.

As it is illustrated, the posterior distributions of the parameters are different from their priors. The posterior distributions of water flow simulation as well as water stress reduction function parameters occupied small portion of their prior values. Thus, the parameters related to the advection process in the solute transport phenomenon have been estimated with high confidence (low uncertainty). These posterior distributions confirmed that the M-H algorithm was able to estimate the water flow parameters of the model with a low uncertainty level [58]. Nevertheless, the posterior distributions of dispersivity (λ) and diffusion coefficient (D_l^w) did not considerably change from the corresponding priors (Figure 3). The posterior distributions of root water uptake reduction parameters (h_{50} and P1) for water stress have been remarkably changed after the M-H algorithm implementation for simulations of ECsw (Figure 4). However, the considerable uncertainty remained in the parameters of root water uptake reduction function for salinity stress (Figure 4) as their posterior distributions covered the prior ranges of the parameters. The unique relative sensitivity of the parameters was expected due to having atmospheric boundary conditions that constituted specific wetting and drying cycles. The relative sensitivity of the parameters to simulate soil salinity dynamics study was lower (higher CV) for solute transport and root water uptake reduction for salinity stress parameters compared to the parameters related to water flow simulation and root water uptake reduction function for water stresses (Table 6). The dispersivity (λ) was found as the least sensitive parameter compared to the others, and similar results were obtained by Skaggs et al., 2013 [59] for soil column experiments. The quantiles of the posterior distribution of the parameters are summarized in Table 7. The 95 CI output uncertainty of the model was obtained from the running model for 1000 parameter sets between the 2.5 and 97.5% quantiles.

Table 7. Quantiles of the posterior distribution of the model parameters.

Parameters	2.50%	25%	50%	75%	97.50%
θ_r	0.69	0.075	0.075	0.078	0.079
θ_s	0.444	0.482	0.497	0.498	0.500
α	0.005	0.007	0.010	0.013	0.015
K_s	77.022	77.411	91.326	91.837	99.127
n	1.296	1.301	1.408	1.411	1.414
l	0.364	0.477	0.555	0.599	0.615
λ	6.182	10.455	16.761	23.547	29.312
Difw	1.023	1.244	1.504	1.757	1.975
a	8.161	9.470	11.047	12.576	13.862
b	2.130	3.270	4.516	5.775	6.884
h_{50}	−802.358	−800.444	−800.408	−800.408	−800.386
P1	2.784	2.915	2.929	2.942	2.974

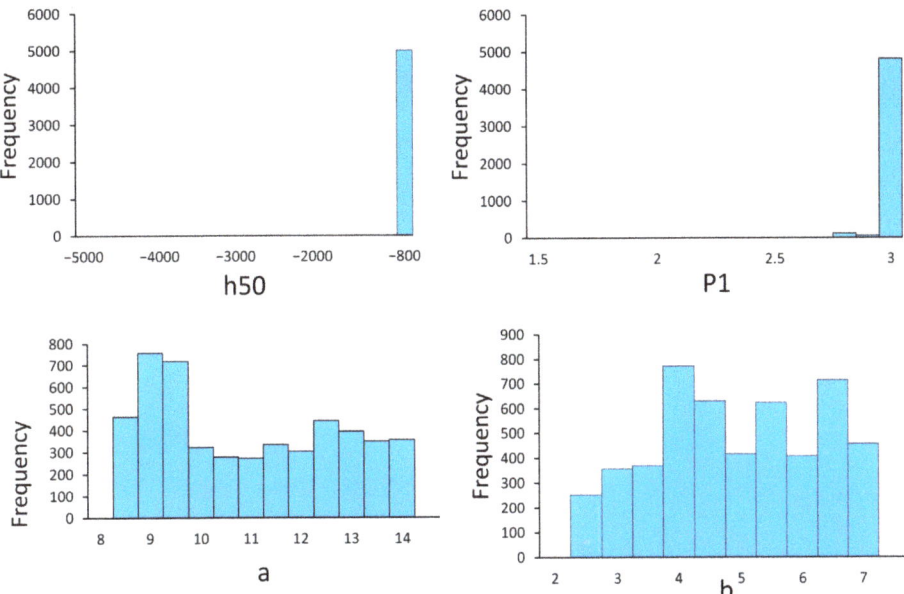

Figure 4. Posterior distributions of root water uptake reduction parameters. The h_{50}, P1, a, and b are the water pressure head for a 50% reduction in root water uptake, empirical parameter, threshold, and slope of root water uptake reduction function for salinity stress, respectively. The *x*-axes of the plots are fixed to prior distribution (range) of the parameters.

3.2. Predictive Uncertainty

Figure 5 shows the 95 CI predictive band associated with the models' parameters to simulate the dynamic of soil salinity at four different soil depths during and out of the growing season. The blue bands are 95 CI of the model outputs, and orange dots are observational points. The red bars in the figure sections are indicators of the end of irrigation events (close to the end of the growing season (a–d)), and the green bars are indicators beginning out of season rainfalls. The majority of the observational data are covered by the 95 CI band, which is an indicator of reaching parameter uncertainty to the model outputs [56]. Furthermore, covering most of the observational data in the 95 CI band shows that a significant portion of uncertainty in model outputs was due to existing uncertainty in the model parameters. However, at 70 cm (Figure 5d) soil depth, some of the observational points were out of the 95 CI band, which showed another source of uncertainty was influential at that specific soil depth. For evaluation of the predictive uncertainty, the r-factor and p-factor were calculated, and the results are presented in Table 5. The model output uncertainty is desirable if the 95 CI band covers 90% of observational data and the r-factor is less than one [55].

Similar to predictive uncertainty graphs (Figure 5), the r- and p-factors results (Table 8) expressed a higher level of uncertainty for simulating soil salt dynamic at 50 and 70 cm soil depth. Furthermore, their corresponding the r-factors and p-factors were over 1 and below 100%, respectively, so the outputs of the model should be categorized as undesirable results.

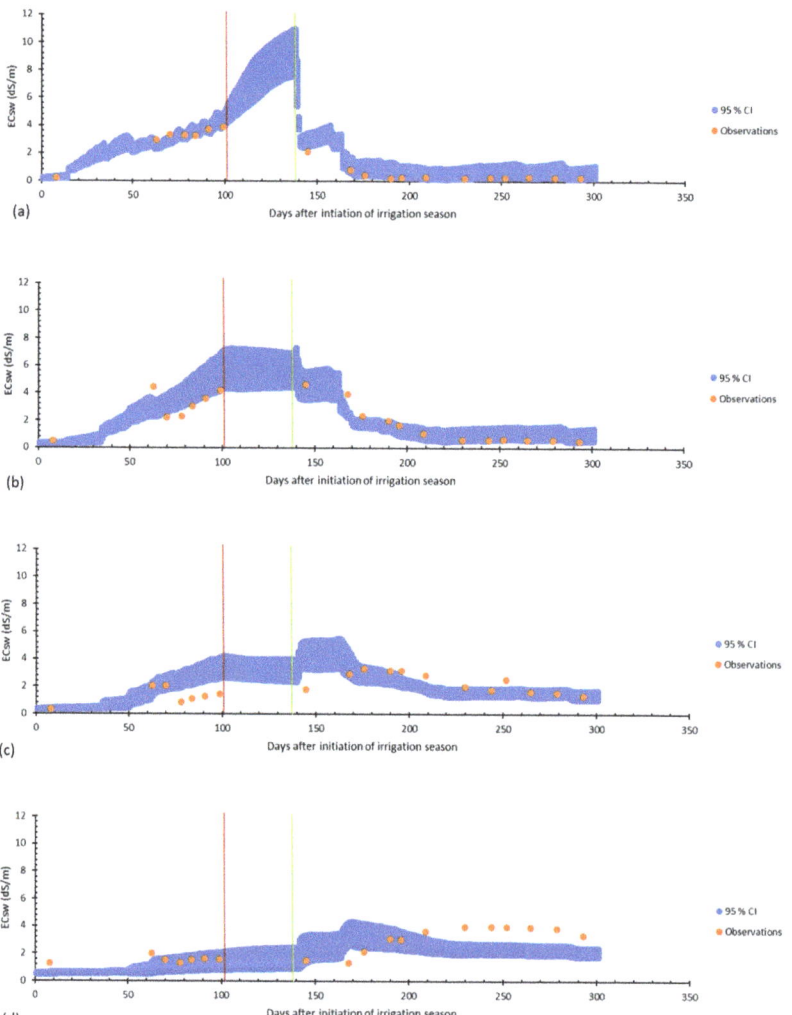

Figure 5. Predictive uncertainty for simulating soil salinity dynamics during and out of growing season at (**a**) 10 (**b**) 30 (**c**) 50 and (**d**) 70 cm soil depth. Orange dots are measured values and blue bands are predictive uncertainty for parameters 95 CI.

Table 8. Evaluation of the predictive uncertainty of the HYDRUS-1D model regarding soil salinity dynamic simulations.

	Soil Depth			
	10 cm	30 cm	50 cm	70 cm
r-factor	0.50	0.74	1.1	1.23
p-factor	95%	95%	74%	60%

As it is depicted in Figure 5, the 95 CI band is smaller during the irrigation season and rainy period out of the growing season. Hence, the predictive uncertainty of the HYDRUS-1D model was significantly higher during dry periods compared to wet periods in terms of simulating soil salinity dynamics.

3.3. Assessment of Simulated Soil Water Content and Salinity

The simulated time series of SWC are presented in Figure 6. The model's SWC outputs were close to observational during the growing season and out of the growing season for two years of the study (2003 and 2004). Based on the statistical indices (Table 9), the overall performance of the HYDRUS-1D model was good regarding the SWC simulation.

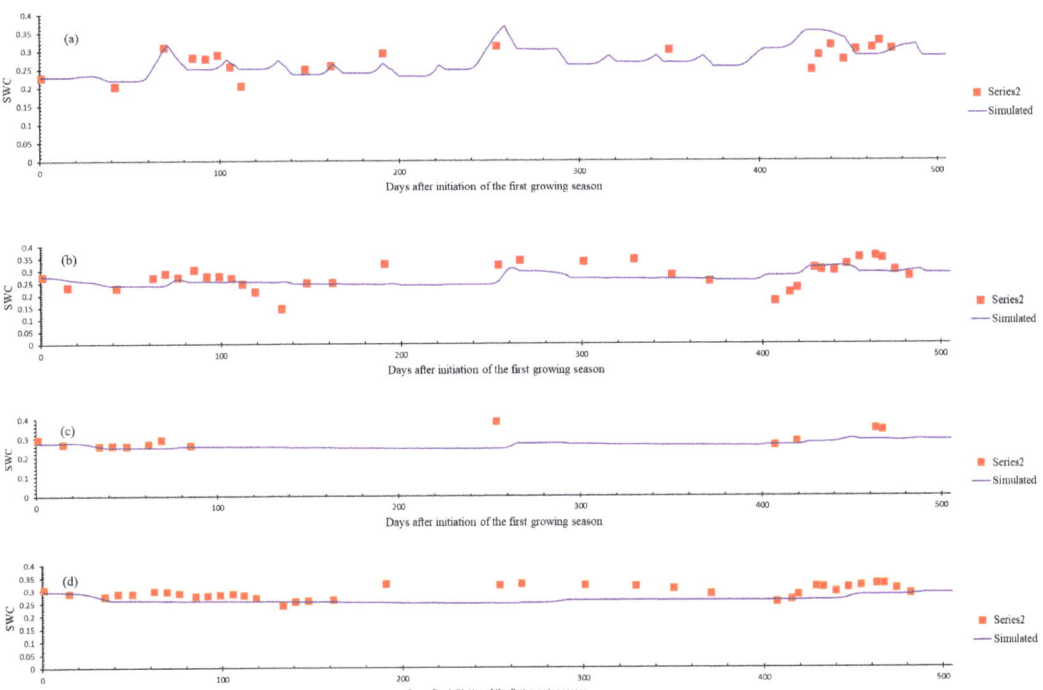

Figure 6. Time series of the simulated soil water content during the growing season and out of the growing season with observational measurements at (**a**) 10, (**b**) 30, (**c**) 50, and (**d**) 70 cm soil depth.

Table 9. Statistical indices to evaluate the performance of the -1D model regarding simulating soil water content.

			Soil Depth	
	10 cm	30 cm	50 cm	70 cm
		Soil water content		
d	0.50	0.70	0.62	0.78
RMSE (cm^3 cm^{-3})	0.05	0.04	0.04	0.02
NRMSE	0.17	0.14	0.13	0.08
		ECsw		
d	0.97	0.92	0.84	0.52
RMSE (dS/m)	0.45	0.62	0.57	1.04
NRMSE	0.23	0.28	0.26	0.36

d = coefficient of agreement, RMSE = root mean square error, NRMSE = normalized root mean square error.

The NRMSE values for SWC simulations were from 0.08 to 0.17 for four different soil depths that fall into good and excellent categories of the model accuracy, and the coefficient of agreement (d) values were from 0.5 to 0.78. The statistical indices and output graphs of the model (Figure 7) regarding the ECsw simulations validated the reliability of the calibrated model in reproducing the dynamics of soil salinity for two years. The d values were from 0.84 to 0.97 for the first three observational depths (10, 30, and 50 cm) and the corresponding NRMSE values were from 0.23 to 0.26, which were in the acceptable range. The highest deviations from the observational data were only found for simulated ECsw at 70 cm soil depth, which might not be desirable. The HYDRUS-1D model parameterized based on ECsw in saline water irrigation conditions was successful in reproducing the soil water movement, solute transport, and root water uptake process for two years during and out of the growing season.

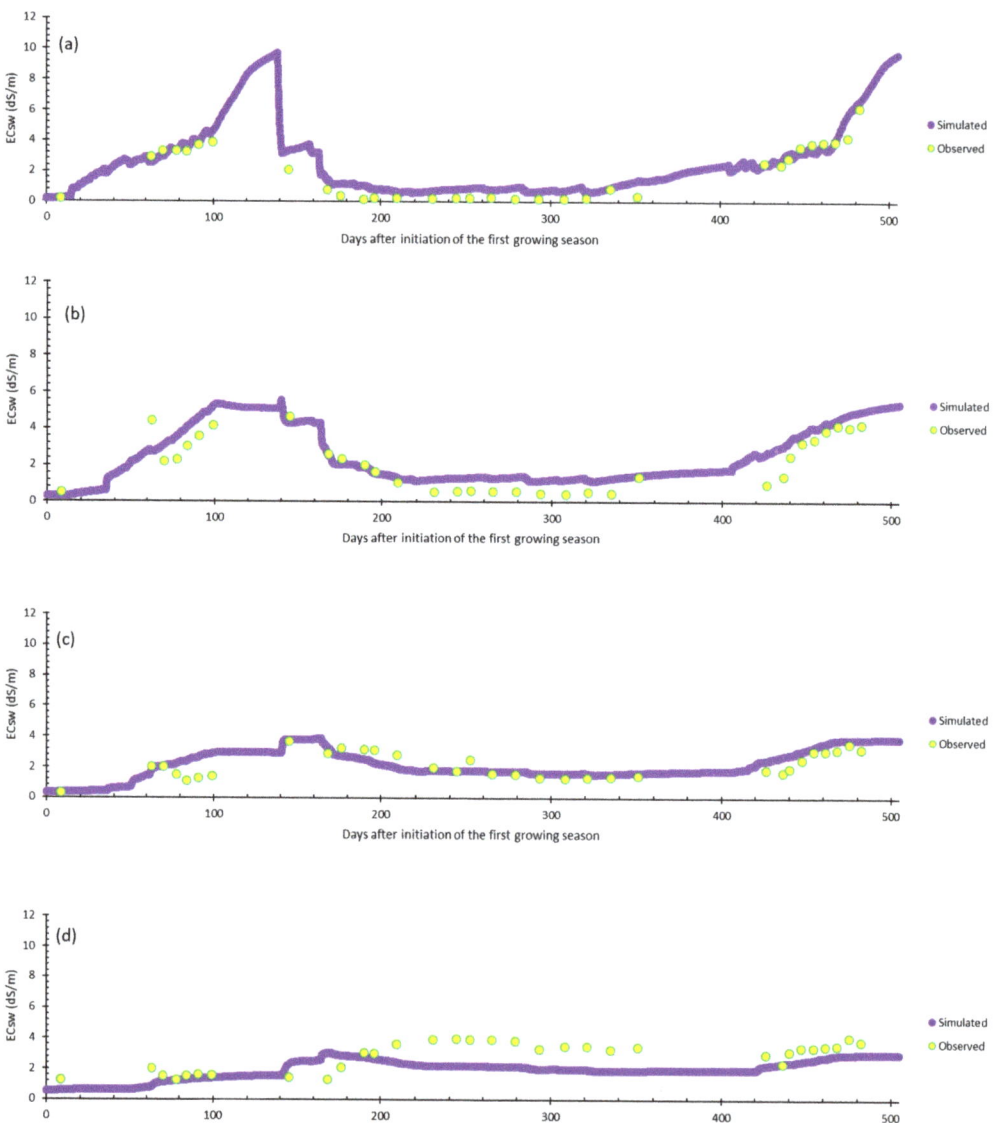

Figure 7. Time series of the simulated ECsw during the growing season and out of the growing season with observational measurements for two years of study at (**a**) 10, (**b**) 30, (**c**) 50, and (**d**) 70 cm soil depth.

4. Discussion

Marginal water quality waters, such as saline water, have been known as alternative sources for agricultural systems to mitigate water scarcity. However, soil salinity is a potential threat in most cases for using these waters as irrigation water. The soil salinity can be controlled through irrigation management methods such as leaching application. The HYDRUS-1D is a well-known numerical model that could be used as a decision support tool to investigate irrigation water management strategies for using saline water as irrigation water. The proper calibration of this model is a common challenge that could be addressed by inverse solutions. The calibration of the models usually comes with some level of uncertainty that needs to be quantified and analyzed. The Markov Chain Monte Carlo (MCMC)-based algorithm, developed by pursuing the Bayesian theorem, has been known to be effective in exploring uncertainty aspects in the models' calibrations. The algorithms that follow the Bayesian theorem integrate prior knowledge about the parameters of the models with available information about the specific study to calibrate the model and quantify the uncertainty by finding posterior distributions. In this study, the MCMC-based Metropolis-Hastings (M-H) algorithm was implemented for the HYDRUS-1D model to seek the posterior distributions of the model's parameters to simulate the dynamics of soil salinity under saline irrigation conditions. The histograms and comparison of statistics of the prior and posterior distribution of the water flow simulation and root water uptake (RWU) reduction function for water stress parameters indicate that observational data included sufficient information to estimate these parameters [26]. The M-H algorithm was able to reasonably estimate residual soil water content (θ_r) at the field scale (CV = 0.04, SD = 0.003), which proved the algorithm's effectiveness in estimating this parameter–which is complicated and time-consuming to be measured. The results indicated a higher level of uncertainty in diffusivity coefficient and dispersivity for solute transport simulations and threshold and slope of the RWU reduction function for salinity stress than the other parameters. A higher uncertainty level in solute transport parameters might be due to the scale of the study or some errors in the models' inputs as an initial condition or boundary condition. Furthermore, the higher level of uncertainty in the threshold and slope parameters of the salinity stress reduction function is presumably due to the lack of sufficient knowledge in the literature for the vegetations of this study threshold to the soil salinity stress. Hence, the selection of our priors for these parameters might have encountered some errors that resulted in the insufficiency of the M-H algorithm in identifying them and reducing the uncertainty level. The posterior distributions of the parameters, which are concentrated in a specific part of the prior distribution, confirm the robustness of the Bayesian statistics concepts–specifically the M-H algorithm–in seeking the posterior distribution of the parameters and reducing the uncertainty. Similar highly concentrated results for seeking posterior distributions of the parameters of the crop models (DSSAT and large-scale crop model) have been reported by He et al., 2009 and Izumi et al., 2009 [58,60]. The results showed a higher level of uncertainty in model outputs during dry periods when no precipitation occurred. One of the hypotheses that could explain this matter is when the soil is exposed to drying (no irrigation or precipitation), the prevailing solute transport condition is more unsaturated than in the wet period, which could increase the complexity of the phenomena predicted by the model. The interactions of evapotranspiration and soil water, and solute fluxes under unsaturated conditions are more difficult to be reproduced by the model. The increase in the unsaturation level of the soil water would make it harder for the model to simulate the water flow and solute transport in the one-dimensional (1D) mode. The predictive uncertainty band covered most of the observational points in all of the observational depths except the 70 cm. This is an indicator of another source of uncertainty in addition to the uncertainty in the model parameters for this soil depth. Among multiple sources of uncertainty, it sounds reasonable to consider the measurement error accountable for these outsider points. However, there was a possibility that if another calibration was done for this soil layer, those points would be 95 CI. The validation of the HYDRUS-1D model, using continuous data for two years of the study, has proved the reliability of the calibration achieved by implementing the M-H algorithm. The model was able to successfully reproduce (Table 9) the ECsw and SWC during and out of the two growing seasons in 2003 and 2004 (Figures 6 and 7). As indicated in the results, there are some detectable deviations in simulating ECsw at 70 cm soil depth. As mentioned in the results, the deviations were expected due to existing uncertainty in the model outputs. The output uncertainty was reflected in the model performance in terms of simulating soil salinity dynamics as the overall NRMSE value was more than 0.3, and R was equal to 0.52 at this soil depth. In this study, it has been proved that even if the target of calibration and uncertainty assessment of the HYDRUS-1D model is soil water salinity (ECsw), a very acceptable performance of the model can still be achieved. This is because the advection term in the advection-dispersion equation is usually the main driving force of the solute transport phenomenon in the soil porous media under saline irrigation conditions. This study's findings authenticated the M-H algorithm's solidity for exploring different aspects of the uncertainty in the HYDRUS-1D model to simulate salinity dynamics under saline water conditions.

The main challenges regarding the accomplishment of the current study to explore the uncertainty aspects of the HYDRUS-1D model for simulating the soil salinity were:

1. The lack of any reported values in the literature regarding the threshold of spontaneous crops to salinity stress.

2. The unavailability of additional data to validate the model performance for more than two years, specifically the recently obtained data that would contain the effects of climate change on the frequency of irrigation events and out-of-growing season precipitations.

To explore further aspects of the uncertainty of the HYDRUS-1D model, the following suggestion might be beneficial for further studies:

1. Testing additive approach for using the water and salinity stress reduction functions for mimicking root water uptake.

2. Investigating the Feddes model as a water stress reduction function [61].

3. Performing SWC and ECsw simulations for row crop cultivations such as corn and cotton.

4. Studying the uncertainty of the HYDRUS-1D model for simulating water and salinity dynamics under crop rotation schemes.

5. Furthermore, an investigation of the HYDRUS-1D uncertainty for reproducing SWC and ECsw under deficit irrigation with marginal quality waters would provide more information about uncertainty in the water and salinity stress reduction function parameters.

5. Conclusions

The uncertainty of the HYDRUS-1D model for simulating the dynamic of salts in the root zone at the field scale was assessed using the Metropolis-Hastings algorithm in the R-Studio environment. In this study, the uncertainty in the model's soil water content, root water uptake and solute transport parameters, and outputs were explored based on measurement data of electrical conductivity of soil water (ECsw). The results of this study indicated a higher level of precision (low uncertainty) in parameters related to water movement simulations comparing with the solute transport parameters (specifically dispersivity and diffusivity). The results of the model's output uncertainty (predictive uncertainty) showed a relatively lower level of uncertainty in ECsw during wet (under irrigation or rainfall) periods of the year compared with dry periods. The majority of output uncertainty in this study originated from parameter uncertainty. Moreover, it has been proved that even if the target of simulations for calibration and uncertainty purposes is ECsw, an excellent performance of the model can be achieved for simulating soil water content. Thus, the HYDRUS-1D outputs are reliable for investigating leaching requirement estimations scenarios to achieve proper irrigation scheduling for saline irrigation water conditions. However, to gain more insight into uncertainty aspects of the HYDRUS-1D model, it is highly recommended to pursue the uncertainty assessment of the model for simulating soil salt dynamic in the root zone of crops such as corn, wheat, or soybean under sprinkle or trickle irrigation systems.

Author Contributions: Conceptualization, F.M. and B.G.; methodology, F.M., A.M. and B.G.; supervision, A.M., J.A. and B.G.; software, F.M. and J.A.; investigation, F.M.; data curation, M.C.G.; resources, J.A. and H.A.; writing—original draft preparation, F.M.; writing—review and editing, F.M., A.M. and J.A. All authors have read and agreed to the published version of the manuscript.

Funding: The APC was funded by Kansas State University.

Institutional Review Board Statement: Not applicable.

Informed Consent Statement: Not applicable.

Data Availability Statement: The data could be available by requesting M.C.G.

Acknowledgments: The group of authors would like to thank Kansas State University for financially supporting the APC for this study. Special thanks to Gyuhyeong Goh, faculty member of the department of statistics at Kansas State University, for the professional comments on this study. In addition, the group of authors would express their gratitude to Maria C. Gonçalves, who is the author of this paper as well, for her remarkable contribution to this project.

Conflicts of Interest: The authors declare no conflict of interest.

References

1. Singh, A. Conjunctive use of water resources for sustainable irrigated agriculture. *J. Hydrol.* **2014**, *519*, 1688–1697. [CrossRef]
2. Chen, L.-J.; Feng, Q.; Li, F.-R.; Li, C.-S. Simulation of soil water and salt transfer under mulched furrow irrigation with saline water. *Geoderma* **2015**, *241–242*, 87–96. [CrossRef]
3. Oster, J.D. Irrigation with poor quality water. *Agric. Water Manag.* **1994**, *25*, 271–297. [CrossRef]
4. Huang, M.; Zhang, Z.; Zhai, Y.; Lu, P.; Zhu, C. Effect of Straw Biochar on Soil Properties and Wheat Production under Saline Water Irrigation. *Agronomy* **2019**, *9*, 457. [CrossRef]
5. Yuan, C.; Feng, S.; Huo, Z.; Ji, Q. Effects of deficit irrigation with saline water on soil water-salt distribution and water use efficiency of maize for seed production in arid Northwest China. *Agric. Water Manag.* **2019**, *212*, 424–432. [CrossRef]
6. Li, J.; Gao, Y.; Zhang, X.; Tian, P.; Li, J.; Tian, Y. Comprehensive comparison of different saline water irrigation strategies for tomato production: Soil properties, plant growth, fruit yield and fruit quality. *Agric. Water Manag.* **2019**, *213*, 521–533. [CrossRef]
7. Hussain, R.A.; Ahmad, R.; Waraich, E.A.; Nawaz, F. Nutrient uptake, water relations, and yield performance lf different wheat cultivars (*Triticum aestivum* L.) under salinity stress. *J. Plant Nutr.* **2015**, *38*, 2139–2149. [CrossRef]
8. Letey, J.; Hoffman, G.J.; Hopmans, J.W.; Grattan, S.R.; Suarez, D.; Corwin, D.L.; Oster, J.D.; Wu, L.; Amrhein, C. Evaluation of soil salinity leaching requirement guidelines. *Agric. Water Manag.* **2011**, *98*, 502–506. [CrossRef]
9. Ayers, R.S.; Westcot, D.W. *Water Quality for Agriculture*; Food and Agriculture Organization of the United Nations: Rome, Italy, 1985; Volume 29.
10. Corwin, D.L.; Grattan, S.R. Are Existing Irrigation Salinity Leaching Requirement Guidelines Overly Conservative or Obsolete? *J. Irrig. Drain. Eng.* **2018**, *144*, 02518001. [CrossRef]
11. Corwin, D.L.; Rhoades, J.D.; Šimůnek, J. Leaching requirement for soil salinity control: Steady-state versus transient models. *Agric. Water Manag.* **2007**, *90*, 165–180. [CrossRef]
12. Simunek, J.; Van Genuchten, M.T.; Sejna, M. The HYDRUS-1D software package for simulating the one-dimensional movement of water, heat, and multiple solutes in variably-saturated media. *Univ. Calif.-Riverside Res. Rep.* **2005**, *3*, 1–240.
13. Zeng, W.; Xu, C.; Wu, J.; Huang, J. Soil salt leaching under different irrigation regimes: HYDRUS-1D modelling and analysis. *J. Arid Land* **2014**, *6*, 44–58. [CrossRef]
14. Yang, T.; Šimůnek, J.; Mo, M.; Mccullough-Sanden, B.; Shahrokhnia, H.; Cherchian, S.; Wu, L. Assessing salinity leaching efficiency in three soils by the HYDRUS-1D and -2D simulations. *Soil Tillage Res.* **2019**, *194*, 104342. [CrossRef]
15. Gonçalves, M.C.; Šimůnek, J.; Ramos, T.B.; Martins, J.C.; Neves, M.J.; Pires, F.P. Multicomponent solute transport in soil lysimeters irrigated with waters of different quality. *Water Resour. Res.* **2006**, *42*, W08401. [CrossRef]
16. Noshadi, M.; Fahandej-Saadi, S.; Sepaskhah, A.R. Application of SALTMED and HYDRUS-1D models for simulations of soil water content and soil salinity in controlled groundwater depth. *J. Arid Land* **2020**, *12*, 447–461. [CrossRef]
17. Phogat, V.; Yadav, A.K.; Malik, R.S.; Kumar, S.; Cox, J. Simulation of salt and water movement and estimation of water productivity of rice crop irrigated with saline water. *Paddy Water Environ.* **2010**, *8*, 333–346. [CrossRef]
18. Slama, F.; Zemni, N.; Bouksila, F.; De Mascellis, R.; Bouhlila, R. Modelling the Impact on Root Water Uptake and Solute Return Flow of Different Drip Irrigation Regimes with Brackish Water. *Water* **2019**, *11*, 425. [CrossRef]

19. Liu, B.; Wang, S.; Kong, X.; Liu, X.; Sun, H. Modeling and assessing feasibility of long-term brackish water irrigation in vertically homogeneous and heterogeneous cultivated lowland in the North China Plain. *Agric. Water Manag.* **2019**, *211*, 98–110. [CrossRef]
20. Helalia, S.A.; Anderson, R.G.; Skaggs, T.H.; Šimůnek, J. Impact of Drought and Changing Water Sources on Water Use and Soil Salinity of Almond and Pistachio Orchards: 2. Modeling. *Soil Syst.* **2021**, *5*, 58. [CrossRef]
21. Tan, X.; Shao, D.; Liu, H. Simulating soil water regime in lowland paddy fields under different water managements using HYDRUS-1D. *Agric. Water Manag.* **2014**, *132*, 69–78. [CrossRef]
22. Karamouz, M.; Meidani, H.; Mahmoodzadeh, D. Inverse unsaturated-zone flow modeling for groundwater recharge estimation: A regional spatial nonstationary approach. *Hydrogeol. J.* **2022**, *30*, 1529–1549. [CrossRef]
23. Wang, X.; Li, Y.; Wang, Y.; Liu, C. Performance of HYDRUS-1D for simulating water movement in water-repellent soils. *Can. J. Soil Sci.* **2018**, *98*, 407–420. [CrossRef]
24. Stafford, M.J.; Holländer, H.M.; Dow, K. Estimating groundwater recharge in the assiniboine delta aquifer using HYDRUS-1D. *Agric. Water Manag.* **2022**, *267*, 107514. [CrossRef]
25. Shafiei, M.; Ghahraman, B.; Saghafian, B.; Davary, K.; Pande, S.; Vazifedoust, M. Uncertainty assessment of the agro-hydrological SWAP model application at field scale: A case study in a dry region. *Agric. Water Manag.* **2014**, *146*, 324–334. [CrossRef]
26. Zeng, W.; Lei, G.; Zha, Y.; Fang, Y.; Wu, J.; Huang, J. Sensitivity and uncertainty analysis of the HYDRUS-1D model for root water uptake in saline soils. *Crop Pasture Sci.* **2018**, *69*, 163–173. [CrossRef]
27. Hartmann, A.; Šimůnek, J.; Aidoo, M.K.; Seidel, S.J.; Lazarovitch, N. Implementation and Application of a Root Growth Module in HYDRUS. *Vadose Zone J.* **2018**, *17*, 170040. [CrossRef]
28. Li, J.; Zhao, R.; Li, Y.; Chen, L. Modeling the effects of parameter optimization on three bioretention tanks using the HYDRUS-1D model. *J. Environ. Manag.* **2018**, *217*, 38–46. [CrossRef]
29. Ang, A.H.S.; Tang, W.H. *Probability Concepts in Engineering: Emphasis on Applications to Civil and Environmental Engineering, 2e Instructor Site*; John Wiley & Sons Incorporated: Hoboken, NJ, USA, 2007; ISBN 047172064X.
30. Moreira, P.H.S.; van Genuchten, M.T.; Orlande, H.R.B.; Cotta, R.M. Bayesian estimation of the hydraulic and solute transport properties of a small-scale unsaturated soil column. *J. Hydrol. Hydromech.* **2016**, *64*, 30–44. [CrossRef]
31. Blasone, R.-S.; Madsen, H.; Rosbjerg, D. Uncertainty assessment of integrated distributed hydrological models using GLUE with Markov chain Monte Carlo sampling. *J. Hydrol.* **2008**, *353*, 18–32. [CrossRef]
32. Vrugt, J.A.; ter Braak, C.J.F.; Diks, C.G.H.; Schoups, G. Hydrologic data assimilation using particle Markov chain Monte Carlo simulation: Theory, concepts and applications. *Adv. Water Resour.* **2013**, *51*, 457–478. [CrossRef]
33. Wang, H.; Wang, C.; Wang, Y.; Gao, X.; Yu, C. Bayesian forecasting and uncertainty quantifying of stream flows using Metropolis–Hastings Markov Chain Monte Carlo algorithm. *J. Hydrol.* **2017**, *549*, 476–483. [CrossRef]
34. Zheng, Y.; Han, F. Markov Chain Monte Carlo (MCMC) uncertainty analysis for watershed water quality modeling and management. *Stoch. Environ. Res. Risk Assess.* **2016**, *30*, 293–308. [CrossRef]
35. Raje, D.; Krishnan, R. Bayesian parameter uncertainty modeling in a macroscale hydrologic model and its impact on Indian river basin hydrology under climate change. *Water Resour. Res.* **2012**, *48*, W08522. [CrossRef]
36. Nguyen, D.H.; Kim, S.-H.; Kwon, H.-H.; Bae, D.-H. Uncertainty Quantification of Water Level Predictions from Radar-based Areal Rainfall Using an Adaptive MCMC Algorithm. *Water Resour. Manag.* **2021**, *35*, 2197–2213. [CrossRef]
37. Brooks, S.; Gelman, A.; Jones, G.; Meng, X.-L. *Handbook of Markov Chain Monte Carlo*; CRC Press: Boca Raton, FL, USA, 2011; ISBN 1420079425.
38. Kunnath-Poovakka, A.; Ryu, D.; Eldho, T.I.; George, B. Parameter uncertainty of a hydrologic model calibrated with remotely sensed evapotranspiration and soil moisture. *J. Hydrol. Eng.* **2021**, *26*, 4020070. [CrossRef]
39. Yang, X.; Li, Y.P.; Liu, Y.R.; Gao, P.P. A MCMC-based maximum entropy copula method for bivariate drought risk analysis of the Amu Darya River Basin. *J. Hydrol.* **2020**, *590*, 125502. [CrossRef]
40. Xu, W.; Jiang, C.; Yan, L.; Li, L.; Liu, S. An adaptive metropolis-hastings optimization algorithm of Bayesian estimation in non-stationary flood frequency analysis. *Water Resour. Manag.* **2018**, *32*, 1343–1366. [CrossRef]
41. Allen, R.G.; Pereira, L.S.; Raes, D.; Smith, M. *Crop Evapotranspiration-Guidelines for Computing Crop Water Requirements-FAO Irrigation and Drainage Paper 56*; FAO: Rome, Italy, 1998; Volume 300, p. D05109.
42. Van Genuchten, M.T. A closed-form equation for predicting the hydraulic conductivity of unsaturated soils. *Soil Sci. Soc. Am. J.* **1980**, *44*, 892–898. [CrossRef]
43. Skaggs, T.H.; van Genuchten, M.T.; Shouse, P.J.; Poss, J.A. Macroscopic approaches to root water uptake as a function of water and salinity stress. *Agric. Water Manag.* **2006**, *86*, 140–149. [CrossRef]
44. Skaggs, T.H.; Shouse, P.J.; Poss, J.A. Irrigating forage crops with saline waters: 2. Modeling root uptake and drainage. *Vadose Zone J.* **2006**, *5*, 824–837. [CrossRef]
45. Ritchie, J.T. Model for predicting evaporation from a row crop with incomplete cover. *Water Resour. Res.* **1972**, *8*, 1204–1213. [CrossRef]
46. Wang, X.; Li, Y.; Chau, H.W.; Tang, D.; Chen, J.; Bayad, M. Reduced root water uptake of summer maize grown in water-repellent soils simulated by HYDRUS-1D. *Soil Tillage Res.* **2021**, *209*, 104925. [CrossRef]
47. Zarate-Valdez, J.L.; Whiting, M.L.; Lampinen, B.D.; Metcalf, S.; Ustin, S.L.; Brown, P.H. Prediction of leaf area index in almonds by vegetation indexes. *Comput. Electron. Agric.* **2012**, *85*, 24–32. [CrossRef]

48. Radcliffe, D.E.; Simunek, J. *Soil Physics with HYDRUS: Modeling and Applications*; CRC Press: Boca Raton, FL, USA, 2018; ISBN 1420073818.
49. Genuchten, M.T.; Hoffman, G.J.; Hanks, R.J.; Meiri, A.; Shalhevet, J.; Kafkafi, U. Management aspect for crop production. In *Soil Salinity under Irrigation*; Springer: Berlin/Heidelberg, Germany, 1984; pp. 258–338. [CrossRef]
50. Zhang, Y.; Arabi, M.; Paustian, K. Analysis of parameter uncertainty in model simulations of irrigated and rainfed agroecosystems. *Environ. Model. Softw.* **2020**, *126*, 104642. [CrossRef]
51. Yang, J.; Reichert, P.; Abbaspour, K.C.; Xia, J.; Yang, H. Comparing uncertainty analysis techniques for a SWAT application to the Chaohe Basin in China. *J. Hydrol.* **2008**, *358*, 1–23. [CrossRef]
52. Zhang, J.; Vrugt, J.A.; Shi, X.; Lin, G.; Wu, L.; Zeng, L. Improving Simulation Efficiency of MCMC for Inverse Modeling of Hydrologic Systems With a Kalman-Inspired Proposal Distribution. *Water Resour. Res.* **2020**, *56*, e2019WR025474. [CrossRef]
53. Maas, E.V.; Hoffman, G.J. Crop salt tolerance—Current assessment. *J. Irrig. Drain. Div.* **1977**, *103*, 115–134. [CrossRef]
54. Grieve, C.M.; Grattan, S.R.; Maas, E.V. Plant salt tolerance. *ASCE Man. Rep. Eng. Pract.* **2012**, *71*, 405–459.
55. Abbaspour, K.C.; Johnson, C.A.; Van Genuchten, M.T. Estimating uncertain flow and transport parameters using a sequential uncertainty fitting procedure. *Vadose Zone J.* **2004**, *3*, 1340–1352. [CrossRef]
56. Bouda, M.; Rousseau, A.N.; Konan, B.; Gagnon, P.; Gumiere, S.J. Bayesian uncertainty analysis of the distributed hydrological model HYDROTEL. *J. Hydrol. Eng.* **2012**, *17*, 1021–1032. [CrossRef]
57. Steduto, P.; Hsiao, T.C.; Raes, D.; Fereres, E. AquaCrop—The FAO crop model to simulate yield response to water: I. Concepts and underlying principles. *Agron. J.* **2009**, *101*, 426–437. [CrossRef]
58. He, J.; Dukes, M.D.; Jones, J.W.; Graham, W.D.; Judge, J. Applying GLUE for estimating CERES-Maize genetic and soil parameters for sweet corn production. *Trans. ASABE* **2009**, *52*, 1907–1921. [CrossRef]
59. Skaggs, T.H.; Suarez, D.L.; Goldberg, S. Effects of soil hydraulic and transport parameter uncertainty on predictions of solute transport in large lysimeters. *Vadose Zone J.* **2013**, *12*, 1–12. [CrossRef]
60. Iizumi, T.; Yokozawa, M.; Nishimori, M. Parameter estimation and uncertainty analysis of a large-scale crop model for paddy rice: Application of a Bayesian approach. *Agric. For. Meteorol.* **2009**, *149*, 333–348. [CrossRef]
61. Homaee, M.; Feddes, R.; Dirksen, C. Simulation of root water uptake. *Agric. Water Manag.* **2002**, *57*, 127–144. [CrossRef]

 *agronomy*

Article

Proper Deficit Nitrogen Application and Irrigation of Tomato Can Obtain a Higher Fruit Quality and Improve Cultivation Profit

Mengying Fan [1], Yonghui Qin [2], Xuelian Jiang [3,*], Ningbo Cui [1,*], Yaosheng Wang [4], Yixuan Zhang [1], Lu Zhao [1] and Shouzheng Jiang [1]

[1] State Key Laboratory of Hydraulics and Mountain River Engineering & College of Water Resource and Hydropower, Sichuan University, Chengdu 610065, China
[2] Heng Yuan Survey and Design of Shandong Co., Ltd, Weifang 261000, China
[3] Key Laboratory of Biochemistry and Molecular Biology in Universities of Shandong, Weifang University, Weifang 261000, China
[4] Institute of Environment and Sustainable Development in Agriculture, Chinese Academy of Agriculture Science, Beijing 100081, China
* Correspondence: jiangxuelian1987@126.com (X.J.); cuiningbo@163.com (N.C.); Tel.: +86-150-0827-5024 (N.C.)

Abstract: Faced with severe global shortage of water and soil resources, studies on the integrated effect of water and nitrogen on tomato cultivation are urgently needed for sustainable agriculture. Two successive greenhouse experiments with three irrigation regimes (1, 2/3, 1/3 full irrigation) and four nitrogen levels (1, 2/3, 1/3, 0 nitrogen) were conducted; plant growth, fruit yield and quality were surveyed; and comprehensive quality and net profit were evaluated. The results show that water and nitrogen deficit decreased plant growth, evapotranspiration and yield while increasing production efficiency and fruit comprehensive quality. An antagonism effect from water and nitrogen application was found in tomato yield, organic acid, solids acid ratio, vitamin C and lycopene, whereas synergistic impact was observed in total soluble solids content. Water deficit had more significant effect on tomato yield and fruit quality parameters compared with that of nitrogen deficiency. Synthesizing the perspectives of yield, quality, resource productivity, market price index and profits, 1/3 full irrigation and 2/3 full nitrogen was the best strategy and could be recommended to farmers as an effective guidance for tomato production.

Keywords: deficit irrigation; nitrogen application; tomato; comprehensive quality; economic evaluation

Citation: Fan, M.; Qin, Y.; Jiang, X.; Cui, N.; Wang, Y.; Zhang, Y.; Zhao, L.; Jiang, S. Proper Deficit Nitrogen Application and Irrigation of Tomato Can Obtain a Higher Fruit Quality and Improve Cultivation Profit. *Agronomy* **2022**, *12*, 2578. https://doi.org/10.3390/agronomy12102578

Academic Editors: Pantazis Georgiou and Dimitris Karpouzos

Received: 12 September 2022
Accepted: 17 October 2022
Published: 20 October 2022

Publisher's Note: MDPI stays neutral with regard to jurisdictional claims in published maps and institutional affiliations.

1. Introduction

Tomato (*solanum lycopersicum* L.), as the most widely cultivated and globally popular vegetable, relies on its savory flavor and rich nutrition [1,2], and higher tomato consumption demonstrates antioxidant, anticancer, antimutagenic and antimicrobial effects on human health [3–5]. The global planting area of tomato reached 5.03×10^6 ha in 2019, with an annual production of 1.81×10^8 t [6]. As planting area and production of tomato are increasing, the fruit quantity can meet the requirements of the market, and consumers pay more attention to fruit quality [7]. Moreover, better fruit quality generally indicates higher economic benefits [8]. In addition, the reduction in agricultural irrigation and fertilization amount, improving water and fertilization productivity, is necessary for sustainable agricultural development in the context of severe water, soil and environmental resource scarcity [9].

Water and nitrogen application is vital to crop yield and quality [10–12]. Deficit irrigation decreased tomato evapotranspiration and yield [13–15] but improved fruit quality, including soluble solids, soluble sugar, vitamin C, polyphenols and lycopene content [16,17]. Gong [18] reported that 50% deficit irrigation decreased tomato evapotranspiration by 16–23%; Lu [19] reported regular deficit irrigation decreased tomato yield by 18.61 t ha^{-1}

on average, increased water use efficiency by 2.33 kg m^{-3} and improved fruit quality. The decline in yield under deficit irrigation was mainly attributed to restrained photosynthesis due to water stress and impaired tomato physiological metabolites [20,21], whereas the increase in quality could be ascribed to the solute concentration caused by water loss in fruit [22]. A proper gradient of irrigation deficit could trade off high water use efficiency, fruit quality and acceptable yield reduction [23]. Plants can only absorb soluble nitrogen-containing ions such as NO_3^- and NH_4^+ in soil through water migration [24]. Nitrogen is an irreplaceable composition of amino acids, proteins, nucleic acids and chlorophyll [25] that affects plant photosynthesis and metabolism directly [26,27] and further affects plant growth and fruitage [28]. Many studies have demonstrated that nitrogen application had positive effect on tomato yield and quality, including vitamin C, sugar–acid ratio, soluble sugar, total soluble solids and total phenols content [29–31]. However, there were other reports showing that nitrogen application rates had no significant effect on processing tomato yield [32] but only increased the aboveground biomass [33]. In addition, it was noted that nitrogen application had a nonnegligible negative impact on increasing nitrate content in tomato fruit [34]. Although multiple studies have reported the effect of water or nitrogen deficit on yield and individual fruit quality parameters, the results vary with tomato breeds, soil textures, climate and agronomic schemas in different experiments [35], and studies on the effect mechanisms of water and nitrogen are still urgent needed.

The integration of water and nitrogen is universally considered to exist in plant growth and fruitage. Zhou [36] found that nitrogen application could partially alleviate the biological stress caused by water stress on tomato plants and enhanced leaf water use efficiency. Nevertheless, irrigation can offset the negative effects of deficit nitrogen on crop productivity [37]. Appropriate sensor-based irrigation and nitrogen sustained high yield and reduced nitrogen leaching in low-holding-capacity soils [38,39], while an inappropriate water–nitrogen deficit level led to a decline in tomato yield and quality simultaneously [40]. Although the effect of the coupling of water and nitrogen is admittedly recognized on plant growth, evidence is still lacking to clarify their inner relationship. Some studies found no significant relationship between water and nitrogen's effects on yield in maize [41], watermelon [42] and tomato [34]. The inner relationship between water and nitrogen and their response threshold, which are indispensable to adjustable irrigation–nitrogen decision making, still remain elusive. Therefore, the marginal productivity efficiency and the comprehensive benefits based on yield, fruit quality, source efficiency and market price were calculated to evaluate the outcome of each application strategy. The aims of the present study are to: (1) investigate the effect of irrigation and nitrogen on tomato growth, yield, quality and water–nitrogen use efficiency; (2) clarify whether there is a synergistic or antagonistic relationship between water and nitrogen's effects on yield and different quality traits; (3) determine the preferable water and nitrogen application strategy based on comprehensive benefit analysis and provide a direct scientific guidance for local tomato cultivation industry.

2. Materials and Methods

2.1. Experimental Site

The greenhouse experiments were conducted from 13 April to 6 July (first season) and 13 August to 28 November (second season) in 2019 at a commercial company located in Gaomi City (latitude 36°38′ N, longitude 112°56′ E, altitude 26.03 m), Shandong Province, in Northern China. The site is located in a monsoon climate with annual precipitation of 646 mm, pan evaporation of 1838 mm, temperature of 11.7 °C, and duration of mean annual sunshine over 2800 h. The greenhouse is 80 m in length and 12 m in width, covering an area of 960 m^2. The experimental soil is clay loam with the average dry bulk density of 1.38 g·cm^{-3}. The total available N, P and K content for 0–1.0 m soil depth initially are 84.3 mg·kg^{-1}, 102 mg·kg^{-1} and 130 mg·kg^{-1}, respectively. The field water capacity (θ_{FC}) and the wilting coefficient for 0–0.8 m depth determined by soil water absorption experiment are 0.33 cm^3·cm^{-3} and 0.14 cm^3·cm^{-3}, respectively.

2.2. Experimental Design

The experimental tomato (*Solanum lycopersicum* L.) variety was Baoli 3 in both seasons. The tomato plants with similar heights were transplanted to the plots on 13 April in the first season and 13 August in the second season. Three growth stages, i.e., seedling stage (from transplant to first fruit set), flowering and fruit development stage (from first fruit set to first fruit maturity) and fruit maturation and harvest stage (from first fruit maturity to uprooting crops after all fruit is harvested), were divided according to local observations. The first season ended on 26 July and the second on 28 November.

Three irrigation levels consisted of full irrigation (W1), 2/3 full irrigation (W2/3) and 1/3 full irrigation (W1/3). Four urea nitrogen levels included full nitrogen (N1), 2/3 full nitrogen (N2/3), 1/3 full nitrogen (N1/3) and no nitrogen (N0) at the whole growth stage. This yielded twelve treatments in a completely randomized block design, and each treatment was replicated thrice. The size of each plot was 2.8 m × 6 m, and the plots were separated by 1 m deep and 2 mm thick acrylic flap. W1 was irrigated to $90 \pm 3\%$ of θ_{FC} when the soil water content within 0.6 m at the seedling stage and 0.8 m in other stages decreased to $75 \pm 3\%$ of θ_{FC}. The nitrogen amount of N1 treatment was consistent with that of local management. The irrigation treatment was irrigated at rate of water requirements based on control treatment W1N1, and the nitrogen treatment was fertilized at rate of N use in W1N1. Furthermore, concerning the relatively low available N content in soil and severe deficit in first season, the same amount of nitrogen application was implemented in second season. The irrigation and fertilization time for deficit treatments was the same as that of W1N1. The description of irrigation and fertilization amount under different treatments is shown in Table 1. The irrigation pattern was dripping irrigation, and each plot had an individual 6 m length branch. Twelve water drippers with a flow of 1.6 L/hour were evenly distributed in the branch, each branch pipe was separately installed with a water meter, and a Venturi fertilizing tank was used to record the irrigation and fertilization amount. Three tomato plants were planted in a plot and uniformly pruned (removing the secondary shoots and only leaving the main stem) in flowering and fruit development stage according to the growth condition.

Table 1. Description of irrigation and nitrogen amount in different treatments.

Num	Treatment	Description (at Whole Season)	First Season		Second Season	
			I	N	I	N
T1	W_1N_1/CK	full irrigation and full nitrogen	260.4 (10)	14.7 (3)	184.3 (10)	19.6 (4)
T2	$W_1N_{2/3}$	full irrigation and 2/3 full nitrogen	260.4 (10)	9.8 (3)	184.3 (10)	15.8 (4)
T3	$W_1N_{1/3}$	full irrigation and 1/3 full nitrogen	260.4 (10)	4.9 (3)	184.3 (10)	12.0 (4)
T4	W_1N_0	full irrigation and no nitrogen	260.4 (10)	0.0 (0)	184.3 (10)	8.2 (1)
T5	$W_{2/3}N_1$	2/3 full irrigation and full nitrogen	180.6 (10)	14.7 (3)	139.4 (10)	19.6 (4)
T6	$W_{2/3}N_{2/3}$	2/3 full irrigation and 2/3 full nitrogen	180.6 (10)	9.8 (3)	139.4 (10)	15.8 (4)
T7	$W_{1/3}N_{1/3}$	2/3 full irrigation and 1/3 full nitrogen	180.6 (10)	4.9 (3)	139.4 (10)	12.0 (4)
T8	W_1N_0	2/3 full irrigation and no nitrogen	180.6 (10)	0.0 (0)	139.4 (10)	8.2 (1)
T9	$W_{1/3}N_1$	1/3 full irrigation and full nitrogen	104.9 (10)	14.7 (3)	87.5 (10)	19.6 (4)
T10	$W_{1/3}N_{2/3}$	1/3 full irrigation and 2/3 full nitrogen	104.9 (10)	9.8 (3)	87.5 (10)	15.8 (4)
T11	$W_{1/3}N_{1/3}$	1/3 full irrigation and 1/3 full nitrogen	104.9 (10)	4.9 (3)	87.5 (10)	12.0 (4)
T12	$W_{1/3}N_0$	1/3 full irrigation and no nitrogen	104.9 (10)	0.0 (0)	87.5 (10)	8.2 (1)
	stage I	seedling stage	13 Apr.–7 May		13 Aug.–6 Sep.	
	stage II	flowering and fruit development stage	8 May–20 Jun.		7 Sep.–13 Oct.	
	stage III	fruit maturation and harvest stage	21 Jun.–26 Jul.		14 Oct.–28 Nov.	

Notes: CK means control treatment; T1 indicates the first treatment and so on for the other treatments; the number in brackets represents the irrigation and fertilization times; I means irrigation amount (mm); N means nitrogen application (g·m^{-2}).

2.3. Measurements

2.3.1. Meteorological Variables

The solar radiation (Ra), air temperature (Ta) and relative humidity (RH) in the two seasons were recorded continuously using a standard automatic weather station (Hobo, Onset Computer Crop, Bourne, MA, USA) installed at the center of the greenhouse. All meteorological mean variables every 30 min were calculated automatically by a data logger, and the daily average value of Ra, Ta and RH are shown in Figure 1.

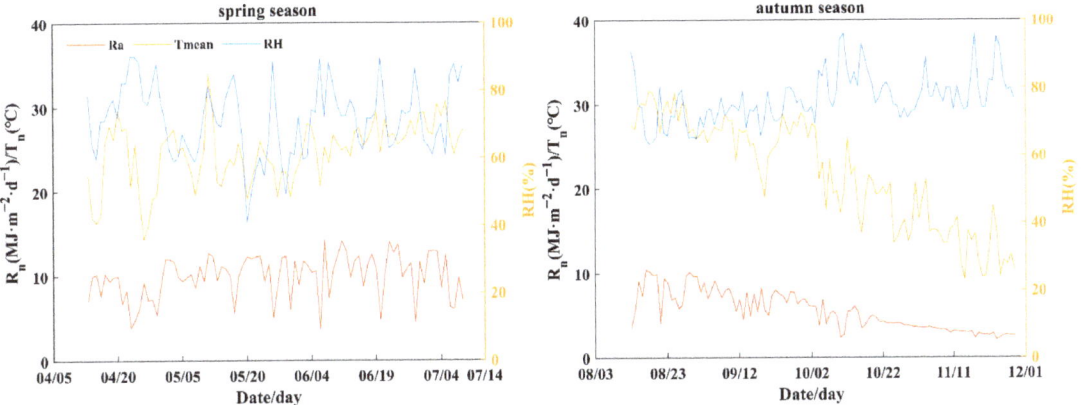

Figure 1. The basic meteorological information during the growth stages.

2.3.2. Evapotranspiration

Crop evapotranspiration (ET) was estimated by soil water balance method [43] as follows:

$$ET = P + I + W - R - D - \Delta W, \qquad (1)$$

where P is precipitation (mm); I is irrigation amount (mm); W is capillary rise to the root zone (mm); R is surface runoff (mm); D is drainage from the root zone (mm); and ΔW is the change in soil water content (mm). ΔW was calculated as follows:

$$\Delta W = H(\theta_i - \theta_{i-1}), \qquad (2)$$

where H is the depth of plant root zone (m); θ_i and θ_{i-1} are the mean water contents in the root zone at time i and $i-1$, respectively.

Since there is no precipitation in greenhouse, P and R can be negligible. The groundwater level was lower than 15 m below the ground surface according to the local observation, so W was also negligible. D can be ignored because the irrigation amount was always within the field water capacity. Thus, Equation 1 is simplified as:

$$ET = I - H(\theta_i - \theta_{i-1}), \qquad (3)$$

2.3.3. Plant Growth

Plant height, stem diameter and leaf index (LAI) were measured at intervals of 7–10 days during the whole growth period. Leaf length and the maximum width were measured, and the leaf area was determined by the sum of the rectangular area of each completely developed leaf (the product of leaf length and maximum width) multiplied by a parameter of 0.64 [36,44]. The LAI was the ratio of leaf area to land area of each plant. Chlorophyll content was measured by a handled chlorophyll analyzer (SPAD502, Spectrum, Aurora, IL, USA, 0.1) every 7–10 days.

2.3.4. Yield, Water and Nitrogen Use Efficiency

Fifteen plants in each plot were randomly selected for measuring the yield, and fruit weight after maturity was measured by an electronic scale. The total yield (Y, $t \cdot ha^{-1}$) and mean single fruit weight was then calculated. Water use efficiency (WUE, $kg \cdot m^{-3}$), irrigation water use efficiency (WUEI, $kg \cdot m^{-3}$) and application nitrogen use efficiency (NUE, $kg \cdot g^{-1}$) were calculated as follows:

$$WUE = Y/ET \times 100, \tag{4}$$

$$WUEI = Y/I \times 100, \tag{5}$$

$$NUE = Y/N \times 0.1, \tag{6}$$

where Y is the yield ($t \cdot ha^{-1}$); I is the irrigation quantity (mm); and N is the nitrogen application amount ($g \cdot m^{-2}$).

2.3.5. Fruit Quality

Fruit quality parameters were measured at fruit maturation and harvest stage. Total soluble solids content (TSS) was measured by a handheld refractometer (PAL-BX/ACID 3, ATAGO, Tokyo, Japan, 0.1 Brix). Organic acid (OA) was titrated with $0.1 \ mol \cdot L^{-1}$ NaOH solution and the solids-acid ratio (SAR) was defined as the ratio of TSS to OA. Vitamin C (VC) content was measured using 2,6-dichloroindophenol titrimetric method (A009-1-1, Nanjing Jiancheng bioengineering institute, China, $0.1 \ ug \cdot ml^{-1}$) [45]. Lycopene content (Lyc) was measured by spectrophotometric method (FT-P6141Z, Fantaibio, China, $0.1 \ ug \cdot mL^{-1}$) [46,47]. Fruit firmness was measured by a hardness tester (GY-4, Handpi, Zhejiang, China, $0.01 \ kg \cdot cm^{-2}$).

2.4. The Calculation of Comprehensive Quality

Tomato quality is an overall result of individual parameters, and the responses of individual fruit quality parameters to irrigation and nitrogen treatments are different, which affects the determination of the treatment that has the best fruit quality. Thus, the comprehensive fruit quality was evaluated using the technique for order preference by similarity to ideal solution (TOPSIS), combined with analytic hierarchy process (AHP), which is briefly outlined below.

(1) Normalize individual fruit quality parameters. The low optimal parameter is converted into high optimal parameter as follows:

$$x_{ij} = \frac{1}{x_{ij,}^{*}} \quad i = 1, \ 2 \cdots m; \ j = 1, \ 2 \cdots n, \tag{7}$$

where x_{ij} is the forward original quality value of i-th treatment and j-th fruit quality parameter. In this study, m = 12 and n = 7; $x_{ij,}^{*}$ is the antidromic original quality value; and only OA was considered to be small optimal index in this study.

Then, x_{ij} is normalized as follows:

$$z_{ij} = \frac{\left| x_{ij} - x_{bestj} \right|}{\sqrt{\Sigma_{i=1}^{n} \left(x_{ij} - x_{bestj} \right)^{2}}} \quad i = 1, \ 2 \cdots m; \ j = 1, \ 2 \cdots n, \tag{8}$$

where z_{ij} is the positively standardized quality value of i-th treatment and j-th fruit quality parameter; x_{bestj} is the best value of j-th parameter among all treatments.

(2) Define the best and worst ideal solutions:

$$Z^{+} = \left(Z_{1}^{+}, Z_{1}^{+}, \ldots Z_{j}^{+}, \ldots Z_{m}^{+} \right), \ Z_{j}^{+} = \max \left\{ z_{1j}, z_{2j}, \ldots, z_{nj} \right\}, \tag{9}$$

$$Z^- = \left(Z_1^-, Z_1^-, \ldots Z_j^-, \ldots Z_m^-\right), \quad Z_j^- = \min\{z_{1j}, z_{2j}, \ldots, z_{nj}\}, \tag{10}$$

where Z^+ is the defining maximum matrix; Z_j^+ is the maximum value of parameter j; Z^- is the defining minimum matrix; and Z_j^- is the minimum value of parameter j.

(3) Calculate the distance using AHP weights:

$$D_i^+ = \sqrt{\sum_{j=1}^{m} \omega_j \left(Z_j^+ - z_{ij}\right)^2}, \quad D_i^- = \sqrt{\sum_{j=1}^{m} \omega_j \left(Z_j^- - z_{ij}\right)^2}, \tag{11}$$

where D_i^+ is the distance between i treatment and the maximum value; D_i^- is the distance between i treatment and the minimum value; ω_j is the weight of index j determined by AHP method [48].

(4) Compute the comprehensive index under different treatments (Qi):

$$Q_i = \frac{D_i^-}{D_i^+ + D_i^-}, \tag{12}$$

2.5. Economic Analysis

The economic benefits are related to both yield and fruit quality, and better fruit quality usually indicates higher sale prices. Thus, an economic profit analysis considering comprehensive quality and yield is necessary to determine the optimal treatment.

Relative sale price considering the comprehensive quality was calculated as:

$$p_i = p_c \left(1 + R\left(\frac{Q_i}{Q_{ck}} - 1\right)\right), \tag{13}$$

where p_i is the price of different fruit quality ($\cdot kg^{-1}$); p_c is the sale price of CK treatment ($\cdot kg^{-1}$), which was defined as 1.5 according to the market sale price in 2019; R is the price index, representing the fluctuation of tomato prices with quality; and Q_i and Q_{ck} are the comprehensive quality index under different irrigation and nitrogen treatment and CK treatment, respectively.

Total cost under different treatment was determined by:

$$C_c = I \times c_w + N \times c_f + C_s, \tag{14}$$

where C_c is the total cost ($); I is the irrigation amount ($m^3 \cdot ha^{-1}$); c_w is the unit price of agricultural water ($\cdot m^{-3}$), which is defined as 0.15 according to the water conservancy company sale price in 2019; N is nitrogen amount applied ($kg \cdot ha^{-1}$); c_f is the unit price of nitrogen fertilizer ($\cdot kg^{-1}$), defined as 2.7 according to the market sale price in 2019; and C_s is the fixed cost under different treatments ($), which does not change due to different treatments, including the cost of greenhouse drip irrigation project, land rent, other facilities costs, $282,750 one-time investment expense for 20 years usage expectation per ha, labor costs (land leveling, tomato interruption, weeding, fertilization, spraying, harvesting and packaging) totaling $652 per ha per year and other consumables costs (fertilizers other than nitrogen, herbicide and insecticide) totaling $580 per ha per year.

Net profit was determined as:

$$C_n = Y_i p_i - C_c, \tag{15}$$

where C_n is the net profit per hectare ($); Y_i is the yield in different treatments ($kg \cdot ha^{-1}$).

The profit change was calculated as:

$$\Delta P_\% = \frac{C_{ni} - C_{nck}}{C_{nck}} \times 100\%, \tag{16}$$

where $\Delta P_\%$ is the change in net profit, %; C_{ni} is the net profit of i treatment; and C_{nck} is the net profit of CK treatment.

2.6. Statistical Analysis

The differences among the treatments were compared by two-way ANOVA and Duncan's multiple range test using SPSS 26.0 (IBM statistics, Armonk, NY, USA). The correlation between independent and dependent variables was determined by the Pearson correlation analysis.

3. Results

3.1. Effect of Deficit Irrigation and Nitrogen on Plant Growth

The variation in plant height, stem diameter, leaf area index and chlorophyll content of tomato under different irrigation and nitrogen treatments in the second season is shown in Figure 2. The plant height increased rapidly in the first two stages and remained stable after reaching its maximum at the beginning of stage III. The stem diameter showed similar change tendency to that of plant height, except for reaching its maximum at the beginning of stage II. Water deficit had no obvious effect on plant height but significantly reduced stem diameter (Figure 2a–f) due to the infinite-growth tomato variety, which may not be very sensitive to water stress in height growth. Nitrogen application improved plant height and stem diameter in low-water treatments (Figure 2c,f), indicating that nitrogen could alleviate the inhibition caused by water deficit as a vital nutritious element of plant growth. Compared with CK, plant height and stem diameter of T12 in stable point (November 10) decreased by 5.88% and 7.62%, respectively.

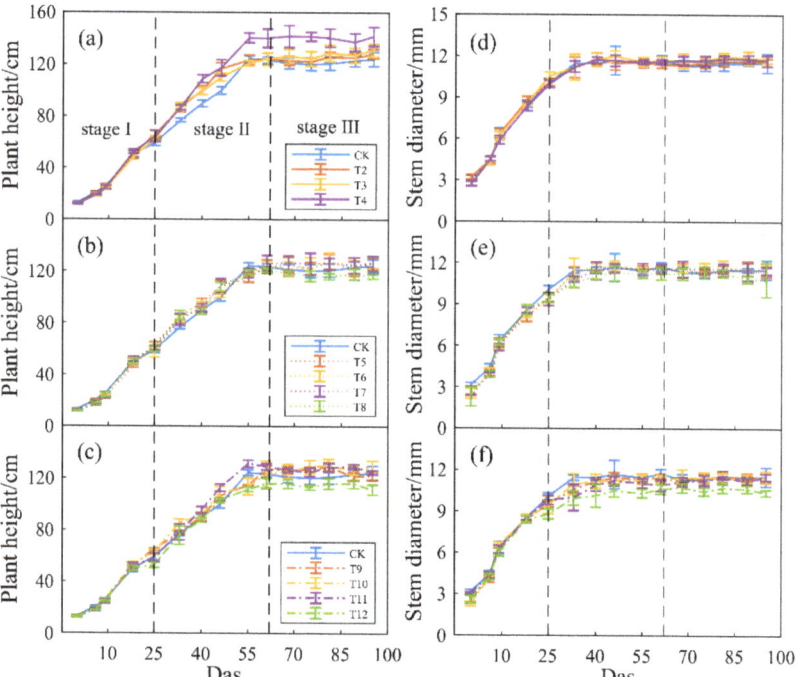

Figure 2. *Cont.*

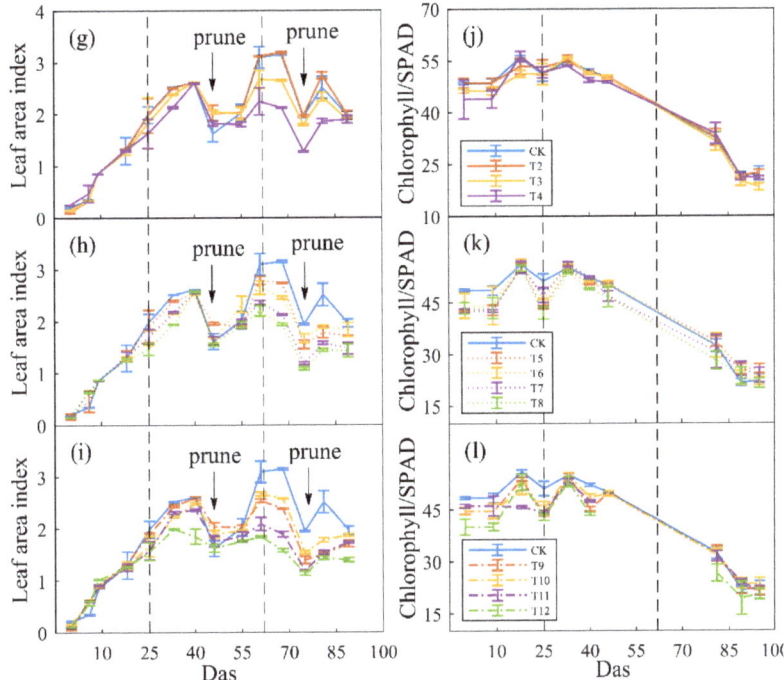

Figure 2. The plant height (h), stem diameter (D), leaf area index (LAI) and chlorophyll content (SPAD) under different water and nitrogen treatments in the second season of 2019. Notes: (**a,d,g,j**) show sufficient irrigation groups of CK, T2, T3 and T4, respectively; (**b,e,h,k**) represent moderate water deficit groups of T5, T6, T7 and T8, respectively; (**c,f,i,l**) indicate severe water deficit groups of T9, T10, T11 and T12, respectively.

Leaf area index (LAI) and chlorophyll content increased at stage I and II and then decreased due to leaf wilting and yellowing (Figure 2g–l). The significant drops on 27 September and 27 October were caused by pruning. Both water and nitrogen deficit reduced LAI observably. Compared with CK, LAI decreased from 26.6% (W2/3) to 33.2% (W1/3) for water deficit treatments and 13.1% (N2/3) to 29.1% (N0) for nitrogen deficit treatments (in maximum point, 20 October). Chlorophyll content decreased by 8.4% (N0) and 9.6% (W1/3) on August 22 at stage I but had no obvious variance after stage I (on 15 September and 10 November, $p > 0.05$).

3.2. Effect of Deficit Irrigation and Nitrogen on Evapotranspiration, Yield, Water and Nitrogen Use Efficiency

The evapotranspiration (ET) at the whole growth stage varied from 300.99 to 173.77 mm in the first season and from 239.80 to 161.49 mm in the second season (Table 2). Both irrigation and nitrogen fertilization decreased ET significantly (Table 2, $p < 0.01$), and the decline in ET reached 38.8% (T9) and 5.7% (T4) in the first season and 23.6% (T9) and 11.6% (T4) in the second season.

The effect of water and nitrogen on tomato yield was significant (Table 2, $p < 0.05$). Compared with CK, yield decreased by 39.1% (T9) and 11.0% (T4) in the first season and 10.1% (T9) and 4.9% (T4) in the second season.

Compared with CK, water use efficiency (WUE) increased first and then decreased with the decline in water supply in the first season but had no significant differences in the second season (Table 2). The decrease in water use efficiency in W1/3 was caused by sharp yield recession, and there was no significant effect of nitrogen on WUE.

Table 2. Evapotranspiration (ET), yield and product efficiency (WUE, WUEI and NUE) under different water and nitrogen treatments in the first and second seasons of 2019.

Treatment		ET (mm)	Yield (t·ha^{-1})	WUE (kg·m^{-3})	WUEI (kg·m^{-3})	NUE (kg·g^{-1})
First season	CK	299.41 a	66.09 a	22.07 ab	25.38 cd	0.45 d
	T2	300.99 a	66.46 a	22.08 ab	25.52 cd	0.68 c
	T3	288.44 ab	65.33 a	22.65 ab	25.09 cd	1.33 a
	T4	282.33 b	58.79 ab	20.82 b	22.58 d	
	T5	245.79 c	68.28 a	27.78 a	37.81 a	0.46 d
	T6	243.26 c	63.84 a	26.24 ab	35.35 ab	0.65 c
	T7	239.89 c	48.23 bc	20.11 b	26.71 bcd	0.98 b
	T8	225.05 d	47.98 bc	21.32 ab	26.57 bcd	
	T9	204.59 e	40.23 c	19.66 b	38.35 a	0.27 e
	T10	190.83 f	38.48 c	20.16 b	36.68 a	0.39 de
	T11	181.28 fg	35.08 c	19.35 b	33.44 abc	0.72 c
	T12	173.77 g	34.32 c	19.75 b	32.72 abc	
Sig test	W	0.000 **	0.000 **	0.022 *	0.000 **	0.000 **
	N	0.000 **	0.019 *	0.278 ns	0.017 *	0.011 *
	W*N	0.326 ns	0.482 ns	0.397 ns	0.608 ns	0.135 ns
Second season	CK	239.80 a	59.92 a	24.99 ab	32.51 cd	0.29 e
	T2	229.59 ab	57.55 ab	25.07 ab	31.23 cd	0.34 de
	T3	225.45 b	54.32 ab	24.09 ab	29.47 d	0.42 bcd
	T4	212.01 c	56.97 ab	26.87 ab	30.91 cd	0.61 a
	T5	206.46 c	49.36 abc	23.91 ab	35.41 bcd	0.25 e
	T6	206.69 c	56.86 ab	27.51 ab	40.79 bcd	0.36 cde
	T7	205.51 c	51.77 abc	25.19 ab	37.14 bcd	0.43 bcd
	T8	183.56 d	42.14 bc	22.96 b	30.23 cd	0.51 ab
	T9	183.16 d	53.85 abc	29.40 ab	61.54 a	0.28 e
	T10	177.75 d	56.79 ab	31.95 a	64.90 a	0.36 cde
	T11	176.77 d	41.46 bc	23.46 b	47.39 b	0.35 de
	T12	161.49 e	38.00 c	23.53 b	43.43 bc	0.46 bc
Sig test	W	0.000 **	0.027 *	0.392 ns	0.000 **	0.134 ns
	N	0.000 **	0.034 *	0.221 ns	0.000 **	0.000 **
	W*N	0.749 ns	0.456 ns	0.384 ns	0.001 **	0.248 ns

Notes: lowercase letters following the data indicate significant differences by Duncan's test at $p < 0.05$ level; * means statistically significant with $p < 0.05$; ** notes statistically extreme significance with $p < 0.01$; ns represents statistically insignificant with $p > 0.05$.

Irrigation water use efficiency (WUEI) significantly increased with the reduction in irrigation amount and deceased with the reduction in nitrogen amount in both seasons. Oppositely, nitrogen use efficiency (NUE) decreased with the reduction in irrigation amount and increased with the reduction in nitrogen amount (Table 2). WUEI was generally higher than WUE, since a plant can use the water in soil that was stored prior to development stage. The maximum of WUEI was 38.35 kg·m^{-3}, found in T9, in the first season and 64.90 kg·m^{-3}, observed in T10, in the second season. T3 in the first season had a much higher NUE of 1.33 kg·g^{-1} than any other treatments. It was interesting that the two-way ANOVA results of water and nitrogen interaction on ET, yield and efficiency were not significant except for WUEI in the second season (Table 2).

3.3. Effect of Deficit Irrigation and Nitrogen on Fruit Quality

The TSS, SAR and VC significantly increased with the increase in irrigation deficit, while SW and OA decreased with the increase in irrigation deficit (Table 3). Compared with CK, T9 increased TSS by 22.9% and 37.0%, SAR by 79.3% and 51.6% and VC by 112.3% and 129.9% in the first and second season, respectively. SW decreased by 14.6% and 12.8%, and OA decreased by 23.8% and 11.1% in the first and second seasons, respectively. Lyc decreased with the decline in water in the second season, but there was no significant variance in the first season. Fn of W1/3 increased 12.7% in the first season but did not change obviously in the second season.

Table 3. Single fruit quality parameters under different water and nitrogen treatments in the first and second seasons of 2019.

Treatment		SW (g)	TSS (°Brix)	OA (%)	SAR (ratio)	VC (mg·kg^{-1})	Lyc (mg·kg^{-1})	Fn (kg·cm^{-2})
First season	CK	93.08 ab	4.85 fg	0.42 a	10.54 d	1.06 b	26.40 a	4.15 abc
	T2	95.86 a	5.38 def	0.38 ab	14.01 cd	1.44 ab	24.00 a	3.45 bc
	T3	95.94 a	5.30 def	0.32 bcd	16.53 bcd	1.39 ab	22.03 a	3.32 bc
	T4	88.25 abc	4.38 g	0.33 bcd	12.93 cd	0.92 b	17.11 a	3.19 bc
	T5	99.69 a	5.06 ef	0.34 abcd	15.83 bcd	1.29 b	23.76 a	3.12 bc
	T6	91.79 abc	5.61 bcde	0.36 abc	15.34 bcd	1.67 ab	20.20 a	3.73 abc
	T7	85.32 abc	5.83 abcd	0.30 bcde	20.00 bc	1.03 b	18.46 a	3.03 c
	T8	84.07 abc	5.55 bcde	0.26 de	21.97 ab	1.08 b	20.70 a	3.24 bc
	T9	79.50 bcd	5.96 abc	0.32 bcd	18.90 bc	2.25 a	20.84 a	3.93 abc
	T10	76.88 cd	6.24 a	0.27 cde	22.59 ab	2.20 a	27.20 a	5.24 a
	T11	69.33 d	6.05 ab	0.33 bcd	18.35 bc	1.54 ab	17.01 a	4.71 abc
	T12	67.00 d	5.42 cde	0.23 e	28.63 a	1.06 b	13.63 a	4.83 ab
Sig test	W	0.000 **	0.000 **	0.001 **	0.000 **	0.012 *	0.715 ns	0.001 **
	N	0.002 **	0.000 **	0.003 **	0.027 *	0.022 *	0.210 ns	0.709 ns
	W*N	0.384 ns	0.034 *	0.254 ns	0.108 ns	0.321 ns	0.773 ns	0.589 ns
Second season	CK	129.56 ab	4.16 f	0.36 bcd	11.76 d	1.34 g	46.08 a	3.18 a
	T2	132.44 ab	5.35 abcd	0.37 bc	15.09 bc	1.40 fg	44.01 ab	3.16 a
	T3	127.35 ab	5.52 abc	0.46 a	12.65 d	1.91 de	40.83 abc	3.76 a
	T4	122.71 ab	5.00 de	0.37 bc	13.76 cd	1.73 efg	30.06 bc	3.79 a
	T5	126.47 ab	5.44 abcd	0.48 a	11.46 d	2.23 bcd	35.10 abc	4.30 a
	T6	108.33 ab	5.55 ab	0.39 b	15.62 abc	2.49 b	37.12 abc	3.36 a
	T7	115.53 ab	5.14 bcde	0.34 bcde	15.15 bc	2.63 b	35.67 abc	3.26 a
	T8	150.01 a	5.07 cde	0.30 e	16.46 ab	1.80 ef	31.39 abc	3.54 a
	T9	113.02 ab	5.70 a	0.32 cde	17.83 a	3.08 a	29.25 bc	3.40 a
	T10	116.37 ab	5.28 abcd	0.31 de	16.94 ab	2.43 b	30.84 bc	3.45 a
	T11	93.50 b	5.47 abcd	0.29 e	17.89 a	2.37 bc	29.97 bc	2.83 a
	T12	84.60 b	4.70 e	0.30 e	16.46 ab	2.00 cde	27.86 c	3.48 a
Sig test	W	0.011 *	0.008 **	0.000 **	0.000 **	0.000 **	0.011 *	0.879 ns
	N	0.819 ns	0.000 **	0.000 **	0.008 **	0.001 **	0.199 ns	0.941 ns
	W*N	0.466 ns	0.000 **	0.000 **	0.003 **	0.000 **	0.755 ns	0.962 ns

Notes: SW, TSS, OA, SAR, VC, Lyc and Fn indicated tomato single weight, total soluble solids, organic acids, solid–acid content ratio, vitamin C content, lycopene content and fruit firmness, respectively. lowercase letters following the data indicate significant differences by Duncan's test at $p < 0.05$ level; * means statistically significant with $p < 0.05$; ** notes statistically extreme significance with $p < 0.01$; ns represents statistically insignificant with $p > 0.05$.

SW declined with the deficit of nitrogen in the first season, but there was no obvious variance in the second season. Compared with N1, nitrogen deficit significantly improved fruit TSS by 8.8% (N2/3 and N1/3), while it decreased fruit TSS by 3.0% under N0 in the first season, and fruit TSS increased by 7.7% with N2/3 and N1/3 while decreasing by 1.4% with N0 under nitrogen deficit in the second season. Similarly, average VC in two seasons of N2/3 increased by 9.7% and decreased by 17.9% under N0. However, nitrogen had no significant effect on Lyc and Fn.

The two-way ANOVA showed that the water more pronouncedly affected the results than the nitrogen treatments, indicating that, compared with nitrogen, water occupied a dominant position in tomato yield and fruit quality, in agreement with the experimental results in northwest China [35,49].

3.4. The Interactive Relationship between Water and Nitrogen Application

Since the two-way ANOVA results indicate that the interaction of water and nitrogen was weak and unclear, marginal values (slope) were calculated to further analyze their relationship, and the results are shown in Figure 3. Tomato marginal yield of W2/3 and W1/3 increased with nitrogen application, and the increasing weakened in W1, indicating that an antagonism effect between water and nitrogen was only found under W1~W2/3 and N1~N1/3 (Figure 3). Although both water and nitrogen promoted tomato yield, their effect intensity was gradually decreased with source supply saturation, which finally caused low efficiency [8]. For OA and SAR, a reciprocal effect from water with nitrogen occurred only under N2/3~N1/3. Water and nitrogen exhibited antagonism on SW, VC and Lyc under N1~N1/3 and synergy on TSS under N1~N1/3 and W1~W2/3. The interaction of water and nitrogen tended to disappear when one of them was extremely deficient, and water took a dominated position in the interaction [49].

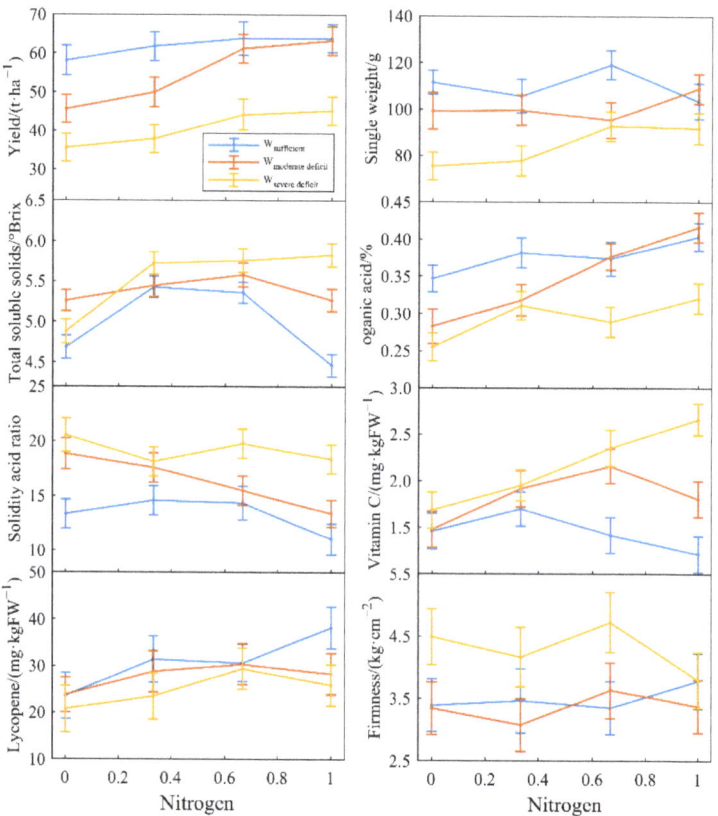

Figure 3. The interaction effect of water and nitrogen on yield, single fruit qualities parameters and comprehensive fruit quality in the first and second seasons of 2019. Notes: the vertical axis is the estimated marginal mean value of each parameter; the horizontal axis is the gradient of nitrogen including full nitrogen, mild stress, medium stress and heavy stress.

3.5. Comprehensive Quality Assessment and Economic Analysis

The weights of each individual quality index calculated by Analytic Hierarchy Process (AHP) are shown in Table 4. The random consistency ratio CR of the judgment matrix was 0.00025 < 0.1, indicating that the calculated weights met the requirements of pairwise comparison consistency. The comprehensive evaluation of the technique for order preference by similarity to ideal solution method (TOPSIS) showed that the individual fruit quality TSS and VC were the main factors affecting overall score in both seasons (Table 4), while the effect of other parameters was restricted by small weights or slight fluctuation of different treatments. Due to the correlation between pivotal quality parameters and comprehensive quality, it was feasible to focus on TSS and VC to evaluate tomato fruit in the absence of quality tests. The assessing results of comprehensive quality in both seasons were roughly consistent. T10 reached the highest quality value (Q) of 0.770 and ranked no. 1 in the first season, and T9 reached Q of 0.599 and 0.641, ranking no. 2 and no. 1 in the first and second seasons, respectively. Treatments with severe water deficit and mild nitrogen stress (T9, T10) attained the best integrated quality.

Table 4. TOPSIS analysis of comprehensive fruit quality under different water and nitrogen treatments in the first and second seasons of 2019.

Treatment Weight		SW 0.110	TSS 0.131	OA 0.088	SAR 0.180	VC 0.173	Lyc 0.198	Fn 0.116	D^+ -	D^- -	Q -	Rank -
First season	CK	0.312	0.255	0.210	0.164	0.207	0.358	0.308	0.169	0.088	0.344	11
	T2	0.321	0.283	0.236	0.218	0.281	0.325	0.256	0.137	0.088	0.392	8
	T3	0.321	0.278	0.275	0.257	0.273	0.299	0.246	0.128	0.086	0.403	7
	T4	0.295	0.230	0.270	0.201	0.180	0.232	0.237	0.178	0.040	0.182	12
	T5	0.334	0.266	0.258	0.246	0.252	0.322	0.231	0.137	0.087	0.388	9
	T6	0.309	0.295	0.245	0.238	0.328	0.274	0.277	0.124	0.090	0.421	6
	T7	0.286	0.306	0.295	0.311	0.202	0.250	0.224	0.143	0.082	0.364	10
	T8	0.281	0.292	0.348	0.341	0.211	0.281	0.240	0.126	0.101	0.445	4
	T9	0.266	0.313	0.280	0.294	0.441	0.282	0.291	0.092	0.137	0.599	2
	T10	0.257	0.328	0.329	0.351	0.431	0.369	0.388	0.051	0.173	0.770	1
	T11	0.232	0.318	0.271	0.285	0.302	0.231	0.349	0.120	0.094	0.438	5
	T12	0.224	0.285	0.395	0.445	0.207	0.185	0.358	0.134	0.141	0.513	3
	A^+	0.334	0.328	0.395	0.445	0.441	0.369	0.388				
	A^-	0.224	0.230	0.210	0.164	0.180	0.185	0.224				
	R	−0.650 *	0.748 **	0.636 *	0.762ns	0.720 **	0.000 ns	0.643 *				
Second season	CK	0.327	0.230	0.277	0.223	0.178	0.377	0.264	0.118	0.080	0.404	10
	T2	0.330	0.296	0.270	0.286	0.186	0.360	0.263	0.103	0.083	0.446	8
	T3	0.313	0.305	0.219	0.240	0.253	0.334	0.312	0.089	0.077	0.464	7
	T4	0.307	0.277	0.270	0.261	0.230	0.246	0.315	0.105	0.057	0.353	12
	T5	0.300	0.301	0.210	0.217	0.296	0.287	0.357	0.090	0.080	0.470	6
	T6	0.264	0.307	0.257	0.296	0.330	0.303	0.279	0.065	0.088	0.575	3
	T7	0.287	0.285	0.297	0.287	0.349	0.291	0.270	0.063	0.093	0.598	2
	T8	0.254	0.281	0.329	0.312	0.239	0.256	0.294	0.096	0.068	0.414	9
	T9	0.296	0.316	0.307	0.338	0.409	0.239	0.282	0.068	0.121	0.641	1
	T10	0.307	0.292	0.319	0.321	0.323	0.252	0.286	0.072	0.093	0.563	4
	T11	0.242	0.303	0.340	0.339	0.315	0.245	0.235	0.087	0.091	0.510	5
	T12	0.211	0.260	0.333	0.312	0.266	0.228	0.289	0.103	0.069	0.401	11
	A^+	0.330	0.316	0.340	0.339	0.409	0.377	0.357				
	A^-	0.211	0.230	0.210	0.217	0.178	0.228	0.235				
	R	−0.189 ns	0.734 **	−0.021 ns	0.399 ns	0.881 **	−0.042 ns	−0.308 ns				

Notes: SW, TSS, OA, SAR, VC, Lyc and Fn indicated tomato single weight, total soluble solids, organic acids, solid–acid content ratio, vitamin C content, lycopene content and fruit firmness, respectively. * means statistically significant with $p < 0.05$; ** notes statistically extreme significance with $p < 0.01$; ns represents statistically insignificant with $p > 0.05$.

The heat map of net profit percentage change compared to CK under different treatments is shown in Figure 4. Yield dominated the comprehensive benefits when the market price sensitivity was low (R0~0.6), and the highest profit ratios found in T5 and T6 were 11% and 19% in the first and second seasons, respectively. Sufficient irrigation and mild nitrogen promised a relative high yield. As the sensitivity of price improved (R0.8~1), quality became more important than yield, and severe deficit treatment with high fruit quality reached the highest profit. The profit ratio of T10 reached 31% in the first season and 33% in the second season. Similar results were also found in grapes [50].

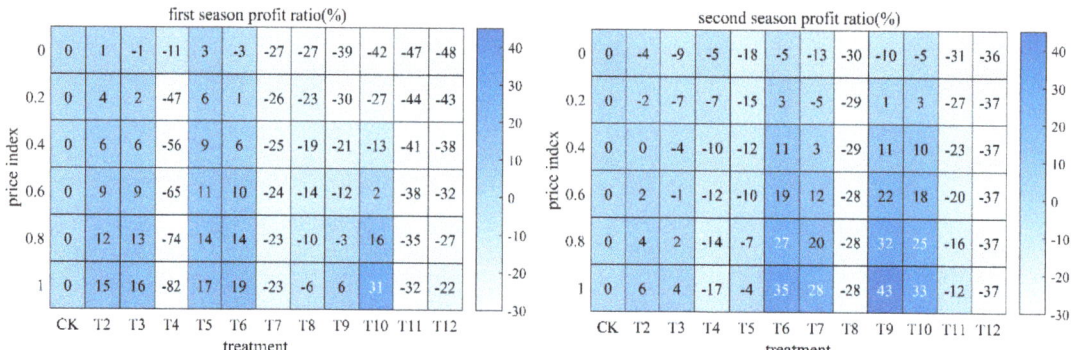

Figure 4. The tomato profit ratio in the first and second seasons of 2019.

4. Discussion

The results show that water and nitrogen application deficit decreased plant height, stem diameter, leaf area index and leaf chlorophyll content (3.1), and this is in agreement with previous studies [24,51]. Nitrogen, as an essential element in synthetic amino acids, protein and chlorophyll, could affect LAI and chlorophyll more seriously than water [27]. Since T12 (W1/3N0) always showed the lowest values in all treatments, the coupling effect of water and nitrogen was more obvious than either of them, which confirmed that under certain circumstances, nitrogen could cripple the inhibition caused by water deficit, and vice versa [40,42].

Evapotranspiration and yield decreased significantly with the decline in irrigation and nitrogen application (Section 3.2). Irrigation deficit could directly reduce root zone soil moisture and plant water use, and nitrogen stress decreased ET through restraining plant leaf development [34,52]. Water deficit decreased yield more seriously than nitrogen (Section 3.2), and the yield in the first season was lower than that of the second season, in line with the results in south Spain [53]. In the current study, the different ET and yield between the two seasons was due to the significant variances in meteorological parameters (Figure 1). The lower temperature and solar radiation in the second season reduced potential plant ET and led to diminished ET (Figure 1 and Table 2), photosynthesis and other meteorological processes, which finally caused a lower yield [54]. Water supply improved NUE, and nitrogen supply enhanced WUE (Section 3.2), which were also observed in previous water–nitrogen deficit experiments [55,56]. On the one hand, as a necessary mineral solvent, water directly participates in nitrogen uptake of plant root and restricts nitrogen use; on the other, nitrogen fertilization can promote plant root biomass, improve its absorptive capacity and enhance water use efficiency [28,57]. In addition to nitrogen, other fertilizers also play an important role in plant growth [58]. As soil–plant–air is a continuous system, the interactions between soil characteristic and fertigation still demand research [59].

Fruit quality parameters, e.g., TSS, SAR and VC, increased with the decline in water supply (Section 3.3). Reduced irrigation decreased water content in tomato fruit, which formed a concentration effect [7]. VC synthesis and the conversion of acid to sugar can be improved by more sugar and less water in fruit [52]. Regarding nitrogen, fruit quality increased first and then decreased with the deficit of nitrogen fertilizer (Section 3.3). Previous studies reported that nitrogen fertilizer provided essential biochemical material and enhanced tomato fruit qualities [7,25,31], while excess nitrogen fertilizer application decreased lycopene and VC content in fruit. The sufficient nitrogen level in this study probably exceeded the actual nitrogen demand according to conventional nitrogen application strategy, and thus, moderate nitrogen deficit treatments had a higher TSS and VC content compared with full nitrogen fertilization.

Although both water and nitrogen affected yield and fruit quality significantly, the integrated relationship between them only existed within a certain threshold (Section 3.4).

The stress adaptive capacity of plants was limited by physiological metabolic process, and water–nitrogen interaction mostly existed in the moderate deficit section. Once the deficit lever of a factor exceeded the threshold, the other influencing factor could probably no longer cooperate with or antagonize the deficit factor, and then, they only would affect plants separately [60–62]. This might explain the non-significance of the interaction effect of water and nitrogen in this study and many earlier studies [34,41,42]. For deeper research, a molecular biology study is needed to determine the variances in hormone signals, key enzyme activities, active genes in a plant when suffering deficit and the inner mechanism of fertigation application. TSS achieved dominance in comprehensive quality evaluation with a high weight and a large fluctuation of water–nitrogen deficit (Section 3.5), which could be the main proxy to judge fruit quality when measurements are limited. In economic analysis, the prices only referred to the local market in Shandong province, China, where the water and labor cost was relatively low and fertilizer cost was relatively high, and the final water and nitrogen application decision may not suitable to other places where the resources prices differ too much. Specific analysis was needed for localization.

5. Conclusions

Deficit irrigation and nitrogen application restrained tomato growth, including plant height, stem diameter and LAI, slightly. Tomato yield and partial quality indicators (SW and OA) decreased with the decline in water and nitrogen. For TSS and VC, water deficit had a promoting effect, while nitrogen deficiency showed an inhibitory impact. Water and nitrogen deficit positively impacted SAR. Water showed greater influence on tomatoes compared with nitrogen, and their integrated relationship was exhibited within the mild deficit threshold of N1~N1/3 and W1~W2/3.

Based on water and nitrogen use efficiency, severe water and nitrogen deficit (W1/3N1/3) was the best scheme; in terms of high fruit quality and net profit, severe water deficit and mild nitrogen deficiency (W1/3N2/3) was optimal. In summary, the 1/3 full irrigation and mild nitrogen deficit (N2/3, N1/3) treatment produced higher profits and is recommended to tomato cultivation industry.

Author Contributions: Conceptualization, X.J. and N.C.; methodology, Y.Q.; software, L.Z. and S.J.; validation, Y.W., Y.Z. and S.J.; formal analysis, M.F. and Y.Z.; investigation, Y.Q.; resources, X.J. and N.C.; data curation, X.J. and M.F.; writing—original draft preparation, M.F., Y.Q. and X.J.; writing—review and editing, M.F.; visualization, M.F.; supervision, Y.W. and N.C.; project administration, X.J., N.C. and L.Z.; funding acquisition, N.C. and Y.W. All authors have read and agreed to the published version of the manuscript.

Funding: This study was financially supported by the Key Research and Development Program of Beijing (Z20111000800000), the National Natural Science Foundation of China (51709203, 51922072, 51779161, 51009101), the Science and Technology Projects of Sichuan (22ZDYF0145, 22QYCX0069, 22QYCX0073, 22QYCX0115) and the Fundamental Research Funds for the Central Universities (2019CDLZ-10, 2020CDDZ-19).

Institutional Review Board Statement: Not applicable.

Informed Consent Statement: Not applicable.

Data Availability Statement: The data presented in this study are available on request from the corresponding author.

Conflicts of Interest: The authors declare no conflict of interest.

References

1. Giuliani, M.M.; Carucci, F.; Nardella, E.; Francavilla, M.; Ricciardi, L.; Lotti, C.; Gatta, G. Combined effects of deficit irrigation and strobilurin application on gas exchange, yield and water use efficiency in tomato (*Solanum lycopersicum* L.). *Sci. Hortic.* **2018**, *233*, 149–158. [CrossRef]
2. Kumar, M.; Tomar, M.; Bhuyan, D.J.; Punia, S.; Grasso, S.; Sá, A.G.A.; Carciofi, B.A.M.; Arrutia, F.; Changan, S.; Radha; et al. Tomato (*Solanum lycopersicum* L.) seed: A review on bioactives and biomedical activities. *Biomed. Pharmacother.* **2021**, *142*, 112018. [CrossRef] [PubMed]
3. Costa-Rodrigues, J.; Pinho, O.; Monteiro, P.R.R. Can lycopene be considered an effective protection against cardiovascular disease? *Food Chem.* **2018**, *245*, 1148–1153. [CrossRef] [PubMed]
4. Li, N.; Wu, X.; Zhuang, W.; Xia, L.; Chen, Y.; Wu, C.; Rao, Z.; Du, L.; Zhao, R.; Yi, M.; et al. Tomato and lycopene and multiple health outcomes: Umbrella review. *Food Chem.* **2020**, *343*, 128396. [CrossRef]
5. Rattanavipanon, W.; Nithiphongwarakul, C.; Sirisuwansith, P.; Chaiyasothi, T.; Thakkinstian, A.; Nathisuwan, S.; Pathomwichai-wat, T. Effect of tomato, lycopene and related products on blood pressure: A systematic review and network meta-analysis. *Phytomedicine* **2021**, *88*, 153512. [CrossRef]
6. FAOSTAT, Food and Agricultural Organization of United Nations, Rome. 2018. Available online: https://www.fao.org/faostat/es/#data (accessed on 16 April 2022).
7. Chen, J.; Kang, S.; Du, T.; Qiu, R.; Guo, P.; Chen, R. Quantitative response of greenhouse tomato yield and quality to water deficit at different growth stages. *Agric. Water Manag.* **2013**, *129*, 152–162. [CrossRef]
8. Valcárcel, M.; Lahoz, I.; Campillo, C.; Martí, R.; Leiva-Brondo, M.; Roselló, S.; Cebolla-Cornejo, J. Controlled deficit irrigation as a water-saving strategy for processing tomato. *Sci. Hortic.* **2019**, *261*, 108972. [CrossRef]
9. Khapte, P.; Kumar, P.; Burman, U.; Kumar, P. Deficit irrigation in tomato: Agronomical and physio-biochemical implications. *Sci. Hortic.* **2019**, *248*, 256–264. [CrossRef]
10. Gu, X.-B.; Li, Y.-N.; Du, Y.-D. Effects of ridge-furrow film mulching and nitrogen fertilization on growth, seed yield and water productivity of winter oilseed rape (*Brassica napus* L.) in Northwestern China. *Agric. Water Manag.* **2018**, *200*, 60–70. [CrossRef]
11. Singh, M.; Singh, P.; Singh, S.; Saini, R.K.; Angadi, S.V. A global meta-analysis of yield and water productivity responses of vegetables to deficit irrigation. *Sci. Rep.* **2021**, *11*, 22095. [CrossRef] [PubMed]
12. Yu, C.; Huang, X.; Chen, H.; Godfray, H.C.J.; Wright, J.S.; Hall, J.W.; Gong, P.; Ni, S.Q.; Qiao, S.C.; Huang, G.R.; et al. Managing nitrogen to restore water quality in China. *Nature* **2019**, *567*, 516–520. [CrossRef] [PubMed]
13. Hooshmand, M.; Albaji, M.; Nasab, S.B.; Ansari, N.A.Z. The effect of deficit irrigation on yield and yield components of greenhouse tomato (*Solanum lycopersicum*) in hydroponic culture in Ahvaz region, Iran. *Sci. Hortic.* **2019**, *254*, 84–90. [CrossRef]
14. Jiang, X.; Zhao, Y.; Wang, R.; Zhao, S. Modeling the Relationship of Tomato Yield Parameters with Deficit Irrigation at Different Growth Stages. *HortScience* **2019**, *54*, 1492–1500. [CrossRef]
15. Lu, J.; Shao, G.; Gao, Y.; Zhang, K.; Wei, Q.; Cheng, J. Effects of water deficit combined with soil texture, soil bulk density and tomato variety on tomato fruit quality: A meta-analysis. *Agric. Water Manag.* **2021**, *243*, 106427. [CrossRef]
16. Dariva, F.D.; Pessoa, H.P.; Copati, M.G.F.; de Almeida, G.Q.; Filho, M.N.D.C.; Picoli, E.A.D.T.; da Cunha, F.F.; Nick, C. Yield and fruit quality attributes of selected tomato introgression lines subjected to long-term deficit irrigation. *Sci. Hortic.* **2021**, *289*, 110426. [CrossRef]
17. Martí, R.; Valcárcel, M.; Leiva-Brondo, M.; Lahoz, I.; Campillo, C.; Roselló, S.; Cebolla-Cornejo, J. Influence of controlled deficit irrigation on tomato functional value. *Food Chem.* **2018**, *252*, 250–257. [CrossRef] [PubMed]
18. Gong, X.W.; Qiu, R.J.; Sun, J.S.; Ge, J.K.; Li, Y.B.; Wang, S.S. Evapotranspiration and crop coefficient of tomato grown in a solar greenhouse under full and deficit irrigation. *Agric. Water Manag.* **2020**, *235*, 106154. [CrossRef]
19. Wu, Y.; Yan, S.; Fan, J.; Zhang, F.; Xiang, Y.; Zheng, J.; Guo, J. Responses of growth, fruit yield, quality and water productivity of greenhouse tomato to deficit drip irrigation. *Sci. Hortic.* **2021**, *275*, 109710. [CrossRef]
20. Agbna, G.H.; Dongli, S.; Zhipeng, L.; Elshaikh, N.A.; Guangcheng, S.; Timm, L.C. Effects of deficit irrigation and biochar addition on the growth, yield, and quality of tomato. *Sci. Hortic.* **2017**, *222*, 90–101. [CrossRef]
21. DU, Y.-D.; Cao, H.-X.; Liu, S.-Q.; Gu, X.; Cao, Y.-X. Response of yield, quality, water and nitrogen use efficiency of tomato to different levels of water and nitrogen under drip irrigation in Northwestern China. *J. Integr. Agric.* **2017**, *16*, 1153–1161. [CrossRef]
22. Lu, B.; Qian, J.; Hu, J.; Wang, P.; Jin, W.; Tang, S.; He, Y.; Zhang, C. The role of fine root morphology in nitrogen uptake by riparian plants. *Plant Soil* **2022**, *472*, 527–542. [CrossRef]
23. Biel, C.; Camprubí, A.; Lovato, P.E.; Calvet, C. On-farm reduced irrigation and fertilizer doses, and arbuscular mycorrhizal fungal inoculation improve water productivity in tomato production. *Sci. Hortic.* **2021**, *288*, 110337. [CrossRef]
24. Maroušek, J.; Maroušková, A. Economic Considerations on Nutrient Utilization in Wastewater Management. *Energies* **2021**, *14*, 3468. [CrossRef]
25. Kusano, M.; Fukushima, A.; Redestig, H.; Saito, K. Metabolomic approaches toward understanding nitrogen metabolism in plants. *J. Exp. Bot.* **2011**, *62*, 1439–1453. [CrossRef] [PubMed]
26. Muttucumaru, N.; Powers, S.J.; Elmore, J.S.; Mottram, D.S.; Halford, N.G. Effects of Nitrogen and Sulfur Fertilization on Free Amino Acids, Sugars, and Acrylamide-Forming Potential in Potato. *J. Agric. Food Chem.* **2013**, *61*, 6734–6742. [CrossRef] [PubMed]

27. Zhang, H.; Liu, X.; Song, B.; Nie, B.; Zhang, W.; Zhao, Z. Effect of excessive nitrogen on levels of amino acids and sugars, and differential response to post-harvest cold storage in potato (*Solanum tuberosum* L.) tubers. *Plant Physiol. Biochem.* **2020**, *157*, 38–46. [CrossRef] [PubMed]
28. Yang, X.; Zhang, P.; Wei, Z.; Liu, J.; Hu, X.; Liu, F. Effects of elevated CO2 and nitrogen supply on leaf gas exchange, plant water relations and nutrient uptake of tomato plants exposed to progressive soil drying. *Sci. Hortic.* **2021**, *292*, 110643. [CrossRef]
29. Cheng, M.; Wang, H.; Fan, J.; Xiang, Y.; Tang, Z.; Pei, S.; Zeng, H.; Zhang, C.; Dai, Y.; Li, Z.; et al. Effects of nitrogen supply on tomato yield, water use efficiency and fruit quality: A global meta-analysis. *Sci. Hortic.* **2021**, *290*, 110553. [CrossRef]
30. Li, Y.; Sun, Y.; Liao, S.; Zou, G.; Zhao, T.; Chen, Y.; Yang, J.; Zhang, L. Effects of two slow-release nitrogen fertilizers and irrigation on yield, quality, and water-fertilizer productivity of greenhouse tomato. *Agric. Water Manag.* **2017**, *186*, 139–146. [CrossRef]
31. Ochoa-Velasco, C.E.; Valadez-Blanco, R.; Salas-Coronado, R.; Sustaita-Rivera, F.; Hernández-Carlos, B.; García-Ortega, S.; Santos-Sánchez, N.F. Effect of nitrogen fertilization and Bacillus licheniformis biofertilizer addition on the antioxidants compounds and antioxidant activity of greenhouse cultivated tomato fruits (*Solanum lycopersicum* L. var. Sheva). *Sci. Hortic.* **2016**, *201*, 338–345. [CrossRef]
32. Geisseler, D.; Aegerter, B.J.; Miyao, E.M.; Turini, T.; Cahn, M.D. Nitrogen in soil and subsurface drip-irrigated processing tomato plants (*Solanum lycopersicum* L.) as affected by fertilization level. *Sci. Hortic.* **2019**, *261*, 108999. [CrossRef]
33. Zotarelli, L.; Dukes, M.; Scholberg, J.; Muñoz-Carpena, R.; Icerman, J. Tomato nitrogen accumulation and fertilizer use efficiency on a sandy soil, as affected by nitrogen rate and irrigation scheduling. *Agric. Water Manag.* **2009**, *96*, 1247–1258. [CrossRef]
34. Wang, C.; Gu, F.; Chen, J.; Yang, H.; Jiang, J.; Du, T.; Zhang, J. Assessing the response of yield and comprehensive fruit quality of tomato grown in greenhouse to deficit irrigation and nitrogen application strategies. *Agric. Water Manag.* **2015**, *161*, 9–19. [CrossRef]
35. Lu, J.; Shao, G.; Cui, J.; Wang, X.; Keabetswe, L. Yield, fruit quality and water use efficiency of tomato for processing under regulated deficit irrigation: A meta-analysis. *Agric. Water Manag.* **2019**, *222*, 301–312. [CrossRef]
36. Zhou, H.P.; Kang, S.Z.; Li, F.S.; Du, T.S.; Shukla, M.K.; Li, X.J. Nitrogen application modified the effect of deficit irrigation on tomato transpiration, and water use efficiency in different growth stages. *Sci. Hortic.* **2020**, *263*, 109112. [CrossRef]
37. Zhang, Q.; Du, Y.; Cui, B.; Sun, J.; Wang, J.; Wu, M.; Niu, W. Aerated irrigation offsets the negative effects of nitrogen reduction on crop growth and water-nitrogen utilization. *J. Clean. Prod.* **2021**, *313*, 127917. [CrossRef]
38. Du, Y.-D.; Zhang, Q.; Cui, B.-J.; Sun, J.; Wang, Z.; Ma, L.-H.; Niu, W.-Q. Aerated irrigation improves tomato yield and nitrogen use efficiency while reducing nitrogen application rate. *Agric. Water Manag.* **2020**, *235*, 106152. [CrossRef]
39. Zotarelli, L.; Scholberg, J.M.; Dukes, M.; Muñoz-Carpena, R.; Icerman, J. Tomato yield, biomass accumulation, root distribution and irrigation water use efficiency on a sandy soil, as affected by nitrogen rate and irrigation scheduling. *Agric. Water Manag.* **2009**, *96*, 23–34. [CrossRef]
40. Li, H.; Liu, H.; Gong, X.; Li, S.; Pang, J.; Chen, Z.; Sun, J. Optimizing irrigation and nitrogen management strategy to trade off yield, crop water productivity, nitrogen use efficiency and fruit quality of greenhouse grown tomato. *Agric. Water Manag.* **2020**, *245*, 106570. [CrossRef]
41. Ran, H.; Kang, S.; Li, F.; Du, T.; Ding, R.; Li, S.; Tong, L. Responses of water productivity to irrigation and N supply for hybrid maize seed production in an arid region of Northwest China. *J. Arid Land* **2017**, *9*, 504–514. [CrossRef]
42. Hong, T.; Cai, Z.; Li, R.; Liu, J.; Li, J.; Wang, Z.; Zhang, Z. Effects of water and nitrogen coupling on watermelon growth, photosynthesis and yield under CO2 enrichment. *Agric. Water Manag.* **2021**, *259*, 107229. [CrossRef]
43. Rana, G.; Katerji, N. Measurement and estimation of actual evapotranspiration in the field under Mediterranean climate: A review. *Eur. J. Agron.* **2000**, *13*, 125–153. [CrossRef]
44. Gong, X.; Qiu, R.; Zhang, B.; Wang, S.; Ge, J.; Gao, S.; Yang, Z. Energy budget for tomato plants grown in a greenhouse in northern China. *Agric. Water Manag.* **2021**, *255*, 107039. [CrossRef]
45. Godana, E.A.; Yang, Q.; Wang, K.; Zhang, H.; Zhang, X.; Zhao, L.; Abdelhai, M.H.; Legrand, N.N.G. Bio-control activity of Pichia anomala supplemented with chitosan against Penicillium expansum in postharvest grapes and its possible inhibition mechanism. *LWT* **2020**, *124*, 109188. [CrossRef]
46. García-Robledo, E.; Corzo, A.; Papaspyrou, S. A fast and direct spectrophotometric method for the sequential determination of nitrate and nitrite at low concentrations in small volumes. *Mar. Chem.* **2014**, *162*, 30–36. [CrossRef]
47. Sharma, S.K.; Le Maguer, M. Kinetics of lycopene degradation in tomato pulp solids under different processing and storage conditions. *Food Res. Int.* **1996**, *29*, 309–315. [CrossRef]
48. Wang, F.; Kang, S.; Du, T.; Li, F.; Qiu, R. Determination of comprehensive quality index for tomato and its response to different irrigation treatments. *Agric. Water Manag.* **2011**, *98*, 1228–1238. [CrossRef]
49. He, Z.; Li, M.; Cai, Z.; Zhao, R.; Hong, T.; Yang, Z.; Zhang, Z. Optimal irrigation and fertilizer amounts based on multi-level fuzzy comprehensive evaluation of yield, growth and fruit quality on cherry tomato. *Agric. Water Manag.* **2020**, *243*, 106360. [CrossRef]
50. Jiang, X.; Liu, B.; Guan, X.; Wang, Z.; Wang, B.; Zhao, S.; Song, Y.; Zhao, Y.; Bi, J. Proper deficit irrigation applied at various stages of growth can maintain yield and improve the comprehensive fruit quality and economic return of table grapes grown in greenhouses. *Irrig. Drain.* **2021**, *70*, 1056–1072. [CrossRef]
51. Lu, B.; Qian, J.; Wang, P.; Wang, C.; Hu, J.; Li, K.; He, X.; Jin, W. Effect of perfluorooctanesulfonate (PFOS) on the rhizosphere soil nitrogen cycling of two riparian plants. *Sci. Total Environ.* **2020**, *741*, 140494. [CrossRef]

52. Wu, Y.; Yan, S.; Fan, J.; Zhang, F.; Zhao, W.; Zheng, J.; Guo, J.; Xiang, Y.; Wu, L. Combined effects of irrigation level and fertilization practice on yield, economic benefit and water-nitrogen use efficiency of drip-irrigated greenhouse tomato. *Agric. Water Manag.* **2021**, *262*, 107401. [CrossRef]

53. Coyago-Cruz, E.; Meléndez-Martínez, A.J.; Moriana, A.; Girón, I.F.; Martín-Palomo, M.J.; Galindo, A.; Pérez-López, D.; Torrecillas, A.; Beltrán-Sinchiguano, E.; Corell, M. Yield response to regulated deficit irrigation of greenhouse cherry tomatoes. *Agric. Water Manag.* **2018**, *213*, 212–221. [CrossRef]

54. Du, M.; Zhang, J.; Wang, Y.; Liu, H.; Wang, Z.; Liu, C.; Yang, Q.; Hu, Y.; Bao, Z.; Liu, Y.; et al. Evaluating the contribution of different environmental drivers to changes in evapotranspiration and soil moisture, a case study of the Wudaogou Experimental Station. *J. Contam. Hydrol.* **2021**, *243*, 103912. [CrossRef] [PubMed]

55. Liu, R.; Yang, Y.; Wang, Y.-S.; Wang, X.-C.; Rengel, Z.; Zhang, W.-J.; Shu, L.-Z. Alternate partial root-zone drip irrigation with nitrogen fertigation promoted tomato growth, water and fertilizer-nitrogen use efficiency. *Agric. Water Manag.* **2020**, *233*, 106049. [CrossRef]

56. Wei, Z.; Fang, L.; Li, X.; Liu, J.; Liu, F. Endogenous ABA level modulates the effects of CO_2 elevation and soil water deficit on growth, water and nitrogen use efficiencies in barley and tomato plants. *Agric. Water Manag.* **2021**, *249*, 106808. [CrossRef]

57. An, T.; Wu, Y.; Xu, B.; Zhang, S.; Deng, X.; Zhang, Y.; Siddique, K.H.; Chen, Y. Nitrogen supply improved plant growth and Cd translocation in maize at the silking and physiological maturity under moderate Cd stress. *Ecotoxicol. Environ. Saf.* **2021**, *230*, 113137. [CrossRef]

58. Maroušek, J.; Maroušková, A.; Zoubek, T.; Bartoš, P. Economic impacts of soil fertility degradation by traces of iron from drinking water treatment. *Environ. Dev. Sustain.* **2021**, *24*, 4835–4844. [CrossRef]

59. Maroušek, J.; Trakal, L. Techno-economic analysis reveals the untapped potential of wood biochar. *Chemosphere* **2021**, *291*, 133000. [CrossRef]

60. Hao, S.; Cao, H.; Wang, H.; Pan, X. The physiological responses of tomato to water stress and re-water in different growth periods. *Sci. Hortic.* **2019**, *249*, 143–154. [CrossRef]

61. Hernandez-Espinoza, L.H.; Barrios-Masias, F.H. Physiological and anatomical changes in tomato roots in response to low water stress. *Sci. Hortic.* **2020**, *265*, 109208. [CrossRef]

62. Sánchez-Rodríguez, E.; del Mar Rubio-Wilhelmi, M.; Blasco, B.; Leyva, R.; Romero, L.; Ruiz, J.M. Antioxidant response resides in the shoot in reciprocal grafts of drought-tolerant and drought-sensitive cultivars in tomato under water stress. *Plant Sci.* **2012**, *188–189*, 89–96. [CrossRef] [PubMed]

 *agronomy*

Article

Effects of Reducing Nitrogen Application Rate under Different Irrigation Methods on Grain Yield, Water and Nitrogen Utilization in Winter Wheat

Jinpeng Li [1], Zhimin Wang [2], Youhong Song [1], Jincai Li [1] and Yinghua Zhang [2,*]

[1] School of Agronomy, Anhui Agricultural University, Hefei 230036, China; jinpeng@ahu.edu.cn (J.L.); y.song@ahau.edu.cn (Y.S.); ljc5122423@126.com (J.L.)
[2] College of Agronomy and Biotechnology, China Agricultural University, No. 2 Yuanmingyuan West Road, Beijing 100193, China; zhimin206@263.net
* Correspondence: yhzhang@cau.edu.cn

Citation: Li, J.; Wang, Z.; Song, Y.; Li, J.; Zhang, Y. Effects of Reducing Nitrogen Application Rate under Different Irrigation Methods on Grain Yield, Water and Nitrogen Utilization in Winter Wheat. *Agronomy* 2022, 12, 1835. https://doi.org/10.3390/agronomy12081835

Academic Editors: Pantazis Georgiou and Dimitris Karpouzos

Received: 12 July 2022
Accepted: 30 July 2022
Published: 2 August 2022

Publisher's Note: MDPI stays neutral with regard to jurisdictional claims in published maps and institutional affiliations.

Abstract: We conducted a two-year field experiment on winter wheat (*Triticum aestivum* L.) from 2016–2018 to compare the effects of reducing nitrogen application rate in spring under three irrigation methods on grain yield (GY), water and nitrogen use efficiency in the North China Plain (NCP). Across the two years, GY of conventional irrigation (CI), micro-sprinkling irrigation (SI) and drip irrigation (DI) decreased by 6.35%, 9.84% and 6.83%, respectively, in the reduced nitrogen application rate (N45) than the recommended nitrogen application rate (N90). However, micro-irrigation (SI and DI) significantly increased GY relative to CI under the same nitrogen application rate, and no significant difference was observed in GY between SI and DI under N45, while SI obtained the highest GY under N90. The difference among different treatments in GY was mainly due to the variation in grain weight. The seasonal evapotranspiration (ET) in N45 was decreased more significantly than N90, and there was no significantly difference in ET among different irrigation methods under N45, but micro-irrigation significantly decreased the ET relative to CI under N90. Micro-irrigation significantly improved water use efficiency (WUE) compared to CI at the same nitrogen application rate. Under N45, compared with CI, WUE in SI and DI increased by 9.09% and 4.70%, respectively; however, the WUE increased by 15.9% and 7.23%, respectively, under N90. Reducing nitrogen application rate did not have a significant impact on WUE under CI, but it did have a substantial negative impact on SI and DI. Nitrogen accumulation in wheat plants at maturity (NAM) in N45 deceased significantly compared with N90 under the same irrigation method. Compared with CI under the same nitrogen application rate, micro-irrigation treatments significantly increased NAM, while SI was the largest. In comparison to N90, under three irrigation methods, N45 significantly increased nitrogen fertilizer use efficiency (N_fUE). The highest N_fUE was attained in SI, followed by DI, while CI was the lowest. Moreover, N45 significantly decreased soil NO_3^--N accumulation (SNC) in three irrigation methods, and micro-irrigation significantly decreased the SNC in deep soil layers compared with CI when nitrogen is applied at the same level. Overall, micro-irrigation with a reduced nitrogen application rate in spring can achieve a relatively higher production of winter wheat while increasing the use efficiency of water and nitrogen and reducing soil NO_3^--N leaching into deep soil layers in the NCP.

Keywords: winter wheat; micro-irrigation; grain yield; nitrogen reduction; water and nitrogen utilization

1. Introduction

North China Plain (NCP) is the main production base of winter wheat in China. The planting area of winter wheat reached 1752.8×10^4 hm^2 in the NCP, which accounts for 71.5% and 79.7% of the national total planting area and output of winter wheat in China, respectively [1]. Therefore, it is crucial to guarantee the long-term production of winter wheat in this area. Although the annual precipitation is 500–950 mm in the

NCP, it mainly concentrates in summer, and the lower precipitation during spring causes drought disasters of winter wheat [2]. The total water consumption for winter wheat production is 433 mm, 413 mm and 373 mm in dry, normal and wet years, respectively [3], and the most generally recognized method of meeting the water requirements of wheat cultivation in the NCP is irrigation. However, the extraction of groundwater for wheat growth and development in the NCP is responsible for around 70% of the need for irrigated water [4,5], and over-exploitation of groundwater had caused continuous ground settlement over the past four decades [6,7], resulting in this region becoming one of the deepest groundwater cones of depression on Earth [8]. Additionally, improper irrigation methods have increased the groundwater resource consumption and reduced the nitrogen fertilizer use efficiency (N_fUE), leading to environmental degradation from water and nitrogen loss [9,10]. Therefore, it is of great significance for the NCP to reduce the nitrogen input and improve the water and nitrogen use efficiency of winter wheat.

Previous studies have reported that optimizing irrigation regimes to promote wheat root growth into deep soil layers and increase water utilization in deep soil is an important means to improve grain yield (GY) and water use efficiency (WUE) of winter wheat [11,12]. However, the unreasonable irrigation methods of wheat often reduce the water absorption from deep soil layers, and the conventional surface irrigation readily causes water and fertilizer to migrate out of the wheat's main root zone, lowering WUE [7,13]. Currently, most farmers in the NCP use flooding irrigation to irrigate their winter wheat fields, and Xu et al. (2018) advocated that irrigation at the jointing and anthesis stages of wheat could achieve a higher GY and WUE [11]. The most common micro-irrigation techniques now applied in wheat production are surface drip irrigation and micro-sprinkling irrigation in the NCP. Studies have shown that micro-irrigation significantly improved crop yield and WUE by ensuring a water supply during the crucial growth stages of winter wheat through reducing the irrigation volume each time and increasing irrigation frequencies compared with the conventional irrigation practice [13,14]. Furthermore, it has been demonstrated that micro-irrigation simultaneously enhanced WUE and N_fUE in wheat by co-locating water and nitrogen fertilizer applications with root distribution of winter wheat in the soil profile [14–16]. Previous studies also indicated that subjecting wheat to micro-irrigation could significantly increase leaf area index and chlorophyll content in leaf during grain filling, extend the leaf function period, enhance photosynthetic rate in plants, prevent premature senescence of wheat and increase grain weight and GY [17,18].

Nitrogen is one of the most important essential elements for crop growth [19]. Studies indicated that crops' growth was highly impacted by the soil's water conditions as well as the nitrogen application rate, and a suitable irrigation amount combined with nitrogen supply could boost the crops' growth and increase GY [20,21]. Presently, the recommended nitrogen application rate for wheat production with a target of high yield and efficient resource utilization is 180–240 kg·ha^{-1} in the NCP, and it is divided into basal dressing and top dressing [22]. The prevailing nitrogen top dressing by farmers for winter wheat in this region is applied artificially combined with irrigation in spring, however, the inadvisable nitrogen application rate for winter wheat production has led to a lower N_fUE [23], and the disposable top dressing accompanied by irrigation easily caused nitrogen to migrate out of the main root zone of wheat and leach into deeper soil layers [14]. Wu et al. (2008) [24] found that the reduction in nitrogen application rate for wheat greatly improved the N_fUE while maintaining a higher yield. Even though an excessive nitrogen application rate may not have a major impact on crop yield, it can have a considerable impact on the amount of nitrogen residue in the soil after wheat harvest and decrease the use efficiency of nitrogen [25,26]. In particular, the rainy season comes soon after wheat harvest in the NCP, which easily exacerbates the residual nitrogen leaching into deeper soil layers. Therefore, the development of an optimized irrigation and fertilization regime is urgently needed to improve the GY, WUE and N_fUE of winter wheat in the NCP. Furthermore, previous studies were focused on the physiological mechanism of micro-irrigation with water and nitrogen integration to achieve high yield and efficient utilization of water and nitrogen in wheat. However, there is insufficient research on the effect of reducing the nitrogen application rate under different irrigation methods on grain yield and water and nitrogen utilization in winter wheat. In this study, we hypothesized that, after increasing irrigation frequency and reducing nitrogen application rate, micro-irrigation with water and nitrogen integration could delay leaf senescence during grain filling and improve dry matter production post-anthesis so as to ensure grain yield, promote the absorption and utilization of soil water and nitrogen, increase nitrogen accumulation in plants and reduce the nitrate leaching to deep soil layers so as to increase WUE and N_fUE. To confirm this hypothesis, a two-year field experiment was conducted to identify the effects of reducing nitrogen application rate under different irrigation methods on (1) grain yield and yield components of winter wheat, (2) leaf area index (LAI), chlorophyll content of flag leaf after anthesis and dry matter accumulation, (3) water and nitrogen utilization in winter wheat. We expect the research results to provide a theoretical basis and technical reference for high winter wheat yield and efficient utilization of water and nitrogen in the NCP.

2. Materials and Methods

2.1. Experimental Site

At the experimental site of the China Agricultural University in Wuqiao County, Cangzhou City, Hebei Province (37°41′02″ N, 116°37′23″ E), an in situ field experiment was carried out from 2016 to 2018 throughout the winter wheat growing seasons. The altitude of the experimental site is 20 m. The field's soil is a light loam consisting of 11.8% clay, 78.1% silt and 10.1% sand. In this two-year trial, summer maize was the previous crop before wheat. Mean bulk densities in 0–100 cm and 100–200 cm soil layers are both 1.43 g·cm^{-3}. The soil total nitrogen, organic matter content, available phosphorus and potassium in the upper 40 cm of the soil layer before sowing were 0.95 g·kg^{-1}, 11.7 g·kg^{-1}, 104.4 mg·kg^{-1} and 29.2 mg·kg^{-1}, respectively; soil pH was 7.5. The nitrate nitrogen (NO_3^--N) content before sowing in the 0–100 cm soil layer was 18.6 mg·kg^{-1} and 12.7 mg·kg^{-1} in 2016–2017 and 2017–2018 winter wheat growing seasons. The total precipitation in the two wheat growing seasons was 95.5 mm and 185.6 mm, respectively. Figure 1 shows the climatic data in different months of the winter wheat growing season during this experiment.

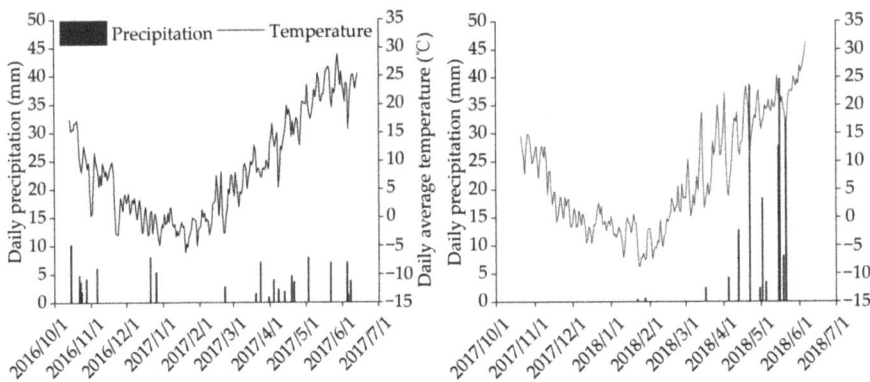

Figure 1. Daily precipitation and average air temperature recorded during the growing seasons of winter wheat from 2016–2018 at the experiment site in this present study.

2.2. Experimental Design

In this experiment, conventional irrigation (CI), micro-sprinkling irrigation (SI) and drip irrigation (DI) were used over the two growing seasons of winter wheat. Wheat growth stages were recorded by the Zadoks scale [27]. Wheat was irrigated with 60 mm in each irrigation event and 120 mm of total irrigation during the jointing (Z31) and anthesis stages (Z61) in CI. SI and DI were carried out with 120 mm of total irrigation and 30 mm in each irrigation event at jointing (Z31), booting (Z45), anthesis (Z61) and filling stages (Z71). CI was carried out using PVC pipe, and SI and DI with hoses [14,28]. The length of the hoses for micro-irrigation treatments (SI and DI) was 30 m. The flow rate of micro-sprinkling hoses was $6.0 \text{ m}^3 \cdot \text{h}^{-1}$, and the sprinkling angle was 80°. The specific details of the layout of micro-sprinkling hose in this experiment field were according to Li et al. [14]. The distance between adjacent drip hoses was 30 cm, and the drip laterals had 30 cm emitter spacing and a flow rate of $2.0 \text{ L} \cdot \text{h}^{-1}$, with a worked pressure of 0.1 MPa. In this experiment, the irrigation water source is fresh water drawn from a well. Three replicates were used in the randomized complete block experimental design. Each experimental plot was 4 m × 30 m. Before sowing, $105 \text{ kg} \cdot \text{ha}^{-1}$ nitrogen, $120 \text{ kg} \cdot \text{ha}^{-1} \text{ P}_2\text{O}_5$ and $90 \text{ kg} \cdot \text{ha}^{-1} \text{ K}_2\text{O}$ were applied as base fertilizer. During the spring season of wheat growth, $45 \text{ kg} \cdot \text{ha}^{-1}$ nitrogen (N45) and $90 \text{ kg} \cdot \text{ha}^{-1}$ nitrogen (N90) were applied using urea (nitrogen content of 46.4%) as top dressing under different irrigation methods. For each irrigation event of SI and DI, a quarter of the top dressing urea was completely dissolved in a fertilization device and applied together with the irrigation, while all nitrogen was artificially spread over the field before irrigation at the jointing stage for CI. One of the most extensively grown varieties in the NCP, the high-yield winter wheat cultivar "Jimai22", was utilized in the experiment. At a planting density of 540 plants per square meter, the wheat was sown on 14 October 2016 and 22 October 2017, and it was harvested on 14 June 2017 and 10 June 2018, respectively. No visible pests or diseases happened in the experimental field during the test period.

2.3. Sampling and Measurements

2.3.1. Water Consumption and Use Efficiency

At sowing and maturity, soil samples were taken from 0 to 200 cm soil depth at 20 cm intervals with a soil corer. The soil water content $(\text{g} \cdot \text{g}^{-1})$ was measured using the oven-drying method. Some fresh soil samples were retained in each soil layer to determine soil $NO_3^- \text{-N}$ content. The difference between the soil water storage (0–200 cm) at sowing and maturity was used to calculate the amount of soil water consumption. The soil water balance equation was used to determine the total seasonal evapotranspiration (ET) during the growth stage of winter wheat [29]:

$$ET = I + P \pm SW - R + CR - D$$

where I (mm) is irrigation, P (mm) is precipitation recorded at the nearby meteorological station, SW (mm) is soil water consumption based on the difference between sowing and maturity, R is surface runoff, CR is capillary rise into the root zone and D is downward flux below the 200 cm soil layer. Due to the lower rainfall during wheat growing seasons in this region, deep soil layer and the large water holding capacity, plus the border of the experimental plots, runoff was rarely observed in the field and it was taken as zero. D can also be ignored in the NCP, including at the experimental site [30]. The soil water consumption percentage for 0–100 cm (SW1) and 100–200 cm (SW2) was the ratio of its water consumption volume to SW. The GY/ET ratio was used to calculate WUE.

2.3.2. Leaf Area Index (LAI) and Chlorophyll Content of Flag Leaves

At anthesis and grain filling stages (20 days after anthesis), the green leaf area of ten wheat plants was measured by a Li-3100 area meter (LI-COR, Inc., Lincoln, NE, USA) from each experimental plot to calculate the leaf area index (LAI) [13]. At these two stages, 20 flag leaf samples were randomly selected from each experimental plot to analyze the chlorophyll content (a + b). Flag leaf chlorophyll (a + b) content was extracted with 95% alcohol for 48 h in the dark, and the optical density (OD) of the alcohol extraction was measured at 649 and 665 nm using a UV-1800 Visible Ultraviolet Spectrophotometer, Shimadzu, Japan [14].

2.3.3. Dry Matter Accumulation (DM) and Grain Yield (GY)

Two 50 cm inner rows of wheat plants from each experimental plot were sampled at ground level at anthesis and maturity stages, then separated into grain and the rest. All samples were dried in an oven at 75 °C to a constant weight. Dry matter translocation (DMT) from the vegetative portions of the grain between anthesis and maturity was calculated as the difference between DM at anthesis and DM at maturity without grain. The contribution of DM pre-anthesis to grain was calculated from the ratio of DMT pre-anthesis to grain at maturity (DMR) [31], and the contribution ratio of DM post anthesis to grain was calculated as the difference by 100-DMR (DMPR). The ratio of grain to total above-ground DM at maturity is defined as the harvest index (HI). To determine spike number, spikes were counted in six 1 m center rows of each plot before harvest. Before harvest, 60 randomly picked spikes from each experimental plot were used to calculate the grain number. Wheat plants from a 3 m^2 area in each plot were harvested and then threshed artificially to determine GY. Actual GY was reported on a 13% moisture basis. By weighing 1000 seeds from each sample and averaging the results of three replicates, the thousand grain weight (TGW) was determined.

2.3.4. Soil Nitrate Nitrogen (NO_3^--N) Residue, Nitrogen Accumulation and Use

Soil NO_3^--N contents were determined using an ultraviolet spectrophotometer, and soil samples were extracted with 0.01 $mol \cdot L^{-1}$ $CaCl_2$ [32]. The calculation method for the accumulated amount of NO_3^--N in the 0–200 cm soil profile is the sum of NO_3^--N accumulation in each layer [33]. The Kjeldahl method was used to determine the total nitrogen content of plants [34]. According to Ruisi et al. (2016), nitrogen accumulation and use were calculated as follows [35]:

$$NAM = DMM \times NC\%,$$

$$N_fUE = GY/Nf$$

where NAM is the nitrogen accumulation in plants at maturity; DMM is dry matter accumulation of plants at maturity; NC is the nitrogen concentration in plants; N_fUE is nitrogen fertilizer use efficiency; Nf is the applied amount of nitrogen fertilizer.

2.4. Data Analysis

Microsoft Excel 2016 (Microsoft, Inc., Redmond, WA, USA) was used for data sorting, SPSS Statistics 22.0 software (IBM, Armonk, NY, USA) was used to analyze the data and the least significant difference test ($p = 0.05$) was used to compare the difference between different irrigation methods and nitrogen application rates in this study. All figures in this paper were generated using Origin Pro 2019 (Origin Lab Corp., Northampton, MA, USA).

3. Results

3.1. Grain Yield and Yield Components

Grain yield (GY) clearly fell from 2017–2018 compared with 2016–2017, which was mostly as a result of a decline in spike number (SN) and thousand grain weight (TGW) (Table 1). However, the two-year study data revealed consistent outcomes. Across the two years, the GY under three irrigation methods was significantly lowered with a reduction in nitrogen application rate, which decreased by 6.35%, 9.84% and 6.83%, respectively, in N45 compared to N90. Under N45, there was no significant difference in GY between SI and DI, and they were both significantly higher than in CI. Under N90, GY in CI was significantly lower than that of DI, and SI yielded the highest GY when compared to those of CI and DI. Notably, there was no significant difference in GY among CIN90, SIN45 and DIN45. SN and grain number per spike (GN) among different treatments in the same year had no significant difference, while there was a large effect on TGW, and the difference in GY under different treatments was mainly caused by the change in TGW.

Table 1. Effects of different irrigation methods and nitrogen application rates on grain yield (GY), spike number (SN), grain number per spike (GN) and thousand grain weight (TGW) of winter wheat.

Year	Treatment	GY (kg·ha^{-1})	SN ($\times 10^4$·ha^{-1})	GN (Grain·Spike^{-1})	TGW (g)
2016–2017	CIN45	7769.6 ± 102.5 d	819.0 ± 11.7 a	27.6 ± 0.25 a	42.3 ± 0.28 e
	CIN90	8503.7 ± 284.5 c	826.0 ± 12.2 a	27.7 ± 0.21 a	43.6 ± 0.32 d
	SIN45	8882.7 ± 102.9 c	819.6 ± 2.2 a	28.2 ± 0.26 a	45.8 ± 0.18 bc
	SIN90	9786.4 ± 120.3 a	830.4 ± 10.4 a	27.8 ± 0.28 a	46.7 ± 0.32 a
	DIN45	8366.0 ± 99.9 c	818.6 ± 9.2 a	27.6 ± 0.21 a	45.5 ± 0.26 c
	DIN90	9111.5 ± 87.8 b	820.9 ± 5.5 a	27.8 ± 0.21 a	46.1 ± 0.09 b
2017–2018	CIN45	6658.2 ± 35.3 d	648.4 ± 5.8 a	29.9 ± 0.15 a	41.3 ± 0.73 e
	CIN90	6940.2 ± 35.2 c	677.8 ± 2.5 a	30.3 ± 0.13 a	42.8 ± 0.19 d
	SIN45	6905.0 ± 35.2 c	665.6 ± 5.4 a	30.4 ± 0.09 a	43.7 ± 0.27 c
	SIN90	7710.7 ± 34.9 a	674.5 ± 6.7 a	30.2 ± 0.02 a	45.4 ± 0.68 a
	DIN45	6881.5 ± 66.7 c	666.9 ± 2.5 a	30.0 ± 0.33 a	43.5 ± 0.12 c
	DIN90	7279.6 ± 89.9 b	668.7 ± 14.2 a	30.1 ± 0.39 a	45.0 ± 1.1 b

CI, conventional irrigation method; SI, micro-sprinkling irrigation method; DI, drip irrigation method; N45 indicates 45 kg·ha^{-1} nitrogen was applied as top dressing; N90 indicates 90 kg·ha^{-1} nitrogen was applied as top dressing. Different letters indicate a significant difference among different irrigation methods at $p < 0.05$ level. All the data are shown as the mean ± standard error ($n = 3$).

3.2. Leaf Area Index (LAI) and Chlorophyll Content Flag Leaf

LAI of winter wheat at anthesis was obviously higher than at the grain filling stage, and it was obviously lower from 2017–2018 than that from 2016–2017 (Figure 2). The two years' experimental results showed that LAI decreased in N45 compared with N90, but no significant difference was presented between N45 and N90 in micro-irrigation treatments in the first growing season. At the anthesis stage, the LAIs of SI and DI were significantly greater than in CI but did not differ significantly from each other under N45. However, irrigation methods had no significant impact on the LAI under N90. At the filling stage, N45 significantly reduced the LAI when compared to N90 under the same irrigation method. The LAI in SI and DI did not differ significantly, while LAI in micro-irrigation was significantly higher than that of CI under N45. Under N90, compared with CI, micro-irrigation significantly improved the LAI, and LAI in SI was significantly higher than that of DI.

Chlorophyll content in flag leaf was obviously decreased when nitrogen application rate was reduced under the same irrigation method, but the results of chlorophyll content in different irrigation methods under the same nitrogen application rate were varied. At the anthesis stage, under N45, the two years' results all showed that the chlorophyll content in CI and SI was not significantly different, while SI significantly improved the chlorophyll content compared to DI. Additionally, under N90, micro-irrigation significantly increased the chlorophyll content compared to CI, and no significant difference in the chlorophyll content was observed between SI and DI. At the filling stage, under N45, the two years' data revealed that CI significantly decreased the chlorophyll content compared to micro-irrigation treatments, and SI had a much greater chlorophyll content than DI. Under N90, compared with micro-irrigation treatments, CI significantly decreased the chlorophyll content, and that of SI was significantly higher than that of DI; however, there was no significant difference between CI and DI from 2017–2018.

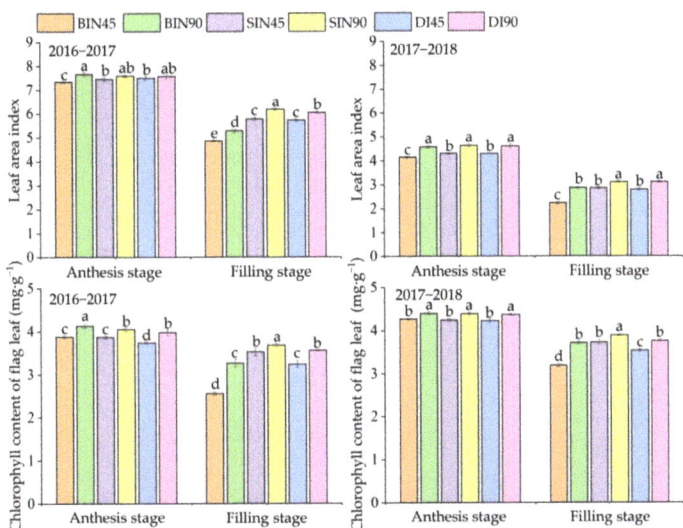

Figure 2. Effects of different irrigation methods and nitrogen application rates on flag leaf chlorophyll content at anthesis and filling stages. Note: CI, conventional irrigation method; SI, micro-sprinkling irrigation method; DI, drip irrigation method; N45 indicates 45 kg·ha^{-1} nitrogen was applied as top dressing; N90 indicates 90 kg·ha^{-1} nitrogen was applied as top dressing. Values followed by the same letter within a column in each year are not significantly different at $p < 0.05$. Vertical bars represent standard errors. All the data are shown as the mean ± standard error ($n = 3$).

3.3. Dry Matter Accumulation and Translocation

Across the two years, dry matter accumulation (DM) at anthesis (DMA) in SI was not significantly impacted by reducing the nitrogen application rate, but it was significantly decreased in CI and DI when nitrogen application was reduced (Table 2). CI treatment significantly increased the DMA compared to SI and DI when nitrogen application was the same, and DMA in micro-irrigation treatments had no significant difference. Under N45, SI and DI significantly decreased DMA compared to CI from 2016–2017; however, no significant difference was observed in DMA between CI and SI from 2017–2018, nor between SI and DI, but DMA in CI was significantly higher than that in SI. At maturity, compared with N90, N45 significantly decreased the total DM (DMM) in both years under the same irrigation method, and under N90, DMM in CI was significantly decreased compared with those of DI and SI, and SI showed the highest DMM. However, compared to CI under N45, the DMM in micro-irrigation treatments was significantly improved, and that of SI was significantly higher than that of DI in the first growing season, while no significant difference was observed between SI and DI in the second year. The contribution ratio of pre-anthesis dry matter translocation to GY (DMR) was significantly higher for N45 than N90 in CI, and the reduction in nitrogen had no significant impact on DMR in SI and DI, but SI significantly decreased the DMR compared with DI. The contribution ratio of post-anthesis dry matter accumulation to grain (DMPR) in different treatments was in opposition to the DMR. In addition, CI significantly decreased harvest index (HI) compared to micro-irrigation treatments under the same nitrogen application rate, whereas SI and DI had no significant effect on HI. Under N45, no significant difference was observed in HI between DI and SI, however, that of DI was significantly lower than that of SI under N90.

Table 2. Effects of different irrigation methods and nitrogen application rates on dry matter accumulation, translocation and harvest index of winter wheat.

Year	Treatment	DMA (kg·ha^{-1})	DMM (kg·ha^{-1})	DMR (%)	DMPR (%)	HI
2016–2017	CIN45	15395.6 ± 130.5 b	20214.9 ± 10.3.3 e	38.6 ± 0.63 a	61.4 ± 0.63 d	0.388 ± 0.002 c
	CIN90	15805.3 ± 81.4 a	21684.0 ± 65.3 d	32.7 ± 0.50 b	67.3 ± 0.50 c	0.403 ± 0.001 b
	SIN45	15055.1 ± 105.1 c	22507.8 ± 100.3 c	17.5 ± 2.39 d	82.5 ± 2.39 a	0.401 ± 0.002 b
	SIN90	15384.0 ± 108.6 b	23794.7 ± 84.8 a	15.7 ± 0.69 d	84.3 ± 0.69 a	0.419 ± 0.001 a
	DIN45	14948.0 ± 87.6 c	21800.4 ± 58.8 d	22.3 ± 0.78 c	77.7 ± 0.78 b	0.405 ± 0.003 b
	DIN90	15409.1 ± 32.9 b	22926.0 ± 99.2 b	22.0 ± 0.67 c	78.0 ± 0.67 b	0.420 ± 0.001 a
2017–2018	CIN45	12402.7 ± 62.0 b	16182.2 ± 71.1 e	43.5 ± 1.39 a	56.5 ± 1.39 d	0.414 ± 0.001 d
	CIN90	12653.3 ± 67.7 a	17488.9 ± 7.3 b	32.9 ± 0.73 c	67.1 ± 0.73 c	0.412 ± 0.001 d
	SIN45	12320.7 ± 99.3 bc	17128.7 ± 91.9 c	33.3 ± 1.01 c	66.7 ± 1.01 c	0.421 ± 0.002 bc
	SIN90	12425.1 ± 53.0 b	17962.4 ± 37.3 a	28.6 ± 0.65 d	71.4 ± 0.65 a	0.432 ± 0.001 a
	DIN45	12248.4 ± 49.6 c	16627.3 ± 63.2 d	37.1 ± 0.56 b	62.9 ± 0.56 c	0.419 ± 0.003 c
	DIN90	12435.8 ± 22.1 b	17476.0 ± 27.2 b	32.0 ± 0.51 c	68.0 ± 0.51 b	0.424 ± 0.003 b

CI, conventional irrigation method; SI, micro-sprinkling irrigation method; DI, drip irrigation method; N45 indicates 45 kg·ha^{-1} nitrogen was applied as top dressing; N90 indicates 90 kg·ha^{-1} nitrogen was applied as top dressing. DMA, dry matter accumulation at anthesis; DMM, dry matter accumulation at maturity; DMR, contribution of dry matter pre-anthesis translocation to grain yield; DMPR, contribution of dry matter post-anthesis accumulation to grain yield. Different letters indicate a significant difference among different irrigation methods at $p < 0.05$ level. All the data are shown as the mean ± standard error ($n = 3$).

3.4. Water Consumption and Utilization

As shown in Figure 3, the total seasonal evapotranspiration (ET) of winter wheat in N45 was significantly lower than that of N90 under the same irrigation method across the two years. The three irrigation methods had no significant effect on ET under N45. Under N90 in the 2016–2017 growing season of winter wheat, no significant difference in ET was observed between CI and DI and between DI and SI, but ET in SI was significantly lower than that in CI; however, CI significantly increased the ET compared with those of DI and SI from 2017–2018. Water use efficiency (WUE) was not significantly impacted by the reduction in nitrogen application rate under CI, but micro-irrigation treatments significantly decreased the WUE in N45 compared to N90. Micro-irrigation treatments significantly improved the WUE compared to CI under the same nitrogen application rate. WUE in SI and DI increased by 9.09% and 4.70%, respectively, compared with that of CI under N45; however, under N90, the WUE increased by 15.9% and 7.23%, respectively. However, WUE in DI was significantly lower than that of SI under the same nitrogen application rate, but SI and DI had no significant effect on WUE from 2017–2018. Under N90, the WUE variation in the various irrigation methods was consistent with the prior year.

3.5. Soil Water Utilization

As shown in Table 3, across the two years, N90 treatments significantly increased soil water consumption (SW) as compared to those of N45 under three irrigation methods. However, under N45, there was no significant variation in SW among the three irrigation methods. Meanwhile, under N90, there was no significant variance in SW between SI and DI, and besides the DI, SW in SI declined significantly over the two years compared to CI. The two years of research indicated that the reduction in nitrogen application rate significantly decreased the soil water consumption from 0–100 cm (SW1), however, CI had a much higher SW1 than micro-irrigation treatments, and SW1 in SI was similar to that in DI. It is worth noting that, compared N90, N45 had no significant impact on the ratio of SW1 to SW in SI and DI, and CI significantly increased SW1 compared to SI and DI from 2016–2017, while there was no significant impact on SW1 under N90 among the three irrigation methods from 2017–2018. In addition, there was no significant difference in soil water consumption in the 100–200 cm soil profile (SW2) between N45 and N90 under SI, and the reduced nitrogen application rate treatments significantly decreased SW2 in the three irrigation methods. Under N45, there was no significant difference in SW2 between SI and DI in the two years, and they were both much higher than in CI. In comparison to CI,

SW2 under N90 increased dramatically with SI and DI from 2016–2017, but the irrigation methods had no significant impact on SW2 from 2017–2018. Most notably, the SW2 to SW ratio in SI and DI was not significantly affected by the reduction in nitrogen application rate, and they significantly increased SW2 compared with CI from 2016–2017, while there was no significant impact on the ratio under N90 among the three irrigation methods from 2017–2018.

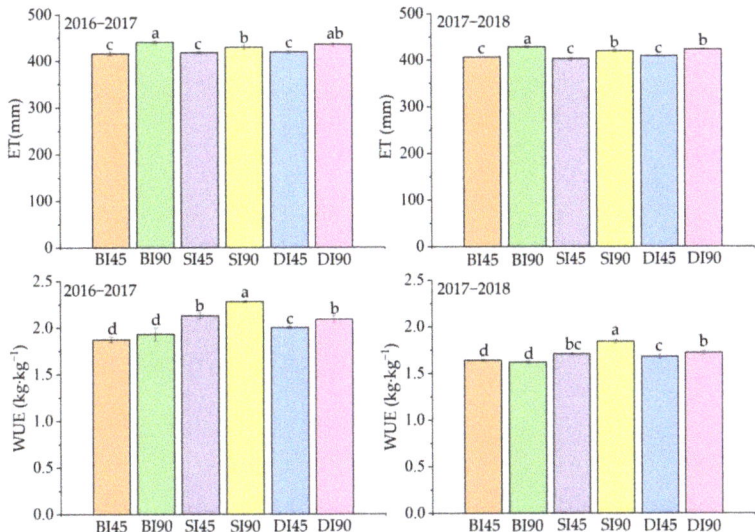

Figure 3. Effects of different irrigation methods and nitrogen application rates on the seasonal evapotranspiration (ET) and water use efficiency (WUE) of winter wheat during the two seasons. Note: CI, conventional irrigation method; SI, micro-sprinkling irrigation method; DI, drip irrigation method; N45 indicates 45 kg·ha^{-1} nitrogen was applied as top dressing; N90 indicates 90 kg·ha^{-1} nitrogen was applied as top dressing. Values followed by the same letter within a column in each year are not significantly different at $p < 0.05$. Vertical bars represent standard errors. All the data are shown as the mean ± standard error ($n = 3$).

3.6. Nitrogen Accumulation and Utilization in Plants

The variation in nitrogen accumulation in winter wheat at maturity (NAM) among different treatments was consistent between the two growing seasons (Figure 4). Compared with N90, NAM was decreased significantly in N45 under the same irrigation method. However, when nitrogen was applied at the same rate, NAM in SI significantly surpassed that of DI and CI, and CI obtained the lowest NAM. Across the two years, N45 significantly increased the nitrogen fertilizer use efficiency (N$_f$UE) compared to N90 under the same irrigation method, but there were some changes in N$_f$UE among the irrigation methods under the same nitrogen application rate. Under N45, CI significantly decreased the N$_f$UE compared with DI from 2016–2017, whereas the N$_f$UE in SI was significantly higher than that in DI. From 2017–2018, N$_f$UE in CI was significantly lower than that of micro-irrigation treatments, and the N$_f$UE in SI was similar to that of DI. SI had a much higher N$_f$UE than DI and CI, and CI had the lowest N$_f$UE under N90 in the two years.

Table 3. Effects of different irrigation methods and nitrogen application rates on soil water consumption of winter wheat during the two seasons.

Year	Treatment	Soil Water Consumption Amount (mm)				SW (mm)
		SW1	Ratio (%)	SW2	Ratio (%)	
2016–2017	CIN45	144.2 ± 1.35 b	72.1 ± 0.61 a	55.8 ± 2.19 d	27.9 ± 0.61 b	200.0 ± 3.51 c
	CIN90	158.9 ± 0.43 a	70.5 ± 0.89 a	66.5 ± 2.69 c	29.5 ± 0.89 b	225.5 ± 2.40 a
	SIN45	134.3 ± 0.86 c	66.4 ± 0.74 b	68.1 ± 2.34 bc	33.6 ± 0.74 a	202.4 ± 3.68 c
	SIN90	140.4 ± 4.97 b	65.7 ± 2.14 b	73.3 ± 5.21 ab	34.3 ± 2.14 a	213.6 ± 4.96 b
	DIN45	135.8 ± 1.16 c	66.7 ± 0.95 b	67.8 ± 2.77 bc	33.3 ± 0.95 a	203.6 ± 2.87 c
	DIN90	143.4 ± 1.39 b	65.3 ± 1.28 b	76.3 ± 2.22 a	34.7 ± 1.28 a	219.7 ± 4.08 ab
2017–2018	CIN45	83.5 ± 3.35 c	85.7 ± 2.51 a	13.9 ± 3.21 d	14.3 ± 2.51 b	97.4 ± 6.24 c
	CIN90	98.5 ± 2.80 a	78.3 ± 1.34 b	27.2 ± 2.86 a	21.7 ± 1.34 a	125.7 ± 5.49 a
	SIN45	75.4 ± 2.32 d	77.7 ± 1.96 b	21.6 ± 2.52 c	22.3 ± 1.96 a	97.0 ± 3.93 c
	SIN90	86.8 ± 1.74 bc	76.7 ± 0.66 b	26.4 ± 1.09 a	23.3 ± 0.66 a	113.2 ± 2.43 b
	DIN45	79.2 ± 1.27 d	76.9 ± 1.84 b	23.7 ± 2.08 bc	23.1 ± 1.84 a	102.9 ± 0.83 c
	DIN90	90.2 ± 1.10 b	77.0 ± 0.46 b	26.9 ± 0.38 a	23.0 ± 0.46 a	117.1 ± 0.78 b

CI, conventional irrigation method; SI, micro-sprinkling irrigation method; DI, drip irrigation method; N45 indicates 45 kg·ha^{-1} nitrogen was applied as top dressing; N90 indicates 90 kg·ha^{-1} nitrogen was applied as top dressing. SW presents the soil water consumption in 0–200 cm soil depth; SW1 presents the soil water consumption in 0–100 cm soil depth; SW2 presents the soil water consumption in 100–200 cm soil depth. Different letters indicate a significant difference among different irrigation methods at $p < 0.05$ level. All the data are shown as the mean ± standard error ($n = 3$).

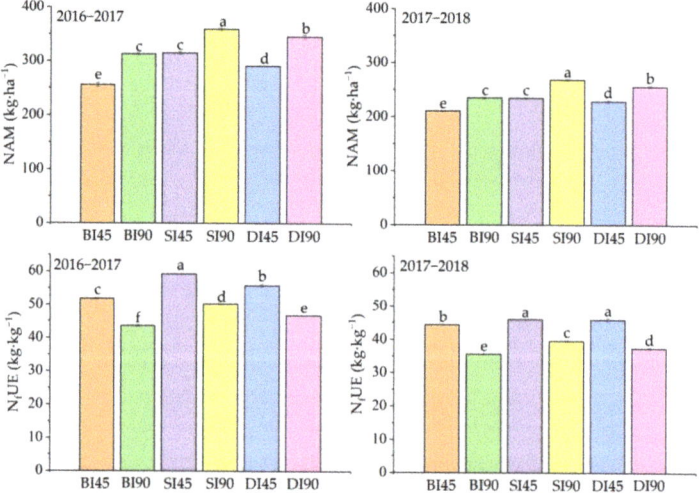

Figure 4. Effects of different irrigation methods and nitrogen application rates on nitrogen accumulation at maturity (NAM) and nitrogen fertilizer use efficiency (N$_f$UE) of winter wheat during the two seasons. Note: CI, conventional irrigation method; SI, micro-sprinkling irrigation method; DI, drip irrigation method; N45 indicates 45 kg·ha^{-1} nitrogen was applied as top dressing; N90 indicates 90 kg·ha^{-1} nitrogen was applied as top dressing. Values followed by the same letter within a column in each year are not significantly different at $p < 0.05$. Vertical bars represent standard errors. All the data are shown as the mean ± standard error ($n = 3$).

3.7. Soil Available Nitrogen Accumulation

Table 4 shows a similar variation in soil nitrate nitrogen (NO$_3^-$-N) accumulation at maturity in 0–200 cm soil depth (SNC) among different treatments from 2016–2017 and 2017–2018, that is, N45 significantly decreased the SNC compared to N90 under the same irrigation method, however, micro-irrigation significantly reduced the SNC compared to CI when nitrogen was applied at the same rate, and no significant difference in SNC was observed between SI and DI. Across the two years, N45 significantly decreased the NO_3^--N residue from 0–100 cm (SNC1) under the same irrigation method, while there was no significant difference in SNC1 among different irrigation methods. In addition, the reduced nitrogen application rate had no significant impact on the ratio of NO_3^--N residue of SNC1 to SNC under CI from 2016–2017 and, compared with N45, N90 significantly increased the ratio of NO_3^--N residue of SNC1 to SNC under the same irrigation method. Under N45, there was no significant difference in the ratio of NO_3^--N residue of SNC1 to SNC among the three irrigation methods from 2016–2017, but SI and DI significantly increased the ratio compared to CI under the same nitrogen application rate in the two growing seasons. N45 significantly decreased the NO_3^--N residue in the 100–200 cm soil profile (SNC2) compared to N90, but under the same nitrogen application rate in the two years, CI significantly increased the SNC2 compared to micro-irrigation treatments, and no significant difference in SNC2 was observed between SI and DI. It was noted that the ratio of SNC2 to SNC in SI and DI was lower than that of CI under the same nitrogen application rate, and there was no significant difference in the ratio between SI and DI.

Table 4. Effects of different irrigation methods and nitrogen application rates on soil NO_3^--N accumulation of winter wheat at maturity.

Year	Treatment	Soil NO_3^--N Accumulation Amount (kg·ha^{-1})				SNC (kg·ha^{-1})
		SNC1	Ratio (%)	SNC2	Ratio (%)	
2016–2017	CIN45	38.0 ± 2.37 b	20.3 ± 1.15 b	148.9 ± 2.37 b	79.7 ± 1.15 a	186.9 ± 1.42 c
	CIN90	64.0 ± 4.77 a	24.4 ± 4.91 b	198.1 ± 4.77 a	75.6 ± 4.91 a	262.1 ± 0.39 a
	SIN45	39.5 ± 6.45 b	22.3 ± 5.39 b	138.0 ± 6.45 c	77.7 ± 5.39 a	177.5 ± 1.41 d
	SIN90	63.2 ± 4.70 a	29.1 ± 0.99 a	154.0 ± 4.70 b	70.9 ± 0.99 b	217.2 ± 3.71 b
	DIN45	41.1 ± 5.84 b	22.8 ± 4.44 b	139.0 ± 5.84 c	77.2 ± 4.44 a	180.1 ± 1.45 d
	DIN90	66.5 ± 1.63 a	30.1 ± 1.90 a	154.5 ± 1.63 b	69.9 ± 1.90 b	221.0 ± 3.48 b
2017–2018	CIN45	92.8 ± 5.51 b	29.3 ± 1.14 c	224.2 ± 2.26 b	70.7 ± 1.14 a	316.9 ± 6.79 c
	CIN90	125.4 ± 2.97 a	33.0 ± 0.82 b	254.2 ± 3.73 a	67.0 ± 0.82 b	379.6 ± 1.74 a
	SIN45	95.1 ± 5.32 b	31.6 ± 0.87 b	206.1 ± 4.39 c	68.4 ± 0.87 b	301.2 ± 9.27 d
	SIN90	128.2 ± 4.65 a	36.5 ± 0.97 a	222.7 ± 2.80 b	63.5 ± 0.97 c	350.9 ± 4.54 b
	DIN45	98.1 ± 4.38 b	31.8 ± 0.58 b	210.5 ± 4.97 c	68.2 ± 0.58 b	308.6 ± 9.03 cd
	DIN90	132.1 ± 3.37 a	37.2 ± 0.66 a	223.0 ± 2.31 b	62.8 ± 0.66 c	355.1 ± 3.96 b

CI, conventional irrigation method; SI, micro-sprinkling irrigation method; DI, drip irrigation method; N45 indicates 45 kg·ha^{-1} nitrogen was applied as top dressing; N90 indicates 90 kg·ha^{-1} nitrogen was applied as top dressing. SNC, soil NO_3^--N accumulation in 0–200 cm soil depth; SNC1, soil NO_3^--N accumulation in 0–100 cm soil depth; SNC2, soil NO_3^--N accumulation in 100–200 cm soil depth. Different letters indicate a significant difference among different irrigation methods at $p < 0.05$ level. All the data are shown as the mean ± standard error ($n = 3$).

4. Discussion

Spike number, grain number per spike and grain weight all affect wheat grain production. Numerous studies have shown that insufficient nitrogen application causes a significant decrease in GY of wheat, which mostly decreased the SN and GN but significantly increased grain weight [36–38]. However, in this study, nitrogen reduction under three irrigation methods had no significant impact on SN and GN, but significantly decreased the TGW, which led to a fall in wheat's GY when compared to the recommended nitrogen application rate (Table 1). Tillering is an important phenological stage for winter wheat, and there is a close correlation between tillering in spring and spike development [39]. The number of tillers produced per plant has been found to be affected by limited nutrients [40]. Nitrogen deficiency has been recognized as an important nutrient factor to limit wheat tiller growth and development [41]. We considered that the excessive nitrogen reduction in earlier studies may have had a negative impact on the development of SN and GN in wheat, and the lower SN and GN were beneficial to the increase in grain weight. In our two-year study, the same nitrogen rate was applied under different treatments at sowing of wheat, and all treatments were applied with nitrogen in spring, therefore, the reduction in nitrogen did not have a negative impact on the SN and GN.

The flag leaf is one of the most important photosynthetic organs of wheat, and it is the basis for obtaining high grain weight and GY for wheat by delaying the leaf senescence and maintaining a higher LAI during the grain filling period [14,42,43]. However, an insufficient nitrogen application rate could easily lead to premature failure of wheat leaves, reducing wheat GY [32]. A prior study revealed that micro-irrigation could greatly increase the production of dry matter by delaying leaf senescence during grain filling, boosting the grain filling rate and raising wheat's TGW and GY [18]. In this study, nitrogen reduction treatments significantly reduced GY under different irrigation methods, and GY in SI was significantly higher than that of DI and CI, and CI had the lowest GY (Table 1). Additionally, no significant difference in GY was found among CIN90, SIN45 and DIN45, and the greater TGW accounted for the increased GY in SI and DI under nitrogen reduction. According to Figure 2, compared to CI, SI and DI significantly increased the chlorophyll content at the grain filling stage while maintaining high levels of LAI under the same nitrogen application rate. Conversely, the chlorophyll content and LAI were significantly decreased when nitrogen was reduced under the same irrigation method, but they were still significantly higher in SI and DI than CIN90, which may be the reason why micro-irrigation treatments could significantly improve the TGW. In addition, nitrogen application during the post-anthesis of winter wheat could improve root activity and increase the photosynthetic rate of flag leaf during the grain filling period [44], which may account for the higher grain weight achieved by micro-irrigation treatments in this present study.

GY of wheat is directly related to dry matter production [45,46]. It is a promising way to improve the DMM and HI of wheat through optimizing irrigation and nitrogen application regime [11]. This study found that micro-irrigation significantly improved the DMM compared with CI under the same nitrogen application rate (Table 2). However, compared with N90, N45 significantly decreased the DMM under the same irrigation method, but micro-irrigation with nitrogen reduction showed a similar DMM to CI with the recommended nitrogen application rate, particularly in SI, which obtained the highest DMM and HI. In addition, grain growth depends on the photosynthesis during the grain filling period and the remobilization of pre-anthesis assimilates stored in vegetable organs to grain [47,48]. In this study, micro-irrigation significantly increased the DM post anthesis and increased the DMPR compared to CI under the same nitrogen application rate, which may be related to the higher chlorophyll content and LAI in the filling period in SI and DI (Figure 2).

Optimized irrigation practices are beneficial to improve the WUE of winter wheat [49,50], and insufficient nitrogen supply may lead to significantly decreased crop yield as well as WUE [32]. This study revealed that micro-irrigation greatly lowered ET when compared with the conventional farmer irrigation method and significantly enhanced wheat GY and WUE [14,51]. Furthermore, enhancing the water uptake and use from deeper soil profiles by the crop is a very promising strategy to increase the WUE and enhance wheat yield when irrigation is limited [12,52,53]. In this study, the ET in CI was significantly lower than in micro-irrigation methods under N90, especially in SI, which may be because SI and DI were beneficial to improving the micro-environment of the wheat canopy, and this favors reducing the ineffective evaporation of soil water in the field [46]. However, under N45, there was no significant change in ET among the various irrigation methods, and the study of this effect is ongoing. Compared with CI under the same nitrogen application rate, micro-irrigation significantly increased WUE of winter wheat, and WUE in SI was the highest. In this investigation, we also discovered that micro-irrigation altered water extraction of deep soil by winter wheat (Table 3, Figure 1). The water consumption of the 0–100 cm soil profile under the same irrigation method was significantly increased by SI and DI in the first growing season of winter wheat, but the deeper (100–200 cm) soil water consumption under N90 in the second year was not significantly different among the three irrigation methods, which may be related to the heavy rainfall during the filling period of winter wheat in this year (Figure 1). Therefore, we suppose that a small amount of micro-irrigation with two nitrogen application rates may have promoted the root to penetrate into the deep soil profile in this present study, and facilitated enhancing the acquisition capacity for deep soil water of winter wheat, thereby improving WUE.

Optimizing nitrogen application is an important means to improve GY and N_fUE of wheat [32]. However, the nitrogen application rate has a great impact on the N_fUE, as does the irrigation regime [16,36,54]. Increasing the irrigation volume usually increases nitrogen leaching, reducing soil's available nitrogen accumulation in the root zone [14,55]. Previous studies showed that soil NO_3^--N is easily leached into the deeper soil layer, especially after a large amount of irrigation for wheat, causing larger NO_3^--N loss, which resulted in the decrease in N_fUE [56,57]. Furthermore, improper agricultural production practices have had a serious impact on the eco-environment, and climate and environmental changes have social consequences affecting people [58]. In addition, the frequent occurrence of droughts over the last two decades has led to in rise in farmers' concerns that field crop production will not be possible without irrigation. The warmer climate will also shorten the growing cycle of all crops [59]. In order to meet the growing demand from an increasing world population, there is a need to increase wheat production. Fertilization and irrigation have a great potential to enhance growth quality, grain yield and yield-related traits of wheat [60–62]. Nevertheless, previous studies found that because of the small irrigation volume and the divided nitrogen application, which encouraged wheat absorption and utilization, the N_fUE in micro-irrigation treatments significantly improved when compared to conventional irrigation methods, and the soil NO_3^--N residue was significantly reduced at maturity [13,14]. In this present study, micro-irrigation significantly improved the plants' nitrogen accumulation at maturity when compared to CI at the same nitrogen application rate, and that of SI was the highest. However, under N45, the N_fUE in micro-irrigation was significantly improved compared to CI (Figure 4), which was mainly due to the significant increase in GY of micro-irrigation treatments (Table 1). In addition, the residual NO_3^--N of micro-irrigation treatments in the soil at maturity was significantly decreased compared to CI, which indicated that the micro-irrigation methods were conducive to reducing the residual amount of soil NO_3^--N, and more nitrogen was absorbed and utilized by the crop (Table 4). In comparison to SI and DI, CI greatly enhanced the soil NO_3^--N leaching into the deeper soil depth with the same nitrogen application rate, which is also one of the causes for the lowering of N_fUE in CI. Furthermore, nitrogen deficit in wheat will stimulate root growth into the deep soil layers to increase nutrient absorption, according to Wang et al. (2014) [52]. In this study, the reason why micro-irrigation improved the WUE and N_fUE compared to CI may be that it led to lower NO_3^--N in the deeper soil layer than in CI from the jointing to booting stage of wheat, and this period is greatly critical for wheat root growth. However, more research will need to be carried out in the future on the effects of different nitrogen application rates on winter wheat root growth and their physiological mechanisms.

5. Conclusions

Compared with CI, using micro-irrigation with integrated water and N fertilizer, and with irrigation and nitrogen application at jointing, booting, anthesis and grain filling of winter wheat, could further reduce the nitrogen application rate and maintain the GY of winter wheat, and improve WUE and N_fUE, particularly in SI. The reason for the higher GY, WUE and N_fUE in micro-irrigation than in CI was because it delayed the leaf senescence during the grain filling period, improved the DM post anthesis and increased the use of water and nitrogen contained in deeper soil layers. Overall, using micro-irrigation technology with reduced nitrogen application rate can guarantee the output and improve the use efficiency of water and N fertilizer in the NCP.

Author Contributions: Methodology, J.L. (Jinpeng Li), Y.Z. and Z.W.; Investigation, J.L. (Jinpeng Li); Data analysis, J.L. (Jinpeng Li); Writing—original draft preparation, J.L. (Jinpeng Li); Writing—review and editing, Y.Z., Y.S. and J.L. (Jincai Li); Funding acquisition, J.L. (Jinpeng Li) and Y.Z. All authors have read and agreed to the published version of the manuscript.

Funding: This study was funded by the National Natural Science Foundation of China (32001474, 31871563), and China Agriculture Research System of MOF and MARA (CARS-03).

Institutional Review Board Statement: Not applicable.

Informed Consent Statement: Not applicable.

Data Availability Statement: Not applicable.

Acknowledgments: The authors thank Yangyang Li, Yulei Zhu and Huihui Liu for guidance in revising this manuscript.

Conflicts of Interest: The authors declare no conflict of interest.

References

1. National Bureau of Statistic of China. *China Statistical Year Book*; China Statistics Press: Beijing, China, 2021.
2. Wu, X.; Wang, P.J.; Huo, Z.G.; Wu, D.R.; Yang, J.Y. Crop drought identification index for winter wheat based on evapotranspiration in the Huang-Huai-Hai Plain, China. *Agric. Ecosyst. Environ.* **2018**, *263*, 18–30. [CrossRef]
3. Ren, P.; Huang, F.; Li, B. Spatiotemporal patterns of water consumption and irrigation requirements of wheat-maize in the Huang-Huai-Hai Plain, China and options of their reduction. *Agric. Water Manag.* **2022**, *263*, 107468. [CrossRef]
4. Yang, X.L.; Chen, Y.Q.; Pacenka, S.; Gao, W.S.; Ma, L.; Wang, G.Y.; Yan, P.; Sui, P.; Steenhuis, T.S. Effect of diversified crop rotations on groundwater levels and crop water productivity in the North China Plain. *J. Hydrol.* **2015**, *522*, 428–438. [CrossRef]
5. Yang, X.L.; Chen, Y.Q.; Pacenka, S.; Gao, W.S.; Zhang, M.; Sui, P.; Steenhuis, T.S. Recharge and groundwater use in the North China Plain for six irrigated crops for an eleven year period. *PLoS ONE* **2015**, *10*, e0115269. [CrossRef] [PubMed]
6. Kong, X.B.; Zhang, X.L.; Lal, R.; Zhang, F.R.; Chen, X.H.; Niu, Z.G.; Han, L.; Song, W. Groundwater depletion by agricultural intensification in China's HHH Plains, Since 1980s. *Adv. Agron.* **2016**, *135*, 59–106.
7. Yang, X.L.; Wang, G.Y.; Chen, Y.Q.; Sui, P.; Pacenka, S.; Steenhuis, T.S.; Siddique, K.H.M. Reduced groundwater use and increased grain production by optimized irrigation scheduling in winter wheat–summer maize double cropping system—A 16-year field study in North China Plain. *Field Crops Res.* **2022**, *275*, 108364. [CrossRef]
8. Pei, H.W.; Scanlon, B.R.; Shen, Y.J.; Reedy, R.C.; Long, D.; Liu, C.M. Impacts of varying agricultural intensification on crop yield and groundwater resources: Comparison of the North China Plain and US High Plains. *Environ. Res. Lett.* **2015**, *10*, 044013. [CrossRef]
9. Liang, H.; Hu, K.L.; Li, B.G.; Liu, H.T. Coupled simulation of soil water-heat- carbon nitrogen process and crop growth at soil-plant-atmosphere continuum system. *Trans. CSAE.* **2014**, *30*, 54–66.
10. Li, T.L.; Xie, Y.H.; Gao, Z.Q.; Hong, J.P.; Li, L.; Meng, H.S.; Ma, H.M.; Jia, J.X. Year-round film mulching system with monitored fertilization management improve grain yield and water and nitrogen use efficiencies of winter wheat in the dryland of the Loess Plateau, China. *Environ. Sci. Pollut. Res. Int.* **2019**, *26*, 9524–9535. [CrossRef] [PubMed]
11. Xu, X.X.; Zhang, M.; Li, J.P.; Liu, Z.Q.; Zhao, Z.G.; Zhang, Y.H.; Zhou, S.L.; Wang, Z.M. Improving water use efficiency and grain yield of winter wheat by optimizing irrigations in the North China Plain. *Field Crops Res.* **2018**, *221*, 219–227. [CrossRef]
12. Ali, S.; Xu, Y.Y.; Ma, X.C.; Ahmad, I.; Manzoor Jia, Q.M.; Akmal, M.; Hussain, Z.; Arif, M.; Cai, T.; Zhang, J.H.; et al. Deficit irrigation strategies to improve winter wheat productivity and regulating root growth under different planting patterns. *Agric. Water Manag.* **2019**, *219*, 1–11. [CrossRef]
13. Li, J.P.; Wang, Y.Q.; Zhang, M.; Liu, Y.; Xu, X.X.; Lin, G.; Wang, Z.M.; Yang, Y.M.; Zhang, Y.H. Optimized micro-sprinkling irrigation scheduling improves grain yield by increasing the uptake and utilization of water and nitrogen during grain filling in winter wheat. *Agric. Water Manag.* **2019**, *211*, 59–69. [CrossRef]
14. Li, J.P.; Xu, X.X.; Lin, G.; Wang, Y.Q.; Liu, Y.; Zhang, M.; Zhou, J.Y.; Wang, Z.M.; Zhang, Y.H. Micro-irrigation improves grain yield and resource use efficiency by co-locating the roots and N-fertilizer distribution of winter wheat in the North China Plain. *Sci. Total Environ.* **2018**, *643*, 367–377. [CrossRef] [PubMed]
15. Li, J.P.; Zhang, Z.; Liu, Y.; Yao, C.S.; Song, W.Y.; Xu, X.X.; Zhang, M.; Zhou, X.N.; Gao, Y.M.; Wang, Z.M.; et al. Effects of micro-sprinkling with different irrigation amount on grain yield and water use efficiency of winter wheat in the North China Plain. *Agric. Water Manag.* **2019**, *224*, 105736. [CrossRef]
16. Liu, S.K.; Lin, X.; Wang, W.Y.; Zhang, B.J.; Wang, D. Supplemental irrigation increases grain yield, water productivity, and nitrogen utilization efficiency by improving nitrogen nutrition status in winter wheat. *Agric. Water Manag.* **2022**, *264*, 107505. [CrossRef]
17. Man, J.G.; Wang, D.; White, P.J. Photosynthesis and drymass production of winter wheat in response to micro-sprinkling irrigation. *Agron. J.* **2017**, *109*, 549–561. [CrossRef]

18. Li, J.P.; Wang, Z.M.; Yao, C.S.; Zhang, Z.; Liu, L.; Zhang, Y.H. Micro-sprinkling irrigation simultaneously improves grain yield and protein concentration of winter wheat in the North China Plain. *Crop J.* **2021**, *9*, 1397–1407. [CrossRef]
19. Wang, H.Y.; Zhang, Y.T.; Chen, A.Q.; Liu, H.B.; Zhai, L.M.; Lei, B.K.; Ren, T.Z. An optimal regional nitrogen application threshold for wheat in the North China Plain considering yield and environmental effects. *Field Crops Res.* **2017**, *207*, 52–61. [CrossRef]
20. Abdou, N.M.; Abdel-Razek, M.A.; Abd El-Mageed, S.A.; Semida, W.M.; Leilah, A.A.A.; Abd El-Mageed, T.A.; Ali, E.F.; Majrashi, A.; Rady, M.O.A. High nitrogen fertilization modulates morpho-physiological responses, yield, and water productivity of lowland rice under deficit irrigation. *Agronomy* **2021**, *11*, 1291. [CrossRef]
21. Agami, R.A.; Alamri, S.A.M.; El-Mageed, T.A.A.; Abousekken, M.S.M.; Hashem, M. Role of exogenous nitrogen supply in alleviating the deficit irrigation stress in wheat plants. *Agric. Water Manag.* **2018**, *210*, 261–270. [CrossRef]
22. Si, Z.Y.; Zain, M.; Mehmood, F.; Wang, G.S.; Gao, Y.; Duan, A.W. Effects of nitrogen application rate and irrigation regime on growth, yield, and water-nitrogen use efficiency of drip-irrigated winter wheat in the North China Plain. *Agric. Water Manag.* **2020**, *231*, 106002. [CrossRef]
23. Zheng, X.J.; Yu, Z.W.; Zhang, Y.L.; Shi, Y. Nitrogen supply modulates nitrogen remobilization and nitrogen use of wheat under supplemental irrigation in the North China Plain. *Sci. Rep.* **2020**, *10*, 3305. [CrossRef]
24. Wu, Y.C.; Zhou, S.L.; Wang, Z.M. Effect of nitrogen fertilizer applications on yield, water and nitrogen use efficiency under limited irrigation of winter wheat in North China Plain. *J. Triticeae Crops* **2008**, *28*, 1016–1620.
25. Gu, B.J.; Ge, Y.; Chang, S.X.; Luo, W.D.; Chang, J. Nitrate in groundwater of China: Sources and driving forces. *Glob. Environ. Chang.* **2013**, *23*, 1112–1121. [CrossRef]
26. Duan, J.Z.; Shao, Y.H.; He, L.; Li, X.; Hou, G.G.; Li, S.N.; Feng, W.; Zhu, Y.J.; Wang, Y.H.; Xie, Y.X. Optimizing nitrogen management to achieve high yield, high nitrogen efficiency and low nitrogen emission in winter wheat. *Sci. Total Environ.* **2019**, *697*, 134088. [CrossRef] [PubMed]
27. Zadoks, J.C.; Chang, T.T.; Konzak, C.F. A decimal code for the growth stages of cereals. *Weed Res.* **1974**, *6*, 415–421. [CrossRef]
28. Man, J.G.; Yu, J.S.; White, P.J.; Gu, S.B.; Zhang, Y.L.; Guo, Q.F.; Shi, Y.; Wang, D. Effects of supplemental irrigation with micro-sprinkling hoses on water distribution in soil and grain yield of winter wheat. *Field Crops Res.* **2014**, *161*, 26–37. [CrossRef]
29. Zhang, X.Y.; Chen, S.Y.; Sun, H.Y.; Shao, L.W.; Wang, Y.Z. Changes in evapotranspiration over irrigated winter wheat and maize in North China Plain over three decades. *Agric. Water Manag.* **2011**, *98*, 1097–1104. [CrossRef]
30. Gao, Z.; Liang, X.G.; Lin, S.; Zhao, X.; Zhang, L.; Zhou, L.L.; Shen, S.; Zhou, S.L. Supplemental irrigation at tasseling optimizes water and nitrogen distribution for high-yield production in spring maize. *Field Crops Res.* **2017**, *209*, 120–128. [CrossRef]
31. Shi, Y.; Yu, Z.W.; Man, J.G.; Ma, S.Y.; Gao, Z.Q.; Zhang, Y.L. Tillage practices affect dry matter accumulation and grain yield in winter wheat in the North China Plain. *Soil Till. Res.* **2016**, *160*, 73–81. [CrossRef]
32. Wang, X.; Shi, Y.; Guo, Z.J.; Zhang, Y.L.; Yu, Z.W. Water use and soil nitrate nitrogen changes under supplemental irrigation with nitrogen application rate in wheat field. *Field Crops Res.* **2015**, *183*, 117–125. [CrossRef]
33. Zhang, J.T.; Wang, Z.M.; Zhou, S.L. Soil nitrate n accumulation under different N-fertilizer rates in summer maize and its residual effects on subsequent winter wheat. *Sci. Agric. Sin.* **2013**, *46*, 1182–1190.
34. Dordas, C.A.; Sioulas, C. Dry matter and nitrogen accumulation, partitioning and retranslocation in safflower (*Carthamus tinctorius* L.) as affected by nitrogen fertilization. *Field Crops Res.* **2009**, *110*, 35–43. [CrossRef]
35. Ruisi, P.; Saia, S.; Badagliacca, G.; Amato, G.; Frenda, A.S.; Giambalvo, D.; Miceli, G.D. Long-term effects of no tillage treatment on soil N availability, N uptake, and ^{15}N-fertilizer recovery of durum wheat differ in relation to crop sequence. *Field Crops Res.* **2016**, *189*, 51–58. [CrossRef]
36. Li, Y.; Huang, G.H.; Chen, Z.J.; Xiong, Y.W.; Huang, Q.Z.; Xu, X.; Huo, Z.L. Effects of irrigation and fertilization on grain yield, water and nitrogen dynamics and their use efficiency of spring wheat farmland in an arid agricultural watershed of Northwest China. *Agric. Water Manag.* **2022**, *260*, 107277. [CrossRef]
37. Wang, L.L.; Palta, J.A.; Chen, W.; Chen, Y.; Deng, X. Nitrogen fertilization improved water-use efficiency of winter wheat through increasing water use during vegetative rather than grain filling. *Agric. Water Manag.* **2018**, *197*, 41–53. [CrossRef]
38. Li, W.Q.; Han, M.M.; Pang, D.W.; Chen, J.; Wang, Y.Y.; Dong, H.H.; Chang, Y.L.; Jin, M.; Luo, Y.L.; Li, Y.; et al. Characteristics of lodging resistance of high-yield winter wheat as affected by nitrogen rate and irrigation managements. *J. Integr. Agr.* **2022**, *20*, 2–21. [CrossRef]
39. Rodriguez, D.; Andrade, F.H.; Goudriaan, J. Effects of phosphorus nutrition on tiller emergence in wheat. *Plant Soil.* **1999**, *209*, 283–295. [CrossRef]
40. Longnecker, N.; Kirby, E.J.M.; Robson, A. Leaf emergence, tiller growth, and apical development of nitrogen-dificient spring wheat. *Crop Sci.* **1993**, *33*, 154–160. [CrossRef]
41. Wang, R.; Wang, Y.; Hu, Y.X.; Dang, T.H.; Guo, S.L. Divergent responses of tiller and grain yield to fertilization and fallow precipitation: Insights from a 28-year long-term experiment in a semiarid winter wheat system. *J. Integr. Agric.* **2021**, *20*, 3003–3011. [CrossRef]
42. Wang, D.; Yu, Z.W.; White, P.J. The effect of supplemental irrigation after jointing on leaf senescence and grain filling in wheat. *Field Crops Res.* **2013**, *151*, 35–44. [CrossRef]
43. Du, X.; Gao, Z.; Sun, X.N.; Bian, D.H.; Ren, J.H.; Yan, P.; Cui, Y.H. Increasing temperature during early spring increases winter wheat grain yield by advancing phenology and mitigating leaf senescence. *Sci. Total Environ.* **2022**, *812*, 152557. [CrossRef]

44. Wu, J.D.; Li, J.C.; Wei, F.Z.; Wang, C.Y.; Zhang, Y.; Sun, G. Effects of nitrogen spraying on the post-anthesis stage of winter wheat under waterlogging stress. *Acta Physiol. Plant.* **2014**, *36*, 207–216. [CrossRef]
45. Ma, S.Y.; Yu, Z.W.; Shi, Y.; Gao, Z.Q.; Luo, L.P.; Chu, P.F.; Guo, Z.J. Soil water use, grain yield and water use efficiency of winter wheat in a long-term study of tillage practices and supplemental irrigation on the North China Plain. *Agric. Water Manag.* **2015**, *150*, 9–17. [CrossRef]
46. Moradi, L.; Siosemardeh, A.; Sohrabi, Y.; Bahramnejad, B.; Hosseinpanahi, F. Dry matter remobilization and associated traits, grain yield stability, N utilization, and grain protein concentration in wheat cultivars under supplemental irrigation. *Agric. Water Manag.* **2022**, *263*, 107449. [CrossRef]
47. Zhang, H.B.; Han, K.; Gu, S.B.; Wang, D. Effects of supplemental irrigation on the accumulation, distribution and transportation of ^{13}C-photosynthate, yield and water use efficiency of winter wheat. *Agric. Water Manag.* **2019**, *214*, 1–8. [CrossRef]
48. Meng, W.W.; Yu, Z.W.; Zhao, J.Y.; Zhang, Y.L.; Shi, Y. Effects of supplemental irrigation based on soil moisture levels on photosynthesis, dry matter accumulation, and remobilization in winter wheat (*Triticum aestivum* L.) cultivars. *Plant Product. Sci.* **2017**, *20*, 215–226. [CrossRef]
49. Xu, C.L.; Tao, H.B.; Tian, B.J.; Gao, Y.B.; Ren, J.H.; Wang, P. Limited-irrigation improves water use efficiency and soil reservoir capacity through regulating root and canopy growth of winter wheat. *Field Crops Res.* **2016**, *196*, 268–275. [CrossRef]
50. Gao, Y.M.; Zhang, M.; Yao, C.S.; Liu, Y.Q.; Wang, Z.M.; Zhang, Y.H. Increasing seeding density under limited irrigation improves crop yield and water productivity of winter wheat by constructing a reasonable population architecture. *Agric. Water Manag.* **2021**, *253*, 106951. [CrossRef]
51. Li, H.R.; Mei, X.R.; Wang, J.D.; Huang, F.; Hao, W.P.; Li, B.G. Drip fertigation significantly increased crop yield, water productivity and nitrogen use efficiency with respect to traditional irrigation and fertilization practices: A meta-analysis in China. *Agric. Water Manag.* **2021**, *244*, 106534. [CrossRef]
52. Wang, C.Y.; Liu, W.X.; Li, Q.X.; Ma, D.Y.; Lu, H.F.; Feng, W.; Xie, Y.X.; Zhu, Y.J.; Guo, T.C. Effects of different irrigation and nitrogen regimes on root growth and its correlation with aboveground plant parts in high-yielding wheat under field conditions. *Field Crops Res.* **2014**, *165*, 138–149. [CrossRef]
53. Jha, S.K.; Gao, Y.; Liu, H.; Huang, Z.D.; Wang, G.S.; Liang, Y.P.; Duan, A.W. Root development and water uptake in winter wheat under different irrigation methods and scheduling for North China. *Agric. Water Manag.* **2017**, *182*, 139–150. [CrossRef]
54. Yan, F.L.; Shi, Y.; Yu, Z.W. Optimized border irrigation improved nitrogen accumulation, translocation of winter wheat and reduce soil nitrate nitrogen residue. *Agronomy* **2022**, *12*, 433. [CrossRef]
55. Yuan, Y.; Lin, F.; Maucieri, C.; Zhang, Y.J. Efficient irrigation methods and optimal nitrogen dose to enhance wheat yield, inputs efficiency and economic benefits in the North China Plain. *Agronomy* **2022**, *12*, 273. [CrossRef]
56. Xu, J.T.; Cai, H.J.; Wang, X.i.Y.; Ma, C.G.; Lu, Y.J.; Ding, Y.B.; Wang, X.W.; Chen, H.; Wang, Y.F.; Saddique, Q. Exploring optimal irrigation and nitrogen fertilization in a winter wheat-summer maize rotation system for improving crop yield and reducing water and nitrogen leaching. *Agric. Water Manag.* **2020**, *228*, 105904. [CrossRef]
57. Yang, X.L.; Lu, Y.L.; YDing Yin, X.F.; Raza, S.; Tong, Y.A. Optimising nitrogen fertilisation: A key to improving nitrogen-use efficiency and minimising nitrate leaching losses in an intensive wheat/maize rotation (2008–2014). *Field Crops Res.* **2017**, *206*, 1–10. [CrossRef]
58. Lemenkova, P. 117 Mapping environmental and climate variations by GMT: A case of Zambia, Central Africa. *Zemljište i biljka* **2021**, *70*, 117–136. [CrossRef]
59. Stričević, R.; Vujadinović-Mandić, M.; Đurović, N.; Lipovac, A. Application of two measures of adaptation to climate change for assessment on the yield of wheat, corn and sunflower by the aquacrop model. *Zemljište i biljka* **2021**, *70*, 41–59. [CrossRef]
60. Ljubičić, N.; Popović, V.; Ćirić, V.; Kostić, M.; Ivošević, B.; Popović, D.; Pandžić, M.; El Musafah, S.; Janković, S. Multivariate interaction analysis of winter wheat grown in environment of limited soil conditions. *Plants* **2021**, *10*, 604. [CrossRef] [PubMed]
61. Ljubičić, N.; Popović, V.; Ivošević, B.; Rajičić, V.; Simić, D.; Kostić, M.; Pajić, M. Spike index stability of bread wheat grown on halomorphic soil. *Selekcija i Semenarstvo* **2022**, *28*, 1–8. [CrossRef]
62. Popovic, V.; Ljubičić, N.; Kostić, M.; Radulović, M.; Blagojević, D.; Ugrenovic, V.; Popovic, D.; Ivosevic, B. Genotype × environment interaction for wheat yield traits suitable for selection in different seed priming conditions. *Plants* **2020**, *9*, 1804. [CrossRef] [PubMed]

agronomy

Article

Irrigation Scheduling and Production of Wheat with Different Water Quantities in Surface and Drip Irrigation: Field Experiments and Modelling Using CROPWAT and SALTMED

Ahmed A. El-Shafei [1,2,3] and Mohamed A. Mattar [1,2,4,*]

1. Prince Sultan Bin Abdulaziz International Prize for Water Chair, Prince Sultan Institute for Environmental, Water and Desert Research, King Saud University, Riyadh 11451, Saudi Arabia; aelshafei1bn.c@ksu.edu.sa
2. Department of Agricultural Engineering, College of Food and Agriculture Sciences, King Saud University, Riyadh 11451, Saudi Arabia
3. Department of Agricultural Engineering, Faculty of Agriculture, Alexandria University, Alexandria 21545, Egypt
4. Agricultural Engineering Research Institute (AEnRI), Agricultural Research Centre, Giza 12618, Egypt
* Correspondence: mmattar@ksu.edu.sa

Citation: El-Shafei, A.A.; Mattar, M.A. Irrigation Scheduling and Production of Wheat with Different Water Quantities in Surface and Drip Irrigation: Field Experiments and Modelling Using CROPWAT and SALTMED. *Agronomy* 2022, 12, 1488. https://doi.org/10.3390/agronomy12071488

Academic Editors: Pantazis Georgiou and Dimitris Karpouzos

Received: 11 May 2022
Accepted: 17 June 2022
Published: 21 June 2022

Publisher's Note: MDPI stays neutral with regard to jurisdictional claims in published maps and institutional affiliations.

Abstract: Water is a key factor in global food security, which is critical to agriculture. The use of mathematical models is a strategy for managing water use in agriculture, and it is an effective way to predict the effect of irrigation management on crop yields if the accuracy of these models is demonstrated. The CROPWAT and SALTMED models were tested in this study, with water quantities applied to surface and drip irrigation (SI and DI) systems to estimate irrigation scheduling and wheat yield. For this purpose, field experiments were conducted for two consecutive years to study the effects of irrigation water levels of 80%, 100%, and 120% crop evapotranspiration (I_{80}, I_{100}, and I_{120}) on the yield and water productivity (WP) of wheat in SI and DI systems. Irrigation treatments affected yield components such as plant height, number of spikes, spike length, and 1000-kernel weight, though they were not statistically different in some cases. In the I_{80} treatment, the biological yield was 12.8% and 8.5% lower than in the I_{100} and I_{120} treatments, respectively. I_{100} treatment under DI resulted in the highest grain yield of a wheat crop. When DI was applied, there was a maximum (22.78%) decrease in grain yield in the I_{80} treatment. The SI system was more water-consuming than the DI system was, which was reflected in the WP. When compared with the WP of the I_{80} and I_{100} treatments, the WP was significantly lower ($p < 0.05$) in the I_{120} treatment in the SI or DI system. To evaluate irrigation scheduling and estimate wheat yield response, the CROPWAT model was used. Since the CROPWAT model showed that increasing irrigation water levels under SI for water stress coefficient (K_s) values less than one increased deep percolation (DP), the I_{120} treatment had the highest DP value (556.15 mm on average), followed by the I_{100} and I_{80} treatments. In DI, I_{100} and I_{120} treatments had K_s values equal to one throughout the growing seasons, whereas the I_{80} treatment had K_s values less than one during wheat's mid- and late-season stages. The I_{100} and I_{80} treatments with DI gave lower DP values of 93.4% and 74.3% compared with that of the I_{120} treatment (on average, 97.05 mm). The I_{120} treatment had the lowest irrigation schedule efficiency in both irrigation systems, followed by the I_{100} and I_{80} treatments. In both seasons, irrigation schedule deficiencies were highest in the I_{80} treatment with DI (on average, 12.35%). The I_{80} treatment with DI had a significant yield reduction (on average, 21.9%) in both seasons, while the irrigation level treatments with SI had nearly the same reductions. The SALTMED model is an integrated model that considers irrigation systems, soil types, crops, and water application strategies to simulate soil water content (SWC) and crop yield. The SALTMED model was calibrated and validated based on the experimental data under irrigation levels across irrigation systems. The accuracy of the model was assessed by the coefficients of correlation (R), root mean square errors (RMSE), mean absolute errors (MAE), and mean absolute relative error (MARE). When simulating SWC, the SALTMED models' R values, on average, were 0.89 and 0.84, RMSE values were 0.018 and 0.019, MAE values were 0.015 and 0.016, and MARE values were 8.917 and 9.133%, respectively, during the calibration and validation periods. When simulating crop yield, relative errors (RE) for the SALTMED model varied between −0.11 and 24.37% for biological yield and 0.1 and 19.18% for grain yield during the calibration period, while in the

validation period, RE was in the range of 3.8–29.81% and 2.02–25.41%, respectively. The SALTMED model performed well when simulating wheat yield with different water irrigation levels under SI or DI.

Keywords: soil water content; water productivity; yield reductions; deficit irrigation

1. Introduction

Water scarcity has become a global problem with a significant impact on agricultural production [1,2]. According to the most recent report [3], irrigation covers more than 20% of global cultivated lands and contributes to more than 40% of global total food production. Agricultural irrigation consumes the most water, but it yields the lowest return per unit of water when compared with other economic sectors [4]. However, traditional irrigation methods, such as flood irrigation, result in less water productivity (WP). There have been many irrigation methods developed to increase WP throughout the world, including furrow and drip irrigation [5]. Furrow irrigation is a finer form of surface irrigation (SI) in which ridge tillage aids root development and water infiltration while reducing deep percolation, causing an increase in WP [6–8]. Furthermore, drip irrigation (DI) has developed rapidly over the past few decades. DI has a distinct advantage over conventional irrigation in reducing water use and regulating salt through reduced evaporation and precise water use, which plays an important role in agricultural production worldwide [9,10]. Deficit irrigation is another water management technique that allows larger agricultural lands to be irrigated with scarce water resources [11,12]. Because crops respond to water stress in different ways at different stages of growth, this technique has a big impact on irrigation scheduling while having a small impact on yield. When using deficit irrigation, an important factor to consider is the timing and degree of water stress to which plants are exposed [13–15]. When crops are irrigated insufficiently, their roots grow deep into the soil and reach the soil water, resulting in significant water savings without reducing crop yield, while increasing WP and increasing net farm income [16–20].

Wheat (*Triticum aestivum* L.) has been adopted mostly as a food crop worldwide and is the world's most widely distributed cereal after maize, ranking eighth in the world. However, growers are concerned about its long-term production and yield in water-stressed conditions [21]. Wheat farmers are currently facing a number of challenges, including water shortages and uncertain water delivery schedules [22]. At all growth stages, wheat needs sufficient soil moisture for normal growth and development, which can be achieved by precise irrigation scheduling that minimizes overwatering [23]. Excessive use of water can lead to waterlogging and nutrients leaching outside the root zone. To improve WP, it is necessary to schedule proper irrigation with an adequate amount of water, as flooding can reduce WP and crop yield [24]. Previous studies [25,26] have suggested that using deficit irrigation to save water in wheat can be beneficial. Several studies have looked at how different irrigation schedules affect wheat growth, yield, and WP [27–30]. Panda et al. [27] reported that wheat WP was highest when irrigation was applied when the available soil moisture was 45% depleted. According to Jalota et al. [28], reducing the number of irrigations to maximize WP and using less irrigation water would conserve wheat grain yield in semi-arid environments. When compared with full wheat-crop irrigation, Wang et al. [29] and Rao et al. [30] found that deficit irrigation improved WP by 11–40%. As a result, developing water-saving farming techniques that decrease irrigation water consumption while increasing WP to produce great and consistent yields similar to those produced with reduced irrigation is critical for achieving sustainable agricultural development [31–33]. Furthermore, decision-makers saved time by using mathematical models to manage irrigation water and forecast production under various conditions [34]. These models are an important tool for scientifically documenting irrigation scheduling with the goal of reducing water consumption and facilitating horizontal agriculture expansion by utilizing limited irrigation water resources.

The use of CROPWAT to decide on an irrigation regime is one of the most popular approaches to assessing irrigation performance and WP in irrigated areas. CROPWAT is an irrigation management and planning application software established by a group of experts [35–38]. CROPWAT model's calculations are based on guidelines for calculating crop water requirements [39] and the yield response to water requirements [40]. The CROPWAT model can be considered as a valuable method for calculating water irrigation requirements for developing irrigation schedules based on the crop parameters and daily soil moisture balance at maximum root depth, which can be used to estimate evapotranspiration using different water source options and irrigation management conditions for a variety of crops under various environmental conditions [41–48].

SALTMED is an integrated model that uses well-known physically based equations to simulate soil water profiles, salinity distribution and nitrogen in the soil, leaching requirements, crop growth and yield, taking into account water application strategies, soil types, irrigation systems, crops, and various water qualities [49,50]. The SALTMED model can be used to assess the future effect on irrigation management and predict water distribution under automatic irrigation scheduling by running different scenarios under different conditions and crop parameters [51]. Some research has been conducted on the SALTMED model, which has shown that it can be used to manage water, crops, and soil under a variety of irrigation applications and environmental conditions [34,52–58].

Therefore, our study aims to (1) compare changes in yield and water productivity of wheat exposed to different water quantities under surface and drip irrigation systems; (2) explore the CROPWAT model's ability to evaluate irrigation scheduling and performance during the wheat-growing stages; and (3) simulate soil water content (SWC), dry matter, and grain yield of wheat using the SALTMED model and compare it with observed data from field experiments.

2. Materials and Methods

2.1. Field Study Site

Field experiments were conducted during the two successive seasons of 2018/2019 and 2019/2020 in the Qarun Kebili region, El Fayoum, Egypt. This region has an altitude of 17 m above sea level at 29°21′55″ N latitude and 30°27′11″ E longitude. The experimental site has a semi-arid climate, as shown in Figure 1, with air temperature (T), relative air humidity (RH), rainfall, and reference evapotranspiration (ET_0). From November to April, the average T and RH concentrations were 16.2 °C and 54.9% in the first growing season, respectively, and 17.9 °C and 58.9% in the second growing season, respectively. The rainfall received in the first growing season was 20 mm, while in the second growing season it was 12.2 mm. In the first and second growing seasons, ET_0 was 490.7 mm and 459 mm, respectively. In Table 1, the physical (i.e., texture, bulk density, field capacity, wilting point, and saturated hydraulic conductivity) and chemical (i.e., electrical conductivity, pH, organic matter, and soluble cations and anions) soil properties of the site were determined at two soil depths (10–30 and 30–60 cm) after eliminating the top soil, by standard procedures according to Page et al. [59] and Klute [60]. The soil of the experimental site had a loamy sand texture. The water used to irrigate the experimental site had a pH of 5 and an electrical conductivity of 2 dS m^{-1} for the experimental period.

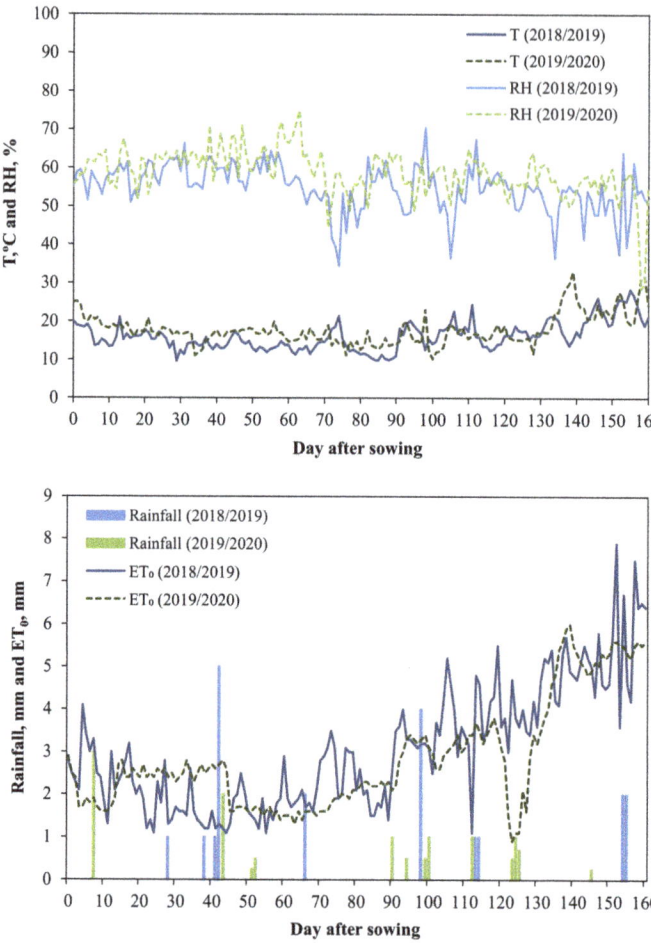

Figure 1. Daily values of climatic data at the experimental site throughout the two growing seasons.

Table 1. Physical and chemical properties of the soil at the experimental site.

Soil's Physical Properties										
Depth (cm)	Particle Size (%)			Texture	ρ_b (g cm^{-3})	FC (%)	WP (%)	θ_s m^3 m^{-3}	TAW m^3 m^{-3}	K_s (mm h^{-1})
	Sand	Silt	Clay							
0–30	72.1	11.9	16.0	Loamy sand	1.57	20.0	12.0	0.44	0.08	32.5
30–60	73.4	12.0	14.6	Loamy sand	1.51	19.5	11.5	0.43	0.08	33.1

Soil's Chemical Properties											
Depth (cm)	EC$_e$ (dS m^{-1})	pH	OM	Soluble Cations (meq L^{-1})				Soluble Anions (meq L^{-1})			
				Ca^{2+}	Mg^{2+}	Na$^+$	K$^+$	CO$_3{}^{2-}$	HCO^{3-}	SO$_4{}^{2-}$	Cl$^-$
0–30	4.42	7.87	0.7	28.7	7.74	9.62	0.46	-	2.98	22.00	21.56
30–60	5.56	7.74	0.8	27.8	5.88	18.88	0.35	-	2.97	22.98	23.93

ρ_b: bulk density; FC: field capacity; WP: wilting point; θ_s: saturated moisture content; TAW: total available water; K_s: saturated hydraulic conductivity; EC$_e$: electrical conductivity; OM: organic matter.

2.2. Experimental Layout and Design

The soil was plowed and leveled after the removal of plant debris prior to the establishment of the experimental layout. Before sowing wheat, all areas where corn was the previous crop received 357 kg ha^{-1} P_2O_5 and 120 kg ha^{-1} K_2O as fertilizer, which was mixed using the disc harrow. After that, the soil was furrowed with 1.2 m spacing, and flatbeds with 1 m width and 0.10–0.15 m height were the result. Wheat (*Triticum aestivum* L. cv. Masr2) seeds were manually sown in the flatbeds at a rate of 110 kg ha^{-1} on 17 November 2018 for the first season and 15 November 2019 for the second season. Four NH_4NO_3 fertilizer doses (286 kg ha^{-1}) were applied at 20, 30, 65, and 45 days after sowing (DAS), with Fe-, Zn-, and Mn- fertilizer sprayed at 53 DAS. Herbicides were sprayed at a rate of 19 g ha^{-1} and 333 g ha^{-1} on 24 and 33 DAS, respectively. The wheat crop was harvested at 160 DAS (i.e., 25 April 2019 and 23 April 2020).

The total study area was divided into four fields, representing replications. Each field contained two irrigation systems that were SI and DI, which represented the main blocks. Each block had three plots which were irrigation water levels, namely, full [100% crop evapotranspiration (ET$_c$), I$_{100}$], deficit [80% ET$_c$, I$_{80}$], and over [120% ET$_c$, I$_{120}$] irrigation. It was maintained at a distance of 1 m between adjacent plots to prevent the potential impact of water leakage. The randomized complete block design (RCBD) was used.

In surface irrigated plots, the water was supplied through perforated PVC pipe with a 63 mm outside diameter along the plot width. In drip-irrigated plots, surface laterals were installed with inline emitters at 50 cm spacing on the lateral line and a 3.5 L h^{-1} flow rate at an operating pressure of 100 kPa. The laterals were placed in the center of the flatbeds and furrow bottoms (i.e., lateral spacing of 60 cm). The irrigation interval time for SI and DI treatments was selected to be 12 and 4 days, respectively. All plots were irrigated based on the crop evapotranspiration (ET$_c$) under standard conditions, which was calculated according to the following equation [39]:

$$ET_c = K_c \times ET_0 \tag{1}$$

where K$_c$ is the crop coefficient and ET$_0$ is the reference evapotranspiration (mm day^{-1}).

On the basis of field observations of crop stages using the FAO-56 data [39], K$_c$ was determined to be 0.35 for the initial stage (up to 20 DAS), 1.15 during the mid-season stage of 51 to 115 DAS, and 0.25 during the late-season stage of 116 to 160 DAS, and the development stage was from 21 to 50 DAS. The Penman–Monteith FAO-56 equation [39] was used to calculate ET$_0$ on a daily basis from the measured climatic data:

$$ET_0 = \frac{0.408\Delta(R_n - G) + \gamma \frac{900}{T_a + 273} u_2(e_s - e_a)}{\Delta + \gamma(1 + 0.34 u_2)} \tag{2}$$

where R$_n$ is net radiation (MJ m^{-2} day^{-1}), G is the soil heat flux (MJ m^{-2} day^{-1}), γ is the psychrometric constant (kPa °C^{-1}), T$_a$ is the mean air temperature at 2 m height (°C), u$_2$ is the wind speed at 2 m height (m s^{-1}), e$_s$ is the saturation vapor pressure (kPa), e$_a$ is the actual vapor pressure (kPa), and Δ is the slope of the saturation vapor pressure–temperature curve at mean air temperature (kPa °C^{-1}).

2.3. Measurement of Soil Water Content

In the irrigation level plots, time domain reflectometry (TDR) probes (Trime FM; IMKO GmbH; Germany-76275 Ettlingen) were installed for continuous monitoring of the SWC over the two growing seasons. In each plot, a 60 cm-long probe was installed after the measurements of the TDR sensors were calibrated. A data logger was used to record SWC data.

2.4. Measurements of Yield Components, Grain Yield, and Water Productivity

Prior to harvesting and at the maturity stage of wheat, sheaves were randomly selected from each treatment area to measure the following data: plant height (cm), spike length (cm), spikes per unit area (m^{-2}), 1000-kernel weight (g), grain yield (GY, Mg ha^{-1}), and biological yield (Mg ha^{-1}).

Plant length and spike length were measured at each plot. The number of spikes was calculated by cutting them from sheaves, counting them, and recalculating the m^2 area. 1000-kernel weight (g) were determined by rubbing out grains from randomly selected 20 plants in each treatment, counting, and weighing them by scales Kern KB 1200-2 [61]. The GY (Mg ha^{-1}) was calculated from each treatment's area by weighting grain samples after air drying and reaching a water content of 14% (g H_2O g^{-1} fresh weight) [62] and then biological yield was measured. To avoid border effects, flatbeds on every side of each plot were not considered at harvest. The harvest index (HI) was determined by dividing grain yield by biological yield. According to Kijne et al. [63], the WP (kg m^{-3}) is defined as the ratio of GY (Mg ha^{-1}) to the amount of applied water (W, m^3 ha^{-1}) (irrigation water + effective rainfall) as follows:

$$WP = \frac{GY}{W} \tag{3}$$

2.5. Maximum Grain Yield

The relationship between GY and W is called the grain water production function (GWPF). As some of the excess applied water is drained or lost, the GWPF becomes curvilinear. It expresses the benefit of applied water in terms of grain yield or biological yield. Helweg's [64] quadratic polynomial function was written as follows:

$$GY = b_0 + b_1W + b_2W^2 \tag{4}$$

where b_0, b_1, and b_2 are fitting coefficients for a specific irrigation system.

When at the maximum GY (GY_{max}) value, the slope of the GWPF against W goes to zero, therefore differentiating Equation (4) and equalizing by zero.

$$\frac{dGY}{dW} = b_1 + 2b_2W = 0 \tag{5}$$

The maximum applied water (W_{max}) was calculated as follows:

$$W_{max} = \frac{-b_1}{2b_2} \tag{6}$$

Then the predicted GY_{max} was calculated by substituting the W_{max} in Equation (4) [56].

$$GY_{max} = b_0 + b_1W_{max} + b_2W_{max}^2 \tag{7}$$

2.6. CROPWAT Model

The United Nations Food and Agriculture Organization (FAO) [35–38] developed the CROPWAT version 8.0 model [65], which is a software package. The CROPWAT model is frequently used for planning and managing irrigation projects based on the method described in Allen et al. [39]. The CROPWAT model was used to calculate crop water requirements and evaluate irrigation schedules for different irrigation strategies in this study. Climatic, crop, and soil variables are included in the CROPWAT model's input data.

- The daily ET_0 values were calculated using the Penman–Monteith FAO-56 equation (Equation (2)), which was based on climatic data from the Central Laboratory for Agricultural Climate's Meteorological Data, as well as the daily rainfall data, for the seasons of 2018/2019 and 2019/2020. The actual crop evapotranspiration was estimated by multiplying ET_0, K_c, and 0.8, 1, or 1.2. The efficiencies of SI and DI were estimated through field investigation, which was about 69% and 93%, respectively.

Irrigation application depth was then estimated at irrigation events for SI and DI, as shown in Figure 2.

- Planting and harvesting dates, duration and water stress coefficient (K_s) of crop growth stages, and root depth were all included in the crop data. In addition, according to Allen et al. [39], the depletion fraction (p) (0.65 for initial and mid-season stages, and 0.57 for late-season stage) were calculated using the following equation:

$$p = p_{ET5} + 0.04(5 - ET_c) \qquad (8)$$

where p_{ET5} is the depletion fraction at $ET_c = 5$ mm/day, which is equivalent to winter wheat as 0.55 [39].

- Total available soil water from measured data (Table 1), maximum rooting depth and maximum rain infiltration rate from FAO, and initial soil moisture depletion from the CROPWAT program were among the soil data.

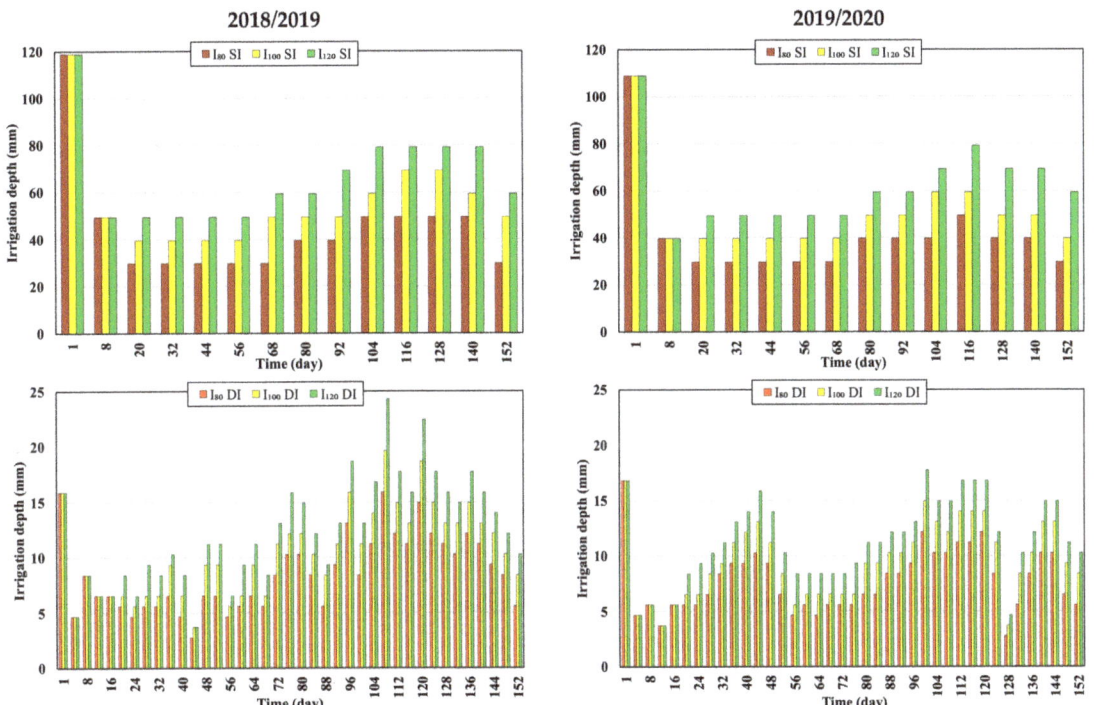

Figure 2. Applied irrigation water at the timing intervals for different treatments in both growing seasons. I_{80} = 80% crop evapotranspiration (ET_c), I_{100} = 100% ET_c, I_{120} = 120% ET_c, SI = surface irrigation, and DI = drip irrigation.

Accordingly, irrigation schedules were developed for 80% ET_c, 100% ET_c, and 120% ET_c with the irrigation systems, namely, SI and DI. The CROPWAT model produces a variety of parameters that can be used to compare irrigation schedules. The output parameters are: root zone depletion (D_r), deep percolation (DP), efficiency of the irrigation schedule (EIS), deficiency of the irrigation schedule (DIS), and yield reduction (Y_R).

Due to the fact that soil water budget parameters are often expressed as depths of water, the D_r is useful since it makes adding and subtracting losses and gains straightforward.

The soil water balance was performed in the schedule module of CROPWAT according to Swennenhuis [65] to estimate the daily D_r (Equation (9)).

$$D_{r,i} = D_{r,i-1} + (ET_{c,i})_{actual} - P_i - I_i + RO_i + DP_i \tag{9}$$

where $D_{r,i}$, and $D_{r,i-1}$ are on days i and i − 1; P_i is the total rainfall over day i; I_i is net irrigation on day i; RO_i is water loss by runoff from the soil surface on day i—since the ends of the plots in SI system were closed in our study, the RO was zero; and DP_i is water loss by deep percolation on day i. If irrigation was used, the D_r was calculated before it was applied.

The readily available water (RAW) is the p fraction of total available water (TAW) that a crop can extract from the root zone without being stressed by water. At a given soil depth, RAW is expressed as a percentage or in mm, as follows:

$$RAW = p \times TAW \tag{10}$$

When daily $D_{r,i}$ is less than RAW_i, daily $K_{s,i} = 1$. Under soil water limiting conditions, $D_{r,i}$ is greater than RAW_i and $K_{s,i} < 1$ and is given by Allen et al. [39] as:

$$K_{s,i} = \frac{TAW_i - D_{r,i}}{TAW_i - RAW_i} \tag{11}$$

Irrigation water reaching the root zone, I_i, is not always advantageously used by the crop due to irrigation losses such as DP_i in our study. Therefore, the EIS evaluates how advantageously the I_i contributions are used by the crop over the growing period, as follows [65]:

$$EIS = \frac{\sum(I_i - DP_i)}{\sum I_i} \times 100 \tag{12}$$

The relationship between seasonal potential water use by crop (ET_c under standard conditions) and seasonal actual water use by crop is expressed by the DIS that was calculated by [65]:

$$DIS = \frac{Seasonal\ (ET_c)_{potential} - Seasonal\ (ET_c)_{actual}}{Seasonal\ (ET_c)_{potential}} \times 100 \tag{13}$$

Due to soil water stress, Y_R was also used in the scheduling performance analysis. Y_R was estimated as a percentage of the maximum crop yield achievable in the case of full satisfaction of crop water needs (GY_{max}) [37], as follows:

$$Y_R = \left(1 - \frac{GY_a}{GY_{max}}\right) = K_y \left(1 - \frac{(ET_c)_{actual}}{(ET_c)_{potential}}\right) \tag{14}$$

where GY_a is the grain yield achievable under actual conditions, and K_y is the yield response factor. For initial, development, mid-season, and late-season stages, as well as the total growing period, K_y is set to 0.4, 0.6, 0.8, 0.4, and 1.0, respectively [40]. As a result, the best irrigation schedules are those that combine an irrigation interval and depth that result in a low DP and a reasonable Y_R.

2.7. SALTMED Model

The SALTMED model version 3.03.21 [49,50] was used for the simulation of SWC, total dry matter (biological yield), and grain yield of wheat by considering irrigation systems and different irrigation water quantities during the two seasons of 2018/2019 and 2019/2020. The data required depends on two main components: the first is for selected application options (global model parameters) and the second is for the interest of the user (field data). The user is not required to provide all of the data in the model tabs. For some applications, the model has multiple options. The user only needs to provide data for the options that are required. The data requirements for the SALTMED model may be directly measured in laboratory and field conditions, or default values may be provided from the SALTMED database for different plant species and soil types. In our study, the data requirements were as follows:

1. Climate data, including the daily data of maximum and minimum temperatures, wind speed, sunshine hours, rainfall, relative humidity, total solar radiation, and net radiation. The Penman–Monteith FAO-56 equation (Equation (2)) was used to calculate the daily ET_0 values.
2. Irrigation management data, including applied irrigation water amounts, dates of irrigation events, and irrigation water quality, were based on field measurement data.
3. Soil parameters, including saturated SWC, initial soil moisture, saturated hydraulic conductivity, and salinity, were based on measurements either in the laboratory or in the field. Soil evaporation coefficient (K_e) values were taken from Allen et al. [39]. The Richards equation was used in the model to simulate two-dimensional water flow in the soil. The analytical functions of van Genuchten [66] in the model were used for determining soil hydraulic properties (i.e., the soil water pressure head and hydraulic conductivity relationships).
4. Crop parameters, including plant height, maximum and minimum root depth, leaf area index, length of the growth stage, and sowing and harvesting dates, were obtained from field measurements. From Allen et al. [39], K_c and fraction cover (F_c) for the initial, middle, and late growth stages were taken. Basal crop coefficient (K_{cb}) values were then estimated as:

$$K_{cb} = K_c - K_e \qquad (15)$$

To simulate crop yield, there are two options in the model: the first by calculating the harvest index and the daily biomass production; and the second, which was used in our study, by using the relative yield index (RY), which is a ratio between the sum of the actual water uptake over the season and the maximum water uptake. Actual yield (AY) can then be calculated as follows:

$$AY = RY \times GY_{max} \qquad (16)$$

The SALTMED model was calibrated and validated for values of SWC, total dry matter (biological yield), and grain yield. The SALTMED model was calibrated for the 2018/2019 growing season by using default values of soil and crop parameters, as well as other measured values of these parameters from the field and laboratory, without any adjustments. Then, to achieve the best agreement between the measured and simulated parameters, a trial-and-error method was used to adjust both soil and crop parameters of the relevant model, namely soil pore-size distribution index, air-entry value, K_c, K_e, F_c, leaf area index, and photosynthesis efficiency. In the validation process, the SALTMED model used data collected during the 2019/2020 growing season to compare observed and simulated SWC, as well as biological and grain yields data of treatments.

2.8. Performance Accuracy Criteria

Five criteria indicators, namely the coefficient of correlation (R), the root mean square error (RMSE), mean absolute error (MAE), mean absolute relative error (MARE), and the relative error (RE), were selected to assess the accuracy of the proposed models. These criteria can be expressed as follows:

$$R = \frac{\sum_{i=1}^{n} (O_i - \overline{O}) (S_i - \overline{S})}{\sqrt{\sum_{i=1}^{n} (O_i - \overline{O})^2 \cdot \sum_{i=1}^{n} (S_i - \overline{S})^2}} \qquad (17)$$

$$\text{RMSE} = \sqrt{\frac{\sum_{i=1}^{n}(S_i - O_i)^2}{n}} \tag{18}$$

$$\text{MAE} = \frac{\sum_{i=1}^{n}|S_i - O_i|}{n} \tag{19}$$

$$\text{MARE} = \frac{1}{n}\left(\sum_{i=1}^{n}\left|\frac{S_i - O_i}{O_i}\right| \times 100\right) \tag{20}$$

$$\text{RE} = \frac{(S_i - O_i)}{O_i} \times 100 \tag{21}$$

where O_i and S_i are observed and simulated values, respectively, $\overline{O}$ and $\overline{S}$ are the average observed and simulated values, respectively, and n is the number of observations.

The degree of correlation between the observed and simulated values is measured by R. RMSE expresses the error in the same units as the variable and measures how close simulated values are to observed values [67]. MAE is a measure of how close the predicted values to the experimental values [68]. An acceptable goodness of fit is indicated by an R value close to 1, and RMSE, MAE, and MARE values close to 0. RE describes bias as a percentage provided by models.

2.9. Statistical Analysis

Using CoStat software (Version 6.303, CoHort, Monterey, CA, USA, 1998–2004) [69], the data from the two growing seasons were subjected to ANOVA analysis following a RCBD with four replicates of each treatment. The significant differences between the two treatment means of the measured parameters of GY, its components, and WP were evaluated using the least significant difference (LSD) method at a 5% significant level [70].

3. Results and Discussion

3.1. Irrigation Water Applied

Irrigation water applied to wheat for the 2018/2019 and 2019/2020 growing seasons was presented in Figure 2. The amount of irrigation water applied in the first growing season was higher than in the second. It is possible that this is due to climatic differences. The air temperature was lower in 2019/2020 than in 2018/2019, while rainfall and relative humidity were higher in 2019/2020 than in 2018/2019 (Figure 1). In both growing seasons, increasing the irrigation level increased the total water applied, as expected. Regarding the irrigation system in Figure 2, the total water applied increased with the SI system in both growing seasons. The highest total irrigation water applied value was obtained in the I_{120} treatment under the SI system, which was 930 mm in the first growing season and 871 mm in the second growing season, while under the DI system it was 492 mm and 450 mm, respectively. In this study, full-irrigated (I_{100}) wheat plants had a similar total irrigation water applied value to those obtained by Moussa and Abdel-Maksoud [71] and Abdelkhalek et al. [72].

3.2. Yield Components and Grain Yield

Table 2 shows the analysis of variance for wheat yield components and grain yield in the 2018/2019 and 2019/2020 growing seasons. Plant height and spike length were not significantly ($p > 0.05$) affected by the systems and levels of irrigation and the interaction between them in both seasons. The values of plant height ranged from 95.25 cm to 100.25 cm in the 2018/2019 season and from 95.75 cm to 100 cm in the 2019/2020 season. While the spike length was between 9.25 cm and 10 cm in the first and second seasons, there was a significant difference in the number of spikes ($p < 0.05$) between different irrigation levels only in the second season, whereas there were no significant differences ($p > 0.05$) between the spikes number values in the first season. The I_{100}- and I_{120}-treated plants in 2019/2020 showed no significant differences in the number of spikes (Figure 3), while the I_{80}-treated number of spikes was significantly reduced by 17.86% and 14.48%, respectively, compared with the I_{100} (396 spikes) and I_{120} (380 spikes) plants. Irrespective of the irrigation systems, the irrigation level treatments showed a significant effect on the 1000-kernel weight and biological yield in both seasons (Table 2). Figure 4 showed that the 1000-kernel weight and biological yield decreased with decreasing or increasing water levels than I_{100}, but the decrease was more pronounced under I_{80} than under I_{120}. In 2018/2019, the 1000-kernel weight values of I_{80} and I_{120} were decreased, compared with I_{100} (47.5 g), by 22.24% and 12.11%, respectively, while the value (38.88 g) for I_{80} was decreased by 17.07% in 2019/2020. The same trend applied to biological yield; the average value for I_{80} and I_{120} was 16.38 Mg ha^{-1} in 2018/2019. The corresponding I_{100} value was 18.13 Mg ha^{-1}. In 2019/2020, the I_{80} treatment had the lowest biological yield of 14.78 Mg ha^{-1} (a 13.99% decrease from the I_{100} value). According to Pandey et al. [73], as the amount of irrigation water applied increased, the growth rate and biological yield of wheat increased as well. Therefore, Table 2 shows that yield components of the wheat had non-significant differences between irrigation systems in both seasons. This is consistent with the findings of Eissa [74] and Noreldin et al. [75], who found that the irrigation system had no significant effect on the wheat plant characteristics studied.

Table 2. Statistical analysis results for grain yield and yield components of wheat under different treatments in both growing seasons.

Treatments	df	Plant Height p-Value	LSD	Number of Spikes p-Value	LSD	Spike Length p-Value	LSD	1000-Kernel Weight p-Value	LSD	Biological Yield p-Value	LSD	Grain Yield p-Value	LSD	Harvest Index p-Value	LSD
2018/2019															
Irr. syst.	1	0.1417	ns	0.2680	ns	0.3037	ns	0.1063	ns	0.0747	ns	0.0188	0.51	0.4502	ns
Irr. lev.	2	0.9522	ns	0.4414	ns	0.1998	ns	<0.001	1.58	0.0239	1.48	0.0008	0.63	0.1491	ns
Irr. syst. × Irr. lev.	2	0.2830	ns	0.4651	ns	0.9103	ns	0.3825	ns	0.2375	ns	0.0405	0.89	0.1187	ns
CV, %		3.78		19.44		6.05		3.53		8.17		8.26		6.86	
2019/2020															
Irr. syst.	1	0.4590	ns	0.2852	ns	0.9403	ns	0.2042	ns	0.0919	ns	0.0476	0.52	0.2376	ns
Irr. lev.	2	0.3757	ns	<0.001	21.99	0.3518	ns	0.0022	3.93	<0.001	0.57	<0.001	0.64	0.0853	ns
Irr. syst. × Irr. lev.	2	0.0829	ns	0.5231	ns	0.4794	ns	0.5828	ns	0.3245	ns	0.4472	ns	0.3028	ns
CV, %		2.74		5.63		6.91		8.57		3.29		8.85		10.83	

Irr. syst.: irrigation systems; Irr. lev.: irrigation levels; df: degrees of freedom; LSD: least significant difference; ns: non-significant; CV: coefficient of variation.

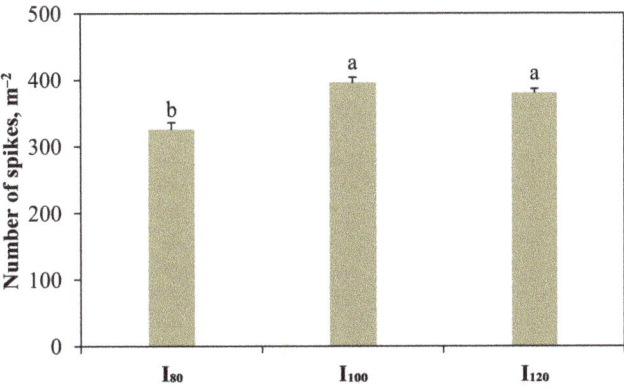

Figure 3. Number of wheat spikes under three irrigation levels in the 2019/2020 growing season (values are averages of two irrigation systems). According to the least significant difference test at $p < 0.05$, the same letters indicate statistically no significant differences. I_{80} = 80% crop evapotranspiration (ET$_c$), I_{100} = 100% ET$_c$, and I_{120} = 120% ET$_c$. Vertical lines give the means ± SE of the mean ($n = 8$).

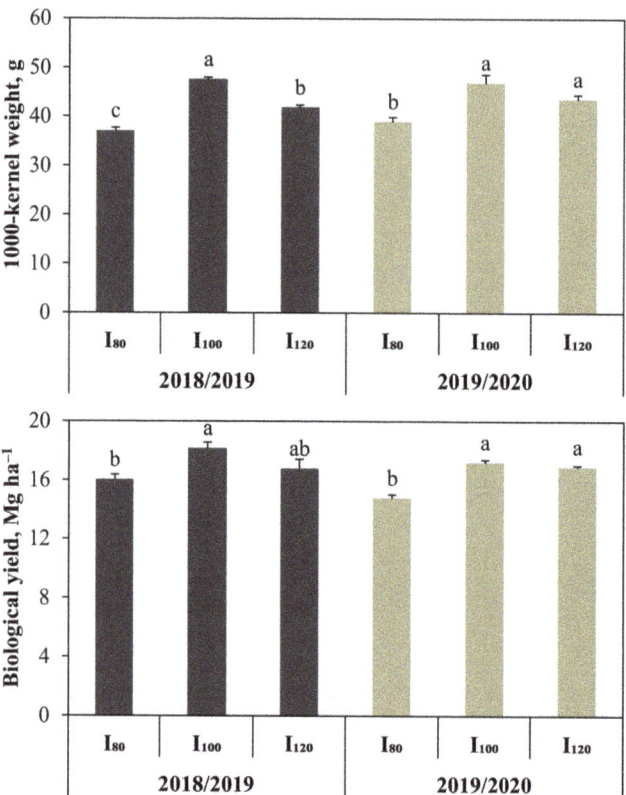

Figure 4. 1000-kernel weight and biological yield in wheat under three irrigation levels in each growing season (values are averages of two irrigation systems). According to the least significant difference test at $p < 0.05$, the same letters within a growing season indicate statistically no significant differences. I_{80} = 80% crop evapotranspiration (ET_c), I_{100} = 100% ET_c, and I_{120} = 120% ET_c. Vertical lines give the means $\pm$ SE of the mean ($n = 8$).

As shown in Table 2, irrigation level treatments had a significant ($p < 0.05$) effect on grain yield regardless of the irrigation system in both seasons. The average values of grain yield under DI were 7.46 and 7.03 Mg ha^{-1}, respectively, in 2018/2019 and 2019/2020, whereas the values under SI were significantly ($p < 0.05$) decreased by 8.50% and 7.50%, compared with those of DI (Figure 5a). Irrespective of the irrigation systems, the value of grain yield for I_{100} (7.89 and 7.52 Mg ha^{-1}) was significantly ($p < 0.05$) the highest, followed by I_{120} (decreasing by 10.05% and 5.29%, respectively) and later by I_{80} (decreasing by 18.34% and 24.58%, respectively) in 2018/2019 and 2019/2020 (Figure 5b). According to Mugabe and Nyakatawa [76], applying 75% of the wheat crop's water requirements reduced yields by 12% in two years. Low yields in the case of deficit irrigation, especially in cases of water limitation, may be offset by increasing production with additional water supply through deficit irrigated areas [77]. Table 2 shows that the interaction between irrigation systems and irrigation levels was significant ($p < 0.05$) in 2018/2019 but not significant ($p > 0.05$) in 2019/2020. In 2018/2019 (Figure 6), the I_{100} with DI treatment (8.21 Mg ha^{-1}) had the highest grain yield, but without significant differences with the I_{120} with DI and I_{100} with SI treatments. The I_{80} with DI treatment (6.34 Mg ha^{-1}) had the lowest value, but without significant differences with the I_{120} with SI and I_{80} with SI treatments. Finally, there were non-significant effects ($p > 0.05$) on the harvest index (Table 2).

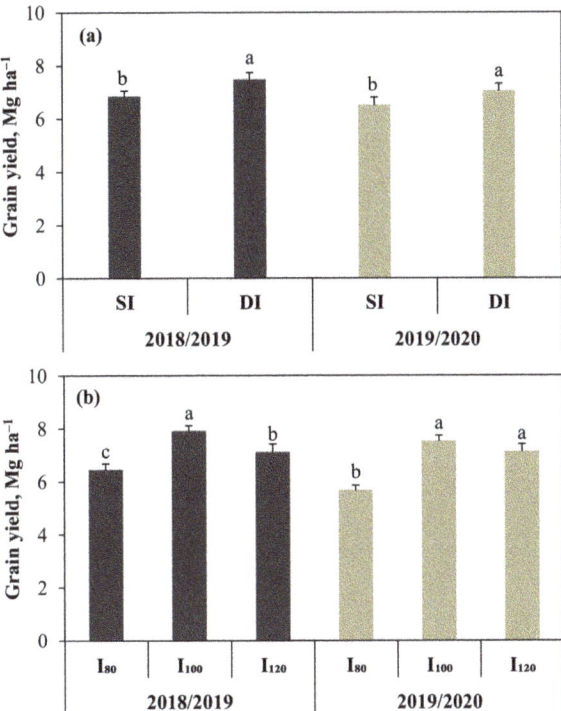

Figure 5. Grain wheat yield in each growing season under (**a**) two irrigation systems (values are averages of three irrigation levels, $n = 12$), and (**b**) three irrigation levels (values are averages of two irrigation systems, $n = 8$). According to the least significant difference test at $p < 0.05$, the same letters within a growing season indicated statistically no significant differences. I_{80} = 80% crop evapotranspiration (ET_c), I_{100} = 100% ET_c, I_{120} = 120% ET_c, SI = surface irrigation, and DI = drip irrigation. Vertical lines give the means $\pm$ SE of the mean.

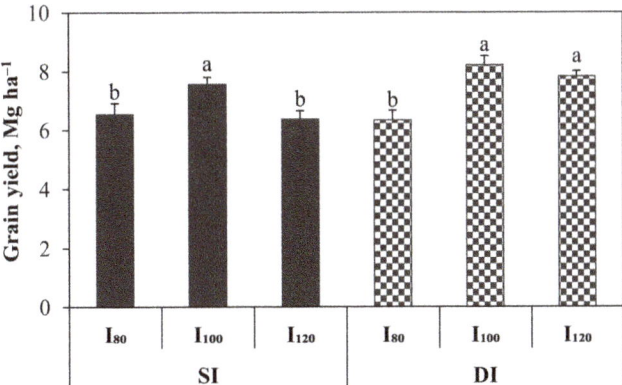

Figure 6. Grain yield in wheat under irrigation levels across irrigation systems in 2018/2019 growing season. According to the least significant difference test at $p < 0.05$, the same letters indicate statistically no significant differences. I_{80} = 80% crop evapotranspiration (ET_c), I_{100} = 100% ET_c, I_{120} = 120% ET_c, SI = surface irrigation, and DI = drip irrigation. Vertical lines give the means $\pm$ SE of the mean ($n = 4$).

3.3. Grain Yield–Water Relationship

From the regression analysis of the crop water production function in Figure 7, it is shown that the GY_{max} values for plants treated with SI were 7.57 and 7.48 Mg ha^{-1}, respectively, in the first and second seasons, and the corresponding calculated W_{max} values were 7738 m^3 and 7379 m^3. While the corresponding values for plants treated with DI were 8.33 and 7.84 Mg ha^{-1}, respectively, the W_{max} values were 4343.9 m^3 and 4086.8 m^3. Thus, it was found that the highest Y_R values (13.52–27.02%) were achieved with deficit irrigation (I_{80}) under the two irrigation systems in both seasons, except in the first season under SI, where the plants treated with I_{120} gave a slightly greater Y_R than those with I_{80}. The least Y_R (0.06–2.77%) was for the full irrigation (I_{100}) plants.

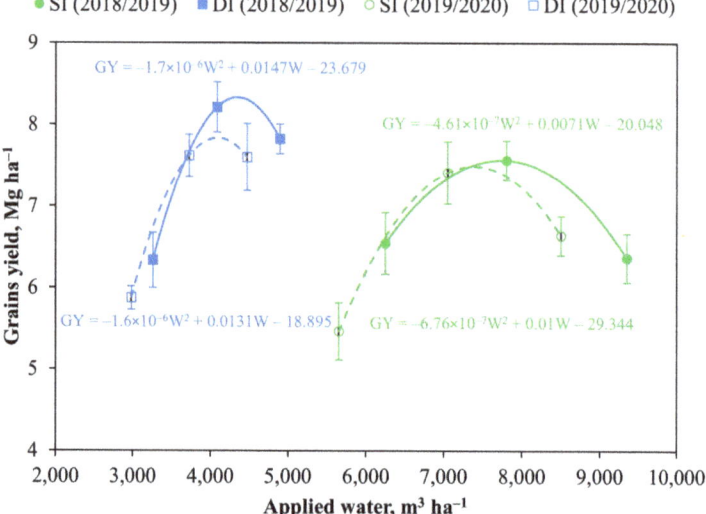

Figure 7. Relationship between grain yield (GY) and applied water (W) under different irrigation systems. SI = surface irrigation, and DI = drip irrigation. Vertical lines give the means ± SE of the mean (*n* = 4).

The results in Figure 8 showed that the SI system was the most water-consuming, followed by the DI system, which was the least water-consuming system. In terms of WP, the DI of the I_{100} treatment was found to be the best. No significant (*p* > 0.05) WP decrease was observed when using the DI of I_{80} treatment, where the WP decreased from 2.01 to 1.94 kg m^{-3} in the first season and from 2.05 to 1.97 kg m^{-3} in the second season for I_{100} and I_{80}, respectively. The I_{120} with either the SI or DI system was the most affected as WP decreased significantly (*p* < 0.05); however, it increased the water amount. Deficit irrigation, according to Geerts and Raes [78] and Pereira et al. [79], can increase WP by reducing the water loss from unproductive evaporation, increasing harvest index, and controlling pests and diseases during crop growth. Due to the relatively small increase in grain yield with increased evapotranspiration, Maurya and Singh [80] reported a decrease in WP with increased irrigation levels.

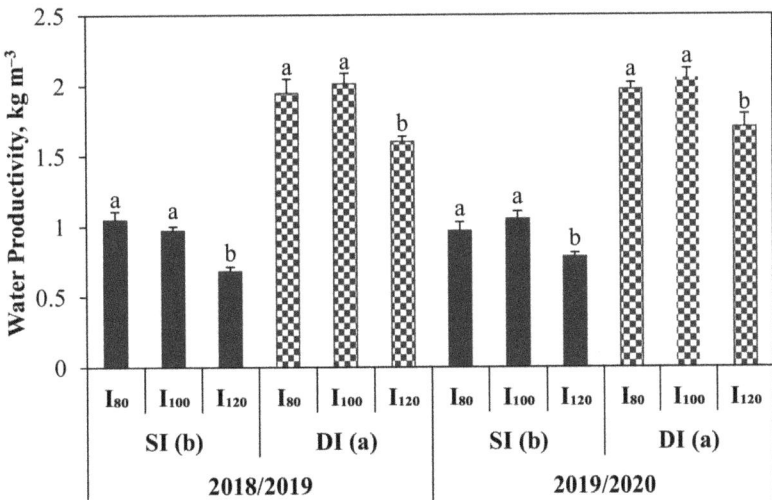

Figure 8. Water productivity in wheat under different irrigation levels across irrigation systems in each growing season. According to the least significant difference (LSD) test at $p < 0.05$, the same letters within an irrigation system indicate statistically no significant differences. Different letters between brackets represent significant differences between irrigation systems based on the LSD test with $p < 0.05$ within growing season. I_{80} = 80% crop evapotranspiration (ET_c), I_{100} = 100% ET_c, I_{120} = 120% ET_c, SI = surface irrigation, and DI = drip irrigation. Vertical lines give the means $\pm$ SE of the mean ($n = 4$).

3.4. CROPWAT Model

Figure 9 shows the soil water balance for wheat in the 2018/2019 and 2019/2020 growth seasons with the CROPWAT model under irrigation levels across irrigation systems. Irrigation scheduling was evaluated by the crop's daily water requirements, the soil's properties (particularly its TAW or water-holding capacity), and the root's effective depth. The TAW evolved in two phases: a filling phase when reserves reached 40 mm at 50 DAS, and a continual stabilization phase from that day through the conclusion of the cycle. The RAW went through a filling phase when reserves reached 20 mm at 50 DAS and then stayed in a steady phase till the cycle ended. The CROPWAT model directly calculates the root growth increase from the first day of vegetation [81]. The same behavior of soil depletion was found in the irrigation-level treatments under SI, where the soil depletion approached the lower limit of RAW at 75 DAS in the 2018/2019 season, while the crop entered stress at 87 DAS in the 2019/2020 season (Figure 9). The crop reached peak stress at 126 DAS and 114 DAS with depletion values of 34.4 mm and 31.7 mm (TAW = 40 mm), respectively, in both seasons. Figure 9 shows that the shape of the depletion curves for irrigation level treatments with DI was very similar in the first 20 days. There were considerable differences in irrigation schedules, as the depletion values for the I_{100} and I_{120} treatments were between the FC and the RAW throughout the growth seasons. The drip-irrigated plot with a water saving of 20% (I_{80}) gave water stress during the mid- and late-season stages, where the maximum depletion was at 106 DAS (31.9 mm) in the 2018/2019 season and 142 DAS (28.7 mm) in the 2019/2020 season. When soil water is extracted through evapotranspiration, depletion increases, and stress occurred when D_r equaled RAW. The D_r exceeded RAW (the water content fell below the threshold), which limited evapotranspiration to less than potential ET_c values, and the crop consumption decreased proportionally to the amount of water retained in the root zone [65]. Accordingly, Figure 10 shows that the lowest K_s values (0.28 and 0.48) in SI treatments were achieved at 126 DAS and 114 DAS (late-season stage). While in DI treatments, the K_s values were not less than one throughout the growing seasons in the I_{100} and I_{120} treatments (Figure 10), while the K_s values were less than one in the I_{80} treatment from 75 DAS (K_s of 0.92) in the 2018/2019 and 87 DAS (K_s of 0.9) in the 2019/2020 (i.e., mid-season) to the end of the season.

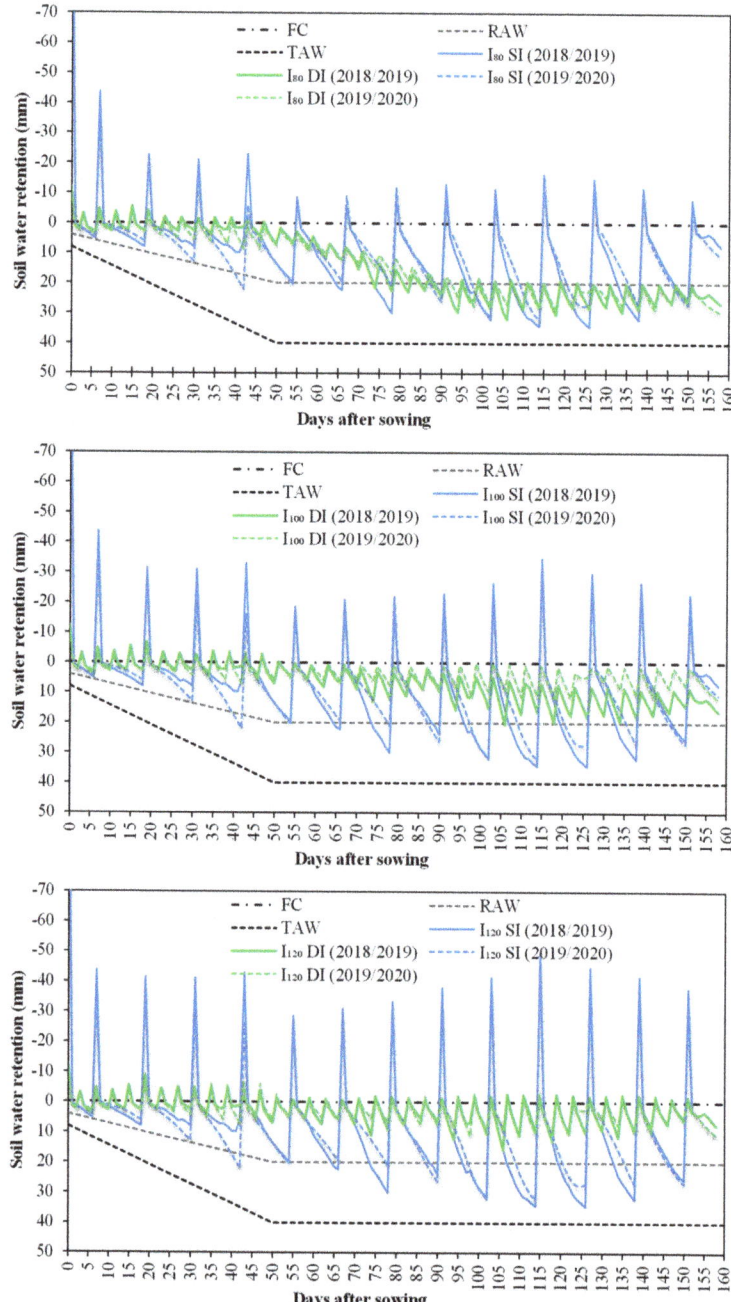

Figure 9. Water balance of wheat during each growth season under different irrigation levels across irrigation systems. I_{80} = 80% crop evapotranspiration (ET_c), I_{100} = 100% ET_c, I_{120} = 120% ET_c, SI = surface irrigation, and DI = drip irrigation.

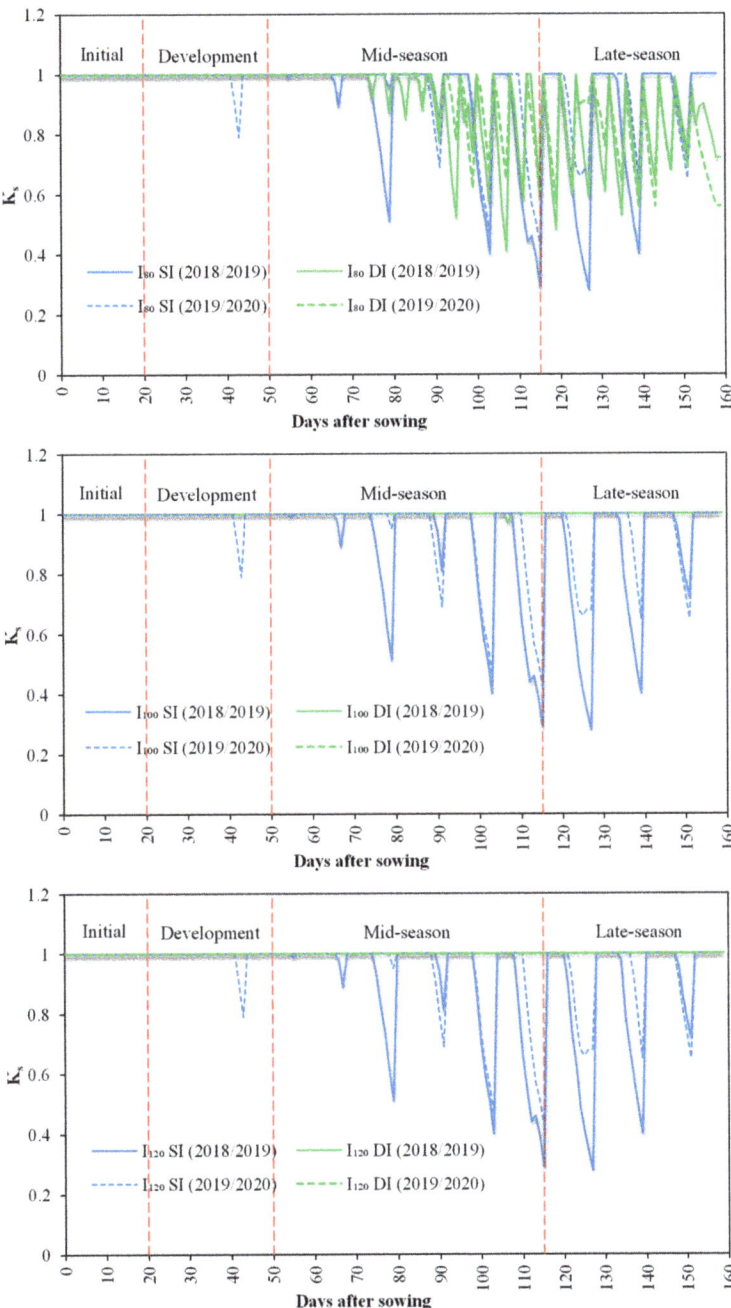

Figure 10. Daily water stress coefficient (K_s) for wheat during each growth season under different irrigation levels across irrigation systems. I_{80} = 80% crop evapotranspiration (ET_c), I_{100} = 100% ET_c, I_{120} = 120% ET_c, SI = surface irrigation, and DI = drip irrigation.

On the other hand, the DP (i.e., irrigation losses) with irrigation levels occurred throughout the seasons. In SI, the I_{120} treatment had the highest DP values (596 and 516.3 mm in both seasons), while the I_{100} and I_{80} treatments gave 26% and 52% lower values in 2018/2019, respectively, and 28.1% and 55.2% in 2019/2020, respectively, compared to the I_{120} treatment (Figure 11). Hence, the EIS values were the highest with the I_{80} treatment (on average 56.7%), followed by the I_{100} and I_{120} treatments (on average 45.5% and 37.8%, respectively).

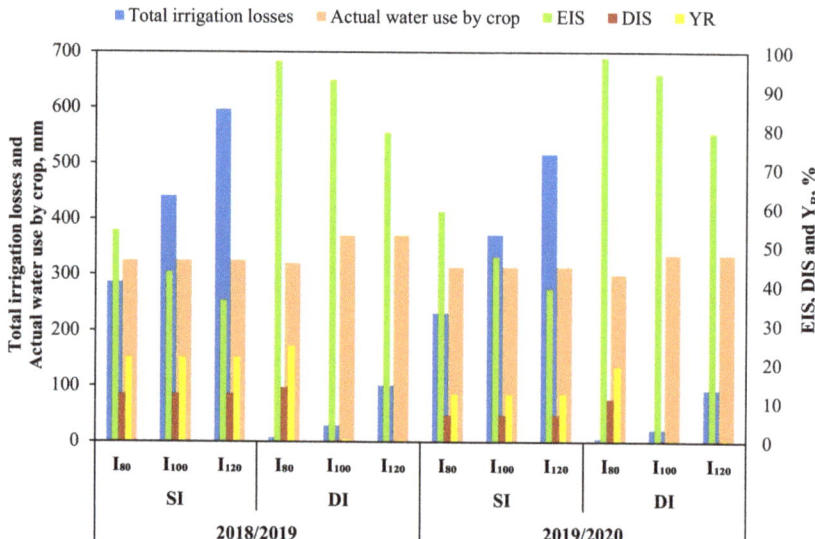

Figure 11. Total irrigation losses, actual water use by crop, efficiency and deficiency of irrigation schedule (EIS and DIS), and yield reduction (Y_R) under different irrigation levels across irrigation systems in each growing season. I_{80} = 80% crop evapotranspiration (ET_c), I_{100} = 100% ET_c, I_{120} = 120% ET_c, SI = surface irrigation, and DI = drip irrigation.

In DI, the DP occurred in I_{80} and I_{100} treatments in about the first 45 DAS, where total DP values were 7.4 and 28.8 mm for I_{80} and I_{100} treatments, respectively, in 2018/2019 and 5.4 and 21 mm, respectively, in 2019/2020 (Figure 11). While I_{120} treatment had DP over both seasons, the total values were 101 mm and 93.1 mm in 2018/2019 and in 2019/2020, respectively. The EIS values for the I_{80} and I_{100} treatments were high at 90%, whereas the I_{120} treatment had an EIS value of 79% in both seasons (Figure 11).

The $(ET_c)_{potential}$ with the CROPWAT model was estimated at 370.8 mm in the 2018/2019 season and 336 mm in the 2019/2020 season. The $(ET_c)_{actual}$ values' SI treatments were higher in the 2018/2019 season (on average, 325 mm) than in the 2019/2020 season (on average, 313.3 mm) (Figure 11). Knežević et al. [81] reported that $(ET_c)_{actual}$ values of winter wheat in Serbia were 345.4 mm and 463.3 mm on soils with a medium and high TAW, respectively, obtained with the CROPWAT model under rainfed conditions. In the I_{80} treatment, the obtained $(ET_c)_{actual}$ values with DI were around 5 mm and 13 mm in the first and second seasons, respectively, which were lower than those obtained with SI. The I_{80} treatment with DI had the highest values of DIS (13.8% and 10.9%) in both seasons. On the contrary, in the I_{100} and I_{120} treatments, the $(ET_c)_{actual}$ values for DI were greater, on average, by 13.8% and 7.1%, than those for SI. Therefore, the I_{100} and I_{120} treatments with DI had the lowest values of DIS (on average, 0.5% and 0.4%, respectively).

Figure 11 indicates the effect of irrigation scheduling on grain yield potential. During the I_{80} treatment with DI in the 2018/2019 and 2019/2020 seasons, tests of irrigation levels across irrigation systems showed significant yield reduction results (24.5 and 19.3%, respectively), while yield reductions across the irrigation level treatments under SI were similar in each season (21.7% in 2018/2019 and 12.2% in 2019/2020). The relative yield obtained with the simulations was compared with the measured yield. The GY_{max} ranged from 7.48 Mg ha^{-1} with SI in the 2019/2020 season to 8.33 Mg ha^{-1} with DI in 2018/2019 (Figure 7). The highest relative yields were achieved with irrigation-level treatments under DI (RE between -0.78% and 7.65%) in both seasons. In SI, the highest relative yield (99%) was estimated in 2019/2020 with I_{120} treatment (RE of -1%), while the lowest relative yield (78.35%) was estimated in 2018/2019 with I_{100} treatment (RE of -21.65%). According to the results of the CROPWAT model in study of Zhou and Zhao [82], an appropriate SI schedule for wheat in sandy soil included more frequent irrigation and lower application depth; this schedule can reduce deep percolation and save approximately 10–20% of irrigation water without affecting crop yield.

The CROPWAT model's outputs show that irrigation is crucial in the middle and late stages of wheat production to avoid yield reductions. Deficit irrigation may be used in areas where water is a limited resource for crop production. Deficit irrigation reduces irrigation water in certain crop growth stages that are thought to be the least sensitive to water stress, without affecting yields, to deal with water issues in areas where supply is limited [83,84]. A well-designed irrigation schedule can improve water productivity over a large area when full irrigation is not possible. However, because of the relationship between ET_c and crop yield, a yield reduction is expected [47,84,85].

3.5. SALTMED Model

The ability of the SALTMED model to represent the experimental data was examined during the periods of calibration (growing season 2018/2019) and validation (growing season 2019/2020) under different irrigation treatments. The SALTMED model was able to simulate SWCs with relatively high accuracy during the calibration period, based on the criteria indices presented in Table 3. In this period, the model for I_{80} in SI had the lowest RMSE, MAE, and MARE values, and vice versa in DI. The differences between measured and simulated SWCs were greatest in the SI for the I_{100} and I_{120} treatments, which had the highest RMSE, MAE, and MARE values. This can be explained by the fact that water infiltrated deeper into the soil during individual irrigation events in full and over-irrigation (I_{100} and I_{120}) treatments compared with water-saving irrigation (I_{80}) treatment. Furthermore, when simulating SWCs under the DI, more precise results were obtained, with lower RMSE, MAE, and MARE values for irrigation-level treatments than with SI. This is due to lower SWC differences in DI as a result of limited irrigation and root water uptake [86].

Table 3. Statistical indices comparing the measured and the SALTMED-simulated soil water content for various irrigation treatments during the calibration (the 2018/2019 growing season) and validation (the 2019/2020 growing season) periods.

Irrigation Systems	Irrigation Levels	2018/2019			2019/2020		
		RMSE	MAE	MARE, %	RMSE	MAE	MARE, %
SI	I_{80}	0.019	0.016	9.0	0.018	0.015	8.5
	I_{100}	0.022	0.02	11.5	0.020	0.018	9.9
	I_{120}	0.022	0.02	11.4	0.021	0.019	10.3
DI	I_{80}	0.018	0.015	9.7	0.017	0.013	8.8
	I_{100}	0.016	0.011	6.6	0.023	0.018	11.3
	I_{120}	0.015	0.009	5.3	0.017	0.010	6.0

Figure 12 shows scatter plots comparing the SALTMED model estimates for SWC to the SWC-measured data. In addition, linear regression is used to evaluate the model statistically. As shown in Figure 12, the results of the model show that points for irrigation level treatment under SI are located above the 1:1 line (perfect line), whereas in the DI, many points are located above and below this line. The fit line equations also show that under SI or DI, the I_{120} treatment had the lowest slope and the highest intercept. This indicates that the I_{120} treatment produced a lower R value.

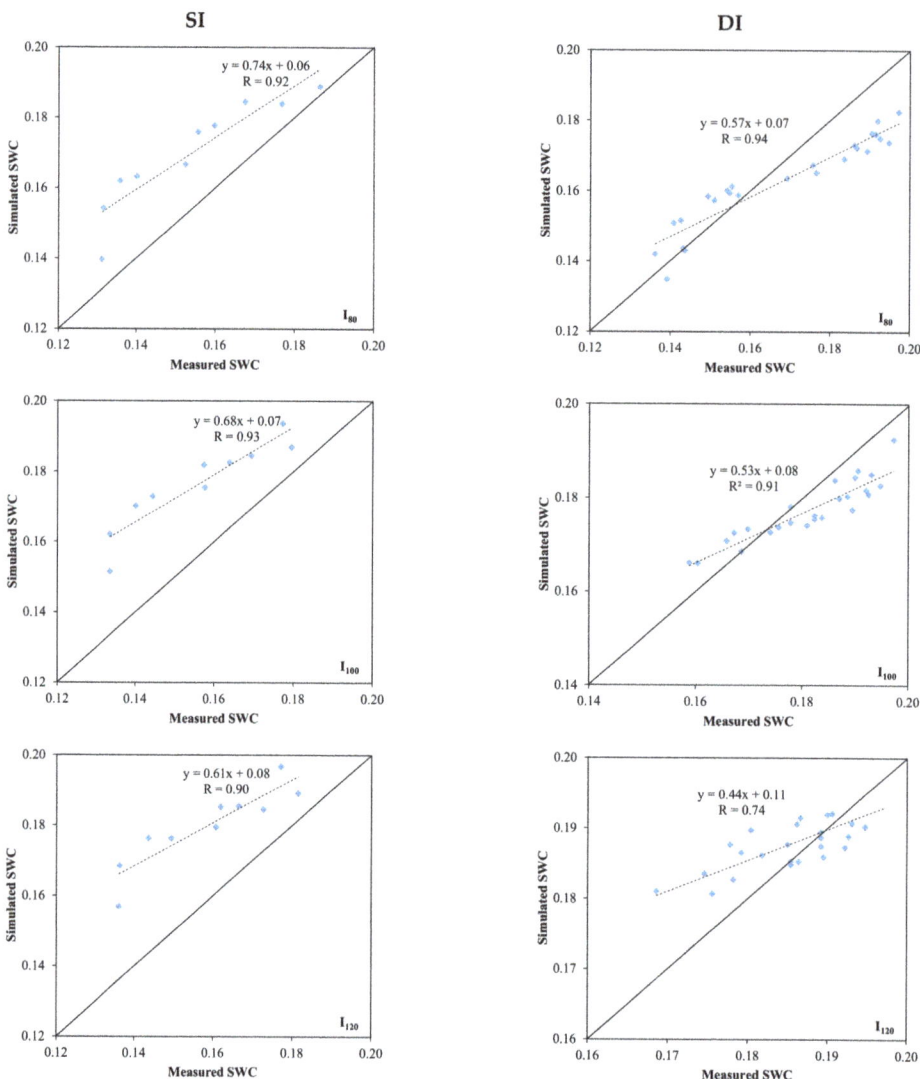

Figure 12. The observed and the SALTMED-simulated soil water content (SWC) for different irrigation levels across irrigation systems during the calibration period (i.e., the 2018/2019 growing season). I_{80} = 80% crop evapotranspiration (ET_c), I_{100} = 100% ET_c, I_{120} = 120% ET_c, SI = surface irrigation, and DI = drip irrigation.

In the validation period (the 2019/2020 growing season), the RMSE, MAE, MARE, and R values varied between 0.017–0.023, 0.01–0.019, 6.0–11.3%, and 0.62–0.95. The lowest RMSE, MAE, and MARE values (Table 3), were found in the SALTMED-simulated SWC values for the I_{80} treatment under SI. In DI, the I_{120} treatment presented the lowest RMSE, MAE, and MARE values. In Figure 13, the R value of the I_{120} treatment under SI or DI was the lowest compared with other treatments. The slope and intercept for the fitted line equation for the I_{80} treatment had the highest and the lowest values of 0.68 and 0.06, respectively, under SI, while these values were 0.66 and 0.05 under DI, according to the scatter plots in Figure 13. This indicates that around the 1:1 line, there was less scatter and more clustering. As a result, the SALTMED model can account for both temporal and spatial variations in SWCs in response to various treatments. According to Aly et al. [56], the SALTMED-simulated SWC values during the growing season of cucumber were very close to the observed values, with R ranging from 0.82 to 0.94 during calibration and 0.76 to 0.91 during validation. Hirich et al. [87] showed that the model could predict SWC into soil layers during the sweet corn growing season, with R values ranging from 0.91 to 0.95 and RMSE values ranging from 0.017 to 0.029 for calibrated data, and R values ranging from 0.91 to 0.96 and RMSE values ranging from 0.027 to 0.062 for validated data. According to Karandish and Simunek [86], the SALTMED model could simulate SWCs with a higher degree of accuracy under water-saving irrigation than under full irrigation.

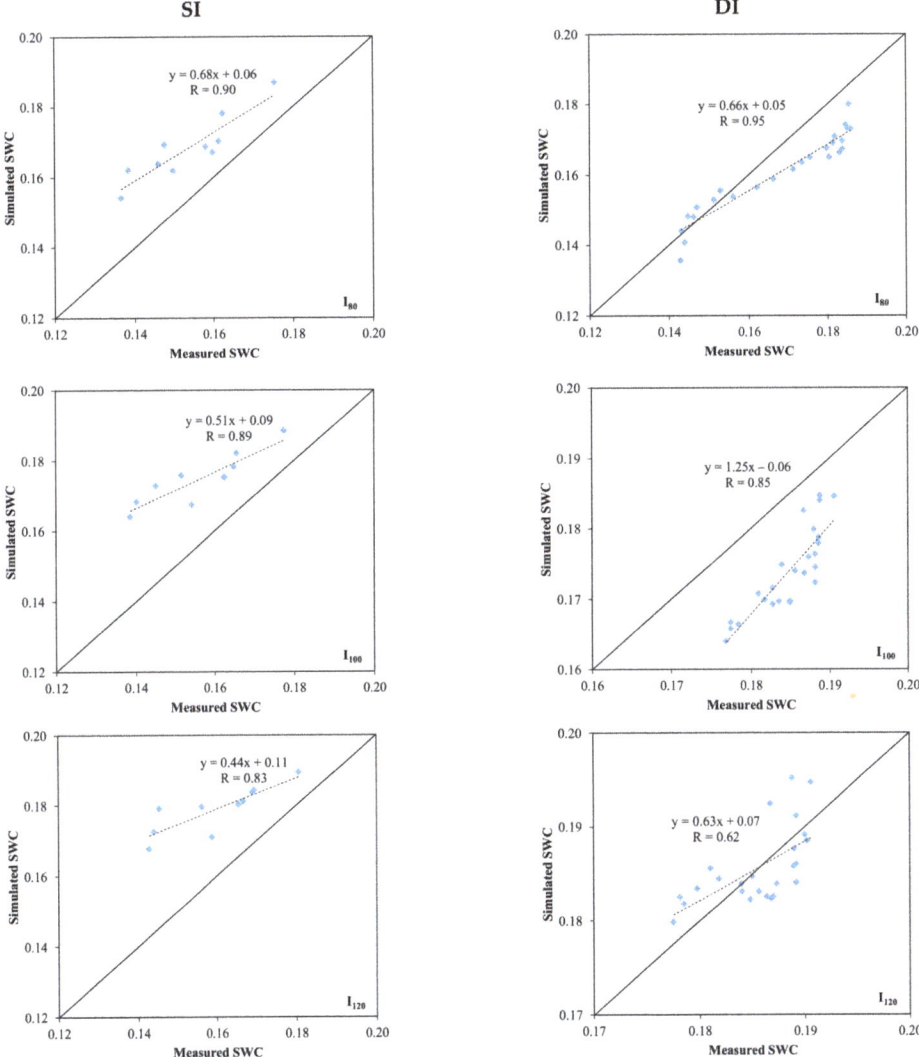

Figure 13. The measured and the SALTMED-simulated soil water content (SWC) for different irrigation levels across irrigation systems during the validation period (i.e., the 2019/2020 growing season). I_{80} = 80% crop evapotranspiration (ET_c), I_{100} = 100% ET_c, I_{120} = 120% ET_c, SI = surface irrigation, and DI = drip irrigation.

When using the SALTMED model, it is important to have a reliable description of the crop's response to applied treatments in addition to simulating soil water. As a result, we tested the SALTMED model's ability to capture temporal variations in biological and grain yields for various treatments during the growing seasons 2018/2019 (calibration period) and 2019/2020 (validation period). During the calibration and validation periods, the SALMED model performed better in the non-water stress (I_{100} and I_{120}) treatments than in the water stress (I_{80}) treatment, as shown in Table 4. The SALTMED model overestimated biological yield by 0.65–24.37% (except for the I_{100} treatment under SI, which underestimated biological yield by 0.11%) and grain yield by 0.13–19.18%, respectively, when compared with observed yields in 2018/2019 season. In the validation period, the SALTMED-simulated values also overestimated observed biological and grain yields by 3.8–29.81% and 2.02–25.41%, respectively. The I_{100} treatment under SI or DI had the lowest RE in the calibration and validation periods. The SALTMED model, according to Karandish and Simunek [86], performed well when simulating maize growth parameters, with a |RE| of 3.5–12%. Our RF range corresponds to the RF range reported by Kaya et al. [88] for quinoa yield (|RE| = 1.2–12.6%), the range reported by Hirich et al. [89] for corn yield (|RE| = 0–29.1%), and the range reported by Ragab et al. [34] for tomato and potato yields (|RE| = 0–21.5%). Over the calibration and validation years (Table 4), close matches between simulated and observed biological yield were observed, with RMSE, MAE, and MARE averaging 2.47, 2.01, and 12.68%, respectively. During the seasons, grain yield followed the same pattern as biological yield. The average RMSE, MAE, and MARE were 0.88, 0.741, and 11.58%, respectively, indicating that the model accurately predicted grain yield at various irrigation water levels. These findings are similar to those of Hirich et al. [87], who discovered that the relationship between observed and simulated yield produced RMSE values of 1.11 in sweet corn. In general, the performance criteria for statistical comparison of the observed and simulated data showed that the SALTMED model was well capable of simulating the yield of a wheat crop.

Table 4. Statistical indices comparing the observed and SALTMED-simulated wheat yields for various irrigation treatments during the calibration (the 2018/2019 growing season) and validation (the 2019/2020 growing season) periods.

Growing Season	Treatments		Biological Yield			Grain Yield		
	Irrigation Systems	Irrigation Levels	Observed	Simulated	RE (%)	Observed	Simulated	RE (%)
2018/2019	SI	I_{80}	15.75	18.20	15.54	6.54	7.55	15.47
		I_{100}	18.00	17.98	−0.11	7.56	7.57	0.13
		I_{120}	15.50	18.37	18.54	6.36	7.58	19.18
	DI	I_{80}	16.25	20.21	24.37	6.34	7.54	18.95
		I_{100}	18.25	18.37	0.65	8.21	8.27	0.68
		I_{120}	18.00	19.04	5.80	7.83	8.28	5.80
	RMSE				2.27			0.83
	MAE				1.74			0.66
	MARE, %				10.80			10.04
2019/2020	SI	I_{80}	14.45	18.22	26.09	5.46	6.85	25.41
		I_{100}	16.88	17.98	6.52	7.41	7.56	2.02
		I_{120}	16.93	18.37	8.53	6.64	7.57	14.01
	DI	I_{80}	15.11	19.61	29.81	5.88	7.14	21.51
		I_{100}	17.49	18.16	3.80	7.62	8.17	7.22
		I_{120}	16.84	18.95	12.55	7.60	8.24	8.48
	RMSE				2.67			0.92
	MAE				2.27			0.82
	MARE, %				14.55			13.11

4. Conclusions

It is important to identify appropriate water irrigation management strategies to reduce the waste of water resources and improve water productivity in irrigation practices. Management strategies are commonly tested through field experiments that are expensive and time-consuming to produce consistent and reliable results. An alternative option for these experiments is to use validated mathematical models. This study discussed the use of irrigation water levels across surface and drip irrigation systems and the application of mathematical models (e.g., CROPWAT and SALTMED) in wheat fields. The study showed that there is a great potential for water savings when using a drip irrigation system, which gives much higher water productivity than a surface irrigation system at the same irrigation water level. Evaluation of the irrigation schedules using the CROPWAT model showed that different irrigation levels need to integrate irrigation application methods. The graphical and statistical comparisons confirmed the ability of the SALTMED model to predict soil water content and simulate the effects of different irrigation water levels on the biological and grain yields of wheat within acceptable limits. The SALTMED model can be an effective tool for identifying the correct irrigation strategy to maximize crop production as well as benefiting from the application of the CROPWAT model to develop accurate irrigation schedules under different management conditions and water supply schemes.

Author Contributions: Conceptualization, investigation, data curation, methodology, formal analysis, software, funding acquisition, writing—original draft, writing—review and editing, A.A.E.-S. and M.A.M. All authors have read and agreed to the published version of the manuscript.

Funding: This research was financially supported by the Deanship of Scientific Research, king Saud University through Vice Deanship of Scientific Research Chairs; Research Chair of Prince Sultan Bin Abdulaziz International Prize for Water.

Institutional Review Board Statement: Not applicable.

Informed Consent Statement: Not applicable.

Data Availability Statement: The data presented in this study are available on request from the corresponding author.

Acknowledgments: The authors extend their appreciation to the Deanship of Scientific Research, king Saud University for funding through Vice Deanship of Scientific Research Chairs; Research Chair of Prince Sultan Bin Abdulaziz International Prize for Water.

Conflicts of Interest: The authors declare no conflict of interest.

References

1. Wu, W.; Ma, B.-L. Assessment of canola crop lodging under elevated temperatures for adaptation to climate change. *Agric. For. Meteorol.* **2018**, *248*, 329–338. [CrossRef]
2. Eck, M.; Murray, A.; Ward, A.R.; Konrad, C. Influence of growing season temperature and precipitation anomalies on crop yield in the southeastern United States. *Agric. For. Meteorol.* **2020**, *291*, 108053. [CrossRef]
3. The World Bank. Water in Agriculture. 2020. Available online: https://www.worldbank.org/en/topic/water-in-agriculture (accessed on 12 December 2021).
4. Monaghan, J.M.; Daccache, A.; Vickers, L.H.; Hess, T.M.; Weatherhead, E.K.; Grove, I.G.; Knox, J.W. More 'crop per drop': Constraints and opportunities for precision irrigation in European agriculture. *J. Sci. Food Agric.* **2013**, *93*, 977–980. [CrossRef] [PubMed]
5. Zhang, T.; Zou, Y.; Kisekka, I.; Biswas, A.; Cai, H. Comparison of different irrigation methods to synergistically improve maize's yield, water productivity and economic benefits in an arid irrigation area. *Agric. Water Manag.* **2021**, *243*, 106497. [CrossRef]
6. Xu, D.; Li, Y. Review on advancements of study on precision surface irrigation system. *J. Hydraul. Eng.* **2007**, *38*, 529–537.
7. Benjamin, J.G.; Havis, H.R.; Ahuja, L.R.; Alonso, C.V. Leaching and water flow patterns in every-furrow and alternate-furrow irrigation. *Soil Sci. Soc. Am. J.* **1994**, *58*, 1511–1517. [CrossRef]
8. Kang, S.; Liang, Z.; Pan, Y.; Shi, P.; Zhang, J. Alternate furrow irrigation for maize production in an arid area. *Agric. Water Manag.* **2000**, *45*, 267–274. [CrossRef]
9. Amayreh, J.; Al-Abed, N. Developing crop coefficients for field-grown tomato (*Lycopersicon esculentum* Mill.) under drip irrigation with black plastic mulch. *Agric. Water Manag.* **2005**, *73*, 247–254. [CrossRef]

10. Wang, F.-X.; Wu, X.-X.; Shock, C.C.; Chu, L.-Y.; Gu, X.-X.; Xue, X. Effects of drip irrigation regimes on potato tuber yield and quality under plastic mulch in arid Northwestern China. *Field Crops Res.* **2011**, *122*, 78–84. [CrossRef]
11. Chaves, M.M.; Zarrouk, O.; Francisco, R.; Costa, J.M.; Santos, T.; Regalado, A.P.; Rodrigues, M.L.; Lopes, C.M. Grapevine under deficit irrigation: Hints from physiological and molecular data. *Ann. Bot.* **2010**, *105*, 661–676. [CrossRef]
12. Sezen, S.M.; Yazar, A.; Daşgan, Y.; Yucel, S.; Akyıldız, A.; Tekin, S.; Akhoundnejad, Y. Evaluation of crop water stress index (CWSI) for red pepper with drip and furrow irrigation under varying irrigation regimes. *Agric. Water Manag.* **2014**, *143*, 59–70. [CrossRef]
13. Jovanovic, Z.; Stikic, R. Strategies for Improving Water Productivity and Quality of Agricultural Crops in an Era of Climate Change. In *Irrigation Systems and Practices in Challenging Environments*; Lee, T.S., Ed.; IntechOpen: Rijeka, Croatia, 2012; pp. 77–102.
14. Loveys, B.R.; Stoll, M.; Davies, W.J. Physiological approaches to enhance water use efficiency in agriculture: Exploiting plant signaling in novel irrigation practice. In *Water Use Efficiency in Plant Biology*; Bacon, M.A., Ed.; University of Lancaster: Lancaster, UK, 2004; pp. 113–141.
15. Yang, H.; Du, T.; Qiu, R.; Chen, J.; Wang, F.; Li, Y.; Wang, C.; Gao, L.; Kang, S. Improved water use efficiency and fruit quality of greenhouse crops under regulated deficit irrigation in northwest China. *Agric. Water Manag.* **2017**, *179*, 193–204. [CrossRef]
16. Chai, Q.; Gan, Y.; Zhao, C.; Xu, H.L.; Waskom, R.M.; Niu, Y.; Siddique, K.H.M. Regulated deficit irrigation for crop production under drought stress. a review. *Agron. Sustain. Dev.* **2016**, *36*, 1–21. [CrossRef]
17. Evett, S.R.; Tolk, J.A. Introduction: Can water use efficiency Be modeled well enough to impact crop management? *Agron. J.* **2009**, *101*, 423–425. [CrossRef]
18. Kato, Y.; Abe, J.; Kamoshita, A.; Yamagishi, J. Genotypic variation in root growth angle in rice (*Oryza sativa* L.) and its association with deep root development in upland fields with different water regimes. *Plant Soil.* **2006**, *287*, 117–129. [CrossRef]
19. Kirda, C. Deficit irrigation scheduling based on plant growth stages showing water stress tolerance. In *Deficit Irrigation Practice*; FAO: Rome, Italy, 2002; pp. 3–10.
20. Stikic, R.; Savic, S.; Jovanovic, Z.; Jacobsen, S.E.; Liu, F.; Jensen, C.R. Deficit irrigation strategies: Use of stress physiology knowledge to increase water use efficiency in tomato and potato. In *Horticulture in 21st Century Series: Botanical Research and Practices*; Sampson, A.N., Ed.; Nova Science Publishers: New York, NY, USA, 2010; pp. 161–178.
21. Pequeno, D.N.L.; Hernández-Ochoa, I.M.; Reynolds, M.; Sonder, K.; Moleromilan, A.; Robertson, R.D.; Lopes, M.S.; Xiong, W.; Kropff, M.; Asseng, S. Climate impact and adaptation to heat and drought stress of regional and global wheat production. *Environ. Res. Lett.* **2021**, *16*, 054070. [CrossRef]
22. Khuhro, W.A.; Kandhro, M.N.; Sadiq, N.; Jakhro, M.I.; Amanullah Naseer, N.S.; Latif, S.A. Impact of irrigation stress on agronomic traits of promising wheat varieties in Tandojam-Pakistan. *Pure Appl. Biol.* **2018**, *7*, 714–720. [CrossRef]
23. Meena, R.P.; Tripathi, S.; Sharma, R.; Chhokar, R.; Chander, S.; Jha, A. Role of precision irrigation scheduling and residue-retention practices on water-use efficiency and wheat (*Triticum aestivum*) yield in north-western plains of India. *Indian J. Agron.* **2018**, *54*, 34–47.
24. Qiu, G.Y.; Wang, L.; He, X.; Zhang, X.; Chen, S.; Chen, J.; Yang, Y. Water use efficiency and evapotranspiration of winter wheat and its response to irrigation regime in the north China plain. *Agric. For. Meteorol.* **2008**, *148*, 1848–1859. [CrossRef]
25. Lobell, D.B.; Ortiz-Monasterio, J.I. Evaluating strategies for improved water use in spring wheat with CERES. *Agric. Water Manag.* **2006**, *84*, 249–258. [CrossRef]
26. Peake, A.S.; Carberry, P.S.; Raine, S.R.; Gett, V.; Smith, R.J. An alternative approach to whole-farm deficit irrigation analysis: Evaluating the risk-efficiency of wheat irrigation strategies in sub-tropical Australia. *Agric. Water Manag.* **2016**, *169*, 61–76. [CrossRef]
27. Panda, R.K.; Behera, S.K.; Kashyap, P.S. Effective management of irrigation water for wheat under stressed conditions. *Agric. Water Manag.* **2003**, *63*, 37–56. [CrossRef]
28. Jalota, S.K.; Sood, A.; Chahal, G.B.S.; Choudhury, B.U. Crop water productivity of cotton (*Gossypium hirsutum* L.)—wheat (*Triticum aestivum* L.) system as influenced by deficit irrigation, soil texture and precipitation. *Agric. Water Manag.* **2006**, *84*, 137–146. [CrossRef]
29. Wang, Q.; Li, F.; Zhang, E.; Li, G.; Vance, M. The effects of irrigation and nitrogen application rates on yield of spring wheat (longfu-920), and water use efficiency and nitrate nitrogen accumulation in soil. *Aust. J. Crop Sci.* **2012**, *6*, 662–672.
30. Rao, S.S.; Regar, P.L.; Tanwar, S.P.S.; Singh, Y.V. Wheat yield response to line source sprinkler irrigation and soil management practices on medium-textured shallow soils of arid environment. *Irrig. Sci.* **2013**, *31*, 1185–1197. [CrossRef]
31. Gao, Z.; Liang, X.-G.; Lin, S.; Zhao, X.; Zhang, L.; Zhou, L.-L.; Shen, S.; Zhou, S.-L. Supplemental irrigation at tasseling optimizes water and nitrogen distribution for high-yield production in spring maize. *Field Crops Res.* **2017**, *209*, 120–128. [CrossRef]
32. Memon, S.A.; Sheikh, I.A.; Talpur, M.A.; Mangrio, M.A. Impact of deficit irrigation strategies on winter wheat in semi-arid climate of Sindh. *Agric. Water Manag.* **2021**, *243*, 106389. [CrossRef]
33. Sadras, V.O.; Angus, J.F. Benchmarking water-use efficiency of rainfed wheat in dry environments. *Aust. J. Agric. Res.* **2006**, *57*, 847–856. [CrossRef]
34. Ragab, R.; Malash, N.; Abdel Gawad, G.; Arslan, A.; Ghaibeh, A. A Holistic Generic Integrated Approach for Irrigation, Crop and Field Management: 2. The SALTMED Model Validation Using Field Data of Five Growing seasons from Egypt and Syria. *Agric. Water Manag.* **2005**, *78*, 89–107. [CrossRef]
35. Doorenbos, J.; Pruitt, W.O. *Guidelines for Predicting Crop Water Requirements*, 2nd ed.; FAO Irrigation and Drainage Paper 24; Food and Agriculture Organization of the United Nations: Rome, Italy, 1977; 156p.

36. Smith, M.; Allen, R.; Monteith, J.; Perrier, L.; Segeren, A. *Report on the Expert Consultation for the Revision of FAO Methodologies for Crop Water Requirements*; FAO/AGL: Rome, Italy, 1991.
37. Smith, M. CropWat. In *A Computer Program for Irrigation Planning and Management*; FAO Irrigation and Drainage Paper 46; Food and Agriculture Organization of the United Nations: Rome, Italy, 1992.
38. Smith, M. *CLIMWAT for CROPWAT, a Climatic Data Base for Irrigation Planning and Management*; FAO Irrigation and Drainage Paper 49; Food and Agriculture Organization of the United Nations: Rome, Italy, 1993; 113p.
39. Allen, R.G.; Pereira, L.S.; Raes, D.; Smith, M. *Crop Evapotranspiration: Guidelines for Computing Crop Water Requirements*; FAO Irrigation and Drainage Paper 56; FAO: Rome, Italy, 1998.
40. Doorenbos, J.; Kassam, A. *Yield Response to Water*; FAO Irrigation and Drainage, Paper 33; FAO, United Nation: Rome, Italy, 1979.
41. Kuo, S.-F.; Ho, S.-S.; Liu, C.-W. Estimation irrigation water requirements with derived crop coefficients for upland and paddy crops in ChiaNan irrigation association, Taiwan. *Agric. Water Manag.* **2006**, *82*, 433–451. [CrossRef]
42. Elamin, A.W.M.; Saeed, A.B.; Boush, A. Water Use Efficiencies of Gezira, Rahad and New Haifa Irrigated Schemes under Sudan Dryland Condition. *Sudan J. Des. Res.* **2011**, *3*, 62–72.
43. Rajaona, A.; Sutterer, N.; Asch, F. Potential of waste water use for Jatropha cultivation in arid environments. *Agriculture* **2012**, *2*, 376–392. [CrossRef]
44. Garg, K.K.; Wani, S.P.; Rao, A.K. Crop coefficients of Jatropha (*Jatropha curcas*) and Pongamia (*Pongamia pinna* ta) using water balance approach. *Wiley Interdiscip. Rev. Energy Environ.* **2014**, *3*, 301–309.
45. Song, L.; Oeumg, C.; Hornbuckle, J. Assessment of rice Water requirement by using CROPWAT model. In Proceedings of the 15th Science Council of Asia Board Meeting and International Symposium, Siem Reap, Cambodia, 15–17 May 2015.
46. Tsakmakis, I.D.; Zoidou, M.; Gikas, G.D.; Sylaios, G.K. Impact of irrigation technologies and strategies on cotton water footprint using AquaCrop and CROPWAT models. *Environ. Process.* **2018**, *5*, 181–199. [CrossRef]
47. Moseki, O.; Murray-Hudson, M.; Kashe, K. Crop water and irrigation requirements of *Jatropha curcas* L. in semi-arid conditions of Botswana: Applying the CROPWAT model. *Agric. Water Manag.* **2019**, *225*, 105754. [CrossRef]
48. Ewaid, S.H.; Abed, S.A.; Al-Ansari, N. Water footprint of wheat in Iraq. *Water* **2019**, *11*, 535. [CrossRef]
49. Ragab, R.A. Holistic Generic Integrated Approach for Irrigation, Crop and Field Management: The SALTMED Model. *Environ. Model. Softw.* **2002**, *17*, 345–361. [CrossRef]
50. Ragab, R.; Malash, N.; Abdel Gawad, G.; Arslan, A.; Ghaibeh, A. A Holistic Generic Integrated Approach for Irrigation, Crop and Field Management: 1. The SALTMED Model and Its Calibration Using Field Data from Egypt and Syria. *Agric. Water Manag.* **2005**, *78*, 67–88. [CrossRef]
51. Marwa, M.A.; El-Shafie, A.F.; Dewedar, O.M.; Molina-Martinez, J.M.; Ragab, R. Predicting the water requirement, soil moisture distribution, yield, water yield of peas and impact of climate change using SALTMED model. *Plant. Arch.* **2020**, *20*, 3673–3689.
52. Montenegro, S.G.; Montenegro, A.; Ragab, R. Improving Agricultural Water Management in The Semi-Arid Region of Brazil: Experimental and Modelling Study. *Irrig. Sci.* **2010**, *28*, 301–316. [CrossRef]
53. Mehanna, H.M.; Sabreen, R.H.P.; El-Hagarey, M.E. Validation of SALTMED model under different conditions of drought and fertilizer for snap bean in delta, Egypt. In Proceedings of the Minta International Conference for Agriculture and Irrigation in the Nile Basin Countries, El-Minia, Egypt, 26–29 March 2012.
54. Pulvento, C.; Riccardi, M.; Lavini, A.; D'andria, R.; Ragab, R. SALTMED Model to Simulate Yield and Dry Matter for Quinoa Crop and Soil Moisture Content under Different Irrigation Strategies in South Italy. *Irrig. Drain.* **2013**, *62*, 229–238. [CrossRef]
55. Rameshwaran, P.; Tepe, A.; Yazar, A.; Ragab, R. The Effect of Saline Irrigation Water on the Yield of Pepper: Experimental and Modeling Study. *Irrig. Drain.* **2015**, *64*, 41–49. [CrossRef]
56. Aly, A.A.; Al-Omran, A.M.; Khasha, A.A. Water management for cucumber: Greenhouse experiment in Saudi Arabia and modeling study using SALTMED model. *J. Soil Water Conserv.* **2015**, *70*, 1–11. [CrossRef]
57. Hassanli, M.; Ebrahimian, H.; Mohammadi, E.; Rahirni, A.; Shokouhi, A. Simulating maize yields when irrigating with saline water, using the AquaCrop, SALTMED, and SWAP models. *Agric. Water Manag.* **2016**, *176*, 91–99. [CrossRef]
58. Al-Omran, A.; Louki, I.; Alkhasha, A.; Abd El-Wahed, M.H.; Obadi, A. Water Saving and Yield of Potatoes under Partial Root-Zone Drying Drip Irrigation Technique: Field and Modelling Study Using SALTMED Model in Saudi Arabia. *Agronomy* **2020**, *10*, 1997. [CrossRef]
59. Page, A.; Miller, R.; Keeney, D. Chemical and microbiological properties. In *Methods of Soil Analysis, Part 2*, 2nd ed.; Agronomy Monogram 9; Agronomy Society of America and Soil Science Society of America: Madison, WI, USA, 1982.
60. Klute, A. *Methods of Soil Analysis*; Part 1 Book Series No. 9; American Society of Agronomy and Soil Science America: Madison, WI, USA, 1986.
61. Jarvan, M.; Edesi, L.; Adamson, A. Effect of sulphur fertilization on grain yield and yield components of winter wheat. *Acta Agric. Scand.* **2012**, *62*, 401–409. [CrossRef]
62. Xie, Y.; Zhang, H.; Zhu, Y.; Zhao, L.; Yang, J.; Cha, F.; Liu, C.; Wang, C.; Guo, T. Grain yield and water use of winter wheat as affected by water and sulfur supply in the North China Plain. *J. Integr. Agric.* **2017**, *16*, 614–625. [CrossRef]
63. Kijne, J.; Barker, R.; Molden, D. Improving water productivity in agriculture. In *Water Productivity in Agriculture: Limits and Opportunities for Improvement*; Kijne, J., Barker, R., Eds.; International Water Management Institute: Colombo, Sri Lanka, 2003; pp. 11–19.
64. Helweg, O. Functions of crop yield from applied water. *Agron. J.* **1991**, *83*, 769–773. [CrossRef]

65. Swennenhuis, J. *CROPWAT Version 8.0 Model*; FAO, Viale delle Terme di Caracalla: Rome, Italy, 2009; Available online: http://www.fao.org/nr/water/infores_databases_cropwat.html (accessed on 12 January 2022).

66. Van Genuchten, M.T. A closed-form equation for predicting the hydraulic conductivity of unsaturated soils. *Soil Sci. Soc. Am. J.* **1980**, *44*, 892–898. [CrossRef]

67. Legates, D.R.; McCabe, G.J., Jr. Evaluating the use of "goodness-of fit" measures in hydrologic and hydroclimatic model validation. *Water Resour. Res.* **1999**, *35*, 233–241. [CrossRef]

68. Willmott, C.J.; Matsuura, K. Advantages of the mean absolute error (MAE) over the root mean square error (RMSE) in assessing average model performance. *Clim. Res.* **2005**, *30*, 79–82. [CrossRef]

69. CoStat Version 6.303 Copyright 1998–2004 CoHort Software798 Lighthouse Ave. PMB 320, Monterey, CA, 93940, USA. Available online: https://cohortsoftware.com/costat.html (accessed on 2 October 2021).

70. Snedecor, G.; Cochran, W. *Statistical Method*, 7th ed.; Ames, I.A., Ed.; The Iowa State University Press: Ames, IA, USA, 1980.

71. Moussa, A.M.; Abdel-Maksoud, H.H. Effect of soil moisture regime on yield and its components and water use efficiency for some wheat cultivars. *Ann. Agric. Sci.* **2004**, *49*, 515–530.

72. Abdelkhalek, A.A.; Darwesh, R.K.; El-Mansoury, M.A.M. Response of some wheat varieties to irrigation and nitrogen fertilization using ammonia gas in North Nile Delta region. *Ann. Agric. Sci.* **2015**, *60*, 245–256. [CrossRef]

73. Pandey, D.S.; Kumar, D.; Misra, R.D.; Prakash, A.; Gupta, V.K. An integrated approach of irrigation and fertilizer management to reduce lodging in wheat (*Triticum aestivum*). *Indian J. Agron.* **1997**, *42*, 86–89.

74. Eissa, M.A. Improving Yield of Drip-Irrigated Wheat under Sandy Calcareous Soils. *World Appl. Sci. J.* **2014**, *30*, 818–826.

75. Noreldin, T.; Ouda, S.; Mounzer, O.; Abdelhamid, M.T. CropSyst model for wheat under deficit irrigation using sprinkler and drip irrigation in sandy soil. *J. Water Land Dev.* **2015**, *26*, 57–64. [CrossRef]

76. Mugabe, F.T.; Nyakatawa, E.Z. Effect of deficit irrigation on wheat and opportunities of growing wheat on residual soil moisture in southeast Zimbabwe. *Agric. Water Manag.* **2000**, *46*, 111–119. [CrossRef]

77. Ali, M.H.; Hoque, M.R.; Hassan, A.A.; Khair, A. Effects of deficit irrigation on yield, water productivity, and economic returns of wheat. *Agric. Water Manag.* **2017**, *92*, 151–161. [CrossRef]

78. Geerts, S.; Raes, D. Deficit irrigation as an on-farm strategy to maximize crop water productivity in dry areas. *Agric. Water Manag.* **2009**, *96*, 1275–1284. [CrossRef]

79. Pereira, L.S.; Oweis, T.; Zairi, A. Irrigation management under water scarcity. *Agric. Water Manag.* **2002**, *57*, 175–206. [CrossRef]

80. Maurya, R.K.; Singh, G.R. Effect of crop establishment methods and irrigation schedules on economics of wheat (*Triticum aestivum*) production moisture depletion pattern, consumptive use and crop water use efficiency. *Indian J. Agric. Sci.* **2008**, *78*, 830–833.

81. Knežević, M.; Perović, N.; Životić, L.; Ivanov, M.; Topalović, A. Simulation of winter wheat water balance with CROPWAT and ISAREG models. *Agric. For.* **2013**, *59*, 41–53.

82. Zhou, H.; Zhao, W. Modeling soil water balance and irrigation strategies in a flood-irrigated wheat-maize rotation system. A case in dry climate, China. *Agric. Water Manag.* **2019**, *221*, 286–302. [CrossRef]

83. English, M. Deficit irrigation. I: Analytical framework. *J. Irrig. Drain. Eng.* **1990**, *116*, 399–412. [CrossRef]

84. Fereres, E.; Garda-Vila, M. Irrigation Management for Efficient Crop Production. In *Crop Science*; Savin, R., Slafer, G.A., Eds.; Springer: New York, NY, USA, 2019; pp. 345–360.

85. Bennett, D.R.; Harms, T.E. Crop yield and water requirement relationships for major irrigated crops in southern Alberta. *Can. Water Resour. J.* **2011**, *36*, 159–170. [CrossRef]

86. Karandisha, F.; Simunek, J. A comparison of the HYDRUS (2D/3D) and SALTMED models to investigate the influence of various water-saving irrigation strategies on the maize water footprint. *Agric. Water Manag.* **2019**, *213*, 809–820. [CrossRef]

87. Hirich, A.; Ragab, R.; Choukr-Allah, R.; Rami, A. The effect of deficit irrigation with treated wastewater on sweet corn: Experimental and modelling study using SALTMED model. *Irrig Sci.* **2014**, *32*, 205–219. [CrossRef]

88. Kaya, C.I.; Yazar, A.; Sezen, S. SALTMED model performance on simulation of soil moisture and crop yield for quinoa irrigated using different irrigation systems, irrigation strategies and water qualities in Turkey. *Agric. Agric. Sci. Procedia* **2015**, *4*, 108–118.

89. Hirich, A.; Choukr-Allah, R.; Ragab, R.; Jacobsen, S.-E.; El Youssfi, L.; El-Omari, H. The SALTMED model calibration and validation using field data from Morocco. *J. Mater Environ. Sci.* **2012**, *3*, 342–359.

Article

Functional Design of Pocket Fertigation under Specific Microclimate and Irrigation Rates: A Preliminary Study

Chusnul Arif [1,*], Yusuf Wibisono [2], Bayu Dwi Apri Nugroho [3], Septian Fauzi Dwi Saputra [4], Abdul Malik [1], Budi Indra Setiawan [1], Masaru Mizoguchi [5] and Ardiansyah Ardiansyah [6]

[1] Department of Civil and Environmental Engineering, IPB University, Kampus IPB Darmaga, Bogor 16680, Indonesia; malik.abede3@gmail.com (A.M.); budindra@apps.ipb.ac.id (B.I.S.)
[2] Department of Bioprocess Engineering, Brawijaya University, Malang 65141, Indonesia; y_wibisono@ub.ac.id
[3] Department of Agricultural and Biosystem Engineering, Gadjah Mada University, Yogyakarta 55281, Indonesia; bayu.tep@ugm.ac.id
[4] Civil Engineering and Management, School of Vocational Sciences, IPB University, Bogor 16680, Indonesia; septianfauzi@apps.ipb.ac.id
[5] Graduate School of Agricultural and Life Sciences, The University of Tokyo, Tokyo 113-8657, Japan; amizo@mail.ecc.u-tokyo.ac.jp
[6] Department of Agricultural Engineering, Jenderal Soedirman University, Purwokerto 53125, Indonesia; ardi.plj@gmail.com
* Correspondence: chusnul_arif@apps.ipb.ac.id

Citation: Arif, C.; Wibisono, Y.; Nugroho, B.D.A.; Saputra, S.F.D.; Malik, A.; Setiawan, B.I.; Mizoguchi, M.; Ardiansyah, A. Functional Design of Pocket Fertigation under Specific Microclimate and Irrigation Rates: A Preliminary Study. *Agronomy* 2022, 12, 1362. https://doi.org/10.3390/agronomy12061362

Academic Editors: Pantazis Georgiou and Dimitris Karpouzos

Received: 7 April 2022
Accepted: 31 May 2022
Published: 5 June 2022

Abstract: Irrigation and fertilization technologies need to be adapted to climate change and provided as effectively and efficiently as possible. The current study proposed pocket fertigation, an innovative new idea in providing irrigation water and fertilization by using a porous material in the form of a ring/disc inserted surrounding the plant's roots as an irrigation emitter equipped with a "pocket"/bag for storing fertilizer. The objective was to evaluate the functional design of pocket fertigation in the specific micro-climate inside the screenhouse with a combination of emitter designs and irrigation rates. The technology was implemented on an experimental field at a lab-scale melon (*Cucumis melo* L.) cultivation from 23 August to 25 October 2021 in one planting season. The technology was tested at six treatments of a combination of three emitter designs and two irrigation rates. The emitter design consisted of an emitter with textile coating (PT), without coating (PW), and without emitter as a control (PC). Irrigation rates were supplied at one times the evaporation rate (E) and 1.2 times the evaporation rate (1.2E). The pocket fertigation was well implemented in a combination of emitter designs and irrigation rates (PT-E, PW-E, PT-1.2E, and PW-1.2E). The proposed technology increased the averages of fruit weight and water productivity by 6.20 and 7.88%, respectively, compared to the control (PC-E and PC-1.2E). Meanwhile, the optimum emitter design of pocket fertigation was without coating (PW). It increased by 13.36% of fruit weight and 14.71% of water productivity. Thus, pocket fertigation has good prospects in the future. For further planning, the proposed technology should be implemented at the field scale.

Keywords: pocket fertigation; water productivity; innovative technology; subsurface irrigation

1. Introduction

Irrigation and fertilization are the main components in determining agricultural production successfully. Climate change causes uncertainty in environmental conditions; thus, optimizing irrigation and fertilization should be adjusted. Suitable adaptation strategies for climate change on irrigation and fertilization could minimize the negative impacts [1]. Water resource availability tends to decrease and become more scarce with the impact of climate change [2]. However, irrigation is often oversupplied, thus resulting in more water loss and reducing water productivity [3]. In addition, excessive use of fertilizers leads to soil damage due to a large amount of soluble nitrate; thus, more nitrogen is wasted before being absorbed by plants [4]. Therefore, it is necessary to develop water-saving and efficient technology in fertilizers. An example of water-saving irrigation technology is subsurface irrigation by the innovative emitter [5]. The technology is very effective in water use because water is supplied directly to the plant roots, reducing evaporation. Several subsurface irrigation technologies have been developed, such as ring-shaped emitter irrigation [6,7] and sheet-pipe technology [8], as well as evapotranspiration irrigation [9]. Unfortunately, the technology still does not consider the use of fertilizers yet.

Both chemical and organic fertilizers should be applied at the right time and in the right amount to avoid the loss and negative impact on the environment. The excessive use of chemical fertilizers and residue in the soil changes the soil's physical and chemical properties, so the soil is easily eroded due to decreased organic content [10]. Furthermore, fertilizers dissolve in water due to rain, and irrigation can cause eutrophication of organic matter accumulation, thus reducing water quality [11]. In addition, long-term use of chemical fertilizers causes a decrease in soil pH [12]. On the other hand, organic fertilizer is more environmentally friendly. However, it is suspected to reduce production, convincing the farmer to consider using it less [13]. In addition, a large amount of organic fertilizer content in the rainwater can make a loss in the nitrate content before being absorbed properly by the crops [4].

This study examines pocket fertigation technology as an innovative idea for water and fertilizer applications. It is developed from a previous emitter irrigation called ring-shaped subsurface irrigation [6,7,14]. This technology uses a ring/disc porous material installed surrounding the roots as an emitter and equipped with a "pocket" for fertilizer storage on the upper side. It is simple, inexpensive, effective, efficient, easy, and fast to construct and manageable by the farmers. All materials used should be available in the local markets and reachable in cost. It is in line with the "farmer-led irrigation development" program [15]. In this sense, the farmers should be capable of planning, constructing, operating, maintaining, repairing, and even developing the irrigation system. This research aims to apply such a type of irrigation technology constructible using locally available materials and easily manageable by the farmers, whether individually or collectively.

By the current technology, water is irrigated through the pocket and then flows directly to the root zone via the emitter. It is expected that water and fertilizer are absorbed by the roots simultaneously. Therefore, it is important to test the performance of the developed technology, particularly for a high economic horticultural product such as melon (*Cucumis melo* L.). Melon is a fruit that has high commercial value in Indonesia with a wide and diverse market range, from traditional markets to modern markets, restaurants, and hotels. Therefore, it can be cultivated because of its competitiveness compared to other commodities. In addition, the fruit by-product can be incubated as a functional food ingredient [16].

The current study was proposed as a preliminary study on the functional design of the pocket fertigation technology. The objective of the study was to evaluate the functional design of the pocket fertigation for melon (*Cucumis melo* L.) production particularly in the emitter design and irrigation aspect. The scope of evaluation aspects consisted of the soil moisture fluctuation, fruit weight, and water productivity under different emitters design and irrigation rates. As an indicator, soil moisture is related to water and nutrient uptake, while crop yield is related to the income obtained by the farmers [17]. In addition, water productivity is related to water use efficiency because it reflects the yield or biomass produced per water used [18].

2. Materials and Methods

2.1. Time, Location, and Soil Properties

The current preliminary study was conducted at lab-scale inside a screenhouse located at Kinjiro Farm with coordinates 6.59° S, 106.77° E, Bogor, West Java, Indonesia. *Glamor*, a variety of melon seeds, was sown on 6 August 2021, planted on 23 August 2021, and harvested on 25 October 2021. The physical characteristics of soils are presented in Table 1.

Table 1. The physical characteristics of planting media soil.

No	Parameter	Value	Unit
1	Dry bulk density	0.77	g/cm^3
2	Particle density	1.92	g/cm^3
3	C-organic	5.73	%
4	Organic content	9.89	%
5	Permeability	5.18	cm/hour
6	Soil texture		
	Sand	17	%
	Silt	59	%
	Clay	24	%
	Soil Texture	Silt Loam	
7	Soil water content at the following soil suction:		
	pF 1	0.476	cm^3/cm^3
	pF 2	0.369	cm^3/cm^3
	pF 2.54	0.294	cm^3/cm^3
	pF 4.2	0.182	cm^3/cm^3

Based on the physical characteristics of the soil, especially the data on soil water content at various pF (soil-water matrix potential) values, a water retention curve was made to determine the saturated and residual soil water contents by the following equation [19]:

$$\theta = \theta_r + \frac{(\theta_s - \theta_r)}{[1 + (\alpha h)^n]^m} \tag{1}$$

where θ is the soil moisture (m^3/m^3) in volumetric water content, θ_s is the saturated soil water content (m^3/m^3), θ_r is the residual soil water content (m^3/m^3), h is the pressure head (cm H_2O), and α, n, and m are constants. The values of θ_s, θ_r, α, n, and m were optimized with a solver in Microsoft Excel (Figure 1). From the optimization results, the values of θ_s and θ_r were 0.485 m^3/m^3 and 0.100 m^3/m^3, respectively.

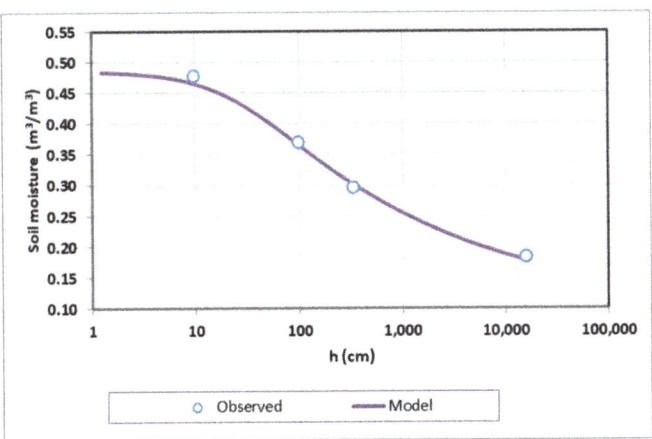

Figure 1. Water retention curve for the type of soil at the study site.

2.2. Experimental Design of the Pocket Fertigation

The experimental design consisted of a combination of emitter types of the pocket fertigation and irrigation rates with six treatments and two replications in total. The pocket fertigation was applied in a pot experiment with a 50 cm diameter in the top and 30 cm diameter in the bottom (Figure 2a). Meanwhile, the design of pocket fertigation is presented in Figure 3. Here, two designs were developed with the same dimensions. As previously mentioned, pocket fertigation has two parts: an emitter and a pocket to store the fertilizer. The emitter material was made from a perforated hose, 14 holes in total, with the interval of the hole being 5 cm. The first design of the emitter was coated with a textile material (PT) and without coating material (PW). The emitter was oval with a longer diameter of 30 cm and a shorter one of 25 cm. The pocket's diameter was 9 cm with a 25 cm height that was created from used plastic bottles with a size of 1500 mL. In this experiment, the emitter was placed 5 cm below the soil surface. For the control, surface irrigation was applied in which the fertilizer was sprinkled on the soil surface (PC).

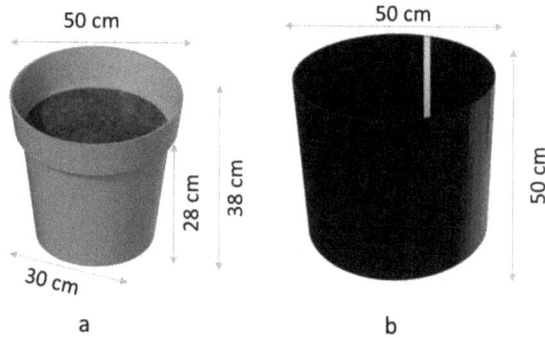

Figure 2. (**a**) The dimensions of pot; (**b**) the dimensions of pan evaporation.

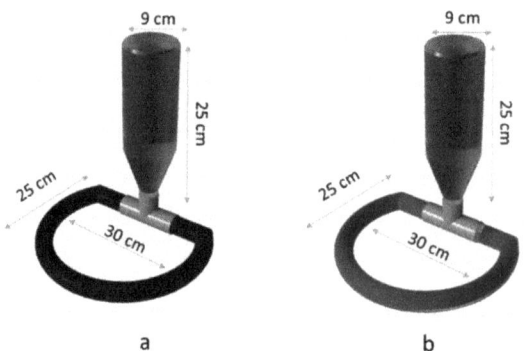

Figure 3. The pocket fertigation design: (**a**) emitter with textile coating (PT), (**b**) emitter without coating (PW).

For the irrigation rate, it is commonly supplied based on crop evapotranspiration (ETc); however, it is difficult to apply by the farmer due to the complicated method. In this research, we used a simple method by pan evaporation to determine the open water evaporation rate on a daily basis. The irrigation water was supplied based on the evaporation rate, i.e., one times the evaporation (E) and 1.2 times the evaporation (1.2E) in all designs of emitters, so there were six treatments in total, i.e., PT-E, PW-E, PC-E, PT-1.2E, PW-1.2E, and PC-1.2E (Figure 4). For the pan evaporation, we used a pan filled with water, 50 cm in diameter and height (Figure 2b). The daily evaporated water was recorded every morning (around 7.00 a.m.). For the leaching process, all treatments were supplied with more water ranging from 2 to 4 L/plant six times at 26, 33, 38, 41, 46, and 51 days after transplanting (DAT). In addition, this watering was also performed to avoid extreme drought in the growing media.

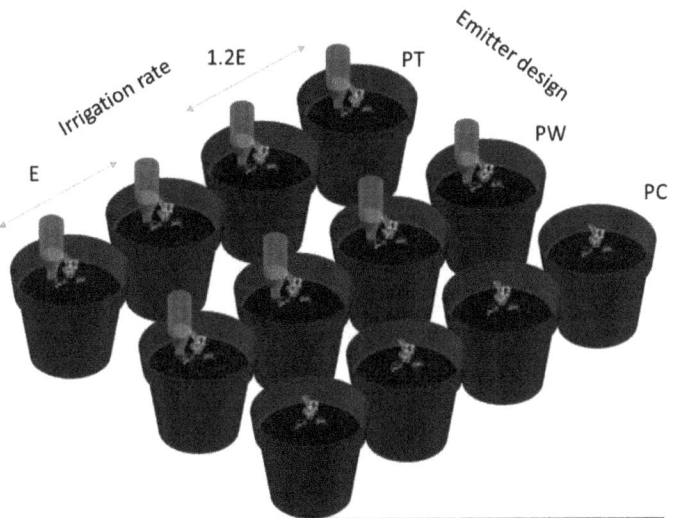

Figure 4. Testing of the pocket fertigation with various emitter designs and irrigation rates.

As we focused on the application of pocket fertigation under different irrigation rates, during the experiment, all treatments were given the same amount and materials content of fertilizer. They were "ABmix" and NPK "Mutiara" fertilizers. The "ABmix" fertilizer contains macro and micro-nutrients. During the planting season, the "ABmix" fertilizer was dissolved with an EC (Electrical Conductivity) value of 4500–5000 µS/cm and the NPK "Mutiara" fertilizer of 20 g/plant at 20 DAT was stored in the pocket.

2.3. Micro-Climate and Soil Moisture Monitoring

The micro-climate inside the screenhouse was measured by an automatic weather station (AWS) connected to the server. It was part of an IoT-based measurement previously developed [20]. There were several weather sensors, i.e., air temperature, relative humidity, wind speed, and solar radiation. Each parameter was measured at 15 min intervals. The micro-climate conditions in the screenhouse fluctuated throughout the cultivation period. However, the daily average, minimum, and maximum air temperatures had a constant trend (Figure 5a). The daily minimum, average, and maximum air temperature values ranged between 22 °C, 28 °C, and 35 °C, respectively. The same thing also occurred with the relative humidity (RH). Although it fluctuated more, the trend was also relatively constant with the average value of RH being approximately was 82% (Figure 5a). Something quite extreme happened on 14 September 2021 (22 DAT). The daily maximum and average air temperatures decreased significantly. On the other hand, RH increased significantly. Here, the daily maximum temperature only reached 27.7 °C with an average of 25.1 °C. Meanwhile, the RH increased and reached a maximum value of 90.1%. In atmospheric pressure, air temperature and RH are inversely proportional, as presented in Figure 5b. The type of greenhouse strongly influences variations in air temperature and RH in the greenhouse used [21]. The air temperature inside the greenhouse should be controlled properly because an increase in air temperature before harvest can reduce fruit sweetness [22]. Many air temperature control systems, including RH control systems, have been developed for optimal plant growth, such as fuzzy control systems [23,24].

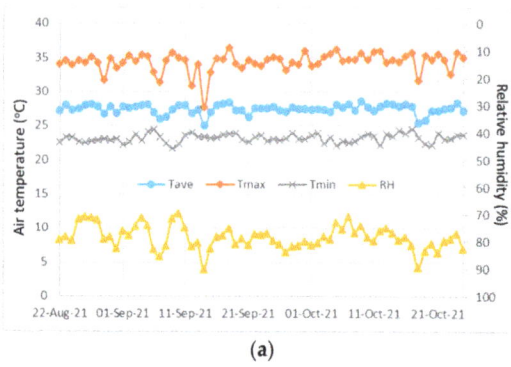
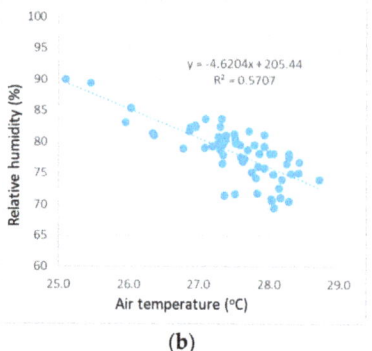

(a)   **(b)**

Figure 5. **(a)** Daily maximum, average, and minimum air temperatures, and relative humidity; **(b)** linear correlation between daily average air temperature and relative humidity.

The weather data (air temperature, relative humidity, wind speed, and solar radiation) were then used to determine the reference evapotranspiration based on the following Penman–Monteith equation [25]:

$$ET_o = \frac{0.408\Delta(R_n - G) + \gamma\frac{900}{T_{ave}+273}u(e_s - e_a)}{\Delta + \gamma(1 + 0.34u)} \tag{2}$$

where ETo is the reference evapotranspiration (mm), Rn is the net radiation ($MJ/m^2/d$), G is the soil heat flux density ($MJ/m^2/d$), T_{ave} is the daily average air temperature (°C), u is the wind speed (m/d), e_s is the saturated vapor pressure (kPa), e_a is the actual vapor pressure (kPa), γ is the psychrometric constant (kPa/°C), and Δ is the slope of the vapor pressure curve (kPa/°C). Rn, G, e_s, e_a, and γ were determined based on observed solar radiation and relative humidity parameters. In addition, to perform the equation, elevation, latitude, and Julian day data were required. The data were compared to evaporation rate that was measured daily as previously explained.

For effectiveness of emitter design, the soil moisture was monitored at a depth of 5 cm below the soil surface and in the middle of the emitter. The 5-TE soil moisture sensor from the Meter Group was used for this purpose. The sensor was placed at a 5 cm soil depth because the emitter of pocket fertigation was kept at this location. The sensor was connected to a ZL datalogger (Meter Group) with a measurement interval of 15 min. From the fluctuations in soil moisture, the actual evapotranspiration between the treatment was estimated and compared.

2.4. Crop Performances and Water Productivity Analysis

The indicators of crop performance were plant growth, fruit weight, and soluble solid content. The soluble solid content represented the sweetness level of fruit. For plant growth parameters, the number of leaves and plant height were measured at the ages of 10, 20, and 30 DAT during the vegetative phase. Meanwhile, in the generative phase (fruit formation), fruit weight and total soluble solid content representing sweetness levels were observed on the harvesting day. The total soluble solid was measured by the Atago Pocket Digital Refractometer in % Brix.

Water productivity was determined based on the product produced per amount of water used based on the definition [26]. As the experiment was conducted inside a screen house and there was no rain, the equation for water productivity is represented as follows:

$$WP_I = \frac{Y}{I}C \tag{3}$$

where Y is the fruit weight (g), I is the total irrigation (mL), C is the conversion factor (in this case, 1000), and WP_I is the water productivity based on total irrigation water (kg weight/m^3 water).

2.5. The Limitation of the Study

The current study only presented the functional design of pocket fertigation. The evaluation scopes were on soil moisture fluctuation, evapotranspiration, and crop and water productivities. As the numbers of pots and screenhouse areas were limited, statistical analysis was limited on the average value and standard deviation. Thus, the values will be compared among the treatments. The proposed technology will be implemented at field scale and it is planned for the next phase of the study.

3. Results

3.1. Evaporation and Evapotranspiration during the Season

Figure 6 shows fluctuations in solar radiation, evaporation, and reference evapotranspiration (ETo) during the growing season. Inside the screenhouse, the solar radiation was relatively low, ranging from 2.1 to 9.8 MJ/m^2/d. The low solar radiation affected the low reference evapotranspiration and pan evaporation (Figure 6). The reference evapotranspiration value ranged from 0.4 to 2.2 mm, while the pan evaporation was from 1 to 4 mm. The pan evaporation value was higher than the reference evapotranspiration because more water evaporated from the water surface than in the soil media when the soil was unsaturated, as found in all treatments. This condition is in line with previous experiments that stated that evaporation increases with the presence of flooded water (unsaturated condition) in the soil and vice versa [27].

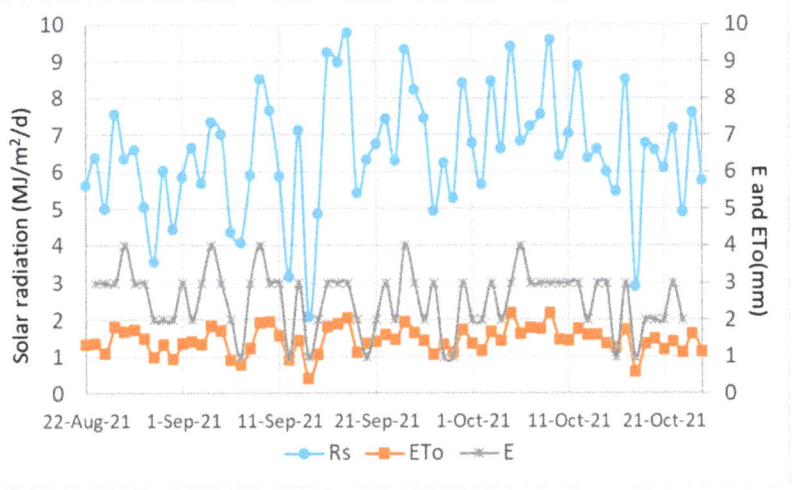

Figure 6. Daily total solar radiation, evaporation (E), and reference evapotranspiration (ETo).

ETo was strongly correlated with solar radiation, represented by high R^2 (>0.85), as shown in Figure 7a. Therefore, solar radiation is the strongest parameter affected on the ETo [20]. The minimum ETo was 0.4 mm when the solar radiation was also a minimum (2.1 MJ/m^2/d). A similar condition also existed for its maximum value, which reached 2.2 mm when the solar radiation was at its maximum level (9.8 MJ/m^2/d). It was indicated that solar radiation had the greatest influence on the evapotranspiration process particularly through the soil surface and plants [28]. The solar radiation also had a positive correlation to evaporation, although it had a lower R^2 compared to the ETo correlation (Figure 7b). Evaporation also correlated (R^2 > 0.48) to ETo, as shown in Figure 7c. It was indicated that evaporation from the water surface and evapotranspiration (evaporation and transpiration) occurred simultaneously. Commonly, evaporation from the water surface (E_{pan}) was higher than that of evaporation from the soil surface, which was measured by a lysimeter (E_{lys}) [29]. Evaporation can be converted to evapotranspiration via the pan coefficient (Kp) [30]. In this study, based on empirical data, Kp was 0.56, indicating that evaporation was approximately 56% higher than the ETo.

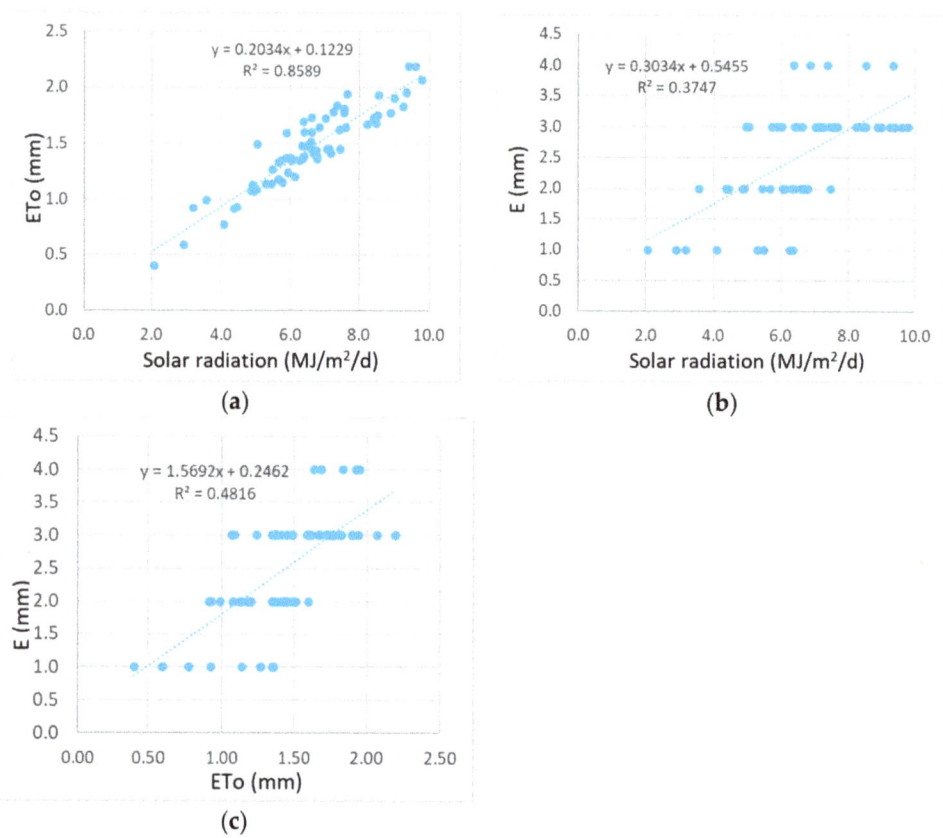

Figure 7. Relationship between (**a**) reference evapotranspiration (ETo) and solar radiation, (**b**) evaporation (E) and solar radiation, and (**c**) evaporation (E) and reference evapotranspiration (ETo).

3.2. Soil Moisture Conditions in Various Irrigation Rates

For in-plant cultivation systems inside the screenhouse or greenhouse, soil moisture is the key to success in horticultural crop production. Thus, it is important to control soil moisture accurately [31]. The soil moisture in PT-E and PT-1.2E fluctuated depending on the irrigation supplied because the plant water requirement for the plants was only supplied from irrigation (Figure 8). The PT-1.2E with a higher irrigation rate had higher soil moisture levels than those in the PT-E. At the PT-1.2E, soil moisture ranged from 0.198 to 0.496 m^3/m^3, while at PT-E, it ranged from 0.116 to 0.437 m^3/m^3. The highest soil moisture level occurred at 41 DAT (3 October 2021) when 4000 mL of irrigation was supplied to PT-E and PT-1.2E treatments. At this time, the soil moisture value was reached at its saturation level in the PT-1.2E. However, the maximum soil moisture in the PT-E treatment was still lower than that of the soil saturation level. At both irrigation rates (E and 1.2E), the soil moisture tended to be at the field capacity level at the beginning of the vegetative phase. Then, water irrigation in large quantities was supplied when the soil moisture level was too low, particularly in the mid-season phase. In the generative phase, the soil moisture condition was maintained in the range of field capacity in both irrigation rates.

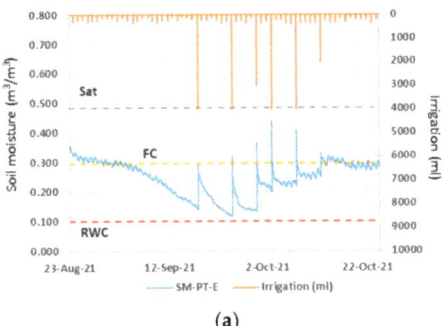
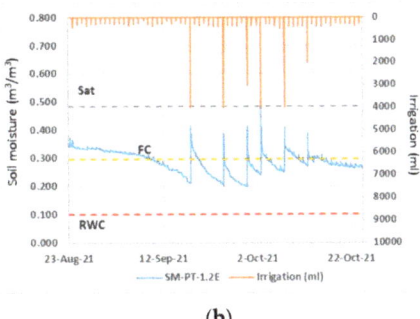

(a)   (b)

Figure 8. The fluctuation in soil moisture and irrigation: (**a**) PT-E treatment, (**b**) PT-1.2E treatment. Note: Sat: saturated water content, FC: field capacity water content, RWC: residual water content.

A similar thing occurred with the PW treatments (Figure 9). The soil moisture level increased rapidly when a large amount of irrigation was supplied. In the PW-E, soil moisture was slightly higher than the field capacity level in the beginning phase until 6 DAT (29 August 2021). Then, the soil moisture decreased below field capacity level until harvest. Here, the soil moisture conditions ranged from 0.147 to 0.339 m^3/m^3. Meanwhile, as more water was supplied, soil moisture in the PW-1.2E was consequently higher than that in the PW-E. At the beginning phase, the soil moisture was at field capacity level until 23 DAT (15 September 2021), and it reached the saturation level when a large amount of water was supplied, particularly at 26 DAT. Hereafter, soil moisture was below the field capacity level. In this treatment, soil moisture ranged from 0.150 to 0.493 m^3/m^3. Overall, the average soil moisture in the PW-E and PW-1.2E was 0.222 and 0.269 m^3/m^3, respectively.

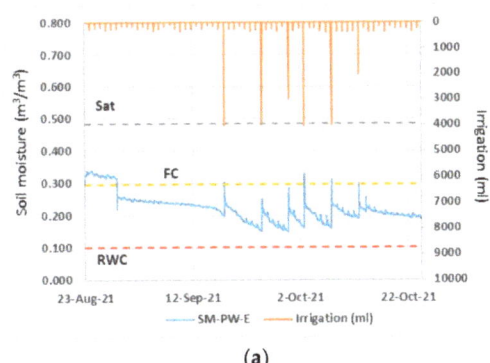
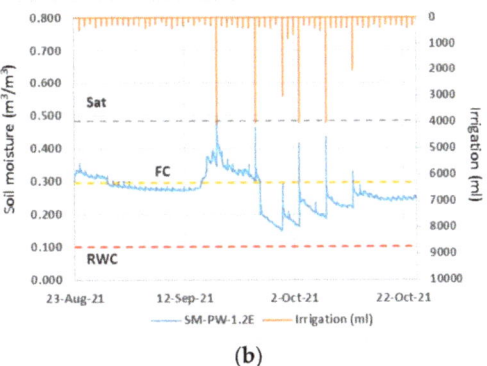

(a)   (b)

Figure 9. The fluctuation in soil moisture and irrigation: (**a**) PW-E treatment, (**b**) PW-1.2E treatment. Note: Sat: saturated water content, FC: field capacity water content, RWC: residual water content.

The fluctuations in soil moisture of the PC treatments are presented in Figure 10. The soil moisture level in the PC-E ranged from 0.112 to 0.426 m^3/m^3, while at the PC-1.2E, it ranged from 0.135 to 0.454 m^3/m^3. For the PC-E, the soil moisture level was below the field capacity level for most of the growing period, except on the specific days (at 26, 33, and 41 DAT) when large amounts of water were applied. Meanwhile, in the PC-1.2E, soil moisture ranged from the field capacity level in the beginning phase to 26 DAT, and then dropped to below field capacity.

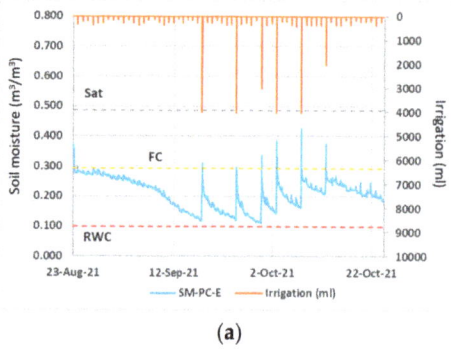

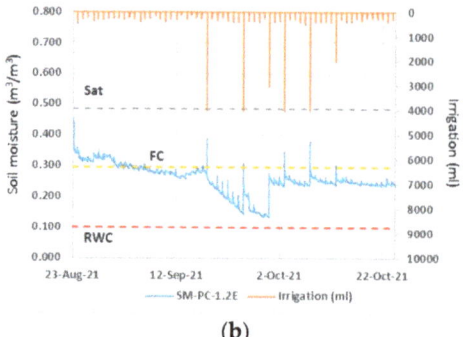

(a)   (b)

Figure 10. The fluctuation in soil moisture and irrigation: (**a**) PC-E treatment, (**b**) PC-1.2E treatment. Note: Sat: saturated water content, FC: field capacity water content, RWC: residual water content.

Table 2 shows the average value of soil moisture levels for each treatment every 10 DAT. Among the two emitter designs (PT and PW) of pocket fertigation, soil moisture tended to be stable with an average level close to the field capacity. PW was more able to maintain soil moisture above the value of 0.200 m^3/m^3 compared to PT. This means the emitter without the coating distributed irrigation water more uniformly and it also reduced actual evapotranspiration by up 13.6%. This indicated that the PW was probably more efficient in water use compared to PT.

Table 2. The maximum, average, minimum soil moisture, and actual evapotranspiration among the treatments.

Parameters	Treatments						Summary	
	PT-E	PT-1.2E	PW-E	PW-1.2E	PC-E	PC-1.2E	Pocket Fertigation *	Control **
Soil moisture (m^3/m^3) at:								
0–10 (DAT)	0.312	0.334	0.295	0.306	0.276	0.326	0.312	0.301
11–20 (DAT)	0.257	0.304	0.240	0.277	0.223	0.283	0.269	0.253
21–30 (DAT)	0.181	0.263	0.214	0.330	0.158	0.250	0.247	0.204
31–40 (DAT)	0.170	0.238	0.178	0.228	0.153	0.187	0.203	0.170
41–50 (DAT)	0.245	0.293	0.202	0.225	0.216	0.252	0.241	0.234
51–62 (DAT)	0.293	0.282	0.207	0.252	0.225	0.245	0.258	0.235
Maximum	0.312	0.334	0.295	0.330	0.276	0.326	0.334	0.326
Minimum	0.170	0.238	0.178	0.225	0.153	0.187	0.170	0.153
Average	0.243	0.285	0.223	0.270	0.209	0.257	0.255	0.233
ETa (mm)	118.8	123.9	98.7	114.9	143.9	107.9	114.1	125.9

* average value of PT-E, PT-1.2E, PW-E, PW-1.2E. ** average value of PC-E and PC 1.2E.

Table 2 also shows that the pocket fertigation was better than the control treatment in retaining soil moisture at a depth of 5 cm. The indicator had a higher soil moisture at the pocket fertigation than that of the control treatment. In addition, pocket fertigation was able to reduce the actual evapotranspiration by 10.32% of the control. The pocket fertigation functioned well, indicated by the higher efficiency of water used. It was seemingly subsurface irrigation that was more effective in distributing water along the root zone than that of surface irrigation. Previous research utilizing a similar emitter type showed that subsurface irrigation can maintain soil moisture in the root zone without causing stress to the plants [7].

3.3. Plant Growth and Their Productivities

The vegetative growth in each treatment is depicted in Figure 11. The highest average number of leaves at 20 DAT was produced by the PT, followed by the PC and PW treatments. However, the PW grew the highest plant height at 20 DAT, followed by the PT and PC treatments. After 30 DAT, pruning of the plants was carried out by maintaining the height of each plant at 200 cm. Overall, the vegetative growth among the treatments was comparable, particularly after 30 DAT.

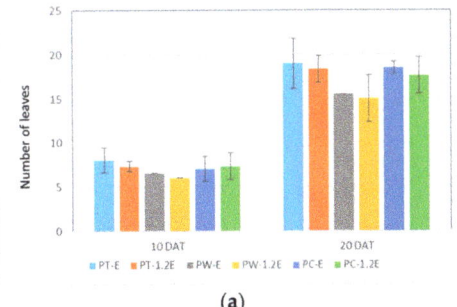
(a)

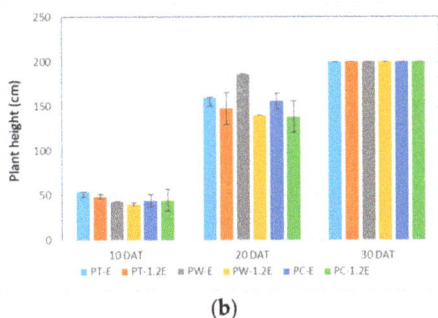
(b)

Figure 11. Plant growth performances among the treatments: (**a**) number of leaves, (**b**) plant height.

According to Table 3, the PW produced a 13.36% bigger average fruit weight than that of the PT. However, the PW produced a 13.07% lower total soluble solid than that of the PT. From the perspective of water used, the PW was more efficient as represented by the higher water productivity by up 14.71%. Therefore, it is recommended to use the pocket fertigation without coating materials. The lower effectivity of the PT is probably due to the clogging problems that occurred by the sedimentation of fertilizers. Thus, this clogging inhibited the distribution of water and fertilizer in the root zone. Clogging is generally a problem that must be overcome when utilizing irrigation systems with low flow rates [32], such as subsurface irrigation.

Table 3. Crop and water productivities among the treatments.

Treatments	Yield (Fruit Weight) (g)	Irrigation (mL)	Total Soluble Solid (%brix)	WP (kg/m^3)
PT-E	733 ± 50.9	34,825	10.5 ± 0.0	21.0
PT-1.2E	925.5 ± 116.7	38,925	9.4 ± 2.4	23.8
PW-E	898 ± 0	34,825	9.3 ± 0	25.8
PW-1.2E	982 ± 5.7	38,975	8.3 ± 0.5	25.2
PC-E	551 ± 0	34,875	10.8 ± 0	15.8
PC-1.2E	1115 ± 0	38,925	7.7 ± 0	28.6
Pocket Fertigation	885 ± 92.6	36,888	9.4 ± 0.8	24.0
Control	833 ± 282.0	36,900	9.3 ± 1.5	22.2
Irrigation rate at E	727.3 ± 141.7	34,842	10.2 ± 0.6	20.9
Irrigation rate at 1.2E	1007.5 ± 79.4	38,942	8.5 ± 0.7	25.9

Note: The presented data are the mean ± SD.

Table 3 shows that better performances were found in the pocket fertigation for fruit weight, total soluble solid, and water productivity compared to the control. It increased the average fruit weight by 6.20% and water productivity by 7.88%. Meanwhile, a higher water irrigation rate at 1.2E produced a bigger fruit weight than that at the E irrigation rate. Fruit weight increased significantly by 38.53% (Table 3). The increasing fruit weight of 1.2E may be contributed by increasing the actual evapotranspiration due to more irrigation water, particularly in the pocket fertigation (Table 2). This reason was supported by a previous study [33]. However, the increase in the fruit weight decreased the sweetness level (total soluble solid), as shown in Figure 12. The heavier melon, the higher water content, and the low dissolved solids may reduce the sweetness level. The results are similar to the previous observation [34,35].

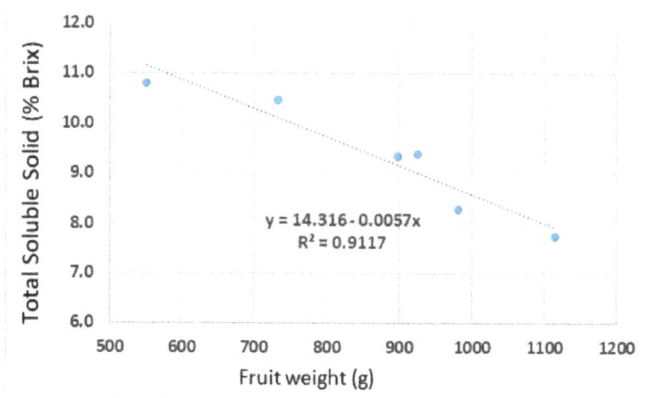

Figure 12. Relationship between total soluble solid and weight of fruit.

4. Discussion

In the context of climate change, water resources for the agriculture sector may become scarce in the future. Therefore, it is important to develop innovative and applicable technologies in utilizing irrigation water more effectively and efficiently, such as the pocket fertigation. Pocket fertigation is easy to produce by the farmers in Indonesia. The basic materials are a hose as the emitter and used bottles to store the fertilizer. In this preliminary study with a limited area, the pocket fertigation was shown to retain soil moisture better than surface irrigation as a control. Maintaining soil moisture implies that more water is stored in the soil, and it can be utilized by plants more optimally. Consequently, the fruit weight was heavier and had higher water productivity (Table 3).

The irrigation water delivery method of the pocket fertigation is similar to drip irrigation in which the emitter is placed below the soil surface near the root zone. Subsurface irrigation, both the pocket fertigation and drip irrigation, proved to be more effective and efficient in the utilization of irrigation water by reducing water loss due to evapotranspiration, as shown in Table 3 and reported in previous studies. As reported by Wang et al. [36], a long-time field experiment of drip irrigation in 2014–2018 showed that irrigation reduced 0.1–23% of evaporation and 7% of evapotranspiration per year. Consequently, the water use efficiency of drip irrigation can be significantly improved under various crop evapotranspiration scenarios [37]. In addition, subsurface irrigation with drip irrigation, combined with fertigation, increased production up to 41% as reported by Rolbiecki et al. [38]. Subsurface irrigation is not only known as effective and efficient in water used, but also more environmentally friendly. The indicates a reduction in greenhouse emissions from the soil under subsurface irrigation, especially N_2O and CO_2 [39,40].

The current developed technology has good prospects in the near future and should be continuously developed. Pocket fertigation is a kind of subsurface irrigation. It has a better performance indicated by the higher effectiveness of water use, and consequently, it can increase water productivity [31]. The performance tests on a field scale are needed not only for melon *(Cucumis melo* L.) but also for other crops. Crop type selection depends on the local climate condition and farmer's preference. Several locations in Indonesia are characterized by dry areas with low rainfall intensity such as East Nusa Tenggara (NTT), a province located in eastern Indonesia [41]. The location lacks water resources, so it is very appropriate to be chosen as the location for field-scale trials.

5. Conclusions

An innovative technology, pocket fertigation, was well implemented in the lab-scale experiment. The pocket fertigation with subsurface irrigation was better than surface irrigation in retaining soil moisture at a 5 cm soil depth. The soil moisture could be maintained at nearly field capacity level. The pocket fertigation was able to reduce the actual evapotranspiration by 10.32%. It also showed better performances in fruit weight production and water productivity. It increased the average fruit weight by 6.20% and water productivity by 7.88%, respectively. Thus, pocket fertigation has good prospects in the future. For further planning, the proposed technology will be implemented at the field scale, particularly in dry areas with minimum water resources.

Author Contributions: Conceptualization, C.A. and B.I.S.; methodology, C.A. and B.I.S.; data collection, C.A., A.M. and S.F.D.S.; writing—original draft preparation, C.A.; writing—review and editing, C.A., B.I.S., Y.W., B.D.A.N., M.M. and A.A. All authors have read and agreed to the published version of the manuscript.

Funding: This research was funded by the Indonesian Collaborative Research Program—WCU (World Class University) scheme by IPB University for the 2021 fiscal year with the number 1376/IT3.L1/PN/2021 dated 23 February 2021 by the project title "Developing Innovative Pocket Fertigation Technology based on Artificial Intelligence and Adaptive to Climate".

Institutional Review Board Statement: Not applicable.

Informed Consent Statement: Not applicable.

Data Availability Statement: The data presented in this study are available upon request from the corresponding author. The data are not publicly available, due to shared ownership between all parties that contributed to the research.

Acknowledgments: We would like to thank and appreciate to reviewers for all valuable comments, critics and suggestions, which helped us to improve the quality of the article. Also, we thank to Ahmad Kohar and Ibrahim for helping us in the field.

Conflicts of Interest: The authors declare no conflict of interest.

References

1. Ventrella, D.; Charfeddine, M.; Moriondo, M.; Rinaldi, M.; Bindi, M. Agronomic Adaptation Strategies under Climate Change for Winter Durum Wheat and Tomato in Southern Italy: Irrigation and Nitrogen Fertilization. *Reg. Environ. Chang.* **2012**, *12*, 407–419. [CrossRef]
2. Karimi, V.; Karami, E.; Keshavarz, M. Climate Change and Agriculture: Impacts and Adaptive Responses in Iran. *J. Integr. Agric.* **2018**, *17*, 1–15. [CrossRef]
3. Arif, C.; Toriyama, K.; Nugroho, B.D.A.; Mizoguchi, M. Crop Coefficient and Water Productivity in Conventional and System of Rice Intensification (SRI) Irrigation Regimes of Terrace Rice Fields in Indonesia. *J. Teknol.* **2015**, *76*, 97–102. [CrossRef]
4. Kottegoda, N.; Sandaruwan, C.; Priyadarshana, G.; Siriwardhana, A.; Rathnayake, U.A.; Berugoda Arachchige, D.M.; Kumarasinghe, A.R.; Dahanayake, D.; Karunaratne, V.; Amaratunga, G.A.J. Urea-Hydroxyapatite Nanohybrids for Slow Release of Nitrogen. *ACS Nano* **2017**, *11*, 1214–1221. [CrossRef]
5. Reskiana, S.; Setiawan, B.I.; Saptomo, S.K.; Mustatiningsih, P.R.D. Uji Kinerja Emiter Cincin. *J. Irig.* **2014**, *9*, 64–74. [CrossRef]
6. Saefuddin, R.; Saito, H.; Šimůnek, J. Experimental and Numerical Evaluation of a Ring-Shaped Emitter for Subsurface Irrigation. *Agric. Water Manag.* **2019**, *211*, 111–122. [CrossRef]
7. Saefuddin, R.; Saito, H. Performance of a Ring-Shaped Emitter for Subsurface Irrigation in Bell Pepper (Capsicum Annum L.) Cultivation. *Paddy Water Environ.* **2019**, *17*, 101–107. [CrossRef]
8. Arif, C.; Setiawan, B.I.; Saptomo, S.K.; Matsuda, H.; Tamura, K.; Inoue, Y.; Hikmah, Z.M.; Nugroho, N.; Agustiani, N.; Suwarno, W.B. Performances of Sheet-Pipe Typed Subsurface Drainage on Land and Water Productivity of Paddy Fields in Indonesia. *Water* **2021**, *13*, 48. [CrossRef]
9. Arif, C.; Saptomo, S.K.; Setiawan, B.I.; Taufik, M.; Suwarno, W.B.; Mizoguchi, M. A Model of Evapotranspirative Irrigation to Manage the Various Water Levels in the System of Rice Intensification (SRI) and Its Effect on Crop and Water Productivities. *Water* **2022**, *14*, 170. [CrossRef]
10. Prasetyo, A.; Utomo, W.H.; Listyorini, E. Hubungan Sifat Fisik Tanah, Perakaran Dan Hasil Ubi Kayu Tahun Kedua Pada Alfisol Jatikerto Akibat Pemberian Pupuk Organik Dan Anorganik (NPK). *J. Tanah Dan Sumberd. Lahan* **2017**, *1*, 27–37.

11. Alfionita, A.N.A.; Patang, P.; Kaseng, E.S. Pengaruh Eutrofikasi Terhadap Kualitas Air Di Sungai Jeneberang. *J. Pendidik. Teknol. Pertan.* **2019**, *5*, 9–23. [CrossRef]
12. Ozlu, E.; Kumar, S. Response of Soil Organic Carbon, PH, Electrical Conductivity, and Water Stable Aggregates to Long-Term Annual Manure and Inorganic Fertilizer. *Soil Sci. Soc. Am. J.* **2018**, *82*, 1243–1251. [CrossRef]
13. Wang, Y.; Zhu, Y.; Zhang, S.; Wang, Y. What Could Promote Farmers to Replace Chemical Fertilizers with Organic Fertilizers? *J. Clean. Prod.* **2018**, *199*, 882–890. [CrossRef]
14. Sumarsono, J.; Setiawan, B.I.; Subrata, I.D.M.; Waspodo, R.S.B.; Saptomo, S.K. Ring-Typed Emitter Subsurface Irrigation Performances in Dryland Farmings. *Int. J. Civ. Eng. Technol.* **2018**, *9*, 797–806.
15. Woodhouse, P.; Veldwisch, G.J.; Venot, J.-P.; Brockington, D.; Komakech, H.; Manjichi, Â. African Farmer-Led Irrigation Development: Re-Framing Agricultural Policy and Investment? *J. Peasant Stud.* **2017**, *44*, 213–233. [CrossRef]
16. Silva, M.A.; Albuquerque, T.G.; Alves, R.C.; Oliveira, M.B.P.P.; Costa, H.S. Melon (*Cucumis Melo* L.) by-Products: Potential Food Ingredients for Novel Functional Foods? *Trends Food Sci. Technol.* **2020**, *98*, 181–189. [CrossRef]
17. Martey, E.; Kuwornu, J.K.M.; Adjebeng-Danquah, J. Estimating the Effect of Mineral Fertilizer Use on Land Productivity and Income: Evidence from Ghana. *Land Use Policy* **2019**, *85*, 463–475. [CrossRef]
18. de Jong, I.H.; Arif, S.S.; Gollapalli, P.K.R.; Neelam, P.; Nofal, E.R.; Reddy, K.Y.; Röttcher, K.; Zohrabi, N. Improving Agricultural Water Productivity with a Focus on Rural Transformation. *Irrig. Drain.* **2021**, *70*, 458–469. [CrossRef]
19. van Genuchten, M.T. A Closed-Form Equation for Predicting the Hydraulic Conductivity of Unsaturated Soils. *Soil Sci. Soc. Am. J.* **1980**, *44*, 892–898. [CrossRef]
20. Arif, C.; Setiawan, B.I.; Saptomo, S.K.; Taufik, M.; Wiranto; Mizoguchi, M. Developing IT Infrastructure of Evaporative Irrigation by Adopting IOT Technology. *IOP Conf. Ser. Earth Environ. Sci.* **2021**, *622*, 012048. [CrossRef]
21. Hou, Y.; Li, A.; Li, Y.; Jin, D.; Tian, Y.; Zhang, D.; Wu, D.; Zhang, L.; Lei, W. Analysis of Microclimate Characteristics in Solar Greenhouses under Natural Ventilation. *Build. Simul.* **2021**, *14*, 1811–1821. [CrossRef]
22. Dufault, R.J.; Korkmaz, A.; Ward, B.K.; Hassel, R.L. Planting Date and Cultivar Affect Melon Quality and Productivity. *HortScience* **2006**, *41*, 1559–1564. [CrossRef]
23. Ben Ali, R.; Bouadila, S.; Mami, A. Development of a Fuzzy Logic Controller Applied to an Agricultural Greenhouse Experimentally Validated. *Appl. Therm. Eng.* **2018**, *141*, 798–810. [CrossRef]
24. Benyezza, H.; Bouhedda, M.; Zerhouni, M.C.; Boudjemaa, M.; Abu Dura, S. Fuzzy Greenhouse Temperature and Humidity Control Based on Arduino. In Proceedings of the 2018 International Conference on Applied Smart Systems (ICASS), Médéa, Algeria, 24–25 November 2018; pp. 1–6.
25. Allen, R.; Pareira, L.; Raes, D.; Smith, M. *FAO Irrigation and Drainage Paper No. 56. Crop Evapotranspiration (Guidelines for Computing Crop Water Requirements)*; Food and Agriculture Organisation of the United Nations: Rome, Italy, 1998.
26. Bouman, B.A.M.; Peng, S.; Castañeda, A.R.; Visperas, R.M. Yield and Water Use of Irrigated Tropical Aerobic Rice Systems. *Agric. Water Manag.* **2005**, *74*, 87–105. [CrossRef]
27. Li, X.; Shi, F. The Effect of Flooding on Evaporation and the Groundwater Table for a Salt-Crusted Soil. *Water* **2019**, *11*, 1003. [CrossRef]
28. Hargreaves, G.H.; Allen, R.G. History and Evaluation of Hargreaves Evapotranspiration Equation. *J. Irrig. Drain. Eng.* **2003**, *129*, 53–63. [CrossRef]
29. Liu, Y.J.; Chen, J.; Pan, T. Analysis of Changes in Reference Evapotranspiration, Pan Evaporation, and Actual Evapotranspiration and Their Influencing Factors in the North China Plain during 1998–2005. *Earth Space Sci.* **2019**, *6*, 1366–1377. [CrossRef]
30. Raghuwanshi, N.S.; Wallender, W.W. Converting from Pan Evaporation to Evapotranspiration. *J. Irrig. Drain. Eng.* **1998**, *124*, 275–277. [CrossRef]
31. de Oliveira, H.F.E.; de Moura Campos, H.; Mesquita, M.; Machado, R.L.; Vale, L.S.R.; Siqueira, A.P.S.; Ferrarezi, R.S. Horticultural Performance of Greenhouse Cherry Tomatoes Irrigated Automatically Based on Soil Moisture Sensor Readings. *Water* **2021**, *13*, 2662. [CrossRef]
32. Zhang, Z.; Liu, S.; Jia, S.; Du, F.; Qi, H.; Li, J.; Song, X.; Zhao, N.; Nie, L.; Fan, F. Precise Soil Water Control Using a Negative Pressure Irrigation System to Improve the Water Productivity of Greenhouse Watermelon. *Agric. Water Manag.* **2021**, *258*, 107144. [CrossRef]
33. Ren, R.; Liu, T.; Ma, L.; Fan, B.; Du, Q.; Li, J. Irrigation Based on Daily Weighted Evapotranspiration Affects Yield and Quality of Oriental Melon. *Sci. Hortic.* **2021**, *275*, 109714. [CrossRef]
34. Chang, Y.H.; Hwang, Y.H.; An, C.G.; Yoon, H.S.; An, J.U.; Lim, C.S.; Shon, G.M. Effects of Non-drainage Hydroponic Culture on Growth, Yield, Quality and Root Environments of Muskmelon (*Cucumis melo* L.). *J. Bio-Environ. Control* **2012**, *21*, 348–353. [CrossRef]
35. Lim, M.Y.; Choi, S.H.; Jeong, H.J.; Choi, G.L. Characteristics of Domestic Net Type Melon in Hydroponic Spring Cultivars Using Coir Substrates. *Korean J. Hortic. Sci. Technol.* **2020**, *38*, 78–86. [CrossRef]
36. Wang, Y.; Li, S.; Qin, S.; Guo, H.; Yang, D.; Lam, H.-M. How Can Drip Irrigation Save Water and Reduce Evapotranspiration Compared to Border Irrigation in Arid Regions in Northwest China. *Agric. Water Manag.* **2020**, *239*, 106256. [CrossRef]
37. Al-Omran, A.; Louki, I.; Alkhasha, A.; Abd El-Wahed, M.H.; Obadi, A. Water Saving and Yield of Potatoes under Partial Root-Zone Drying Drip Irrigation Technique: Field and Modelling Study Using SALTMED Model in Saudi Arabia. *Agronomy* **2020**, *10*, 1997. [CrossRef]

38. Rolbiecki, R.; Sadan, H.; Rolbiecki, S.; Jagosz, B.; Szczepanek, M.; Figas, A.; Atilgan, A.; Pal-Fam, F.; Pańka, D. Effect of Subsurface Drip Fertigation with Nitrogen on the Yield of Asparagus Grown for the Green Spears on a Light Soil in Central Poland. *Agronomy* **2022**, *12*, 241. [CrossRef]

39. Hamad, A.A.A.; Wei, Q.; Wan, L.; Xu, J.; Hamoud, Y.A.; Li, Y.; Shaghaleh, H. Subsurface Drip Irrigation with Emitters Placed at Suitable Depth Can Mitigate N_2O Emissions and Enhance Chinese Cabbage Yield under Greenhouse Cultivation. *Agronomy* **2022**, *12*, 745. [CrossRef]

40. Edwards, K.P.; Madramootoo, C.A.; Whalen, J.K.; Adamchuk, V.I.; Mat Su, A.S.; Benslim, H. Nitrous Oxide and Carbon Dioxide Emissions from Surface and Subsurface Drip Irrigated Tomato Fields. *Can. J. Soil Sci.* **2018**, *98*, 389–398. [CrossRef]

41. Fisher, R.; Bobanuba, W.E.; Rawambaku, A.; Hill, G.J.E.; Russell-Smith, J.; Fisher, R.; Bobanuba, W.E.; Rawambaku, A.; Hill, G.J.E.; Russell-Smith, J. Remote Sensing of Fire Regimes in Semi-Arid Nusa Tenggara Timur, Eastern Indonesia: Current Patterns, Future Prospects. *Int. J. Wildland Fire* **2006**, *15*, 307–317. [CrossRef]

 *agronomy*

Article

Crop Sequencing to Improve Productivity and Profitability in Irrigated Double Cropping Using Agricultural System Simulation Modelling

Ketema Zeleke [1,2,*] and Jeff McCormick [1,2]

[1] School of Agricultural, Environmental and Veterinary Sciences, Charles Sturt University, Wagga Wagga, NSW 2650, Australia; jmccormick@csu.edu.au

[2] Gulbali Institute for Agriculture, Water and Environment, Charles Sturt University, Wagga Wagga, NSW 2650, Australia

* Correspondence: kzeleke@csu.edu.au

Citation: Zeleke, K.; McCormick, J. Crop Sequencing to Improve Productivity and Profitability in Irrigated Double Cropping Using Agricultural System Simulation Modelling. *Agronomy* 2022, 12, 1229. https://doi.org/10.3390/agronomy12051229

Academic Editors: Pantazis Georgiou and Dimitris Karpouzos

Received: 30 April 2022
Accepted: 18 May 2022
Published: 20 May 2022

Publisher's Note: MDPI stays neutral with regard to jurisdictional claims in published maps and institutional affiliations.

Abstract: Land and water are two major inputs for crop production. Simulation modelling was used to determine crop sequences that maximise farm return. Crop yield was determined for different irrigation scheduling scenarios based on the fraction of available soil water (FASW). Farm returns ($ ML^{-1} and $ ha^{-1}) were evaluated for seven crop sequences. Three irrigation water price scenarios (dry, median, wet) were considered. The yield of summer crops increased with irrigation. For winter crops, despite increase in irrigation, the yield would not increase. The optimum irrigation (ML ha^{-1}) was: soybean 8.2, maize 10.4, wheat 2.5, barley 3.1, fababean 2.5, and canola 2.7. The water productivity curve of summer crops has a parabolic shape, increasing with FASW, reaching a maximum value at FASW 0.4–0.6, and then decreasing. The water productivity of winter crops decreases as FASW increases following a power function. Gross margins are positive when water is cheap ($60 ML^{-1}) and when water has a median price ($124 ML^{-1}). When water is expensive ($440 ML^{-1}), positive gross margin would be obtained only for the continuous wheat scenario. Deficit irrigation of summer crops leads to significant yield loss. Supplemental irrigation of winter crops results in the highest gross margin per unit of water.

Keywords: APSIM; Australia; gross margin; double cropping; irrigation; water price

1. Introduction

The rapid growth of the world population and pressure on land and environmental resources has amplified the need to increase food and fibre production with minimal resource input. Multiple cropping or increasing cropping intensity is one means of increasing global crop production [1]. Crop diversification increases food production sustainability by supressing pests, absorbing climatic shocks, reducing fertiliser use, and reducing business risks [2]. In the Australian grain crop production system, winter crops are normally grown under a dryland/rainfed environment. When and where water is available, supplemental irrigation is applied as pre-irrigation in autumn and at the reproductive growth stage in spring. However, with the changing climate and competing needs such as environmental watering requirements, the amount of water available for irrigation is declining. In the Riverina region of south-eastern New South Wales (NSW), the farm-land holdings are also relatively small compared to other parts of the country. As a result, there is a need to optimise the limited amount of available irrigation water and land resources.

Irrigation farmers in the region have water entitlements, which are an ongoing share of surface and ground water resources in their catchment. However, they do not always obtain 100% of their entitlement as this depends on the rainfall in the season and water already available in the dams. Depending on the reliability of obtaining the full amount of entitlement, there are two kinds of water securities: high security water and general security water. High security water entitlement holders obtain 100% of their entitlement unless there is severe drought. This is the water for permanent plantings such as horticulture and viticulture. General security water is for annual crops. Most of the water in the region is as general security entitlement. When there is low or no general security water allocated, farmers buy water on the water market. The price of water varies from year to year depending on the amount of water available in the dams and the prevailing rainfall. As a result, the gross margin per unit of irrigation water also varies with the price of water.

In recent years, low water allocation brought about by a combination of climatic and environmental policy-related factors have constrained Riverina irrigators' production capacity. In the past 20 years, the amount of water in the region has been around 40% less than the long-term average (Figure 1). As a result, average seasonal allocations were less than 50% of the entitlement with high year to year variability. Innovative on-farm water management practices and planning are required to maintain a profitable irrigation industry in a future climate of reduced and variable irrigation water supply [3]. Some of the approaches that could help farmers adapt to reduced water allocations by increasing water and land productivity are: partial (deficit) irrigation, changes in crop rotations, crop species or varieties, and changes in water allocation to winter crops relative to summer crops [4–6]. In this study, different winter–summer crop sequences were evaluated in terms of gross margin per unit of water applied and per unit of cultivated land area. Crop yields were simulated using Agricultural Production Systems sIMulator (APSIM v.7.10) [7].

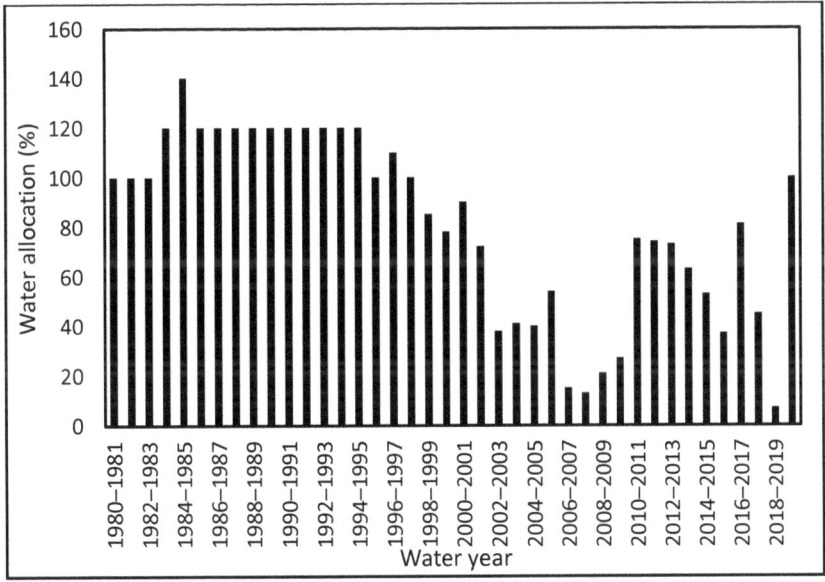

Figure 1. Historical annual percentage allocation of irrigation water entitlement in the Murrumbidgee Irrigation Area, Riverina, NSW (data from https://www.industry.nsw.gov.au/water/allocations-availability/water-accounting/historical-available-water-determination-data, accessed on 20 November 2021).

2. Materials and Method

2.1. General Description of the Study Area

The study area is at Leeton in the Riverina region in NSW, Australia. The Riverina region of southern NSW has a mixed (crop and livestock) farming system. There are two large irrigation areas, Murrumbidgee and Coleambally, in the region. The Murrumbidgee river, the third longest in the country, originates from two large dams, Blowering and Burrinjuck, in the Snowy mountain of the Great Dividing Range. Water is diverted to the two irrigation areas using the two weirs, Berembed and Gogeldrie, on the Murrumbidgee river. The major irrigated crops in the region are horticulture, grapevine, cotton, rice, legumes, and maize. The climate of the region (south-eastern NSW) is partly semi-arid and partly temperate. It has a cold winter and hot summer with monthly temperature variation shown in Figure 2.

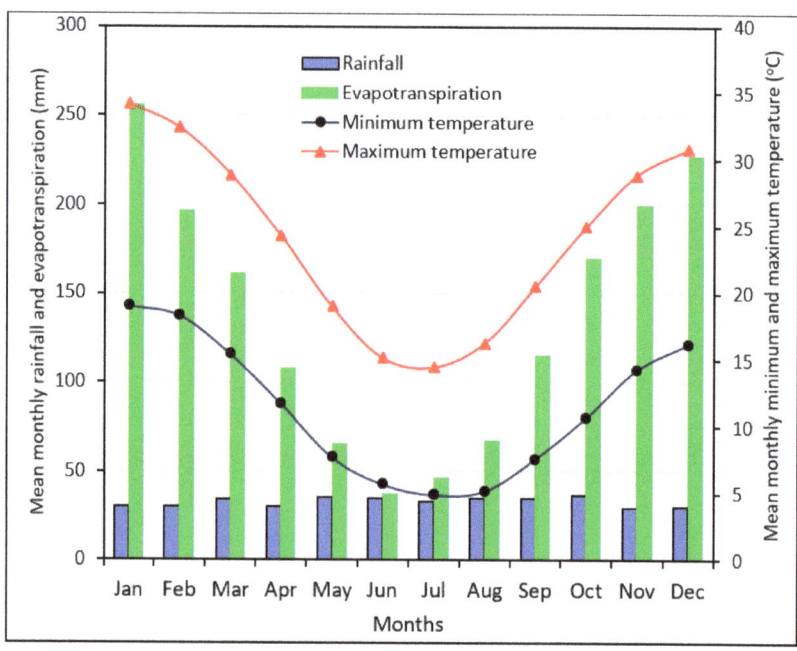

Figure 2. Mean monthly rainfall, temperature and evapotranspiration at Leeton station, Riverina region, NSW Australia (1989–2018).

The annual mean minimum temperature is 11 °C while the mean maximum temperature is 24 °C. It has high evapotranspiration except during the winter season. The mean annual rainfall is 395 mm (monthly average 33 mm) and mean annual evapotranspiration is 1652 mm (http://www.bom.gov.au/ (accessed on 14 September 2021)).

2.2. Crops Used in the Study

Two summer crops (maize and soybean) and four winter crops (barley, canola, fababean, and wheat) commonly grown in the Australian wheat-belt were chosen for this study. In the Riverina region, in the summer season, after the winter crop harvest, fields are either left fallow or sown to a summer crop (double-cropping), depending on irrigation water availability. Introduction of summer crops in a winter-dominant cropping system is important for farming system sustainability to manage diseases and weeds. However, determining the crop sequence or which crop follows which requires careful planning. When double cropping is used, the time of harvest of one crop may overlap with the time of sowing of the next crop. Careful planning and choice of variety is required to ensure smooth transition. The availability of short-season varieties in recent years has enabled timely sowing of a crop and harvest without incurring substantial yield penalty. In order to meet crop water demand during the critical reproductive growth stages of crop development, where possible, supplemental irrigation of winter crops is used [8].

In the region, maize (*Zea mays* L.) is grown for silage and grain production. Depending on the variety, the length of growing season from planting to harvest is 130 to 150 days. The average maize irrigation requirement is 8–9 ML ha^{-1} with an average grain yield of 10.2 t ha^{-1} [9]. In southern NSW, the optimum time of sowing of maize is from early

October to mid-November. Riverina is one of the two areas where soybean is widely grown in NSW. The soybean (*Glycine max* L.) irrigation requirement is 6–8 ML ha^{-1}. Its optimum sowing time is mid-November to mid-December. For barley (*Hordeum vulgare* L.), sowing time is important to avoid risk of frost damage and drought stress during and after anthesis [10]. Its recommended sowing time is from mid-May to mid-June. Canola (*Brassica napus* L.) is a profitable break crop for weed and disease control. Its best sowing time is from the fourth week of April to the second week of May. Fababeans (*Vicia faba* L.) is an important break crop in crop rotation for disease control and nitrogen fixation [11]. The recommended sowing window is from the fourth week of April to the second week of May [12]. The optimum sowing time of wheat (*Triticum aestivum* L.) is not fixed but varies with location, season and variety [13]. Supplemental irrigation is used for winter crops during critical growth stages [14].

2.3. Simulation Setup

2.3.1. Weather and Soil Data

The Agricultural Production Systems sIMulator (APSIM) version 7.10 was used to simulate water-limited potential yield of the crops [7]. The details of the model can be found at www.apsim.info (accessed on 4 August 2021). Briefly, APSIM links the specific crop module (CROP), the soil water module (SOILWAT), the nitrogen module (SOILN), and residue module (RESIDUE) and irrigation. The model calculates biomass accumulation from solar radiation interception and adjusts for water and nitrogen stresses. Empirical coefficients are used to partition the biomass into different organs. The model can be used to study the effect of environmental factors and management decisions on resource use, crop growth, and yield. APSIM requires climate, soil, crop, and management decisions' data to simulate crop yield. The SILO patched point daily climate dataset (1989–2020) of the Yanco Agricultural Institute Station ($-34.6222°$ latitude and $146.4326°$ longitude) was used ([15], https://legacy.longpaddock.qld.gov.au/silo/ (accessed on 23 July 2021)). The soil of the study area is Brown Chromosol with a moderate water holding capacity (126 mm/m), the hydrologic characteristics of which, as used in APSIM, are shown in Table 1 [16]. The APSIM model has already been calibrated and validated for the crops used in this study: wheat [17,18], soybean [19], fababean [20,21], maize [22], barley [23], and canola [24,25].

Table 1. Hydrologic properties of the Brown Chromosol soil in Murrumbidgee Irrigation Area, NSW Australia (http://www.apsim.info) (accessed on 4 August 2021).

Soil Depth (cm)	Bulk Density (g cm^{-3})	Wilting Point (LL15) * (cm^3 cm^{-3})	Field Capacity (DUL) + (cm^3 cm^{-3})	Saturation Moisture Content (cm^3 cm^{-3})	Plant Available Water Capacity, PAWC (mm)
0–15	1.47	0.101	0.265	0.414	24.6
15–30	1.44	0.247	0.375	0.427	19.2
30–60	1.43	0.244	0.380	0.430	40.8
60–90	1.50	0.244	0.354	0.404	32.7
90–120	1.58	0.228	0.325	0.375	29.1
120–150	1.59	0.224	0.319	0.366	26.7
150–160	1.49	0.224	0.324	0.408	17.7

* LL15 is the soil water content at 15 bar pressure, which is the lower limit of the plant available water. + DUL (drainable upper limit) is the soil water content at field capacity.

2.3.2. Crop Sequences

Seven winter–summer crop sequences were evaluated under different water allocation/water price scenarios and irrigation amounts (Table 2). These were selected and adapted based on Napier et al. [26]. The simulation was done for five seasons (three winter and two summer—starting with winter season and ending with winter season). Crop cultivars and sowing dates used in the APSIM simulation were: soybean (cv. dragon), 15 Nov; maize (generic—early maturing), 15 Nov; fababean (cv. fiord), 15 May; wheat (cv. suntop), 01 May; barley (cv. scope), 15 May; canola (generic—early), 01 May. The soil water, nitrogen and organic matter were reset at the time of sowing in the long-term simulations.

Table 2. Winter–summer crop sequence scenarios used in the simulation.

Rotation 1—R1	F-S-F-M-F	Fallow–Soybean–Fallow–Maize–Fallow
Rotation 2—R2	F-S-F-S-F	Fallow–Soybean–Fallow–Soybean–Fallow
Rotation 3—R3	W-F-Fb-M-F	Wheat–Fallow–Fababean–Maize–Fallow
Rotation 4—R4	W-S-W-S-W	Wheat–Soybean–Wheat–Soybean–Wheat
Rotation 5—R5	W-F-W-F-W	Wheat–Fallow–Wheat–Fallow–Wheat
Rotation 6—R6	B-S-B-S-B	Barley–Soybean–Barley–Soybean–Barley
Rotation 7—R7	C-M-Fb-F-C	Canola–Maize–Fababean–Fallow–Canola

2.3.3. Soil Water Deficit

Irrigation would be applied when the soil moisture is at a certain fraction of available soil water, hereafter designated as FASW. FASW varies between 0 and 1: 1 when the soil water is close to field capacity or drained upper limit and 0 when the soil water is close to permanent wilting point or drained lower limit. FASW is computed as:

$$FASW = \frac{\theta - LL15}{DUL - LL15} \tag{1}$$

where FASW = fraction of available soil water (0–1); θ = soil water content (cm^3 cm^{-3}); DUL = drained upper limit or field capacity (cm^3 cm^{-3}); $LL15$ = soil water content ((cm^3 cm^{-3}) at wilting point or 1.5 MPa soil water potential). In this study, irrigation scheduling at the following FASWs was investigated: 0.1, 0.2, 0.3, 0.4, 0.5, 0.6, 0.7, 0.8, and 0.9. For example, FASW 0.1 means irrigation is applied when the moisture remaining in the soil is 10% of the available soil water. This is a dry scenario, which means the crop is under water stress. On the other hand, in FASW 0.9, irrigation is applied while the soil moisture is still 90% of full capacity (a wet scenario). For all crop sequence scenarios and soil moisture deficit levels, the variations in the total seasonal irrigation, crop yield, and water productivity were determined.

Gross margins ($ ML^{-1} and $ ha^{-1}) were determined for different winter–summer crop sequences, irrigation water price scenarios, and soil moisture deficit levels. Irrigation scheduling is determined by the amount of water available in the soil and the rate at which the crop uses this water. For most crops, crop yield is not affected if irrigation is applied before 50% of the available soil moisture (FASW 0.5) is depleted. For gross margin analysis, three soil moisture deficit scenarios were considered. In the first scenario, irrigation would be applied when the soil water is depleted to 20% of the available water (FASW 0.2) (dry scenario). The other scenarios are FASW 0.5 (medium scenario) and FASW 0.8 (wet scenario). The furrow irrigation water application efficiency was set at 75%.

2.4. Water Allocation Scenarios

Water price highly varies from year to year depending on dam storage and rainfall in the catchments. As a result, the gross margin was determined for different percentiles of water price obtained from historical water allocation prices of the Murrumbidgee river catchment (https://www.awe.gov.au/abares/research-topics/water/water-market-outlook (accessed on 24 August 2021)). The 25 percentile irrigation water price was $60 ML^{-1}, the median (50 percentile) irrigation water price was $124 ML^{-1}, and the 75 percentile irrigation water price was $440 ML^{-1}. The annual average water prices for different water-years were as shown in Figure 3. The generalized price-allocation function is shown in Equation (2):

$$P = 437 \times e^{-0.025A} \tag{2}$$

where P is the market water price ($ ML^{-1}); A is the seasonal allocation (%). A similar relationship was reported by [3].

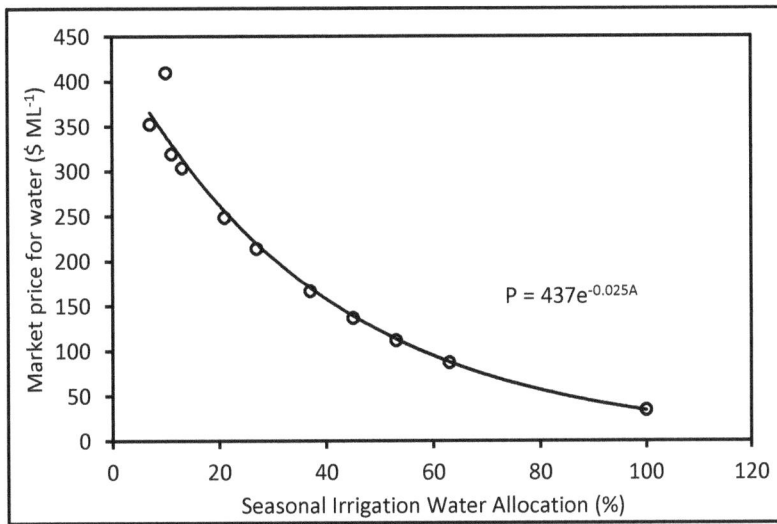

Figure 3. Water price as a function of seasonal water allocation in the Murrumbidgee valley of NSW, Australia.

2.5. Gross Margin Analysis

Gross margin was calculated using the Decision Support Tool developed during the "Correct Crop Sequencing" Grain Research and Development Cooperation (GRDC) project (https://www.dpi.nsw.gov.au/agriculture/budgets/costs/cost-calculators/correct-crop-sequencing-decision-support-tool (accessed on 25 September 2021)).

Gross margin for a given cropping season was calculated as:

$$\text{Gross margin (\$ ha}^{-1}) = \text{Gross income (\$ ha}^{-1}) - \text{Variable cost (\$ ha}^{-1}) \tag{3}$$

$$\text{Gross income (\$ ha}^{-1}) = \text{Grain yield (t ha}^{-1}) * \text{Price (\$ t}^{-1}) \tag{4}$$

$$\text{Variable cost (\$ ha}^{-1}) = \text{Cost of water (\$ ha}^{-1}) + \text{Cost of other inputs and farm operations (cultivation, sowing, fertiliser, spraying, harvest) (\$ ha}^{-1}) \tag{5}$$

$$\text{Cost of water (\$ ha}^{-1}) = \text{Water use (ML ha}^{-1}) * \text{Water price (\$ ML}^{-1}) \tag{6}$$

$$\text{Gross margin per unit of water (\$ ML}^{-1}) = \text{Gross margin (\$ ha}^{-1})/\text{Water use (ML ha}^{-1}) \tag{7}$$

The data required to run this model are grain yield, grain yield price, water use, cost of water, and costs related to other farm operations (cultivation, sowing, fertiliser, spraying, harvest). It calculates gross margins per unit area ($ ha^{-1}) and per unit of water ($ ML^{-1}) for different crop sequences. The default variable cost (excluding cost of water) used in the decision support tool was adopted for this simulation study as it is from the same site as this simulation study. The variation of the variables costs over time were not considered. The long-term average price for the Riverina area, as obtained from Grain Price Australia Listings igrainPlus (https://www.igrain.com.au/) (accessed on 14 August 2021) was used in the Gross Margin Decision Support Tool. Accordingly, the average prices per tonne were: soybean ($535), fababean ($374), barley ($291), canola ($559), wheat ($324), and maize ($384). Water price varied highly from year to year. For example, it was as high as $1349 in December 2007, $665 in September 2008, and $759 in January 2020. It was also as low as $6 in April 2011, $4 in May 2012, and $6 in June 2017 (https:

//www.awe.gov.au/abares/research-topics/water/water-market-outlook) (accessed on 4 August 2021). As a result, the gross margin was analysed for three water price (\$ ML^{-1}) scenarios: 25 percentile (\$60) (dry), median (\$124) (normal), and 75 percentile (\$440) (wet).

Crop yield and crop water use were simulated using APSIM [7] for typical varieties, sowing dates, and long-term climate data for Yanco station, Leeton, NSW.

3. Results and Discussion

3.1. Yield of Summer and Winter Crops at Different Soil Moisture Deficit Levels

The simulated irrigation water requirement and corresponding grain yield at different soil moisture deficit levels is presented in Figure 4. The optimum irrigation, the amount of irrigation that maximises yield per unit of water applied, is: soybean, 820 mm (8.2 ML ha^{-1}); maize, 1040 mm (10.4 ML ha^{-1}); wheat, 250 mm (2.5 ML ha^{-1}); barley, 310 mm (3.1 ML ha^{-1}); fababean, 250 mm (2.5 ML ha^{-1}); and canola, 270 mm (2.7 ML ha^{-1}). In this study area, Gaydon et al. [3] estimated the irrigation requirement of 8 ML ha^{-1} for soybean and 3 ML ha^{-1} for barley. The relationship between grain yield, irrigation, and soil water deficit levels is different for summer (soybean, maize) and winter (canola, fababean, wheat, barley) crops. Figure 4a shows that if soybean is scheduled to be irrigated when the soil moisture is only slightly depleted (FASW 0.8), the total amount of irrigation is high. However, if irrigation is applied when the soil moisture is depleted to a level that is only 20% of the available soil water (FASW 0.2), the total amount of irrigation decreases significantly. The grain yield also follows a similar trend. However, the rate of yield increase slows when irrigation is scheduled in the high soil moisture range. Above FASW 0.5, the rate of yield increase is only 5%. However, when FASW is less than 0.5, the rate of yield increase is 20%. Figure 4b also shows that maize has a similar pattern. It has a slightly quicker increase in yield in the early part of the curve and plateaus once FASW > 0.45. The average rate of yield increase is 56% for FASW < 0.45 and it was only 3% once FASW > 0.45. This pattern is different for winter crops; there is only a slight yield increase as the amount of irrigation increases. Overall, the highest yield is obtained at the lowest soil moisture deficit level. If irrigation was to be applied only after the soil moisture was highly depleted (FASW 0.2), less irrigation would be required.

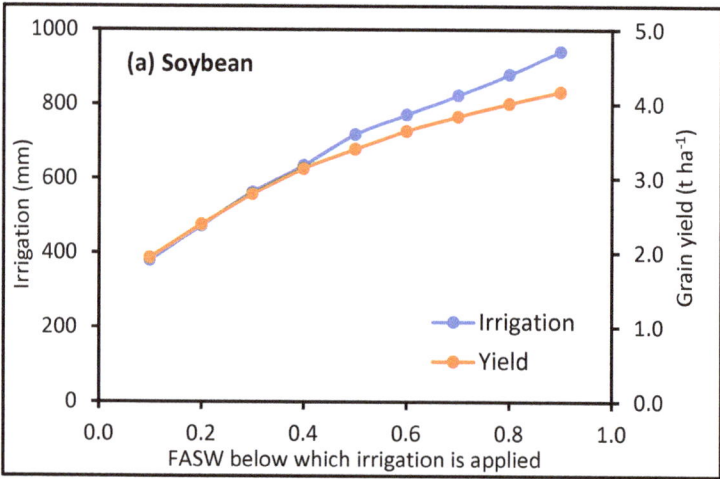

Figure 4. *Cont.*

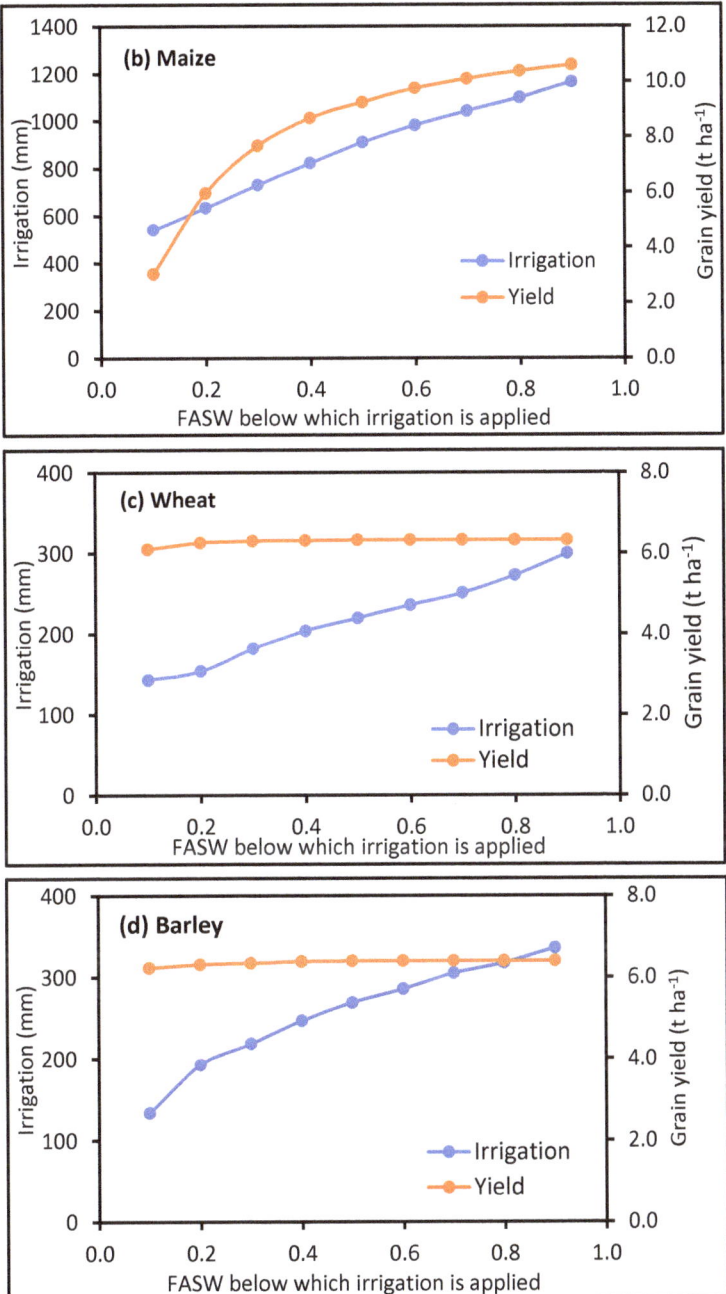

Figure 4. *Cont.*

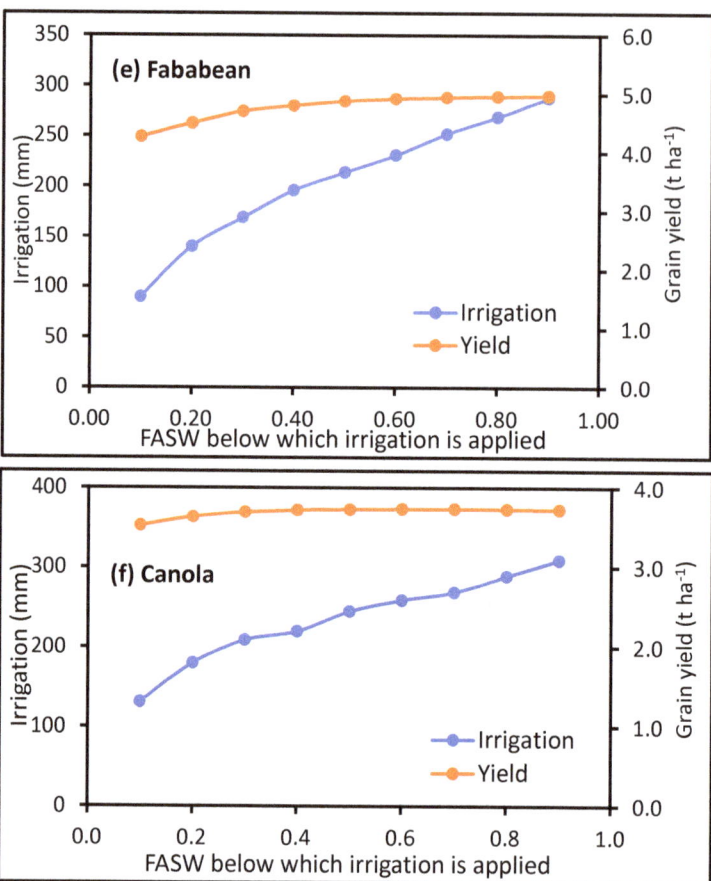

Figure 4. The amount of irrigation and corresponding grain yield of summer and winter crops for different soil water deficit levels in the Riverina region, NSW, Australia.

The FASW below which crop yields are affected depends on the crop species, growth stage, and environmental condition. For example, for shallow-rooted crops it is 0.25–0.40, for deep rooted-crops it is 0.50, and for deep-rooted crops with dense rooting systems it is 0.60–0.65. FAO recommends that the commonly used average FASW 0.50 should be increased by 15% when the reference evapotranspiration ETo <3 mm day^{-1} and decreased by 15% when ETo > 8 mm day^{-1} [27].

The relationship between the amount of irrigation and grain yield is presented in Figure 5. From Figure 5a (soybean), it can be seen that when irrigation was reduced from 942 to 636 mm (32%), yield dropped by 25%, and when irrigation was reduced from 564 to 381 mm (32%), yield dropped by 31%. For maize (Figure 5b), when irrigation was reduced from 1165 to 824 mm (29%), yield dropped by only 18%. However, when irrigation was reduced from 732 to 542 mm (26%), yield dropped by 60%. This shows that for soybean, the rate of crop yield loss per unit reduction in irrigation amount is the same at a higher and lower irrigation ranges. When a higher amount of irrigation is applied, some of the applied water is lost by deep percolation below the root zone [3]. There is high solar radiation and temperature during the summer period, implying no yield-limiting environmental factors. As a result, unlike winter crops, for summer crops, when a higher amount of irrigation water is applied, correspondingly a higher yield can be obtained. For soybean, the amount of irrigation that maximizes yield per unit of water is 8.2 ML ha^{-1} resulting in 3.8 t ha^{-1} yield, and for maize, it is 10.4 ML ha^{-1} irrigation resulting in 10.1 t ha^{-1} yield.

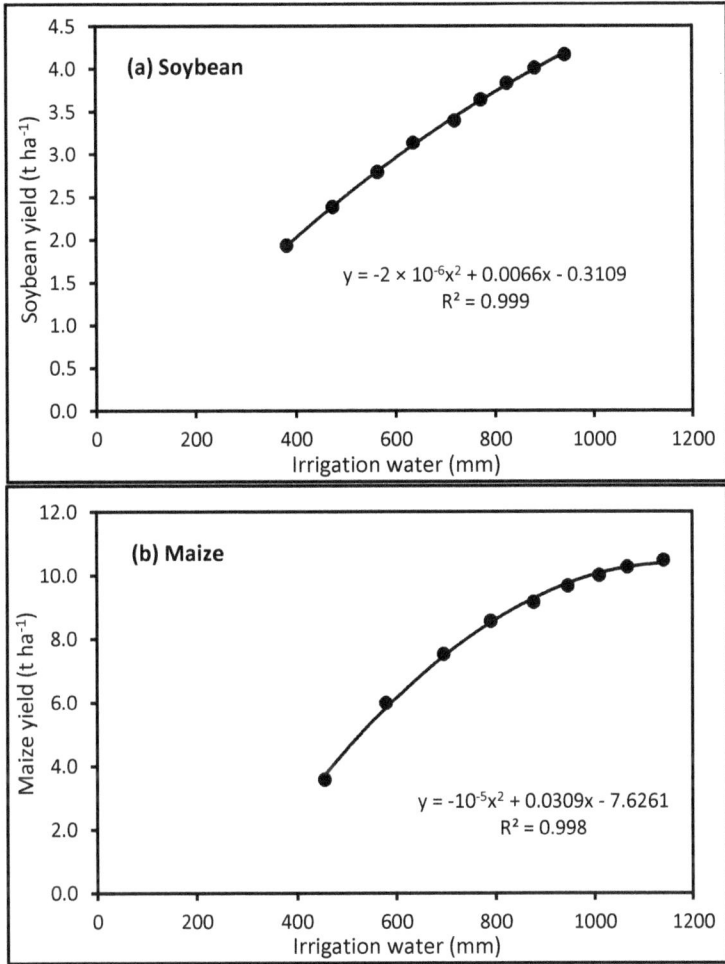

Figure 5. Simulated water production function of soybean and maize and fitted polynomial function in the Riverina region of NSW, Australia.

When the available soil moisture is highly depleted and the plant is unable to easily extract water, plant water stress and crop yield loss result. Different crops have different water stress tolerance [28]. The stress tolerance level also depends on the season, summer or winter. Due to low evaporative demand during the winter period, the plants are not significantly affected by the low soil moisture levels. During the summer period, however, due to high evaporative demand, the plants would not be able to withstand high soil water depletion. The total amount of irrigation is different for different soil moisture deficit levels. If irrigation was to be applied only after the soil moisture is highly depleted (long irrigation interval), the overall amount of irrigation required would be lower. However, irrigating while the soil moisture is still high results in more frequent and higher amount of irrigation. Crop water use is high when the soil moisture is high because crops can easily extract soil water. As a result, the total seasonal irrigation decreases as irrigation is applied less frequently, that is, waiting until the soil water is significantly depleted. However, this happens at the expense of crop yield.

3.2. Water Productivity of Summer and Winter Crops at Different Soil Moisture Deficit Levels

Crop water productivity was calculated as a ratio of crop yield and the total amount of water applied (rainfall plus irrigation). This was done for the water deficit levels varying from 0.1 to 0.9. The response of water productivity to different irrigation trigger soil moisture levels differs for summer and winter crops (Figure 6). The variation of water productivity with the soil water deficit levels was different for summer and winter crops. For summer crops, water productivity increases as the soil moisture level at which irrigation is applied increases. However, once it reached a maximum value at about FASW 0.5, it decreases. However, for winter crops, it starts at a higher value at FASW 0.1 and starts decreasing. This is attributed to less evapotranspiration during the winter crop growing season. Compared to the summer crops, the variation in water productivity with soil moisture depletion levels is not that high. For summer crops, if irrigation is applied only after the soil moisture drops to a low level (such as FASW 0.2), water productivity is low. It increases as the FASW increases and reaches a maximum at about FASW 0.4–0.6 before it decreases again; this is well-represented by a second-degree polynomial equation (Figure 6a,b). For winter crops, the highest water productivity is when FASW is low (e.g., FASW 0.2). From this highest point, it decreases as the FASW increases following a trend line represented by a power function (Figure 6c–f).

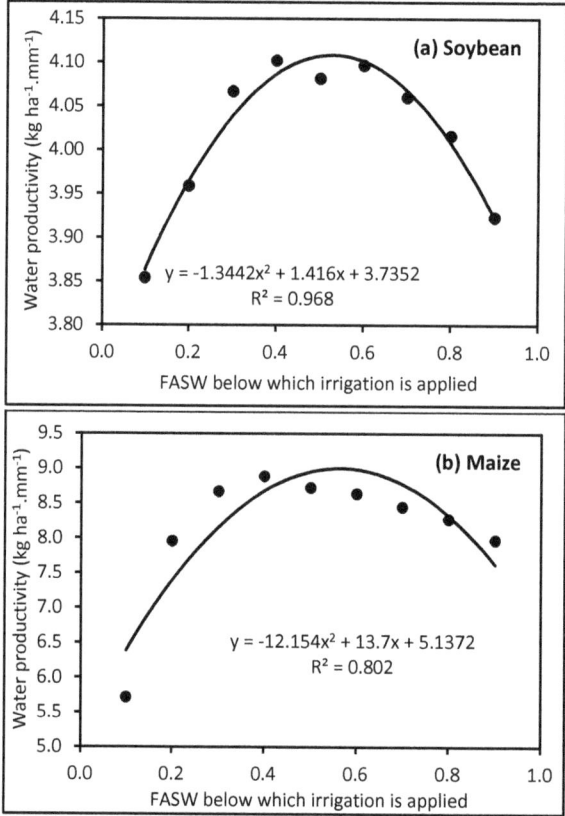

Figure 6. *Cont.*

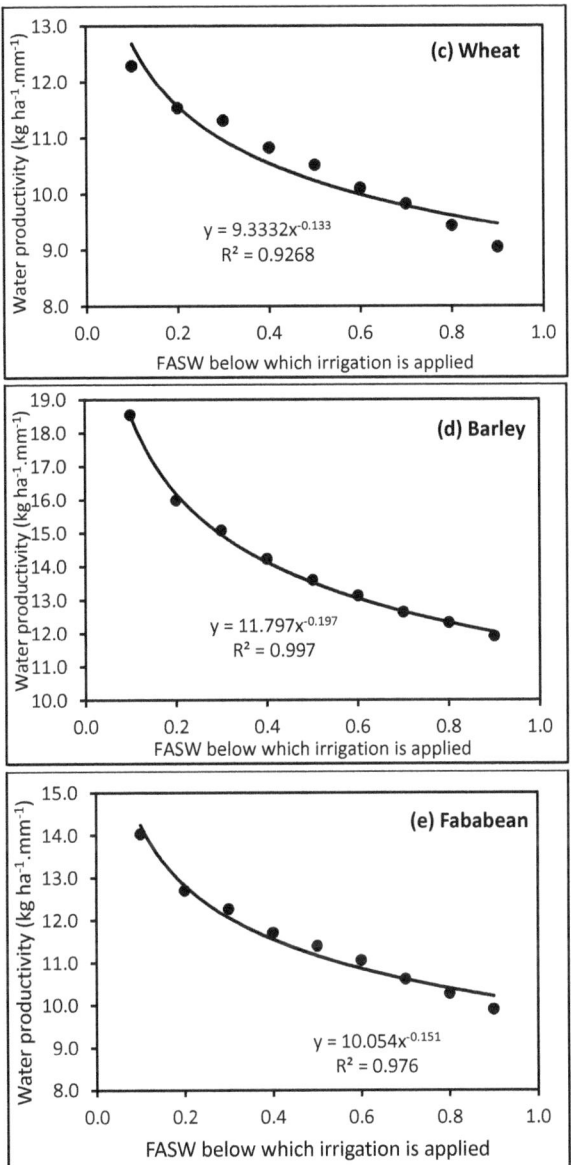

Figure 6. *Cont.*

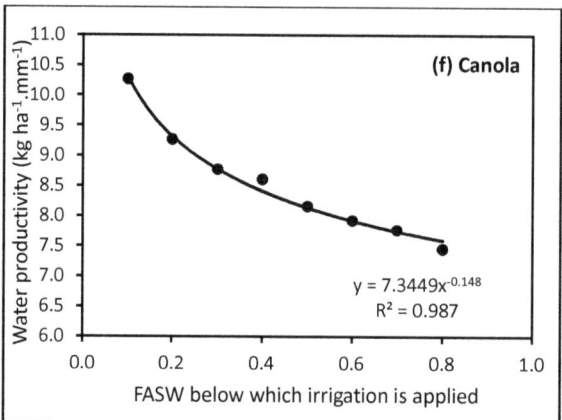

Figure 6. Water productivity of summer and winter crops at different soil water deficit levels in the Riverina region of NSW, Australia.

Summer crops have higher water requirements compared to winter crops due to high evaporative demand during the summer season. If irrigation is applied before the plant available water is depleted, the crop grows at an optimum level and yield will not be affected. However, depending on the crop type, growth stage, and crop water demand, crop yields can be affected when irrigation is applied after the soil water is depleted below a certain level called the management allowed deficit or critical soil moisture content [29]. Supplemental or deficit irrigation is practiced when irrigation is applied after the soil moisture is depleted below the critical soil moisture level. If the soil water is allowed to be depleted below the critical level before irrigation is applied, the practice is called regulated deficit irrigation [30].

3.3. Gross Margins for Different Crop Sequences

Gross margin ($ ML^{-1} and $ ha^{-1}) for different water prices, soil water deficits, and crop sequences is presented in Figure 7. When irrigation water is limited, irrigation scheduling depends on the prevailing seasonal conditions and the value of water. It can be seen that when irrigation is applied at the commonly used FASW 0.5, positive gross margins were obtained under all crop sequence scenarios and the 25 percentile ($60 ML^{-1}) and 50 percentile ($124 ML^{-1}) water price scenarios. When water is expensive (75 percentile—$440 ML^{-1}), all, except the continuous wheat scenario, resulted in negative gross margins. Obviously, the gross margin (both per ML and per ha) was the highest when water was cheap ($60 ML^{-1}). When water was cheap ($60 ML^{-1}), continuous wheat (WFWFW) resulted in the highest gross margin per unit of water used ($536 ML^{-1}) and the lowest was for the FSFMF sequence ($153 ML^{-1}). Per unit of cultivated land area, the two non-fallow crop sequences (WSWSW and BSBSB) resulted in the highest return, $5010 ha^{-1} and $5175 ha^{-1}, respectively. When water was expensive ($440 ML^{-1}), the biggest loss was for the soybean-only sequence (FSFSF) scenario ($-$257 ML^{-1} and $-$3696 ha^{-1}).

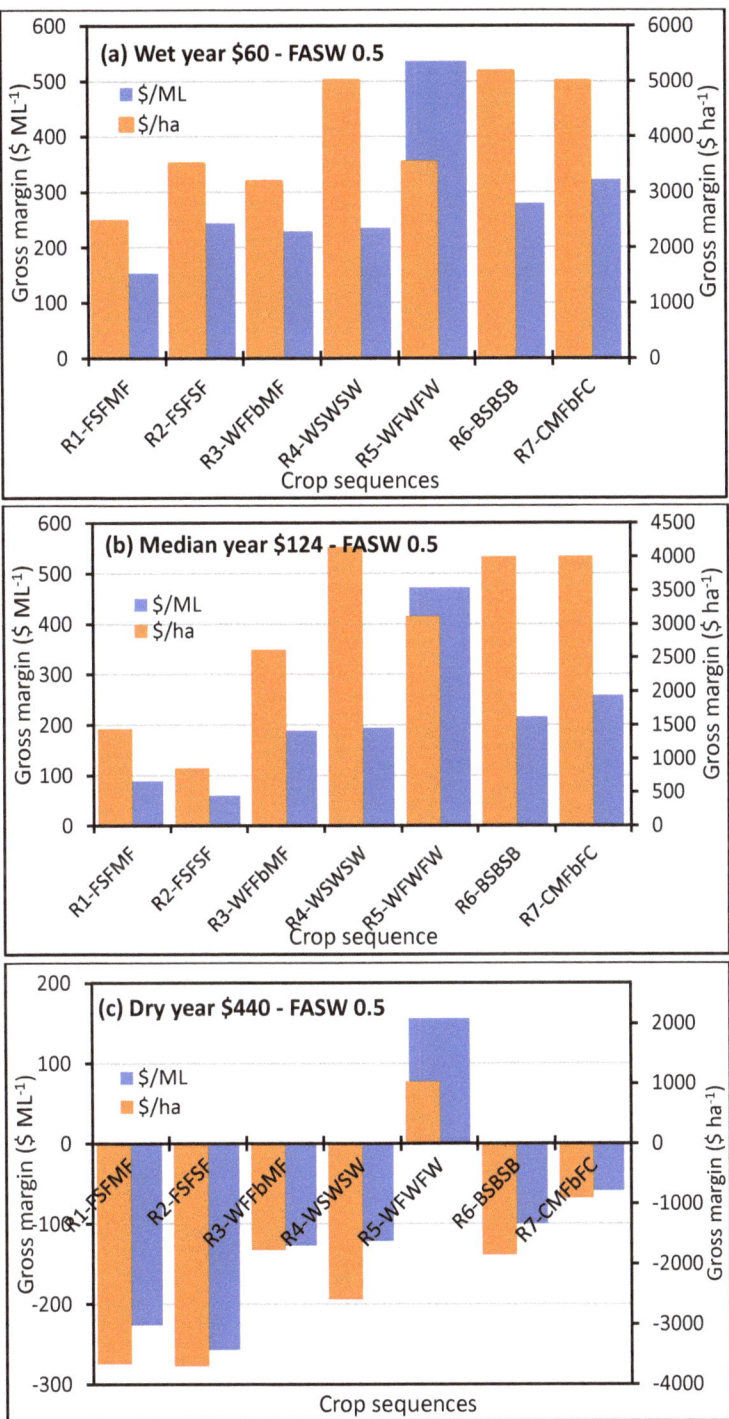

Figure 7. *Cont.*

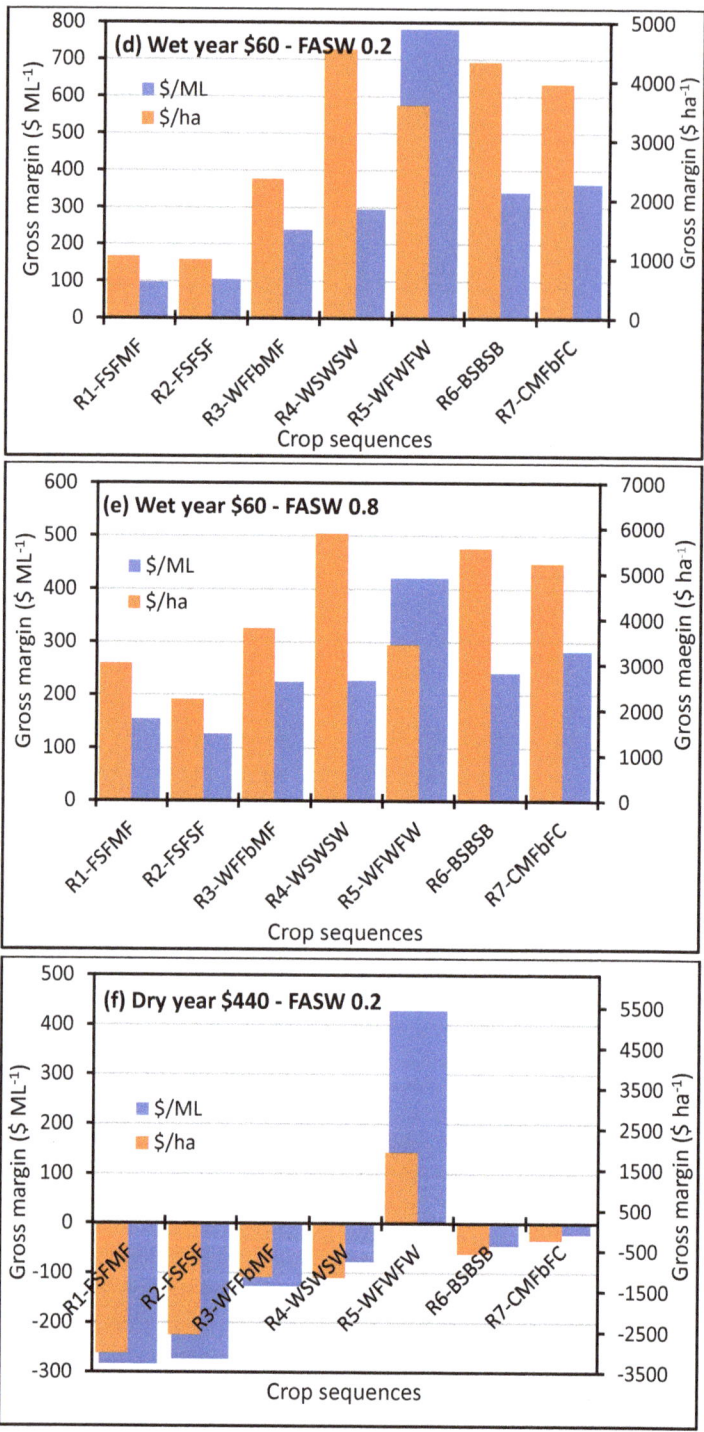

Figure 7. *Cont.*

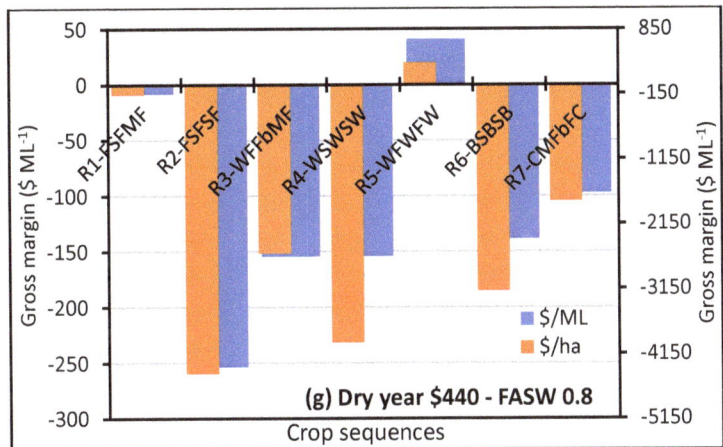

Figure 7. Total gross margin per unit of water and per cultivated land for different water prices ($) and available soil water fractions (FASW) as affected by crop sequences. Wet, medium, and dry refer to years with above average, average, and below average rainfall, respectively.

Comparing the three deficit levels (FASW 0.2, FASW 0.5, FASW 0.8) under the cheap water scenario ($60 ML^{-1}), the highest gross margin per ML was obtained for the WFWFW crop sequence. The gross margin for the three soil water deficit levels FASW 0.2, FASW 0.5, and FASW 0.8 was $781 ML^{-1}, $536 ML^{-1}, and $420 ML^{-1}, respectively. For all of the three water deficit scenarios, the highest per unit area gross margin would be obtained for crop sequences where there is neither summer nor winter fallow (WSWSW and BSBSB). The highest per unit area gross margin was obtained for the FASW 0.8. For example, considering the WSWSW scenario, the gross margin per cultivated land area was 5881 $ ha^{-1}, 5010 $ ha^{-1}, and 4542 $ ha^{-1}, for the FASW 0.8, FASW 0.5, and FASW 0.2 scenarios, respectively.

When water is expensive ($440 ML^{-1}), under all three soil moisture deficit levels, only the WFWFW crop sequence resulted in a positive gross margin. For this crop sequence, under the expensive water scenario, the highest gross margin (as $ ML^{-1}) was when FASW was 0.8. The gross margin as $ ML^{-1} for the three soil moisture deficit scenarios FASW 0.2, FASW 0.5, and FASW 0.8 was 429 $ ML^{-1}, 156 $ ML^{-1}, and 41 $ ML^{-1}, respectively. Per unit of cultivated land area it also followed a similar pattern, 1930 $ ha^{-1}, 1028 $ ha^{-1}, and 335 $ ha^{-1} for FASW 0.2, FASW 0.5, and FASW 0.8, respectively. The biggest loss in gross margin as $ ML^{-1} was for the only-soybean scenario FSFSF. The gross margin losses were −$272 ML^{-1}, −$257 ML^{-1}, and −$253 ML^{-1}, for FASW 0.2, FASW 0.5, and FASW 0.8, respectively. The biggest gross margin loss per ha was also for the soybean-only scenario FSFSF. The gross margin losses were −4457 $ ha^{-1}, −3696 $ ha^{-1}, and −2561 $ ha^{-1} for FASW 0.8, FASW 0.5, and FASW 0.2, respectively.

When water is plentiful and cheap and land is not limited, full irrigation of winter and summer crops that results in high return per unit of water and land area can be practiced. However, when water is limited, partial/deficit irrigation results in better return in $ per ML. A high gross margin per unit area ($ ha^{-1}) is obtained when both summer and winter crops are sown (i.e., no fallow). Crop intensification can minimize the expansion of agricultural land, although its viability depends on attainable crop yield [2].

3.4. Gross Margin under Different Water Price and Soil Water Deficit Scenarios

One of the factors determining the gross margin of an irrigated farm is the price of water, which varies with the amount of rainfall received by the catchments and runoff into reservoirs. There is significant year to year rainfall variability in Australia. Water trade was instituted in Australia to move water from where it has a lower value to where it can be used at its highest value (e.g., for permanent horticulture). The Australian water market is highly complex and occurs across catchments and state boundaries. For this study, 15 years' water price data in the Murrumbidgee catchment were used. Accordingly, the 25 percentile water price ($) was $60 ML^{-1}, the median water price was $124 ML^{-1}, and the 75 percentile water price was $440 ML^{-1}. The gross margin was determined for the three water prices and different irrigation scheduling criteria. Irrigation could be applied at different soil moisture deficit levels between field capacity (drained upper limit) and permanent wilting point (drained lower limit). From Figure 8a it can be seen that having two summer crops (soybean and maize) with winter fallows in between has positive gross margins ($ ML^{-1} and $ ha^{-1}) for 25 percentile and median water prices with the gross margin of the $60 water price being higher than that of the median price $124. When water is cheap, all crop sequence scenarios resulted in positive gross margins (Figure 8a–g). At a higher water price of $440 ML^{-1}, the gross margins were negative. At an intermediate water price ($124 ML^{-1}), all crop sequences resulted in positive gross margins for an FASW of 0.5. The highest gross margin ($ ML^{-1} and $ ha^{-1}) was for the winter crop-only sequence R5-WFWFW, while the lowest was for the summer crop-only sequence R2-FSFSF. A water price of $60 ML^{-1} and FASW 0.8 resulted in the highest yield and profit when winter crops are gown. However, for a summer crop-only sequence or when there is at least one season summer crop, partial irrigations (FASW 0.2 and 0.5) resulted in better gross margins. In seasons with low water allocation and when water is expensive, maximizing crop yields does not necessarily lead to the highest whole-farm gross margins.

When seasonal allocation is low and water is expensive, farmers need to concentrate on fully irrigated winter cropping. This can be seen from Figure 8, where at 440 $ ML^{-1} all of the crop sequence scenarios resulted in negative gross margins both per unit of water and per unit of land area. The only scenario that resulted in a positive gross margin was R5-WFWFW (Figure 4e), where the gross margin at all three soil moisture deficit levels was positive. The gross margin was highest for the full irrigation (FASW 0.8) and lowest for the deficit irrigation scenario (FASW 0.2). The 50% deficit (FASW 0.5) resulted in the intermediate gross margin.

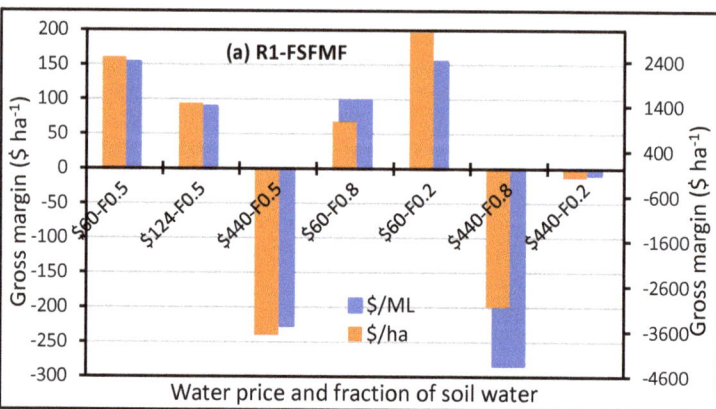

Figure 8. *Cont.*

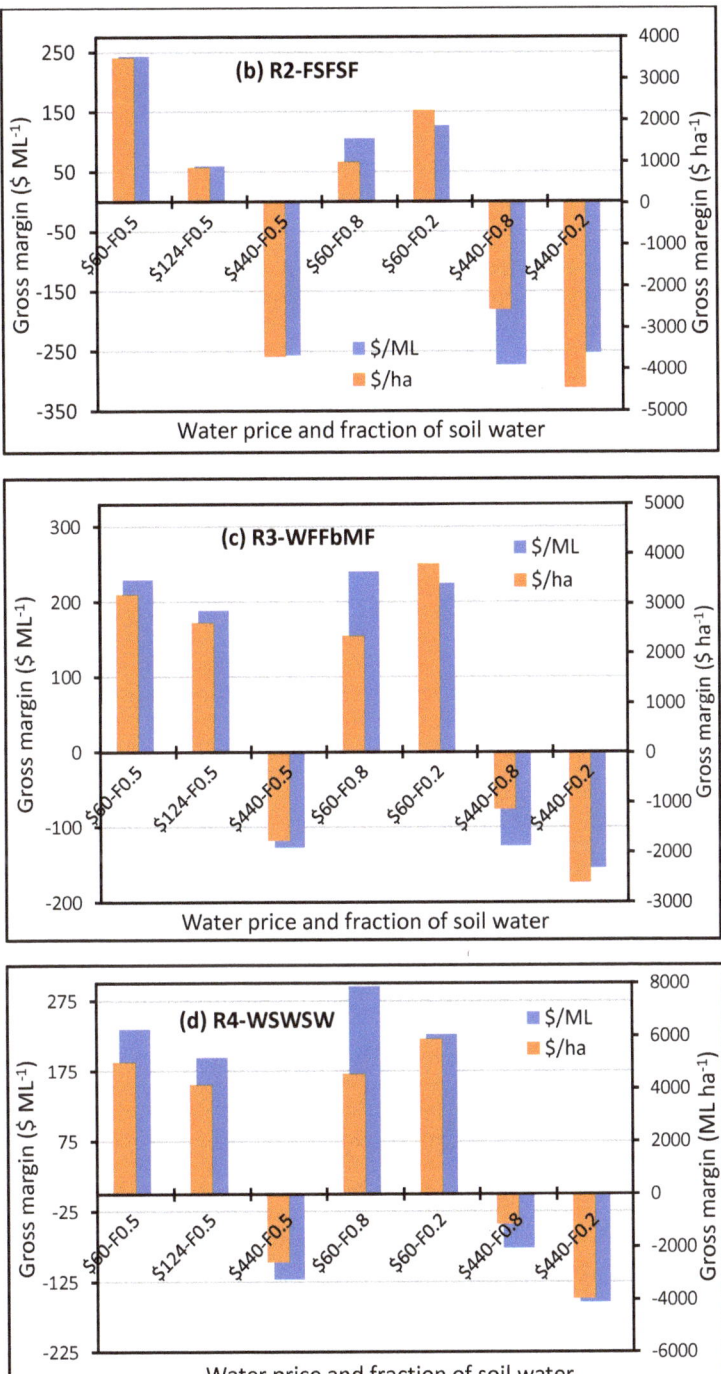

Figure 8. *Cont.*

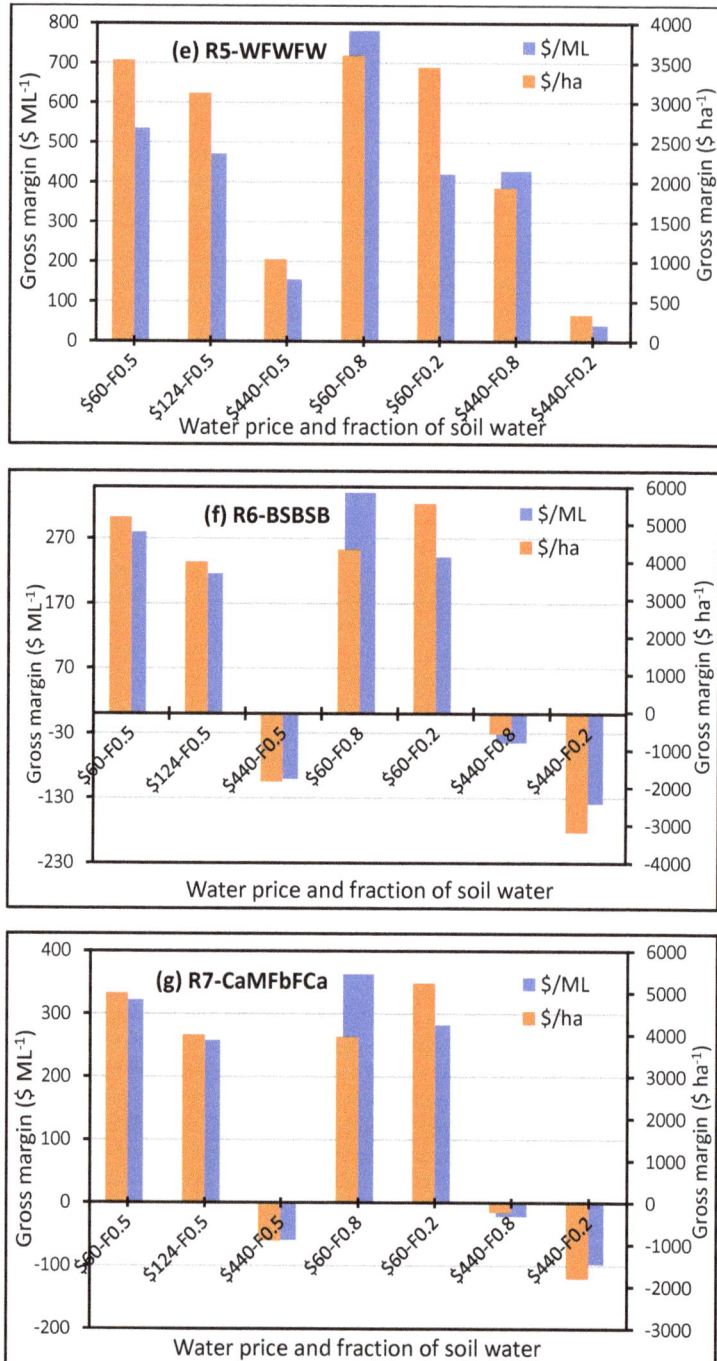

Figure 8. Total gross margin per unit of water and per unit cultivated land for different water prices ($) and available soil water fractions (F).

Historically, irrigation allocation was close to 100%. However, in recent years the allocation has been much lower. This requires maximising crop yield and profit per unit of water. For winter crops, the greatest yield return per unit of water is when about 200–300 mm (2–3 ML ha^{-1}) of irrigation water is applied. Above this amount, the rate of return of yield for each unit of irrigation water applied decreases (diminishing rate of return). The relationship between crop production and transpiration is linear [31]; hence this decreased slope shows that there are other yield-limiting factors. For winter crops, solar radiation and temperature are the liming factors. Even if a higher amount of irrigation water is applied, the crops do not grow in proportion to the amount of water applied. However, in this study, for summer crops the yield continued to increase as the amount of irrigation was increased.

4. Conclusions

This study evaluated different crop sequence, irrigation scheduling and water price scenarios to improve the whole-farm gross margin. APSIM, the agricultural system model used in this study, is a vital tool in prioritizing the use of limited resources such as water and land. The analyses demonstrated that when water supply is not limited and full irrigation is practiced, irrigation needs to be triggered with only small soil water deficits, leading to smaller but frequent irrigations, generally resulting in a higher total irrigation. When full irrigation is not possible due to limited water supply, irrigation should be applied only after the soil water is highly depleted, leading to larger but fewer irrigation events and, generally, a lower amount of total irrigation. These strategies can be used to maximise long-term profitability.

Summer and winter crops show different relationships between water productivity and soil water deficit levels. Summer crops have a parabolic relationship with water productivity, increasing, reaching a maximum, and then decreasing as the amount of irrigation increases. Winter crops, however, have the highest water productivity at the lowest irrigation level.

In the current environment, and more so in the future water-limited environment, supplemental irrigation of winter crops will have more whole farm return compared to fully irrigated intensive winter–summer cropping systems. When water is limited, it would be better to practice supplemental irrigation of a winter crop rather than full irrigation of a summer crop. In hot and dry environments such as the Riverina region, deficit irrigation of summer crops is not a viable option due to high evaporative demand. However, this needs to be decided based on the relative price of summer and winter crops as well. In order to maximise long-term average returns, farm management strategies that vary on a season-by-season basis, based on resource availability, cost, and commodity prices, are required.

The interaction between relative prices of summer and winter crops and different irrigation water allocations was not considered. The gross margin comparison as influenced by the fluctuation in commodity prices needs to be investigated in any future study. Farm return was evaluated only from the water and land value perspective; for example, addition of nitrogen by legumes was not considered.

Author Contributions: Conceptualization, methodology, software: K.Z.; investigation, visualization, writing: K.Z. and J.M. All authors have read and agreed to the published version of the manuscript.

Funding: This research received no external funding.

Institutional Review Board Statement: Not applicable.

Informed Consent Statement: Not applicable.

Acknowledgments: Charles Sturt University for paying the publication processing fee.

Conflicts of Interest: There is no conflict of interest to declare.

References

1. Ray, D.K.; Foley, J.A. Increasing global crop harvest frequency: Recent trends and future directions. *Environ. Res. Lett.* **2013**, *8*, 44041. [CrossRef]
2. Waha, K.; Dietrich, J.P.; Portmann, F.T.; Siebert, S.; Thornton, P.K.; Bondeau, A.; Herrero, M. Multiple cropping systems of the world and the potential for increasing cropping intensity. *Glob. Environ. Chang.* **2020**, *64*, 102131. [CrossRef] [PubMed]
3. Gaydon, D.; Meinke, H.; Rodriguez, D. The best farm-level irrigation strategy changes seasonally with fluctuating water availability. *Agric. Water Manag.* **2012**, *103*, 33–42. [CrossRef]
4. Fereres, E.; Soriano, M.A. Deficit irrigation for reducing agricultural water use. *J. Exp. Bot.* **2007**, *58*, 147–159. [CrossRef] [PubMed]
5. Howell, T.A. Enhancing Water Use Efficiency in Irrigated Agriculture. *Agron. J.* **2001**, *93*, 281–289. [CrossRef]
6. Lorite, I.; Mateos, L.; Orgaz, F.; Fereres, E. Assessing deficit irrigation strategies at the level of an irrigation district. *Agric. Water Manag.* **2007**, *91*, 51–60. [CrossRef]
7. Holzworth, D.P.; Huth, N.I.; Devoil, P.G.; Zurcher, E.J.; Herrmann, N.I.; McLean, G.; Chenu, K.; van Oosterom, E.J.; Snow, V.; Murphy, C.; et al. APSIM—Evolution towards a new generation of agricultural systems simulation. *Environ. Model. Softw.* **2014**, *62*, 327–350. [CrossRef]
8. Zhang, X.; Wang, Y.; Sun, H.; Chen, S.; Shao, L. Optimizing the yield of winter wheat by regulating water consumption during vegetative and reproductive stages under limited water supply. *Irrig. Sci.* **2013**, *31*, 1103–1112. [CrossRef]
9. Serafin, L.; Hertel, K.; Moore, N. *Summer Crop Management Guide*; NSW Department of Primary Industries: Sydney, NSW, Australia, 2019. Available online: https://www.dpi.nsw.gov.au/__data/assets/pdf_file/0011/1187750/SCMG-web-FINAL-5Nov.pdf (accessed on 18 October 2021).
10. Dodig, D.; Kandić, V.; Zorić, M.; Nikolić-Đorić, E.; Nikolić, A.; Mutavdžić, B.; Perović, D.; Šurlan-Momirović, G. Comparative kernel growth and yield components of two- and six-row barley (Hordeum vulgare) under terminal drought simulated by defoliation. *Crop Pasture Sci.* **2018**, *69*, 1215–1224. [CrossRef]
11. Felton, W.L.; Marcellos, H.; Alston, C.; Martin, R.J.; Backhouse, D.; Burgess, L.W.; Herridge, D.F. Chickpea in wheat-based cropping systems of northern New South Wales. II. Influence on biomass, grain yield, and crown rot in the following crop. *Aust. J. Agric. Res.* **1998**, *49*, 401–407. [CrossRef]
12. Matthews, P.; McCaffery, D.; Jenkins, L. *Winter Crop Variety Sowing Guide 2017*; NSW Department of Primary Industries: Sydney, NSW, Australia, 2017. Available online: https://www.dpi.nsw.gov.au/__data/assets/pdf_file/0017/1302173/nsw-dpi-wcvsg-20 21-web.pdf (accessed on 24 September 2021).
13. Kirkegaard, J.A.; Hunt, J.R. Increasing productivity by matching farming system management and genotype in water-limited environments. *J. Exp. Bot.* **2010**, *61*, 4129–4143. [CrossRef] [PubMed]
14. Sissons, M.; Ovenden, B.; Adorada, D.; Milgate, A. Durum wheat quality in high-input irrigation systems in south-eastern Australia. *Crop Pasture Sci.* **2014**, *65*, 411–422. [CrossRef]
15. Jeffrey, S.J.; Carter, J.O.; Moodie, K.B.; Beswick, A.R. Using spatial interpolation to construct a comprehensive archive of Australian climate data. *Environ. Model. Softw.* **2001**, *16*, 309–330. [CrossRef]
16. ApSoil. A Database of Soil Characteristics. 2013. Available online: http://www.apsim.info/Products/APSoil.aspx (accessed on 24 August 2021).
17. Lilley, J.M.; Bell, L.W.; Kirkegaard, J. Optimising grain yield and grazing potential of crops across Australia's high-rainfall zone: A simulation analysis. 2. Canola. *Crop Pasture Sci.* **2015**, *66*, 349–364. [CrossRef]
18. Zeleke, K.; Nendel, C. Analysis of options for increasing wheat (*Triticum aestivum* L.) yield in south-eastern Australia: The role of irrigation, cultivar choice and time of sowing. *Agric. Water Manag.* **2016**, *166*, 139–148. [CrossRef]
19. Archontoulis, S.V.; Miguez, F.E.; Moore, K.J. A methodology and an optimization tool to calibrate phenology of short-day species included in the APSIM PLANT model: Application to soybean. *Environ. Model. Softw.* **2014**, *62*, 465–477. [CrossRef]
20. Turpin, J.E.; Robertson, M.J.; Haire, C.; Bellotti, W.D.; Moore, A.D.; Rose, I. Simulating fababean development, growth, and yield in Australia. *Aust. J. Agric. Res.* **2003**, *54*, 39–52. [CrossRef]
21. Zeleke, K.; Nendel, C. Growth and yield response of faba bean to soil moisture regimes and sowing dates: Field experiment and modelling study. *Agric. Water Manag.* **2019**, *213*, 1063–1077. [CrossRef]
22. Peake, A.S.; Robertson, M.J.; Bidstrup, R.J. Optimising maize plant population and irrigation strategies on the Darling Downs using the APSIM crop simulation model. *Aust. J. Exp. Agric.* **2008**, *48*, 313–325. [CrossRef]
23. Liu, K.; Harrison, M.T.; Hunt, J.; Angessa, T.T.; Meinke, H.; Li, C.; Tian, X.; Zhou, M. Identifying optimal sowing and flowering periods for barley in Australia: A modelling approach. *Agric. For. Meteorol.* **2010**, *282–283*, 107871. [CrossRef]
24. Robertson, M.J.; Kirkegaard, J.A. Water-use efficiency of dryland canola in an equi-seasonal rainfall environment. *Aust. J. Agric. Res.* **2005**, *56*, 1373–1386. [CrossRef]
25. Zeleke, K.; Luckett, D.; Cowley, R. The influence of soil water conditions on canola yields and production in Southern Australia. *Agric. Water Manag.* **2014**, *144*, 20–32. [CrossRef]
26. Napier, T.; Gaynor, L.; Johnston, D.; Morris, G.; Rollin, M. Crop Sequencing for Irrigated Double Cropping—Murrumbidgee Valley Site. GRDC Update. 2016, pp. 95–99. Available online: https://grdc.com.au/resources-and-publications/grdc-update-papers/tab-content/grdc-update-papers/2016/07/crop-sequencing-for-irrigated-double-cropping-murrumbidgee-valley-site (accessed on 8 September 2021).

27. Allen, R.; Pereira, L.S.; Raes, D.; Smith, M. *Crop Evapotranspiration: Guidelines for Computing Crop Water Requirements*; FAO Irrigation and Drainage Paper No 56; FAO: Rome, Italy, 1998; 300p.
28. Osakabe, Y.; Osakabe, K.; Shinozaki, K.; Tran, L.-S.P. Response of plants to water stress. *Front. Plant Sci.* **2014**, *5*, 86. [CrossRef] [PubMed]
29. Soothar, R.K.; Singha, A.; Soomro, S.A.; Chachar, A.-U.; Kalhoro, F.; Rahaman, A. Effect of different soil moisture regimes on plant growth and water use efficiency of Sunflower: Experimental study and modeling. *Bull. Natl. Res. Cent.* **2021**, *45*, 121. [CrossRef]
30. Chai, Q.; Gan, Y.; Zhao, C.; Xu, H.-L.; Waskom, R.M.; Niu, Y.; Siddique, K.H. Regulated deficit irrigation for crop production under drought stress. A review. *Agron. Sustain. Dev.* **2015**, *36*, 3. [CrossRef]
31. Perry, C.; Steduto, P.; Allen, R.G.; Burt, C.M. Increasing productivity in irrigated agriculture: Agronomic constraints and hydrological realities. *Agric. Water Manag.* **2009**, *96*, 1517–1524. [CrossRef]

 *agronomy*

Article

Agronomic Performance of Grain Sorghum (*Sorghum bicolor* (L.) Moench) Cultivars under Intensive Fish Farm Effluent Irrigation

Ildikó Kolozsvári [1], Ágnes Kun [1,*], Mihály Jancsó [1], Andrea Palágyi [2], Csaba Bozán [1] and Csaba Gyuricza [3]

[1] Research Center for Irrigation and Water Management, Institute of Environmental Sciences, Hungarian University of Agriculture and Life Sciences, Anna-Liget Str. 35., H-5540 Szarvas, Hungary; kolozsvari.ildiko@uni-mate.hu (I.K.); jancso.mihaly@uni-mate.hu (M.J.); bozan.csaba@uni-mate.hu (C.B.)
[2] Cereal Research Non-Profit Ltd., Alsókitötő sor 9., H-6726 Szeged, Hungary; andrea.palagyi@gabonakutato.hu
[3] Institute of Agronomy, Hungarian University of Agriculture and Life Sciences, Páter Károly Str. 1., H-2100 Gödöllő, Hungary; gyuricza.csaba@uni-mate.hu
* Correspondence: kun.agnes@uni-mate.hu; Tel.: +36-70-684-1404

Citation: Kolozsvári, I.; Kun, Á.; Jancsó, M.; Palágyi, A.; Bozán, C.; Gyuricza, C. Agronomic Performance of Grain Sorghum (*Sorghum bicolor* (L.) Moench) Cultivars under Intensive Fish Farm Effluent Irrigation. *Agronomy* **2022**, *12*, 1185. https://doi.org/10.3390/agronomy12051185

Academic Editors: Pantazis Georgiou and Dimitris Karpouzos

Received: 28 March 2022
Accepted: 10 May 2022
Published: 14 May 2022

Publisher's Note: MDPI stays neutral with regard to jurisdictional claims in published maps and institutional affiliations.

Abstract: The growing global water shortage is an increasing challenge for the agricultural sector, which aims to produce sufficient quantity and quality of food and animal feed. In our study, effluent water from an intensive African catfish farm was irrigated on grain sorghum plants in four consecutive years. In our study the effects of the effluent on the N, P, K, Na content of the seeds, on the phenological parameters (plant height, relative chlorophyll content), the green mass, and on the grain yield of three varieties ('Alföldi 1', 'Farmsugro 180' and 'GK Emese') were investigated. Five treatments (Körös River (K) water and effluent (E) water: 30 and 45 mm weekly irrigation water dose; non-irrigated control) were applied with micro-spray irrigation. Compared to non-irrigated plants, effluent water did not negatively affect the N, P, K and Na contents of the grain crop. In terms of phenological parameters, the quality of the irrigation water (150–230 cm) had no negative effect on any of the measured parameters compared to the control (133–187 cm) values. In terms of biomass in 2020, grain yields were 89–109 g/plant with variety Alföldi 1, 64–91 g/plant with variety Farmsugro 180, and 86–110 g/plant with variety GK Emese.

Keywords: sorghum; effluent water; irrigation; growth response; mineral content; biomass

1. Introduction

The growing global demand for energy and the high use of fossil fuels are a matter of distress in both the long and short term, as these energy sources are not renewable [1,2]. There may be a shortage of these non-renewable energy sources in the future, which could result in economic and political conflicts between energy-scarce nations. Therefore, there is an increasing urgency to research for renewable energy sources that can meet humanity's energy needs in the long term [3,4]. Hungary is poor in fossil fuels, but half of its area is under arable cultivation, and its agro-ecological characteristics also favor biomass production. For this reason, energy produced as biomass as an alternative energy source may be the main perspective in the future [5]. Areas that do not allow the successful cultivation of other crops can be used to for energy crops. At the same time, they meet the growing conditions for some woody or herbaceous energy plants. The sorghum plant may be a perfect candidate for the production of low-cost biofuels in the future, as its abiotic stress tolerance, diverse genetics, and reliable seed production all contribute to this property [6].

During changes in climatic conditions, drought periods and uneven rainfall distribution become more frequent. It has been described several studies that sorghum has excellent

drought tolerance, with dry-land regions growing more than maize. In drought conditions, sorghum grain absorbs nutrients more efficiently than maize. However, the sorghum crop grown under non-irrigated conditions does not exceed the irrigated crop [7–9]. Declining freshwater supplies and pollution are global problems [10]. According to a study by Mekonnen and Hoekstra [11], approximately four billion people live in water scarcity worldwide, and an estimated five hundred million people live in areas with grave water crisis. For this reason, one of the most significant resources today is water. Nowadays, the biggest challenge is to provide irrigation water for agriculture in the context of the increasingly frequent drought phenomenon. As a consequence of climate change periodically and regionally, there may be a phenomenon in which the surface freshwater supply is insufficient to meet irrigation water demand [12]. The need for irrigation water can be solved by making more optimal use of the available irrigation water. However, in some situations, it may be necessary to use municipal wastewater or agricultural effluent water [13]. Municipal water sources contain lower concentrations of potential pollutants compared to industrial wastewater [14]. During the use of reused water in agriculture, environmental changes that may have a positive or negative property should be monitored [15]. Another source is the irrigation utilization of the effluent of intensive aquaculture systems. Moreover, the effluent is usually rich in organic matter; therefore, the fertilizer doses applied to the production area can also be reduced [16]. At the same time, nutrient accumulation caused by large amounts of organic and inorganic metabolites and residual fish feed should be taken into account when placing effluents in natural recipients [17].

The importance of growing sorghum is increased by the fact that it does not require as intensive plant protection and nutrient replenishment as maize [18]. It is less sensitive to the quality of the area and can be grown successfully in places where other crops make little or no profit in an average year. The uses of sorghum are diverse. Sorghum also plays a significant role in human consumption; in terms of production area, it ranks fifth after maize, rice, wheat and barley among the cereal crops [19]. Sorghum is a plant of physiological type C_4 with high productivity and good drought tolerance [20,21]. Species have good drought tolerance due to their original habitat conditions; indeed, sorghum's gene center is the steppe and savannah region of Africa [22]. It has a water demand of 500–580 mm/year and a transpiration coefficient of 150–250 l/kg dry matter [8]. It can be used as a multi-purpose energy crop in both human food and feed production, although it can also be grown for energy purposes [23,24].

The aim of our study was to determine the growth rate of the sorghum varieties that were irrigated with effluent water from an intensive fish farm with a higher Na content, and to define the biomass production. Our further objective was to determine the concentration of N, P, K and Na elements which were accumulated in plant parts and its effect on the macronutrient content in the soil.

2. Materials and Methods

2.1. Site Description and Climatic Conditions

The field experiment was set up at the Lysimeter Research Station (46°51′49″ N 20°31′39″ E, Szarvas, Hungary) of the Hungarian University of Agriculture and Life Sciences (MATE), Institute of Environmental Sciences (IES), Research Center for Irrigation and Water Management (ÖVKI).

Szarvas is located in one of the warmest and driest areas of Hungary. The climate of the Great Hungarian Plain is characterized by large annual and daily temperature fluctuations, late spring and early autumn frosts, high sunlight, relatively low air humidity, and extremely capricious rainfall conditions. In all four experimental years, the spring was characterized by drought in which the average monthly precipitation did not exceed 49.7 mm (Figure 1). We measured the lowest precipitation in 2019, during which the average annual precipitation was 516.4 mm. In terms of temperature, the warmest year was observed in 2019, where the average annual temperature was 27.1 °C.

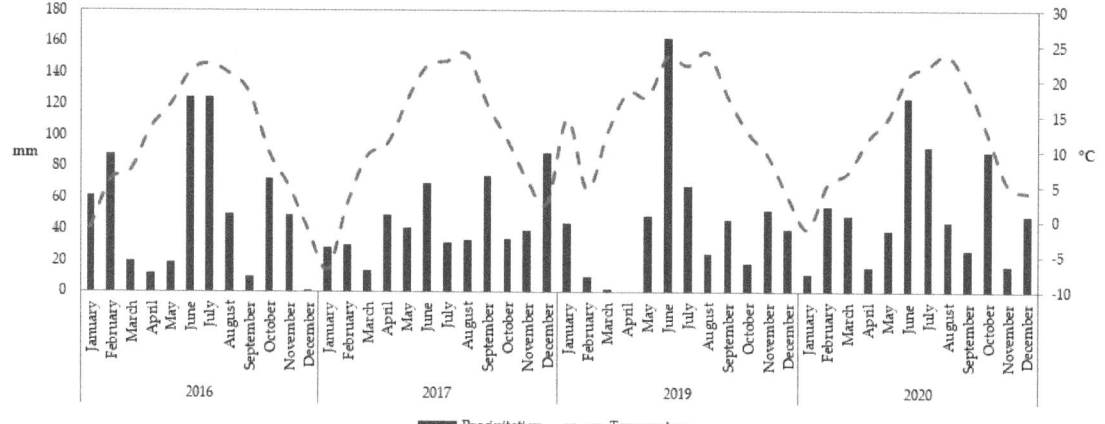

Figure 1. Average mean temperature and precipitation data for the 2016–2020 experimental years.

The soil parameters were analyzed before the start of the study and at the end of the experiments. The soil of the small-plot experiment was Vertisol. The pH of the soil was neutral, total carbonate content and total organic carbon content were low; however, the nutrient supply was high for phosphorus and potassium and moderate for nitrogen (Table 1). Based on the recorded electrical conductivity (EC) and sodium (Na) concentration values, the soil is not saline (Table 1).

Table 1. Soil parameters of the experimental field before the start of the experiment in the 0–30 cm and 30–60 cm soil layers (2016, Szarvas, Hungary).

	Depth of the Soil Layer	
Soil Parameter	**0–30 cm**	**30–60 cm**
pH (KCL)	7.23 ± 0.06	7.15 ± 0.05
Texture	clay loam	clay loam
EC (μS/cm)	410 ± 28	458 ± 30
Total carbonate content (m/m%)	1.96 ± 0.97	1.41 ± 0.51
Total organic carbon content (m/m%)	1.21 ± 0.08	1.33 ± 0.09
KCL-NO_2^- + NO_3^--N (mg/kg)	3.47 ± 0.56	4.42 ± 0.63
AL-P_2O_5 (mg/kg)	2350 ± 607	3013 ± 395
AL-K_2O (mg/kg)	627 ± 137	957 ± 195
AL-Na (mg/kg)	45.0 ± 11.0	56.2 ± 13.6

Comments: EC—specific electrical conductivity of saturated soil paste, KCL—extraction with potassium chloride solution, AL—extraction with ammonium lactate solution.

2.2. The Plant Material and Experimental Design

In the present study, the growth parameters of three registered grain sorghum cultivars ('Alföldi 1', 'Farmsugro 180', 'GK Emese') of the Cereal Research Non-Profit Ltd. (Szeged, Hungary) were monitored using different amounts and qualities of irrigation water. The sowing time was set up when the average soil temperature (at 5 cm deep) reached 12–13 °C in each experimental year (late April or early May). The row spacing was 70 cm and a stocking density was 190–230 thousand plants/hectare (114–138 plants per plot). Each plot contained 4 rows (1 m), and the measurements were performed in the middle two rows (in 6 replicates) in each case. Accordingly, the size of one sample area was 3 m long and 2.1 m wide.

In our experiment, two different types of irrigation water were used (Table 2). One of these was a surface freshwater from the local oxbow lake of the Körös River (46°51′38.6″ N 20°31′28.0″ E, Szarvas, Hungary). The second was an untreated effluent water which was collected from the direct outflow tank of an intensive African catfish farm. The amount of water applied during the experiment is shown in Table 3.

Table 2. Types and average quality parameters of irrigation water used in the irrigation experiment of grain sorghum varieties (Szarvas, Hungary).

	EC	NH$_4$-N	N	P	K	Na	SAR
	(μS/cm)	(mg/L)	(mg/L)	(mg/L)	(mg/L)	(mg/L)	
Effluent water	1307.0	22.5	29.2	3.9	6.4	275.5	12.1
Körös River oxbow lake water	371.0	0.6	2.1	0.2	3.7	31.2	1.2

Comment: SAR—Sodium adsorption ratio.

Table 3. Date of sowing and the available amount of water (irrigation and precipitation) during the growing seasons of grain sorghum in 2016–2020.

	Date of Sowing	Irrigation Water Doses (mm)	Number of Irrigation	Amount of Water Applied by Irrigation (mm)	Precipitation during the Growing Season (mm)	Amount of Additional Irrigation (Körös River) during Germination (mm)	Total Amount of Available Water (mm)
2016	4 May	30 / 45	5	150 / 225	296	120	566 / 641
2017	2 May	30 / 45	6	180 / 270	144	80	404 / 494
2019	7 May	30 / 45	8	240 / 360	208	40	488 / 608
2020	27 April	30 / 45	4	120 / 180	288	90	498 / 558

In that fish farm, the continuous water supply was provided by a flow-through system, and the water was obtained from a geothermal reservoir to ensure minimum (16 °C) water temperature and quality needs for the African catfish. The average daily effluent from the fish farm exceeds 1000 m^3 per day [16]. During the irrigation experiment, two doses of irrigation water (30 and 45 mm) were set on a weekly basis and applied with a micro sprinkler irrigation system. Five treatments were set up, one non-irrigated control (C), two surface water irrigated treatments (K30 and K45), and two effluent treatments (E30 and E45). For each variety, six replicates were set. In the first four weeks after sowing, the plants were irrigated with Körös River water to supplement the precipitation in all treatments to promote germination and initial growth. It had a uniform water condition of 30 mm on a weekly basis, which was ensured by the total amount of precipitation and irrigation water in the Körös. Subsequently, differentiated irrigation was implemented.

2.3. Assay of Phenologycal Parameters and Mineral Content

The phenological measurements of the plants were recorded weekly during the growing seasons. The plant height was measured with a measuring rod at the intersections of the upper two leaves. The Soil Plant Analyses Development (SPAD) index was measured with the Chlorophyll Meter SPAD-502 (Konica Minolta Inc., Tokyo, Japan) on 3 plants per sampling point, on the most advanced leaf, at 4 points per leaf. The SPAD measurement was distributed proportionally along the length of the leaf plate at two points on the right and left sides of the leaf plate. For the determination of biomass, the whole above-ground part of the plant was sampled, when the moisture content of sorghum grains dropped below 20 m/m%. We also measured wet green weight and the weight of grains (in both cases we worked with six replicates).

We performed the studies based on our previously published study of Kolozsvári et al. [16]. The analysis of the soil samples and mineral content of different plant parts was carried out at the end of the growing season assayed by the Hungarian and International Organization for Standardization (ISO) methods. Sodium, phosphorus, and potassium were extracted with nitric acid + hydrogen peroxide and their concentrations were measured using inductively coupled plasma-optical emission spectrometry (ICP-OES) (according to Hungarian standard MSZ 08 1783 28-30:1985). The ISO 5983-2:2009 standard method was used to determine nitrogen. In the analytical studies, we worked with three repetitions.

2.4. Statictical Analyses

IBM SPSS Statistics 25.0 software was used for statistical evaluation. The significant differences between different irrigation treatments and cultivars were determined by one-way analysis of variance ANOVA, where the Tukey's test was considered significant at $p \leq 0.05$. Pearson correlation was used in correlation analysis.

3. Results

3.1. Changes in Soil Parameters during the Experiment

Soil properties were examined in two soil layers. In the sub-soil layer (30–60 cm), there was no significant difference between the different treatments, expect the sodium content (Table 4). The properties of the upper soil layer were changed for five parameters due to irrigation or irrigation water quality. The pH values were significantly higher in the effluent water treatment than in the K30, K45 and control treatments. The highest EC values were measured in the control, and there were no significant differences between the treatments. In case of two macronutrients, phosphorus and potassium, there were less available amounts in the soils in E30 and E45 treatments than in surface water or non-irrigated treatments (Table 4). The AL-Na content was lowest in the control treatment; for the irrigated soil samples, the value in E45 was significantly higher than others.

Table 4. Soil parameters of the experimental area sampled in the final year of the irrigation experiment. Average soil parameters data are presented from five treatments. Results are means $\pm$ SD, $n = 6$. Different letters show significant differences among irrigation water qualities for the four cultivation years, corroborating to the Tukey's test at $p \leq 0.05$.

2020	E30	E45	K30	K45	C	
			0–30 cm			p^1
pH $_{(KCl)}$	7.28 ± 0.04 b	7.30 ± 0.06 b	7.22 ± 0.02 ab	7.25 ± 0.04 ab	7.18 ± 0.03 a	*
EC (µS/cm)	355 ± 38 a	352 ± 38 a	411 ± 38 ab	402 ± 38 ab	464 ± 38 b	**
Total carbonate content (m/m%)	2.31 ± 0.67	2.02 ± 0.93	1.90 ± 0.23	2.35 ± 0.23	1.43 ± 0.24	n.s.
Total organic carbon (m/m%)	1.09 ± 0.08	1.08 ± 0.05	1.18 ± 0.05	1.16 ± 0.04	1.23 ± 0.07	n.s.
KCL-NO$_2^-$ + NO$_3^-$-N (mg/kg)	3.01 ± 0.45	3.45 ± 0.81	3.44 ± 0.95	3.85 ± 1.04	3.42 ± 0.64	n.s.
AL-P$_2$O$_5$ (mg/kg)	1243 ± 262 a	1247 ± 159 a	1880 ± 285 ab	1543 ± 179 bc	2300 ± 225 c	**
AL-K$_2$O (mg/kg)	350 ± 75 a	333 ± 45 a	475 ± 67 ab	460 ± 31 ab	509 ± 41 b	**
AL-Na (mg/kg)	113.7 ± 18.9 ab	122.9 ± 21.3 b	89.2 ± 21.0 ab	82.1 ± 17.9 a	86.8 ± 23.3 a	**
			30–60 cm			p^1
pH $_{(KCl)}$	7.13 ± 0.10	7.08 ± 0.03	7.16 ± 0.06	7.17 ± 0.07	7.12 ± 0.05	n.s.
EC (µS/cm)	458 ± 38	476 ± 38	464 ± 38	452 ± 38	451 ± 38	n.s.
Total carbonate content (m/m%)	1.92 ± 1.16	1.41 ± 0.74	1.74 ± 0.52	1.91 ± 1.10	1.40 ± 0.47	n.s.
Total organic carbon (m/m%)	1.30 ± 0.18	1.42 ± 0.06	1.29 ± 0.07	1.30 ± 0.14	1.40 ± 0.06	n.s.
KCL-NO$_2^-$ + NO$_3^-$-N (mg/kg)	3.55 ± 0.32	3.37 ± 0.31	3.22 ± 0.92	3.26 ± 0.66	3.53 ± 0.36	n.s.
AL-P$_2$O$_5$ (mg/kg)	2100 ± 447	2447 ± 716	2373 ± 234	2127 ± 318	3010 ± 828	n.s.
AL-K$_2$O (mg/kg)	612 ± 114	607 ± 94	635 ± 93	727 ± 66	640 ± 68	n.s.
AL-Na (mg/kg)	122.2 ± 14.7 ab	127.0 ± 18.2 b	88.9 ± 20.0 a	95.6 ± 24.2 ab	89.9 ± 20.0 a	**

Comment: p^1: represent p-value. * $p < 0.05$, ** $p < 0.01$, n.s.—not significant. EC—specific electrical conductivity of saturated soil paste, KCL—extraction with potassium chloride solution, AL—extraction with ammonium lactate solution.

3.2. Results of the Main Macroelement Content of Sorghum Plant Parts

3.2.1. Changing of Nitrogen Content in Different Plant Parts

Figure 2a shows the nitrogen (N) analyses of the grains of the 'Alföldi 1' variety during the experimental four years. The grains show a decreasing trend every year. The highest value is shown by the control samples (2.4 m/m%), while the lowest nitrogen content was in the K45 (1.8 m/m%) treatment in 2016. In the 2017 growing year, the samples were characterized by a balanced nitrogen content. The results obtained in the next two years converge, where it can be seen that the control (C) values were the lowest and the samples irrigated with effluent irrigation water had the highest nitrogen content. There was only a significant difference in the first year, where the E45 treatment ($p = 0.005$) had a significantly lower N element content compared to the C and K30 treatments.

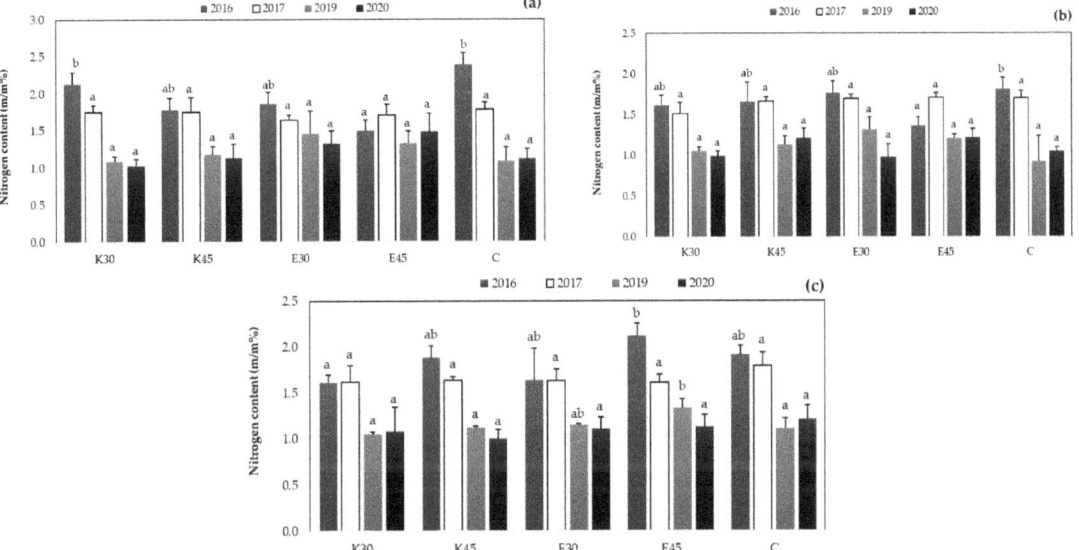

Figure 2. Nitrogen content of sorghum grains from 2016 to 2020. (**a**) The nitrogen content of 'Alföldi 1' variety; (**b**) the nitrogen content of 'Farmsugro 180' variety; (**c**) the nitrogen content of 'GK Emese' variety. The mean nitrogen content data are presented in five treatments. The results are means ± SD, $n = 3$. The different letters establish significant differences between irrigation water qualities for the four vegetation periods, corroborating to the Tukey's test at $p \leq 0.005$.

The Figure 2b shows the development of nitrogen content in the grain yield of 'Farmsugro 180'. In the first two growing years, almost the same N-element content was measured in the grains, during which the C treatment had the highest concentration (1.7–1.8 m/m%). In addition, a decrease can be observed in 2019 and 2020. No significant difference can be described; however, it can be observed that higher N content was measurable for irrigated treatments. The significant difference was characteristic of the first year of cultivation, where we measured a significantly lower N content in the E45 treatment ($p = 0.035$) compared to the C treatment.

In the 'GK Emese' variety (Figure 2c), it can be stated that the highest N content in the grains occurred in the first year of cultivation. The highest values were measured for the E45 treatment (2.1 m/m%) and the lowest for the K30 sample (1.0 m/m%). In 2017, there was some decrease in plant samples for all treatments. Almost the same course can be observed in the last two growing years. We measured a significant difference in 2016 and 2019. The K30 treatment ($p = 0.047$) had significantly less N content compared to the E 45 treatment with the highest value. In the latter case, we also measured a significantly higher nitrogen content in the plant sample of the 45 mm irrigated effluent treatment compared to the K30 ($p = 0.005$), C ($p = 0.018$) and K45 ($p = 0.029$) samples.

3.2.2. Changing of Phosphorus Content in Sorghum Plant Part

Figure 3a shows the phosphorus (P) content measured in the grains of the 'Alföldi 1' grain sorghum variety. The phosphorus values were between 2700 and 3700 mg/kg dry matter (d.m.) during the four years of cultivation. In 2016, treatment C had the highest concentration (3700 mg/kg d.m.), while K30 treatment had the lowest concentration (3190 mg/kg d.m.). In the second year of cultivation, the highest phosphorus concentration was measured in the C treatment plant sample, and the lowest in the E45 treatment. In the experimental years of 2019 and 2020, a small decrease was observed, especially for treatment C, where the values were around 3000 and 2700 mg/kg d.m. We measured significant differences between the treatments, except for the last year. In 2016, K30 ($p = 0.028$) and E45 ($p = 0.046$) treatments had significantly less phosphorus content compared to the C treatment with the highest value. In the second year, we also measured significantly lower values for E45 ($p = 0.003$), E30 ($p = 0.047$), and K45 ($p = 0.049$) samples compared to the C treatment. In 2019, the phosphorus value of the E30 treatment proved to be the highest, where we measured a significantly lower value than the C sample ($p = 0.045$).

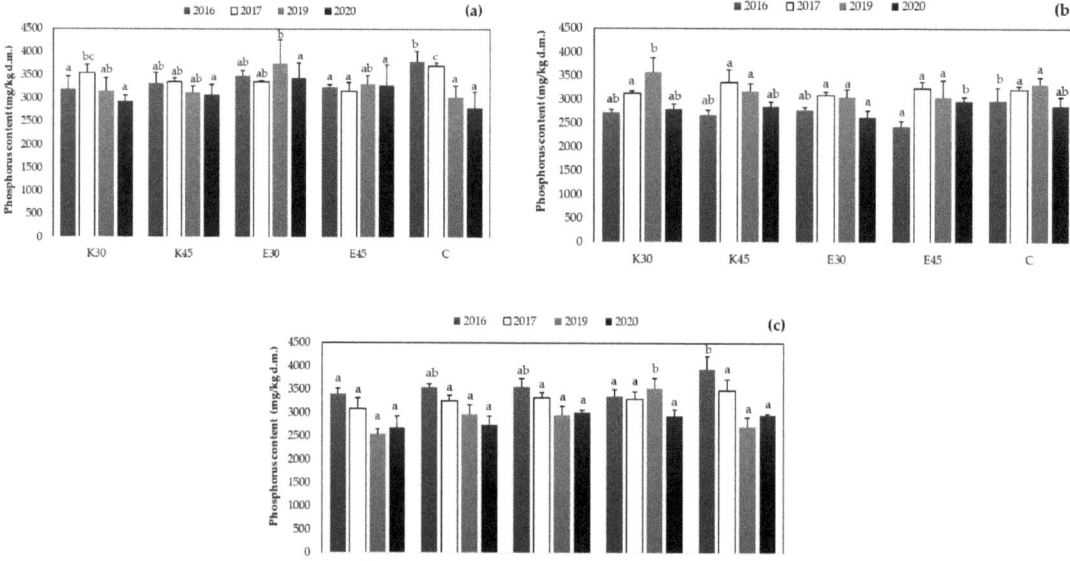

Figure 3. Phosphorus content values of sorghum grains from 2016 to 2020. (**a**) The phosphorus content of 'Alföldi 1' variety; (**b**) the phosphorus content of 'Farmsugro 180' variety; (**c**) the phosphorus content of 'GK Emese' variety. The mean phosphorus content data are presented in five treatments. The results are means ± SD, $n = 3$. The different letters establish significant differences between irrigation water qualities for the four vegetation periods, corroborating to the Tukey's test at $p \leq 0.005$.

Lower phosphorus values were measured for the 'Farmsugro 180' variety in the first experimental year compared to the other years (Figure 3b). The highest value was recorded by C treatment (2987 mg/kg d.m.), while the lowest was recorded by E45 treatment (2430 mg/kg d.m.). In the following two experimental years, the values show an upward trend. However, a repeated decline in phosphorus levels can be observed in 2020. The significant differences were detected between the applied treatments, with the exception of the year 2017. In 2016, we measured a significantly lower concentration for the E45 ($p = 0.004$) sample compared to the value of treatment C. In 2019, the K30 ($p = 0.043$) sample had a significantly higher phosphorus concentration. Compared to the E45 sample with the highest value, the last year was characterized by significantly lower phosphorus content in E30 ($p = 0.048$).

In Figure 3c, the phosphorus value of the grains of the 'GK Emese' cultivar ranged from 2540 to 3950 mg/kg d.m. In the first year of cultivation, E45 treatment had the lowest concentration, while C treatment had the highest concentration. The difference between the measured values exceeded 600 mg/kg d.m. The year 2017 was characterized by equalized values. In the last two years, a decrease in phosphorus levels was observed, where the K30 treatment had the lowest value and the E45 treatment had the highest concentration. There was no significant difference between the treatments in 2017 and 2020. However, in 2016, treatments E45 ($p = 0.015$) and K30 ($p = 0.023$) contained significantly less phosphorus capable of C treatment with the highest P levels. Furthermore, we measured significantly more phosphorus in the E45 treatment in 2019.

3.2.3. Changing of Potassium Content in Sorghum Plant Part

In the first year of cultivation, the potassium (K) levels of the grains of the 'Alföldi 1' variety ranged from 4020 to 5000 mg/kg d.m. (Figure 4a). The K30 treatment had the lowest value and the E45 treatment had the highest value. In 2017, a decrease in potassium levels was observed for all treatments. Subsequently, in the last two growing years, the values showed a nearly identical trend, where we measured higher potassium levels with K30 treatment. We detected a significant difference between the treatments in the first two experimental years. In 2016, E45 ($p = 0.044$) treatment had significantly more potassium content than the others. Furthermore, E30 ($p = 0.016$) had significantly less K element content compared to the C treatment with the highest value in 2017.

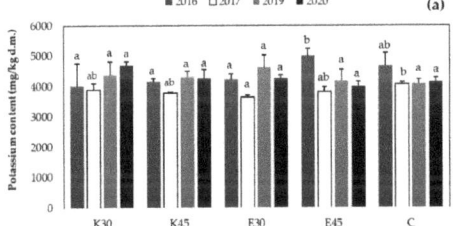

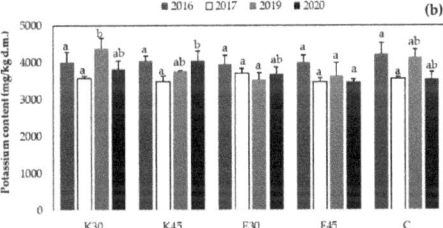

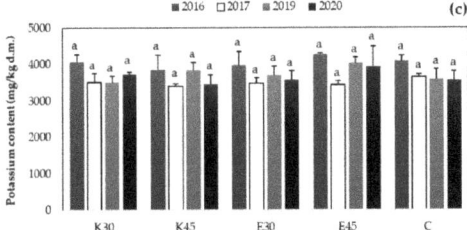

Figure 4. Potassium content values of sorghum grains from 2016 to 2020. (**a**) The potassium content of 'Alföldi 1' variety; (**b**) the potassium content of 'Farmsugro 180' variety; (**c**) the potassium content of 'GK Emese' variety. The mean potassium content data are presented five treatments. The results are means ± SD, $n = 3$. The different letters establish significant differences between irrigation water qualities for the four vegetation periods, corroborating to the Tukey's test at $p \leq 0.005$.

In the case of the 'Farmsugro 180' cultivar, it can also be described that the samples were characterized by elevated K levels in the first year. It is also typified by balanced values (Figure 4b). In 2017, potassium levels ranged from 3490–3580 mg/kg d.m. The samples from 2019 and 2020 showed significant differences. The E30 (p = 0.010) and E45 (p = 0.021) treatments had significantly less K content compared to the highest value of K30 in 2019. In the last year, E45 (p = 0.029) had significantly less K content compared to the highest value of K45 treatment.

In case of 'GK Emese', we detected a higher K level in the samples, which was between 3843 and 4266 mg/kg d.m. in the first year of cultivation (Figure 4c). In 2017, the potassium content in the grains was similar to the other varieties as some decrease can be observed. The K content of the last two years was balanced, with nearly the same values, where treatment C showed a decreased potassium content (3580 and 3560 mg/kg d.m.). The highest values were measured for the E45 treatment in both years. In the evaluation of the K-level measured in the grains of the 'GK Emese' variety, no significant difference between the treatments was detected in any of the examined cultivation years.

3.2.4. Changing of Sodium Content in Sorghum Plant Part

In 2016, the sodium (Na) content detected in the grains of the 'Alföldi 1' variety proved to be the lowest, where its measured values ranged from 29 to 34 mg/kg d.m (Figure 5a). Some decreases were observed in the 2017 and 2019 cultivation years; however, even the highest Na content does not exceed 44 mg/kg d.m. In the last experimental year, a significant increase in Na levels was observed for all treatments. In this case, treatments E45 and C exceed 100 mg/kg d.m. The effects of the treatments were compared within that cultivation year, where no significant differences were detected.

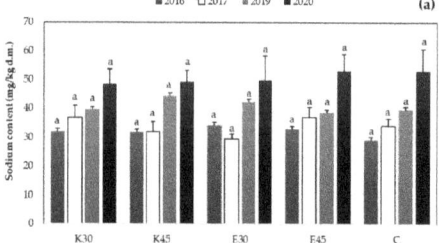
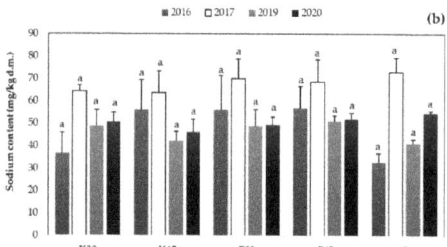

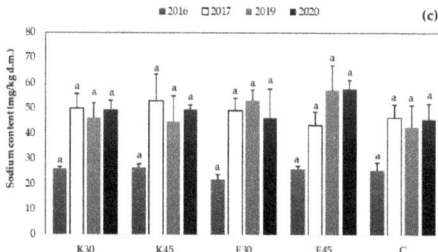

Figure 5. Sodium content values of sorghum grains from 2016 to 2020. (**a**) The sodium content of 'Alföldi 1' variety; (**b**) the sodium content of 'Farmsugro 180' variety; (**c**) the sodium content of 'GK Emese' variety. The mean sodium content data are presented five treatments. The results are means $\pm$ SD, n = 3. The different letters establish significant differences between irrigation water qualities for the four vegetation periods, corroborating to the Tukey's test at $p \leq 0.005$.

In the case of the 'Farmsugro 180' cultivar, the Na content measured in the grains was higher than in the case of the 'Alföldi 1' cultivar. In the first year of cultivation, treatment C had the lowest value of 32 mg/kg d.m., while treatments K45, E30 and E45 had the highest values (56 mg/kg d.m.) (Figure 5b). The following year was marked by growth. The highest sodium values in the control samples were 73 mg/kg d.m. and the lowest was 63 mg/kg d.m. for K45 treatment. In 2019, a repeated decrease was observed in the samples, where they ranged from 41 to 49 mg/kg d.m. In 2020, a remarkable increase was observed between 92 and 109 mg/kg d.m. There were no significant differences for this variety when comparing the treatments in that year.

The level of Na content measured in the grains was similar for the 'GK Emese' cultivar (Figure 5c). In 2016, a balanced sodium level was measured for each treatment, during which it did not exceed 26 mg/kg d.m. In the second year of cultivation, samples irrigated with water from the Körös oxbow lake represented higher Na levels. In 2019, however, samples irrigated with effluent showed higher values (53 and 57 mg/kg d.m.). It is also observed in this variety that the Na content in the grains increased notably in 2020, especially in the case of the E45 sample, where it reached 115 mg/kg d.m. In this case, there was no significant difference between the treatments in the statistical evaluation.

3.3. Phenological Results

3.3.1. Development of Relative Chlorophyll Content

Table 5 shows the average chlorophyll values. In the experimental year of 2016, the highest SPAD value in the leaf parts was measured in the case of the C treatment in the 'Alföldi 1' variety, while the lowest was detected in the case of the E45 treatment. A decrease was observed for all treatments in the following year. The SPAD value developed similarly in 2019 and 2020, with the highest measured at E45 and the lowest at K30. There was no significant difference between treatments for the first two years. In the third and the fourth year, E30 ($p = 0.000$), K30 ($p = 0.000$), and C ($p = 0.001$) treatments had significantly lower SPAD values compared to the highest E45 treatment. By 2020, the chlorophyll content of the leaves was also significantly lower in the E30 ($p = 0.001$), K30 ($p = 0.002$) and C ($p = 0.001$) treatments compared to the E45 value.

Table 5. Chlorophyll values over the four experimental years. Average chlorophyll data are presented from five treatments. Results are means $\pm$ SD, $n = 30$. Different letters recommend significant differences among irrigation water qualities for the four cultivation years, corroborating to the Tukey's test at $p \leq 0.05$.

Variety of Grain Sorghum	Treatments	Average SPAD Values			
		2016	2017	2019	2020
'Alföldi 1'	K30	50.7 ± 5.8 a	51.7 ± 6.2 a	42.8 ± 6.7 a	42.5 ± 6.1 a
	K45	52.6 ± 7.5 a	52.1 ± 11.6 a	48.5 ± 9.0 bc	49.9 ± 8.8 bc
	E30	52.6 ± 8.0 a	49.0 ± 6.7 a	44.0 ± 7.2 ab	44.5 ± 7.7 ab
	E45	50.0 ± 6.1 a	49.1 ± 6.7 a	52.4 ± 4.5 c	53.2 ± 4.7 c
	C	52.8 ± 9.1 a	49.8 ± 7.8 a	44.3 ± 9.2 ab	45.3 ± 9.4 ab
'Farmsugro 180'	K30	46.3 ± 7.3 a	49.8 ± 5.7 a	44.8 ± 5.1 b	45.6 ± 5.5 b
	K45	50.0 ± 9.0 a	51.0 ± 5.8 a	39.1 ± 5.4 a	41.0 ± 5.6 a
	E30	49.5 ± 10.1 a	50.7 ± 6.2 a	38.9 ± 5.0 a	41.1 ± 4.6 a
	E45	47.7 ± 8.8 a	52.5 ± 6.9 a	47.1 ± 5.1 b	48.0 ± 4.9 b
	C	47.6 ± 10.2 a	49.8 ± 7.4 a	46.8 ± 5.0 b	47.7 ± 4.6 b
'GK Emese'	K30	51.9 ± 7.0 a	49.2 ± 5.7 a	45.7 ± 6.3 b	45.5 ± 6.2 b
	K45	56.5 ± 7.3 a	50.1 ± 7.3 a	40.0 ± 7.3 a	37.8 ± 8.0 a
	E30	52.4 ± 7.2 a	47.3 ± 7.6 a	41.0 ± 6.9 ab	38.0 ± 10.2 a
	E45	53.1 ± 7.1 a	48.6 ± 6.8 a	44.8 ± 6.8 b	45.0 ± 7.5 b
	C	55.8 ± 7.3 a	48.4 ± 7.5 a	42.8 ± 6.7 ab	43.0 ± 6.6 ab

Comments: SPAD—Soil Plant Analyses Development.

Regarding the 'Farmsugro 180' variety, it can be observed that in all cases except the first year, E45 treatment had the highest SPAD value, especially in 2017, where it reached 52.5. Of the four experimental years, the lowest mean chlorophyll values were detected in 2019 (38.9–47.1), while the highest was measured in 2017 (49.8–52.5). During the one-way statistical evaluation, there was no significant difference between treatments in the first two years. In 2019 and 2020, E30 and K45 ($p = 0.000$) treatments had significantly lower chlorophyll values compared to E45 treatment with the highest SPAD data.

For the 'GK Emese' sorghum variety, the highest chlorophyll value was measured in 2017 out of the four experimental years (51.9–56.5), while the lowest was in 2020 (38.8–45.5). It was characteristic of the first two cultivation years that the treatments irrigated with the water of the Körös River had a higher SPAD value, especially the K45 treatments. However, in 2019 and 2020, plants with K30 treatment had the highest chlorophyll value. In this case, it can be described that there was no significant difference between the treatments during the first two years of the statistical analysis. In the year 2019, the K45 ($p = 0.012$) treatment had a significantly lower SPAD value compared to the value measured at the highest K30 treatment. In 2020, E45 ($p = 0.003$) and K45 ($p = 0.002$) treatments also had significantly lower chlorophyll values compared to the highest K30 treatment.

3.3.2. Determination of Growth Parameter during the Seasons

In the case of the 'Alföldi 1' variety, it can be observed that the height values measured during the first year exceeded the measurements of the following years (Figure 6a). For the E45 treatment, we measured the highest plants where they reached 199 cm. Plants with the E30 treatment grew to the lowest one (155 cm). In the following year, a decrease in height was observed for all treatments. Height values from 2019 to 2020 were detected in nearly the same range. As in previous years, the plants grew smaller with E30 treatment and the largest with E45 treatment (156–169 cm). The significant difference between the treatments was detected only in the first year of cultivation, where sorghum plants grew significantly smaller ($p \leq 0.01$) than all other treatments compared to the E45 treatment with the highest value.

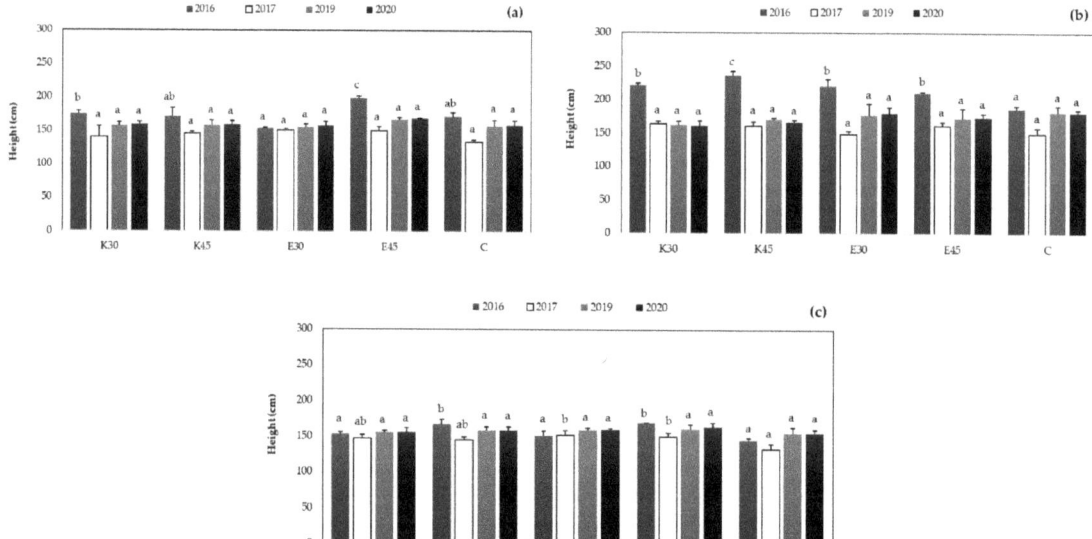

Figure 6. Plant height data measured at the last evaluation time of sorghum grains from 2016 to 2020. (**a**) Height values of 'Alföldi 1' variety; (**b**) height values of 'Farmsugro 180' variety; (**c**) height values of 'GK Emese' variety. The mean height data are presented five treatments. The results are means ± SD, $n = 3$. The different letters establish significant differences between irrigation water qualities for the four vegetation periods, corroborating to the Tukey's test at $p \leq 0.005$.

For 'Farmsugro 180' (Figure 6b), higher plant heights were also detected in the first year of cultivation, with the lowest plants in treatment C (187 cm) and the highest in treatment K45 (236 cm). In the following years, a decrease was observed with all treatments. In 2017, the height of the 'Farmsugro 180' variety was between 149 and 164 cm. In 2019 and 2020, an increase of 10% was observed, during which the height ranged from 161 to 182 cm. In 2016, a significant difference was found between the treatments, where plants of other treatments grew significantly lower ($p \leq 0.05$) compared to K45 treatment.

The 'GK Emese' variety is also characterized by the finding that the height values measured in 2016 exceeded those in other years (Figure 6c). The measured values were between 140–170 cm in the first year and 133–152 cm in the second year. In all four years of cultivation, treatment C was found to be the lowest and treatment E45 to be the highest. In the last two years, a decrease can be observed, where the measured values were between 156–162 cm. In the first two years there was a significant difference between treatments. In 2016, C ($p = 0.001$) E30 ($p = 0.008$) and K30 ($p = 0.016$) treatments proved to be significantly lower than E45. In the second year, significantly lower values were detected for C treatment ($p = 0.018$) compared to the highest E30 treatment.

3.4. Development of Biomass Product over the Four Experimental Years

3.4.1. Development of the Green Mass of the Three Sorghum Cultivars during the Experimental Years

The development of the wet green mass of grain sorghum varies from variety to variety. In the case of the 'Alföldi 1' variety, the 2017 growing year proved to be the most productive in terms of wet green mass, during which the E45 treatment reached 476 g/plant (Table 6). In the same year, the lowest weight was measured for treatment C, which was 349 g/plant. In addition, the green mass of the first year of cultivation proved to be the lowest, where the E45 treatment (365 g/plant) had the highest biomass and E30 treatment had the lowest (296 g/plant) value. In the one-way analysis of variance, there was no significant difference when comparing treatments over the years.

Table 6. Development of wet green mass at harvest between 2016–2020 growing seasons. Average green mass weight data are presented from five treatments. Results are means ± SD, $n = 6$. Different letters recommend significant differences among irrigation water qualities for the four cultivation years, corroborating to the Tukey's test at $p \leq 0.05$.

Variety of Grain Sorghum	Applied Treatments	2016		2017		2019		2020	
		Green Mass Weight (g/Plant) Mean ± SD	*p*-Value	Green Mass Weight (g/Plant) Mean ± SD	*p*-Value	Green Mass Weight (g/Plant) Mean ± SD	*p*-Value	Green Mass Weight (g/Plant) Mean ± SD	*p*-Value
'Alföldi 1'	K30	301 ± 9 a		422 ± 79 a		315 ± 58 a		379 ± 104 a	
	K45	302 ± 49 a		451 ± 92 a		357 ± 54 a		420 ± 123 a	
	E30	296 ± 62 a	n.s	436 ± 100 a	n.s.	386 ± 81 a	n.s.	402 ± 31 a	n.s.
	E45	365 ± 108 a		476 ± 117 a		329 ± 93 a		376 ± 173 a	
	C	320 ± 80 a		349 ± 89 a		397 ± 70 a		269 ± 55 a	
'Farmsugro 180'	K30	583 ± 81 b		427 ± 48 a		331 ± 36 a		397 ± 80 a	
	K45	528 ± 96 ab		489 ± 105 a		308 ± 52 a		383 ± 62 a	
	E30	539 ± 91 ab	*	460 ± 65 a	n.s.	346 ± 37 a	n.s.	422 ± 107 a	n.s.
	E45	433 ± 84 a		458 ± 71 a		330 ± 33 a		461 ± 60 a	
	C	540 ± 58 ab		411 ± 74 a		321 ± 60 a		448 ± 53 a	
'GK Emese'	K30	244 ± 70 a		427 ± 98 a		435 ± 74 ab		347 ± 86 a	
	K45	229 ± 47 a		418 ± 74 a		531 ± 83 abc		303 ± 65 a	
	E30	256 ± 20 a	n.s	401 ± 75 a	n.s.	573 ± 96 c	**	335 ± 70 a	n.s.
	E45	209 ± 25 a		421 ± 61 a		555 ± 59 bc		401 ± 140 a	
	C	227 ± 27 a		339 ± 67 a		421 ± 63 a		330 ± 56 a	

Comments: * $p < 0.05$, ** $p < 0.01$, n.s.—not significant.

For the 'Farmsugro 180' cultivar, the highest green weight was detected in the first year of cultivation, reaching 583 g/plant for the K30 treatment, while the E45 treatment with the lowest weight value 433 g/plant. The plants of 2019 had the lowest weight, where their measured values ranged from 308 to 346 g/plant. A significant difference between the treatments was detected in the first year of cultivation ($p \leq 0.05$).

In the case of the 'GK Emese' cultivar, the year 2016 had the lowest plant weight, where the highest measured values did not exceed 256 g/plant. However, in 2019, these values increased for all treatments, with the lowest C treatment reaching 421 g/plant and the E30 treatment having the highest plant weight of 573 g/plant, although there was no significant difference between treatments during the statistical annual evaluation.

3.4.2. Improvement of the Grain Yield of the Three Sorghum Cultivars during the Experimental Years

In the course of the change in the weight of the grain yield of the 'Alföldi 1' variety, the smallest grains yield was detected in the year 2019, which was between 67–91 g/plant (Table 7). The highest grain yield values were measured in 2016, where the C treatment had the lowest grain yield of 82 g/plant and the E30 samples had the highest grain yield of 138 g/plant. During the one-way analysis of variance, there was a significant difference between treatments only in 2016 ($p \leq 0.001$).

Table 7. Development of wet grains yield at harvest between 2016–2020 growing seasons. Average grains yield data are presented from five treatments. Results are means $\pm$ SD, $n = 6$. Different letters recommend significant differences among irrigation water qualities for the four cultivation years, corroborating to the Tukey's test at $p \leq 0.05$.

Variety of Grain Sorghum	Applied Treatments	2016		2017		2019		2020	
		Grain Yield (g/Plant) Mean $\pm$ SD	p-Value	Grain Yield (g/Plant) Mean $\pm$ SD	p-Value	Grain Yield (g/Plant) Mean $\pm$ Std. Deviation	p-Value	Grain Yield (g/Plant) Mean $\pm$ Std. Deviation	p-Value
'Alföldi 1'	K30	125 ± 16 bc		123 ± 18 a		68 ± 17 a		102 ± 38 a	
	K45	128 ± 21 c		115 ± 9 a		83 ± 21 a		109 ± 53 a	
	E30	138 ± 17 c	***	116 ± 17 a	n.s.	91 ± 16 a	n.s.	105 ± 7 a	n.s.
	E45	100 ± 14 ab		120 ± 24 a		67 ± 11 a		106 ± 42 a	
	C	82 ± 10 a		104 ± 22 a		83 ± 11 a		89 ± 46 a	
'Farmsugro 180'	K30	75 ± 6 a		94 ± 11 a		88 ± 16 a		67 ± 13 ab	
	K45	86 ± 2 b		105 ± 25 a		88 ± 13 a		57 ± 12 a	
	E30	88 ± 3 b	***	90 ± 19 a	n.s.	96 ± 17 a	n.s.	64 ± 23 ab	*
	E45	87 ± 3 b		81 ± 8 a		93 ± 16 a		81 ± 21 ab	
	C	70 ± 6 a		86 ± 13 a		94 ± 12 a		91 ± 21 b	
'GK Emese'	K30	140 ± 13 a		109 ± 27 a		80 ± 18 a		94 ± 22 a	
	K45	129 ± 26 a		106 ± 17 a		75 ± 18 a		77 ± 13 a	
	E30	110 ± 52 a	n.s.	112 ± 15 a	n.s.	77 ± 15 a	n.s.	86 ± 19 a	n.s.
	E45	127 ± 15 a		107 ± 17 a		87 ± 9 a		110 + 46 a	
	C	100 ± 19 a		98 ± 13 a		73 ± 19 a		94 ± 15 a	

Comments: * $p < 0.05$, *** $p < 0.001$, n.s.—not significant.

For the 'Farmsugro 180' variety, the grain yield was the lowest in the last experimental year, during which the measured value was only 57 g/plant for the K45 treatment. However, the highest values were detected in 2017, during which the grain yield of sorghum was the lowest for the E45 (81 g/plant) treatment and the highest for the K45 (105 g/plant) treatment. There was no verifiable significant difference between the treatments in any of the years.

Compared to the other cultivars, we measured a higher amount of grain yield in the case of the 'GK Emese' every year (Table 7), during which the lowest values measured in 2019 exceeded it (77–110 g/plant). The year 2016 had the highest weight value, with the lowest value for treatment C (100 g/plant) and the highest for treatment K30 (140 g/plant), although there was no significant difference between treatments during statistical evaluation.

4. Discussion

The irrigation experiment applied to the grain sorghum varieties took place between 2016 and 2020. With population growth and swift urbanization, the agricultural sector is under increasing pressure as freshwater supplies for crop production declining in all parts of the world [25]. The use of wastewater and effluents of industrial or agricultural origin is an essential element in the protection of our water resources. According to Qi et al. [26], effluents from aquaculture systems have a rich organic matter content which can be advantageously used in crop production. At the same time, it increases soil fertility, improves cultivation success and reduces fertilizer costs [27]. Irrigation of the higher salinity effluent with us gave similar results as Guimarães et al. [28] described in their research that the cultivation of the sorghum forage could be solved by irrigation with saline effluent water.

The high Na^+ and HCO_3^- concentration in irrigation water is known to be responsible for soil salinization. In sodic soils, ionic exchange between Na^+ and H^+ causes the dissociation of water in soil solution, leading to increasing concentrations of NaOH in the soil solution and the soil pH may increase to values above 10.5 [29,30]. The negative relationship between basic respiration and pH in salt-affected soils [31] could be another reason of the alkalization of the soil irrigated with effluent water (Table 4). In case of total carbonate, total organic carbon and N values, there were no significant differences due to the treatments (Table 4).

According to our results in the non-irrigated treatment, the highest EC value was measured (in surface soil layer). Strong correlation was found between EC and P and K content of the soil (Pearson correlation coefficients 0.824 and 0.823, respectively, sig. < 0.01) in the surface layer, but there was no correlation between them in the subsoil layer. We assume the EC differences occur because of the more available nutrients (P, K) in the soil at 0–30 cm depth.

The impact of irrigation and water quality on the available phosphorus content of the soil was proved in the surface soil layer where the lowest mean P content were calculated in E30 and E45, despite the P content of the effluent and the river water (Tables 1 and 4). We assume the disintegration of soil aggregates was due to soil salinization was the significant role of the released colloid-sized clay particles in P fixation; however, further studies are needed to prove our assumption. According to Arienzo et al. [32], potassium availability is strongly affected by the pH of the wastewater, as well as by the pH of the receiving soil. Normally, potassium availability is sustained for most plants in neutral or slightly acidic soils. In this study, the pH was significantly lower in the soil irrigated with Körös River water, which may have caused higher K content in the control treatment.

One of the acidic extractants, the ammonium lactate (AL, pH = 3.7) solution, introduced by Egner et al. [33], is commonly used in Europe. When the soil is treated with AL extraction solution, the soluble substance enters the solution partly through dissolution and partly through ion exchange, and AL extraction solution could decompose the carbonates also. The higher sodium concentration of the soils irrigated with effluent (Table 4) indicates the start of the sodication process.

The increase in the nitrogen content of sorghum plants is directly proportional to the higher crude protein content [34], which may mean a more nutritious feed for the animals. Although, the lower nitrogen content affects the physiology of the plant processes in which the macronutrient content of the grain changes, in particular the uptake of Ca, Mg and S [34]. For all three cultivars, significantly higher nitrogen values were measured in the grain yield in the first two growing years. In certain years of the experiment, it was found for each variety that the higher N content of the effluent irrigation water could be well utilized in grain yields.

The P demand of plants is high during the development of vegetative organs, but it is also significant during crop production. The seeds are the phosphorus-containing plant organs [35]. Nitrogen and phosphorus are antagonists of each other in terms of their physiological effects, where N stimulates the growth of vegetative organs, while phosphorus stimulates the appearance of generative organs and crop ripening [36]. Regarding phosphorus, there was no significant difference between the varieties and the irrigation treatments. On the other hand, the 'Alföldi 1' and 'GK Emese' varieties were able to make slightly better use of the higher P content of effluent irrigation water, especially in the last two years.

Potassium is an essential element for growth and one of the most frequently occurring cations in plant organs. Unlike other elements such as nitrogen, phosphorus, magnesium, calcium and sulfur, potassium is not incorporated into organic matter. Over time, the K content of older organs showed a decreasing trend. [37]. In the experiment, sorghum plants had high K levels of grain yield between 3500 and 5000 mg/kg d.m. There was no significant difference between the varieties. The Na^+/K^+ ratio is considered to be the basis of the salt tolerance of plants, which increases in direct proportion to the increase in salinity [38]. According to Ahmad et al. [39] and Iqbal et al. [40] studies, effluent irrigation water with higher salinity did not reduce the accumulation of K^+ in plant organs.

High salinity in plants cause hyperionic and hyperosmotic stress effects, as well as limited growth. Sodium is not essential even for extreme salt-tolerant plants, requiring only small amounts of C_4 and CAM plants [41,42]. Due to this, in addition to salt stress, sorghum is able to maintain its photosynthetic activity and dry matter production [43]. The sodium content of the grain crop was the lowest in the first year of cultivation. There was a difference in the accumulation of the cultivars, where a higher Na level was detected in the case of 'Farmsugro 180', while the lowest sodium content was measured in 'GK Emese'. The level of Na in the grain yield of sorghum also shows an upward trend between the years, but it occurred to different degrees for the three sorghum cultivars. However, this value has not yet been shown to be toxic to the dose. In a vegetation period—in proportion to the amount of annual irrigation—41–66 g/m^2 Na was applied in the case of 30 mm effluent irrigation and 49–99 g/m^2 Na in the case of 45 mm effluent.

Sixto et al. [44] have shown that a decrease in vegetative growth parameters can be observed in plants as a function of increasing salinity. In plants exposed to salt stress, a decrease in shoot, stem and root development, fresh and dry stem and root mass, leaf area and number of leaf, and relative chlorophyll amount and yield were observed [45–48]. For all three varieties, the average SPAD value of the leaves was lower in the last two growing years. There is a linear relationship between the nitrogen content of the crop and the chlorophyll value, where a positive correlation ($r = 0.737$, Pearson correlation) was observed during the study.

In the case of plant height, it can be stated that the highest plants (149–236 cm) were detected in the first year of cultivation, which can be explained mainly by the maximum amount of total water (precipitation + irrigation). Subsequently, a decrease was observed for all three cultivars (133–181 cm), depending on the total annual water volumes, as plant height is primarily affected by precipitation and temperature. In the experiment, the 'Farmsugro 180' fell short of its average height of 180–220 cm except for the first year. However, the measured height data of 'Alföldi 1' (140–16 cm), and mainly the 'GK Emese' (130–150 cm), corresponded to their characteristic height, which means that they were well adapted to the experimental stress conditions. This trend is also observed in the weight of grain yield. In addition, several studies have reported that higher salinity in irrigation water reduces plant mass, crop, and biomass product [49,50].

Hussein et al. [51] showed that higher Na concentrations of irrigation water had a negative effect on the growth profile of sorghum. In the 2017 growing year, the amount of irrigation water presented a positive correlation ($r = 0.026$, Pearson correlation) for both green mass and grain yield. In the case of both green mass and grain yield, it was observed that lower biomass values were detected in the last year of cultivation. The sorghum is a moderately salt tolerant plant [52], and no yield reduction is expected at irrigation water with EC of 4.5 dS/m and soil salinity up to 6.8 dS/m. According to our soil EC values (Table 4), it is not proven that salinity could cause the decrease; however, a detailed analyses of soil exchangeable sodium percentage would be justified to further investigate

Agronomy **2022**, *12*, 1185

the effluent water impact on these sorghum cultivars. Nevertheless, a decrease occurred in all treatments, and hence cannot be linked to water quality with absolute certainty. For example, the sensitivity of 'Farmsugro 180' should be emphasized, during which the value of grain yield in the case of samples irrigated with the water of the Körös River in the last year was only between 57–67 g/plant.

5. Conclusions

Irrigation of fish farm effluent water with a higher Na content may provide an alternative solution for regions with water scarcity; however, the possibility of its long-term use should be considered as it may cause salinization of the soil.

During the chemical analysis of plant parts compared to non-irrigated plants, effluent water irrigation did not negatively affect the N, P, K, and Na contents of the grain crop. Furthermore, based on the plant height and SPAD values, it can be concluded that compared to the control values, the applied irrigation waters did not have a negative effect on the two parameters mentioned above.

In summary, in the short term, in water-scarce or unfavorable soil areas, a good alternative could be to irrigate the effluent water of the intensive African catfish farm we studied on a grain sorghum plantation.

Author Contributions: Conceptualization, I.K., Á.K., A.P., C.B., C.G. and M.J.; Data curation, I.K. and Á.K., Formal analysis, I.K.; Funding acquisition, C.B.; Supervision, C.G., Methodology, I.K., Á.K., C.B. and M.J.; Writing—original draft preparation, I.K. and Á.K.; Writing—review and editing, I.K., Á.K. and M.J. All authors contributed critically to the drafts and gave final approval for publication. All authors have read and agreed to the published version of the manuscript.

Funding: This work was supported by the Hungarian Ministry of Agriculture under Project no. OD001, and by the National Research, Development and Innovation Office (GINOP-2.3.3-15-2016-00042 project). We gratefully acknowledge the staff of ÖVKI Lysimeter Station.

Data Availability Statement: The data presented in this study are available on request from the corresponding author. The data are not public, as this study is part of a forthcoming PhD dissertation.

Conflicts of Interest: The authors declare no conflict of interest.

References

1. Heaton, E.A.; Clifton-Brown, J.; Voigt, T.B.; Jones, M.B.; Long, S.P. Miscanthus for Renewable Energy Generation: European Union Experience and Projections for Illinois. *Mitig. Adapt. Strateg. Glob. Chang.* **2004**, *9*, 433–451. [CrossRef]
2. Zhao, Y.L.; Dolat, A.; Steinberger, Y.; Wang, X.; Osman, A.; Xie, G.H. Biomass Yield and Changes in Chemical Composition of Sweet Sorghum Cultivars Grown for Biofuel. *Field Crops Res.* **2009**, *111*, 55–64. [CrossRef]
3. Ghatak, H.R. Biorefineries from the Perspective of Sustainability: Feedstocks, Products, and Processes. *Renew. Sustain. Energy Rev.* **2011**, *15*, 4042–4052. [CrossRef]
4. Owusu-Sekyere, E.; Scheepers, M.E.; Jordaan, H. Economic Water Productivities Along the Dairy Value Chain in South Africa: Implications for Sustainable and Economically Efficient Water-Use Policies in the Dairy Industry. *Ecol. Econ.* **2017**, *134*, 22–28. [CrossRef]
5. Simon, S.; Wiegmann, K. Modelling Sustainable Bioenergy Potentials from Agriculture for Germany and Eastern European Countries. *Biomass Bioenergy* **2009**, *33*, 603–609. [CrossRef]
6. Sadia, B.; Saeed Awan, F.; Saleem, F.; Razzaq, A.; Irshad, B. Sorghum an Important Annual Feedstock for Bioenergy. In *Biomass Bioenergy-Recent Trends Future Chall*; El-Fatah Abomohra, A., Ed.; IntechOpen: London, UK, 2019; ISBN 978-1-78923-987-4. Available online: https://www.researchgate.net/publication/337694786_Sorghum_an_Important_Annual_Feedstock_for_Bioenergy (accessed on 13 May 2022).
7. Staggenborg, S.A.; Dhuyvetter, K.C.; Gordon, W.B. Grain Sorghum and Corn Comparisons: Yield, Economic, and Environmental Responses. *Agron. J.* **2008**, *100*, 1600–1604. [CrossRef]
8. Assefa, Y.; Staggenborg, S.A.; Prasad, V.P.V. Grain Sorghum Water Requirement and Responses to Drought Stress: A Review. *Crop Manag.* **2010**, *9*, 1–11. [CrossRef]
9. Plénet, D.; Cruz, P. Maize and Sorghum. In *Diagnosis of the Nitrogen Status in Crops*; Lemaire, G., Ed.; Springer: Berlin/Heidelberg, Germany, 1997; pp. 93–106. ISBN 978-3-642-64506-8.
10. Gosling, S.N.; Arnell, N.W. A Global Assessment of the Impact of Climate Change on Water Scarcity. *Clim. Chang.* **2016**, *134*, 371–385. [CrossRef]
11. Mekonnen, M.M.; Hoekstra, A.Y. Four Billion People Facing Severe Water Scarcity. *Sci. Adv.* **2016**, *2*, e1500323. [CrossRef]

12. O'Connor, G.A.; Elliott, H.A.; Bastian, R.K. Degraded Water Reuse: An Overview. *J. Environ. Qual.* **2008**, *37*, S-157–S-168. [CrossRef]
13. Oster, J.D. Irrigation with Poor Quality Water. *Agric. Water Manag.* **1994**, *25*, 271–297. [CrossRef]
14. Kestemont, P. Different Systems of Carp Production and Their Impacts on the Environment. *Aquaculture* **1995**, *129*, 347–372. [CrossRef]
15. WHO. *A Regional Overview of Wastewater Management and Reuse in the Eastern Mediterranean Region*; CEHA, 2005. Available online: https://apps.who.int/iris/bitstream/handle/10665/116463/dsa759.pdf (accessed on 13 May 2022).
16. Kolozsvári, I.; Kun, Á.; Jancsó, M.; Bakti, B.; Bozán, C.; Gyuricza, C. Utilization of Fish Farm Effluent for Irrigation Short Rotation Willow (*Salix alba* L.) under Lysimeter Conditions. *Forests* **2021**, *12*, 457. [CrossRef]
17. Lin, Y.-F.; Jing, S.-R.; Lee, D.-Y.; Wang, T.-W. Nutrient Removal from Aquaculture Wastewater Using a Constructed Wetlands System. *Aquaculture* **2002**, *209*, 169–184. [CrossRef]
18. Tsuchihashi, N.; Goto, Y. Cultivation of Sweet Sorghum (*Sorghum bicolor* (L.) Moench) and Determination of Its Harvest Time to Make Use as the Raw Material for Fermentation, Practiced during Rainy Season in Dry Land of Indonesia. *Plant Prod. Sci.* **2004**, *7*, 442–448. [CrossRef]
19. Paterson, A.H. Genomics of Sorghum. *Int. J. Plant Genom.* **2008**, *2008*, 1362451. [CrossRef]
20. Mace, E.S.; Tai, S.; Gilding, E.K.; Li, Y.; Prentis, P.J.; Bian, L.; Campbell, B.C.; Hu, W.; Innes, D.J.; Han, X.; et al. Whole-Genome Sequencing Reveals Untapped Genetic Potential in Africa's Indigenous Cereal Crop Sorghum. *Nat. Commun.* **2013**, *4*, 2320. [CrossRef]
21. Teetor, V.H.; Duclos, D.V.; Wittenberg, E.T.; Young, K.M.; Chawhuaymak, J.; Riley, M.R.; Ray, D.T. Effects of Planting Date on Sugar and Ethanol Yield of Sweet Sorghum Grown in Arizona. *Ind. Crops Prod.* **2011**, *34*, 1293–1300. [CrossRef]
22. Murray, S.C.; Rooney, W.L.; Hamblin, M.T.; Mitchell, S.E.; Kresovich, S. Sweet Sorghum Genetic Diversity and Association Mapping for Brix and Height. *Plant Genome* **2009**, *2*, 48–62. [CrossRef]
23. Vasilakoglou, I.; Dhima, K.; Karagiannidis, N.; Gatsis, T. Sweet Sorghum Productivity for Biofuels under Increased Soil Salinity and Reduced Irrigation. *Field Crops Res.* **2011**, *120*, 38–46. [CrossRef]
24. Al-Jaloud, A.A.; Hussain, G.; Al-Saati, A.J.; Karimulla, S. Effect of Wastewater Irrigation on Mineral Composition of Corn and Sorghum Plants in a Pot Experiment. *J. Plant Nutr.* **1995**, *18*, 1677–1692. [CrossRef]
25. Flörke, M.; Schneider, C.; McDonald, R.I. Water Competition between Cities and Agriculture Driven by Climate Change and Urban Growth. *Nat. Sustain.* **2018**, *1*, 51–58. [CrossRef]
26. Qi, D.; Yan, J.; Zhu, J. Effect of a Reduced Fertilizer Rate on the Water Quality of Paddy Fields and Rice Yields under Fishpond Effluent Irrigation. *Agric. Water Manag.* **2020**, *231*, 105999. [CrossRef]
27. Abdelraouf, R.E.; Abou-Hussein, S.D.; Badr, M.A.; El-Tohamy, N.M. Safe and Sustainable Fertilization Technology with Using Fish Water Effluent as a New Bio-Source for Fertilizing. *Acta Hortic.* **2016**, *1142*, 41–48. [CrossRef]
28. Guimarães, M.J.M.; Simões, W.L.; Tabosa, J.N.; dos Santos, J.E.; Willadino, L. Cultivation of Forage Sorghum Varieties Irrigated with Saline Effluent from Fish-Farming under Semiarid Conditions. *Rev. Bras. Eng. Agrícola Ambient.* **2016**, *20*, 461–465. [CrossRef]
29. Sou/Dakouré, M.Y.; Mermoud, A.; Yacouba, H.; Boivin, P. Impacts of Irrigation with Industrial Treated Wastewater on Soil Properties. *Geoderma* **2013**, *200–201*, 31–39. [CrossRef]
30. Wang, M.; Chen, S.; Chen, L.; Wang, D.; Zhao, C. The Responses of a Soil Bacterial Community under Saline Stress Are Associated with Cd Availability in Long-Term Wastewater-Irrigated Field Soil. *Chemosphere* **2019**, *236*, 124372. [CrossRef]
31. Yang, C.; Wang, X.; Miao, F.; Li, Z.; Tang, W.; Sun, J. Assessing the Effect of Soil Salinization on Soil Microbial Respiration and Diversities under Incubation Conditions. *Appl. Soil Ecol.* **2020**, *155*, 103671. [CrossRef]
32. Arienzo, M.; Christen, E.W.; Quayle, W.; Kumar, A. A Review of the Fate of Potassium in the Soil–Plant System after Land Application of Wastewaters. *J. Hazard. Mater.* **2009**, *164*, 415–422. [CrossRef]
33. Egner, H.; Riem, H.; Domingo, W. Untersuchungen Über Die Chemische Bodenanalyse Als Grundlage Für Die Beurteilung Des Nährstoffzustandes Der Böden. II. Chemische Extraktionsmethoden Zur Phosphor Und Kaliumbestimmung. *K. Lantbr. Ann.* **1960**, *26*, 199–215.
34. Campos, F.S.; Araújo, G.G.L.; Simões, W.L.; Gois, G.C.; Machado Guimarães, M.J.; da Silva, T.G.F.; Rodrigues Magálhaes, A.L.; Oliveira, G.F.; de Almeida Araujo, C.; Silva, T.S.; et al. Mineral and Fermentative Profile of Forage Sorghum Irrigated with Brackish Water. *Commun. Soil Sci. Plant Anal.* **2021**, *52*, 1353–1362. [CrossRef]
35. Malhotra, H.; Vandana; Sharma, S.; Pandey, R. Phosphorus Nutrition: Plant Growth in Response to Deficiency and Excess. In *Plant Nutrients and Abiotic Stress Tolerance*; Hasanuzzaman, M., Fujita, M., Oku, H., Nahar, K., Hawrylak-Nowak, B., Eds.; Springer: Singapore, 2018; pp. 171–190. ISBN 978-981-10-9043-1.
36. Gordon, W.B.; Whitney, D.A. Effects of Phosphorus Application Method and Rate on Furrow-irrigated Ridge-tilled Grain Sorghum. *J. Plant Nutr.* **2000**, *23*, 23–34. [CrossRef]
37. Marschner, H.; Marschner, P. *Mineral Nutrition of Higher Plants*, 3rd ed.; Elsevier: London, UK; Academic Press: Waltham, MA, USA, 2012; ISBN 978-0-12-384905-2.
38. Chhipa, B.; Lal, P. Na/K Ratios as the Basis of Salt Tolerance in Wheat. *Aust. J. Agric. Resour. Econ.* **1995**, *46*, 533. [CrossRef]
39. Ahman, S.; Islam Khan, N.; Iqbal, M.Z.; Hussain, A.; Hassan, M. Salt Tolerance of Cotton (*Gossypium hirsutum* L.). *Asian J. Plant Sci.* **2002**, *1*, 715–719. [CrossRef]

40. Iqbal, N.; Ashraf, M.Y.; Javed, F.; Martinez, V.; Ahmad, K. Nitrate Reduction and Nutrient Accumulation in Wheat Grown in Soil Salinized with Four Different Salts. *J. Plant Nutr.* **2006**, *29*, 409–421. [CrossRef]
41. Rao, D.L.N. The Effects of Salinity and Sodicity upon Nodulation and Nitrogen Fixation in Chickpea (*Cicer arietinum*). *Ann. Bot.* **2002**, *89*, 563–570. [CrossRef]
42. Rout, N.P.; Shaw, B.P. Salt Tolerance in Aquatic Macrophytes: Possible Involvement of the Antioxidative Enzymes. *Plant Sci.* **2001**, *160*, 415–423. [CrossRef]
43. Calone, R.; Sanoubar, R.; Lambertini, C.; Speranza, M.; Vittori Antisari, L.; Vianello, G.; Barbanti, L. Salt Tolerance and Na Allocation in *Sorghum bicolor* under Variable Soil and Water Salinity. *Plants* **2020**, *9*, 561. [CrossRef]
44. Sixto, H.; Grau, J.M.; Alba, N.; Alía, R. Response to Sodium Chloride in Different Species and Clones of Genus *Populus* L. *Forestry* **2005**, *74*, 93–104. [CrossRef]
45. Shannon, M.C.; Grieve, C.M. Tolerance of Vegetable Crops to Salinity. *Sci. Hortic.* **1998**, *78*, 5–38. [CrossRef]
46. Chookhampaeng, S. The Effect of Salt Stress on Growth, Chlorophyll Content Proline Content and Antioxidative Enzymes of Pepper (*Capsicum annuum* L.) Seedling. *Eur. J. Sci. Res.* **2011**, *49*, 103–109.
47. Sevengor, S.; Yasar, F.; Kusvuran, S.; Ellialtioglu, S. The Effect of Salt Stress on Growth, Chlorophyll Content, Lipid Peroxidation and Antioxidative Enzymes of Pumpkin Seedling. *Afr. J. Agric. Res.* **2011**, *6*, 4920–4924. [CrossRef]
48. Padilla, F.M.; de Souza, R.; Peña-Fleitas, M.T.; Gallardo, M.; Giménez, C.; Thompson, R.B. Different Responses of Various Chlorophyll Meters to Increasing Nitrogen Supply in Sweet Pepper. *Front. Plant Sci.* **2018**, *9*, 1752. [CrossRef] [PubMed]
49. Almodares, A.; Sharif, M.E. Effects of Irrigation Water Qualities on Biomass and Sugar Contents of Sugar Beet and Sweet Sorghum Cultivars. *J. Environ. Biol.* **2007**, *28*, 213–218. [PubMed]
50. Katerji, N.; Van Hoorn, J.W.; Hamdy, A.; Mastrorilli, M.M.; Mou, E. Osmotic Adjustment of Sugar Beets in Response to Soil Salinity and Its Influence on Stomatal Conductance, Growth and Yield. *Agric. Water Manag.* **1997**, *34*, 57–69. [CrossRef]
51. Hussein, M.M.; Abdel-Kader, A.A.; Kady, K.A.; Youssef, R.A.; Alva, A.K. Sorghum Response to Foliar Application of Phosphorus and Potassium with Saline Water Irrigation. *J. Crop Improv.* **2010**, *24*, 324–336. [CrossRef]
52. Ayers, R.S.; Westcot, D.W. *Water Quality for Agriculture*; FAO Irrigation and Drainage Paper; Food and Agriculture Organization of the United Nations: Rome, Italy, 1985; ISBN 978-92-5-102263-4.

Article

Hydrochemical Assessment of Water Used for Agricultural Soil Irrigation in the Water Area of the Three Morava Rivers in the Republic of Serbia

Radmila Pivić [1,*], **Jelena Maksimović** [1], **Zoran Dinić** [1], **Darko Jaramaz** [1], **Helena Majstorović** [2], **Dragana Vidojević** [3] and **Aleksandra Stanojković-Sebić** [1]

[1] Institute of Soil Science, Teodora Drajzera 7, 11000 Belgrade, Serbia; jelena.maksimovic@soilinst.rs (J.M.); soils.dinic@gmail.com (Z.D.); soilsjaramaz@gmail.com (D.J.); soils.stanojkovicsebic@gmail.com (A.S.-S.)
[2] Institute "Tamiš" Pančevo, 26000 Pančevo, Serbia; helenabane@gmail.com
[3] Ministry of Environmental Protection, Serbian Environmental Protection Agency, 11000 Belgrade, Serbia; dragana.vidojevic@sepa.gov.rs
* Correspondence: drradmila@pivic.com

Citation: Pivić, R.; Maksimović, J.; Dinić, Z.; Jaramaz, D.; Majstorović, H.; Vidojević, D.; Stanojković-Sebić, A. Hydrochemical Assessment of Water Used for Agricultural Soil Irrigation in the Water Area of the Three Morava Rivers in the Republic of Serbia. *Agronomy* **2022**, *12*, 1177. https://doi.org/10.3390/agronomy12051177

Academic Editors: Pantazis Georgiou and Dimitris Karpouzos

Received: 11 April 2022
Accepted: 9 May 2022
Published: 13 May 2022

Publisher's Note: MDPI stays neutral with regard to jurisdictional claims in published maps and institutional affiliations.

Abstract: The assessment of the suitability and status of irrigation water quality from the aspect of its potential negative impact on soil salinization and mapping of spatial distribution within the area of the three Morava rivers, which includes the South, West, and Great Morava basins, was the purpose of this research. A total of 215 samples of irrigation water were tested, and their quality was evaluated based on the analysis of the following parameters: pH, electrical conductivity (EC), total dissolved salt (TDS), sodium adsorption ratio (SAR), and content of SO_4^{2-}, Cl^-, HCO_3^-, CO_3^{2-}, Mg^{2+}, Ca^{2+}, Na^+, and K^+. The results showed that the average content of ions was as follows: $Ca^{2+} > Mg^{2+} > Na^+ > K^+$ and $HCO_3^- > SO_4^{2-} > Cl^- > CO_3^{2-}$. The assessment of irrigation water suitability was determined by calculating the following indices: percentage sodium (Na %), residual sodium carbonate (RSC), permeability index (PI), magnesium hazard (MH), potential salinity (PS), Kelley's index (KI), total hardness (TH), irrigation water quality index (IWQI). Based on Wilcox's diagram, the USSL diagram, and the Doneen chart, it was concluded that most of the samples were suitable for irrigation. Using multivariate statistical techniques and correlation matrices in combination with other hydrogeochemical tools such as Piper's, Chadha's, and Gibbs diagrams, the main factors associated with hydrogeochemical variability were identified.

Keywords: hydrochemical characterization; irrigation water quality; irrigation suitability; soil; hydrochemical facies; GIS

1. Introduction

The irrigation of cultivated plants on agricultural soil involves the use of water with the appropriate physical, chemical, and biological properties, so it is very important to examine the quality of water used for its intended purpose to assess the impact on soil and plants. Intensification of irrigation depends primarily on the provision of the required amount of water of adequate quality [1,2]. According to the report of the Republic Bureau of Statistics (RBS), in 2021 [3], 52.236 hectares of agricultural soil in the Republic of Serbia were irrigated, capturing about 92.574 thousand m³ of water, which was mostly pumped from watercourses, about 84.3%, while the remaining quantities were collected from groundwaters, lakes, reservoirs, and other sources.

The area of study, the three Morava rivers, covers an area of agricultural production, within which the application of irrigation is expanding, so the examination of the quality of irrigation water is important for its intensification. This was recognized by the governing body, the Ministry of Agriculture, Water Management and Forestry of the Republic of Serbia, and the Agricultural Soil Administration, which enabled researchers to assess the quality of irrigation water to intensify agricultural production and prevent a negative impact on soil degradation.

Different sources of water are used for irrigation: rivers, streams, natural and artificial reservoirs and lakes, and groundwater from different depths (tube wells and dug wells). Information on irrigation water quality is of critical significance for understanding the changes in product quality and for making necessary modifications in water management [4]. Ayers and Wescot [5] stated the importance of water quality assessment for irrigation. It is a prerequisite for planning, designing, and operating irrigation systems [6].

Surface water quality according to [7], is mostly determined by the quality and scope of industrial, agricultural, and other anthropogenic activities in the basins of a particular area.

Anthropogenic impacts and natural processes can affect the quality of surface water and threaten its use as drinking water, and for use in industry, agriculture, and other purposes [8–10].

When assessing the quality of irrigation water, it is very important to evaluate whether it can harm the characteristics of the soil on which it is used. Ghazaryan et al. [11] stated that salinization involves the accumulation of soluble salts in the soil profile and is a process of soil degradation, mainly anthropogenic, which affects highly productive irrigated agricultural ecosystems in semi-arid and arid regions and can negatively affect agricultural production and the sustainable development of agricultural regions. Salinization reduces the production capacity of the soil and degrades the chemical and physical properties of the soil. Given the problem of food shortages worldwide, one of the biggest problems of agriculture is the reduction and control of soil salinity [12].

On irrigated soils, salinization is the main cause of conditions limiting agricultural production and lack of yield and is one of the harmful effects on the environment. Singh et al. [13] list the three most common worldwide problems of irrigation water used being of inadequate quality to include salinity, reduced permeability, and increased specificity of ionic toxicity.

Poor quality irrigation waters can have negative effects on heavy, clayey soils, while the same water can be used satisfactorily on sandy and/or permeable soils [14].

The quality of irrigation water is defined by the type and concentration of dissolved salts and solids [15]. Based on the results of the chemical analyses of surface waters based on hydrochemical parameters, it is possible to obtain adequate data on water types, different geochemical processes, and water classifications [16]. The interaction of the surface water chemistry and geochemical characteristics offers a valuable context for trend analysis, identification of unique environmental problems, and the exchange of information on water sources, geochemical processes, water quality, and water susceptibility to contamination [13,17].

There are several methods and classifications for assessing the quality of irrigation water based on the assessment of analyzed hydrochemical parameters, each of which cannot be considered applicable to all conditions in crop production because each depends on the soil characteristics, plant tolerance, precipitation regime, drainage conditions, watering methods, water accessibility to plants and climate. Different methods and different hydrochemical indices are used to assess the suitability of water for irrigation. Hem [18] states that reliable results can be obtained by analyzing the chemistry of all ions, rather than the individual parameters of irrigation water. Irrigation water quality is determined based on sodium adsorption ratio (SAR), sodium percentage (%N), magnesium ratio (MR), residual sodium carbonate (RSC), permeability index (PI), and Kelley's index (KI) [11,19–21].

By determining the irrigation water quality index (IWQI), which combines several indicators and expresses the quality of irrigation water in the form of a single value as proposed by several authors [22–25], it is possible to obtain more reliable evaluations.

Graphical representations of the irrigation water quality assessment and its application ability are defined by physicochemical parameters [26], using graphical techniques such as the [27], Wilcox Diagram [28], and Doneen's chart [29]. Gibbs diagram [30], as a method for defining the main geochemical control processes that affect the chemical composition of surface waters [31], is an applicable and frequently used method for defining the main geochemical control processes that affect the chemical composition of surface water and groundwater, i.e., irrigation water. In addition, by applying multivariate statistical techniques in combination with other hydrogeochemical tools such as Piper's and Chadha's diagrams, the main factors associated with hydrogeochemical variability have been identified. The application of cluster analysis (CA)-multivariate statistical analysis [32,33] is often used to assess water quality. It is a multivariate technique for classifying the physico-chemical parameters into water classes according to the relationship between the chemical properties of the surface water [34].

Within this paper, the ion content in irrigation water samples, collected from available sources at observation sites (surface and groundwater), i.e., along with agricultural areas where irrigation is used or planned, the water characteristics and the relationship between each component and each type of ion were analyzed. The results have a significant effect on understanding the characteristics of the current situation of the regional hydrochemistry of the analyzed irrigation water and the impact of the examined parameters on the soil, primarily in terms of regional salinization treatment and the impact on agricultural production.

This study aimed to assess the suitability and status of the irrigation water quality and to provide a graphical representation of the applied classifications of the tested irrigation water samples within the three Morava river basin districts, which include the basins of the Južna (English: South), Zapadna (English: West), and Velika (English: Great) Morava rivers. Given that the data of this type of research have not been systematized, processed, and presented so far, the present results will contribute through the adoption of adequate assessment methods that will provide a basis for monitoring and establishing irrigation water quality.

2. Materials and Methods

2.1. Description of the Study Area

2.1.1. Location, Hydrological Setting, and Climate

The area of the three Morava rivers is of great importance for the national economy—agriculture, industry, energy, and other human activities. In geographical terms, the basin of the three Morava rivers' water area lies between 42°04′ and 44°82′ northern latitude and 19°19′ and 23°14′ eastern longitude (Figure 1).

The surface area of the water of the three Morava rivers in the Republic of Serbia, which includes the basins of the Južna, Zapadna, and Velika Morava, is about 36.207 km^2. It belongs to the Black Sea basin and spreads over the most fertile and most densely populated area of central Serbia.

Given the small difference in latitude between the southernmost and northernmost points, it could be expected that the general climatic conditions change very little at the examined locality. However, the influence of climatic parameters, primarily the relief and the degree of continentality, determines the diversity of the climate. The climate of the area is moderately continental. Climatic conditions for agricultural production are generally favorable, especially the thermal potential.

Air temperature is one of the most important climatic elements, based on which the insight into the thermal conditions in an area is obtained. Temperatures have been rising steadily since the coldest January when an average minimum temperature of -2.5 °C and an average maximum of 5.2 °C were recorded. The warmest months are July and August with an average minimum temperature of 15.4 °C and 15.0 °C, and an average maximum temperature of 29.3 °C and 29.4 °C, respectively. The highest amplitude is in August, at 14.4 °C, while the annual temperature amplitude is 11.3 °C. The average annual air temperature is 11.8 °C. The annual actual insolation of the observed locality is 2198 h, and the average annual relative air humidity is 72%.

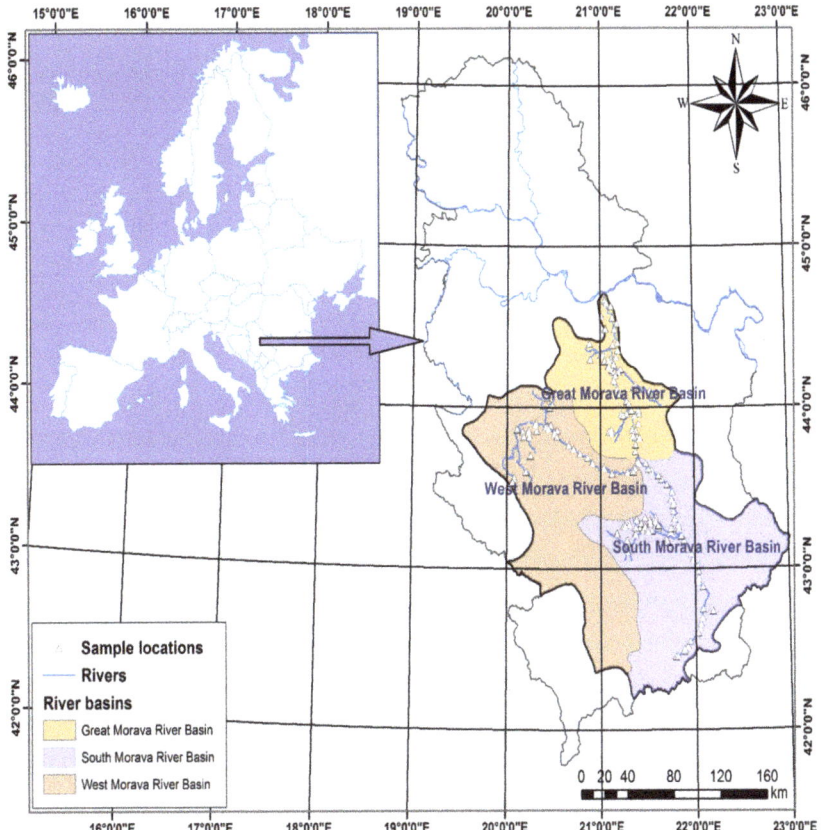

Figure 1. Study area with locations of sampling points.

Pomoravlje and its surroundings have a very low annual precipitation of about 665 mm, which is a value close to those of arid areas. The distribution of precipitation throughout the seasons is not favorable for the development of crops that have a higher demand for water in summer. Precipitation is not evenly distributed over the months. The most rain falls in June and May. The months of February and October have the least precipitation when on average 5 to 6% of the total annual precipitation falls.

Droughts during summer and autumn can cause the soil to dry to a depth of 2–4 m. Exceptions are areas with groundwater at shallow depths but in very limited areas.

The wind has a significant effect on evapotranspiration. It can currently modify the weather situation depending on the amount of moisture it carries. Ground air currents are mostly conditioned by an orography. In the warmer part of the year, winds from the northwest and west prevail. During the colder part of the year, the east and southeast wind, called Košava, dominates. For the summer period, it is important to emphasize the appearance of southern winds when there is a great drying of the soil, especially if the average temperatures are high.

Data on the reference evapotranspiration (ETo), calculated using the Penman–Monteith method, showed that the highest value of ETo was registered in July and averaged 4.8 mm day^{-1}.

In the study area, the largest deficit occurs in August and occasionally in July. The values of peak water requirements are relatively uniform with an average of 4.2 mm day^{-1}.

2.1.2. Geological and Hydrogeological Setting

Within the study area, three different relief units are observed on which very different soils are formed. The terrain of the Zapadna Morava basin has a very heterogeneous geological structure, with very different relief units: river valleys, basins, the coast of Šumadija, and the foothills of the mountains and mountainous area. The basic morphological characteristic of the Južna Morava basin is its great fragmentation. The Južna Morava valley is a typical composite valley, consisting of several valleys separated by gorges. In the valleys filled with loose lake and river sediments, the slopes of the Južna Morava river are less pronounced than in the gorges built of more resistant rocks, various types of crystal-like slates through which recently formed eruptive rocks broke in places. From the steep and mostly deforested mountainsides, the flow of atmospheric water is fast and high, which makes the Južna Morava river and its tributaries have the characteristics of torrents.

In the basin of the Velika Morava river, there are very diverse rocks on which different soils are formed. In the river valleys, huge amounts of alluvial and deluvial sediments were deposited in the quaternary. Neogene sediments predominate on the lake surfaces, and shists, sandstones, limestones, and other compact rocks dominate in the mountainous areas. Certain types of soil are formed on quaternary sediments, primarily on alluvial deposits and loess. Neogene sediments are of great importance for soil formation in the area of lake surfaces.

Shists and compact rocks are important for the formation of shallow mountain soils. Figure 2 shows a geological map of the study area.

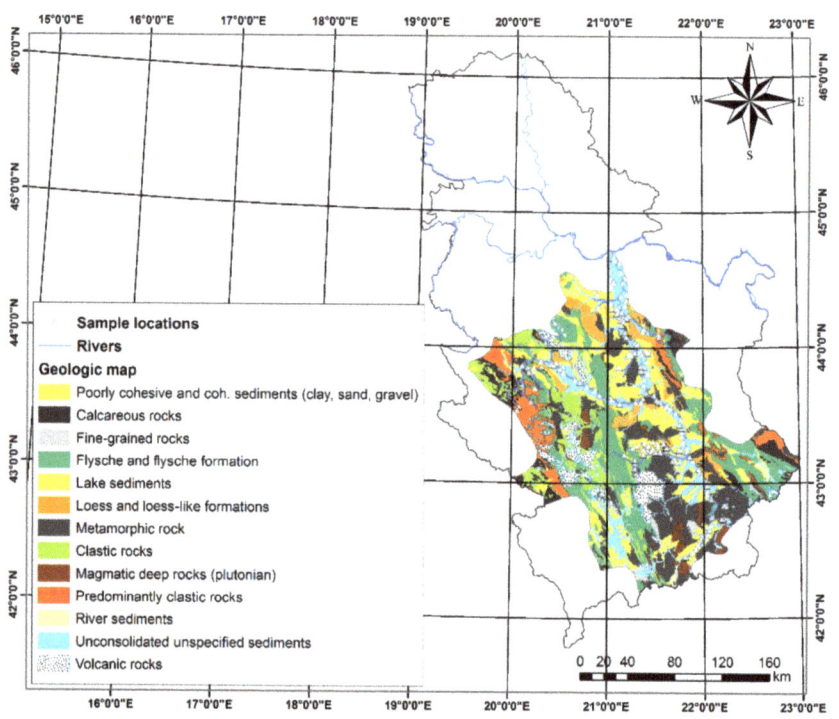

Figure 2. Geological map of the study area.

2.1.3. Soil Use Type

Soils suitable for irrigation are primarily alluvial soils along the three Morava rivers and meadow soils that are heavier in texture than alluvium [35,36]. The water physical properties of the soil along the three Morava rivers are of very heterogeneous composition. Represented are applied gravel, sandy gravel, sandy, loamy, and clay composition. All of these soils can be irrigated with varying amounts of water. The basic soil types in the river basin are Fluvisols, Luvisols, Eutric Luvisols, Eutric Cambisols and Eutric Vertisols [37]. These soils have a high potential for fertility, and irrigation is one of the factors that contribute. Irrigation provides increased yields of the crops grown in the river basin. Figure 3 shows a pedological map of the study area.

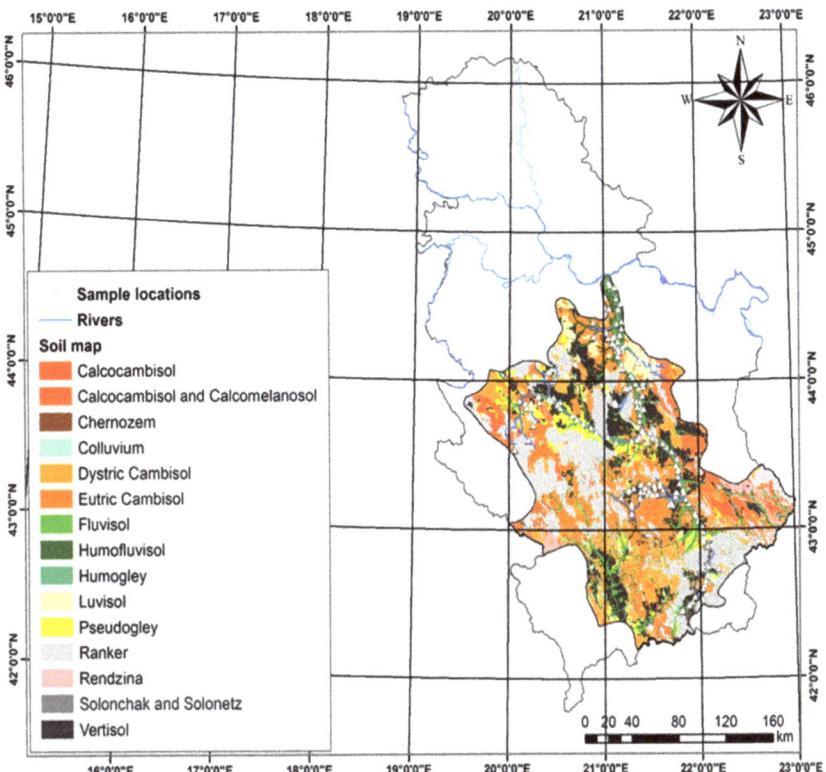

Figure 3. Pedological map of the study area.

2.2. Sampling and Collection of Water Samples

A sampling of 215 water samples for irrigation (surface water samples and groundwater samples) within the irrigated agricultural areas or agricultural areas where irrigation is planned was conducted in the period 2014 to 2019, in phases as a result of the study commissioned by the Ministry of Agriculture, Water Management and Forestry of the Republic of Serbia. Water samples were collected according to the methodology described in the professional literature. The sampling locations were determined with a Garmin GSP map 62s GPS device with UTM coordinate recording. Samples were taken in polyethylene sampling bottles with a volume of 2000 mL to determine physicochemical parameters. Before use, they were washed with distilled water, and on the spot three times with sampled water. They were marked, sealed, adequately stored at temperatures up to 4 °C, and transported to the laboratory.

2.3. Laboratory Analysis

Analyses of the sampled water for irrigation were conducted in the laboratory of the Institute of Soil Science, Belgrade. The measured parameters were determined by the following methods: pH—potentiometric [38]; electrical conductivity (EC)—conductiometric [39]; total dissolved solids (TDS)—gravimetric; CO_3^{2-}, HCO_3^-, Cl^-—volumetric; K^+, Na^+—flame photometric [40];

sodium adsorption ratio (SAR)—calculation [41]; Ca^{2+}, Mg^{2+}—spectrophotometric [42]; SO_4^{2-}—gravimetric [43]. After analysis of ion concentrations, the charge balance error (*CBE*) was calculated to ensure suitably high quality, and the standard error for each sample was calculated using Equation (1) [44].

$$CBE = \frac{\sum cations - \sum anions}{\sum cations + \sum anions} * 100\%$$

(1)

CBE values with a limit of 5% were considered acceptable [45]. All *cations* and *anions* were expressed in meq L^{-1}.

2.4. Data Analysis

Statistical description (range, mean and standard error) of physicochemical parameters was determined using SPSS version 22 (SPSS Inc., Chicago, IL, USA) and shown in Table 1. The relationships between the main physicochemical parameters were processed using Microsoft Excel to identify geochemical processes and control mechanisms that affect the quality of irrigation water.

Table 1. Irrigation water quality parameters and their proposed limiting values.

q_i	EC (μS m^{-1})	SAR ((mmol L^{-1})$^{0.5}$)	Na (meq L^{-1})	Cl (meq L^{-1})	HCO$_3$ (meq L^{-1})
85–100	[200–750)	[2–3)	[2–3)	[1–4)	[1–1.5)
60–85	[750–1500)	[3–6)	[3–6)	[4–7)	[1.5–4.5)
35–60	[1500–3000)	[6–12)	[6–9)	[7–10)	[4.5–8.5)
0–35	EC < 200 or EC $\geq$ 3000	SAR < 2 or SAR $\geq$ 12	Na < 2 or Na $\geq$ 9	Cl < 1 or Cl $\geq$ 10	HCO$_3$ < 1 or HCO$_3$ $\geq$ 8.5

Analysis of the obtained data of the irrigation water analysis results was estimated using Gibbs diagram, USSL diagram, Wilcox's diagram, Doneen's chart, Piper's diagram, and Chadha's diagram, using Microsoft Excel 2016 (Redmond, Washington, DC, USA).

Multivariate statistical analysis, including correlation and cluster analysis (CA), was applied to assess the quality of irrigation water, using the Ward method to describe the similarities between the two clusters to identify different geochemical groups with a similar content of physicochemical parameters in the tested samples. Cluster analysis (Dendrograms) was performed using SPSS version 22 (SPSS Inc., Chicago, IL, USA).

2.5. Spatial Distribution

GIS software was used as a platform for geostatistical analysis of spatial data processing. Cartographic data processing was performed using ESRI ArcGIS Desktop 10.7.1 (Esri, Redlands, CA, USA).

2.6. Irrigation Water Quality Evaluation

The following parameters were determined: sodium absorption ratio, sodium hazard, residual sodium carbonate, permeability index, magnesium ratio, Kelley ratio, potential salinity, and irrigation water quality index.

These indicators were obtained using the following formulas:

$$SAR = \frac{Na^+}{\sqrt{\frac{Ca^{2+} + Mg^{2+}}{2}}}$$

(2)

$$Na\% = \frac{(Na^+ + K^+) * 100}{Ca^{2+} + Mg^{2+} + Na^+ + K^+}$$

(3)

$$RSC = \left(CO_3^{2-} + HCO_3^-\right) - \left(Ca^{2+} + Mg^{2+}\right)$$

(4)

$$PI = \frac{\left(Na^+ + \sqrt{HCO_3^-}\right) * 100}{Ca^{2+} + Mg^{2+} + Na^+ + K^+} \tag{5}$$

$$MH = \frac{Mg^{2+}}{Ca^{2+} + Mg^{2+}} * 100 \tag{6}$$

$$KI = \frac{Na^+}{Ca^{2+} + Mg^{2+}} \tag{7}$$

$$PS = Cl^- + \frac{1}{2}SO_4^{2+} \tag{8}$$

The calculation of the irrigation water quality index (IWQI), developed by Meireles et al. [21], for the calculation of the mentioned index values, individual data on the values were used, as follows: EC, Na^+, Cl^-, SAR and HCO_3^-, by Equation (9)

$$IWQI = \sum_{i=1}^{n} q_i \, w_i \tag{9}$$

where:

The values of the limit values q_i are determined by the following Equation (10), where the corresponding parameters are shown in Table 1.

$$q_i = q_{max} - \left(\frac{\left[\left(x_{ij} - x_{inf}\right) * q_{imap}\right]}{x_{amp}} \right) \tag{10}$$

where:

q_{max} is the upper value of the corresponding class of q_i;

X_{ij} represents the data points of the parameters (observed value of each parameter);

X_{inf} refers to the lower limit value of the class to which the observed parameter belongs;

q_{imap} represents the class amplitude for q_i classes;

x_{imap} corresponds to the class amplitude to which the parameter belongs.

For the calculation of W_i, the following Equation (11) is used:

$$W_i = \frac{\sum_{j=1}^{k} F_j \, A_{ij}}{\sum_{j=1}^{k} \sum_{i=1}^{n} F_j \, A_{ij}} \tag{11}$$

where W_i and F are the comparative weights of the IWQI physicochemical characteristics, and component i is a constant value; The parameter i that can be described by factor j is denoted by Aij. The number of physicochemical parameters used in the IWQI ranges from 1 to n, while the number of factors chosen in the IWQI ranges from 1 to k, and where n represents the number of parameters considered, in this case, 5, values q_i in Table 1 were multiplied by the corresponding weight W_i of each parameter listed in Table 2, according to [23].

Table 2. The weights of the IWQI parameters.

Parameters	W_i
[EC]	0.211
[Na]	0.204
[HCO$_3$]	0.202
[Cl]	0.194
[SAR]	0.189
Total	1

3. Results and Discussion

3.1. Irrigation Water Suitability Indicators

Several indicators indicate the suitability of irrigation water. Within this study, the ion content in irrigation water samples was considered, and the water characteristics and the relationship between each component and each ion type were analyzed.

The results have a significant impact on the understanding of the characteristics of the current situation of regional surface water hydrochemistry and the impact of the examined parameters on agriculture, providing support for data for the treatment of regional salinization.

The results of the statistical analysis (min, max, mean, standard deviation value) of the physical and chemical parameters of the analyzed samples of irrigation water are shown in Table 3.

Table 3. Physicochemical water characteristics.

Parameters	Unit	Range	Mean		STDEV
pH	/	6.20–8.90	7.73	±	0.47
EC	μSm^{-1}	20–2260	650.91	±	370.02
TDS	$mg\,L^{-1}$	50–2800	497.42	±	359.76
CO_3^{2-}	$meq\,L^{-1}$	0.00–14.70	4.47	±	2.49
HCO_3^-	$meq\,L^{-1}$	0.00–3.60	0.83	±	0.56
Cl^-	$meq\,L^{-1}$	0.10–6.12	1.04	±	0.80
SO_4^{2-}	$meq\,L^{-1}$	0.05–6.02	1.05	±	0.82
Ca^{2+}	$meq\,L^{-1}$	0.46–11.03	3.85	±	1.84
Mg^{2+}	$meq\,L^{-1}$	0.16–10.91	2.54	±	1.71
K^+	$meq\,L^{-1}$	0.01–12.4	1.60	±	2.15
Na^+	$meq\,L^{-1}$	0.01–4.30	0.48	±	0.53

The pH values of the analyzed samples ranged from 6.2 to 8.9 (mean of 7.7). The EC of the samples varied in the ranges of 20.0–2260.0 μSm^{-1}, with an average value of 650.9 μSm^{-1}. TDS values of the samples varied in a wide range from 50.0–2800.0 $mg\,L^{-1}$, with a mean value of 497.4 $mg\,L^{-1}$. The results showed that the average content of ions was as follows: $Ca^{2+} > Mg^{2+} > Na^+ > K^+$ and $HCO_3^- > SO_4^{2-} > Cl^- > CO_3^{2-}$. The results of the hydrochemical properties of the tested samples are presented below, based on which the performance of the irrigation water suitability was obtained (Table 3).

For a critical assessment of the irrigation water suitability, Table 4 shows the various indicators and associated classifications along with the number and percentage of water samples belonging to each class.

Table 4. Irrigation water suitability.

Classification Pattern	Sample Range				Categories	Ranges	Description	Number of Samples	Sample (%)
	Min.	Max	Mean	STDEV					
Sodium absorption ratio (SAR) [27]	0.01	10.34	0.93	1.27	Excellent	0–10	Don't have sodium hazard	214	99.53
					Good	10–18	Low sodium hazard	1	0.47
					Fair	18–26	Harmful for almost all types of soils		
					Poor	>26	Unsuitable for irrigation		
Percent sodium (% Na) [28]	0.49	51.89	18.80	9.86	Excellent	0–20	Excellent for irrigation	136	63.26
					Good	20–40	Good for irrigation	72	33.49
					Permissible	40–60	Permissible for irrigation	7	3,26
					Doubtful	60–80	Doubtful for irrigation		
					Unsuitable	>80	Unsuitable for irrigation		
Residual sodium carbonate (RSC) [27]	−8.79	14.16	−1.08	2.62	Good	<1.25	Generally safe for irrigation	197	91.63
					Medium	1.25–2.5	Marginal as an irrigation source		
					Bad	>2.5	Generally not suitable for irrigation without improvement	18	8.37

Table 4. *Cont.*

Classification Pattern	Sample Range				Categories	Ranges	Description	Number of Samples	Sample (%)
	Min.	Max	Mean	STDEV					
Permeability index (PI) [29]	0.07	1.19	0.48	0.17	Class-I	>75	Good for irrigation	201	93.49
					Class-II	25–75	Suitable for irrigation	9	4.19
					Class-III	<25	Unsuitable for irrigation	5	2.33
Electrical conductivity (EC, $\mu S\,cm^{-1}$) [28]	20.00	2260.00	650.91	370.02	Good	250–750	Medium salinity water	135	62.79
					Permissible	750–2250	High-salinity water	79	36.74
					Doubtful	2250–5000	Doubtful for irrigation	1	0.47
					Unsuitable	>5000	Unsuitable for irrigation		
Total dissolved salts (TDS, mg L^{-1}) [27]	50.00	280000	497.42	359.76	Excellent	<150	Low salinity hazard	8	3.72
					Good	150–500	Permissible for irrigation	134	62.33
					Fair	500–1500	Doubtful for irrigation	69	32.09
					Poor	>1500	Unsuitable for irrigation	4	1.86
Magnesium Hazard (MH) [46]	4.94	77.68	38.35	12.55	MH	<50%	Suitable	173	85.12
					MH	>50%	Unsuitable	32	14.88
Kelly's Index (KI) [47]	0.00	4.42	0.29	0.42	KI	<1	Suitable	202	93.95
					KI	>1	Unsuitable	13	6.05
Potential Salinity (PS) (meq L^{-1}) [29]	0.28	7.32	1.57	1.06	PS	<3.0	Excellent to good	197	91.63
					PS	3.0–5.0	Good to injurious	15	6.98
					PS	>5.0	Injurious to unsatisfactory	3	1.4
Total Hardness (TH) (meq L^{-1}) [48]	0.70	17.17	6.38	3.11	TH	0–60	Soft	215	100
					TH	61–120	Moderate		
					TH	121–180	Hard		
					TH	>181	Very hard		
IWQI [23]	52.96	99.42	89.55	9.19	ClassI	85–100	Excellent	164	76.27
					ClassII	70–85	Good	41	19.07
					ClassIII	55–70	Poor	7	3.26
					ClassIV	40–55	Very poor	3	1.4
					ClassV	0–40	Unsuitable		

3.1.1. Electrical Conductivity (EC)

Conductivity is a measure of the ability of an aqueous solution to carry an electric current. Increasing levels of conductivity and *cations* are the products of decomposition and mineralization of organic materials [49]. Natural waters have values of electrical conductivity typically less than unity. Measurement of the electrical conductivity is performed at a specific temperature and it corresponds to the presence of dissolved salts. These are most often sodium chloride (table salt) and may be represented, sodium sulphate, calcium chloride, calcium sulfate, magnesium chloride, and others. Some of the large numbers of different elements dissolved in water favor the plant and their presence is useful, but sometimes these useful items can become harmful if their concentration is too high.

EC and sodium concentration are very important in the classification of irrigation water. Salts, in addition to directly affecting plant growth, also affect soil structure, permeability, and aeration, which indirectly affect plant growth [14].

The presence of a charged particle excess limits the quality of irrigation water. The EC values of the tested samples were in the range of 20–2260 μSm^{-1} (Table 3) and the graph is shown in Figure 4.

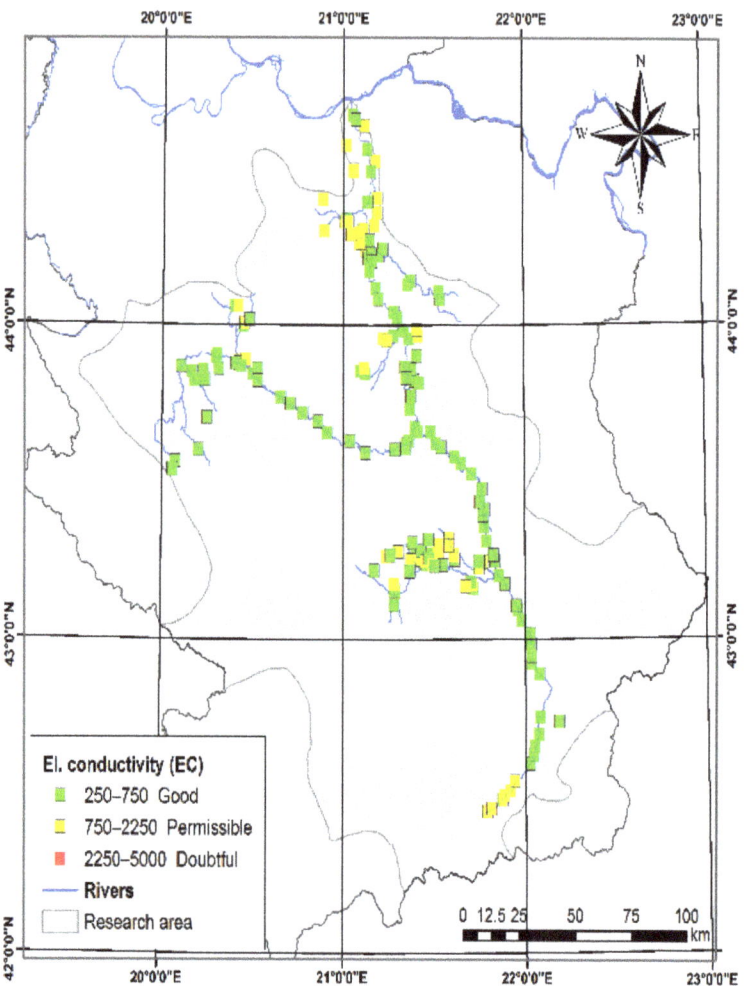

Figure 4. Spatial distribution maps of EC.

3.1.2. Total Dissolved Solids (TDS) and Total Hardness (TH)

Assessment of total dissolved solids (TDS) is very important for understanding the status of pollutants present in irrigation water. The suitability of using water with a TDS value of less than approximately 500 mg L^{-1} is often considered good, while a TDS level greater than approximately 1500 mg L^{-1} is not acceptable for irrigation [27].

In the tested irrigation water samples, the range for TDS was from 50 to 2800 mg L^{-1} (Table 3). Among the analyzed samples (Table 4), 3.72% of irrigation water samples were in the excellent category (TDS less than 150 mg L^{-1}), 62.33% of samples were in the good category (TDS in the range of 150–500 mg L^{-1}) below very low, 32.09% of the samples were in the fair category (TDS in the range 500–1500 mg L^{-1}), and the remaining 1.86% were in the poor category (TDS > 1500 mg L^{-1}).

The TDS zoning map (Figure 5) shows that TDS values increased from South to Northwest, which may be due to anthropogenic factors and the geological characteristics of the aquifer in the study area. Somewhat higher values were also noticeable in the Southeastern part, which can also be explained by anthropogenic and probably mining activity.

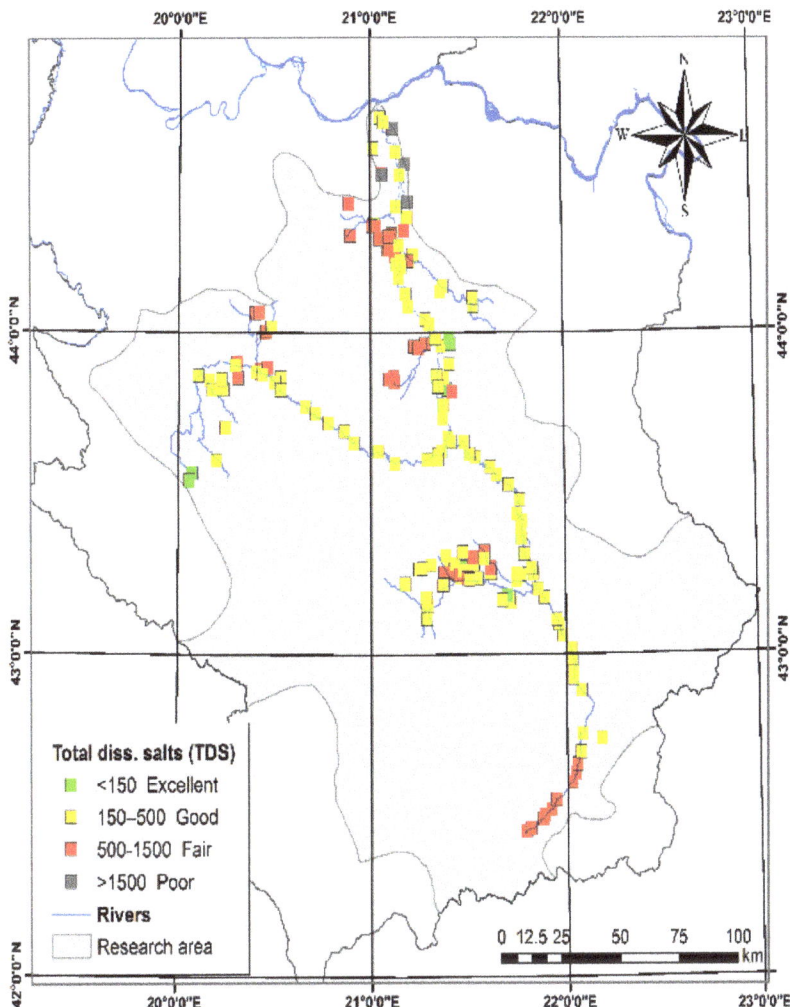

Figure 5. Spatial distribution maps of TDS.

Water hardness (WH) increases due to the increase in the content of alkaline earth elements such as calcium and magnesium [50]. The presence of the minerals calcite and dolomite are the main reasons for the increase in the concentration of Ca^{2+} and Mg^{2+} in groundwater [51,52]. Classification of groundwater done by Durfor and Becker [48] based on WH is given in Table 4. All tested samples belong to the soft category. The spatial distribution of the examined parameter is shown in Figure 6.

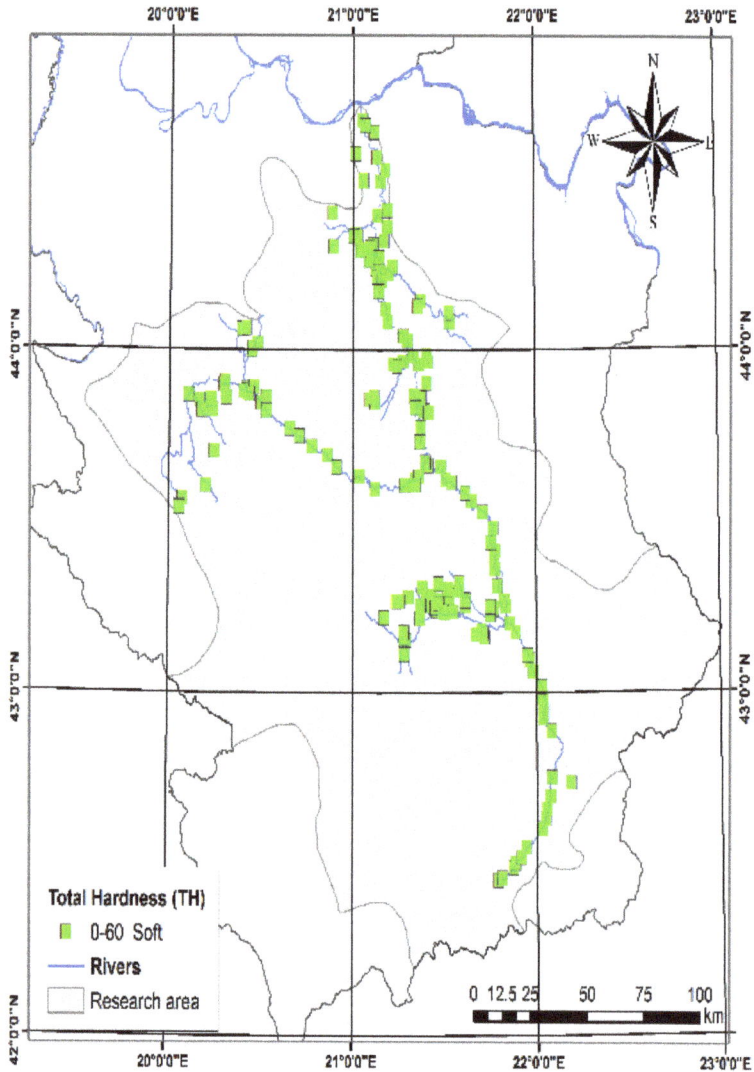

Figure 6. Spatial distribution maps of TH.

3.1.3. Sodium Absorption Ratio (SAR)

SAR represents the relative activity of Na^+ in soil cation exchange reactions and is used to estimate the degree of alkalization of irrigation water [5]. The SAR concept is used to detect the likely danger of sodium [53]. This parameter was originally proposed by Richards [27]. If the water used for irrigation has a high sodium content and a low calcium content, the cation exchange complex may become saturated with sodium. This can worsen soil structure due to the dispersion of clay particles [14].

Irrigation with water with a high SAR can lead to the formation of an impermeable layer, which leads to reduced soil permeability, internal drainage, and air circulation, or deterioration of the soil structure [54].

Irrigation with water with a high SAR can lead to the formation of an impermeable layer of irrigated soil, which leads to reduced soil permeability, internal drainage and air circulation, and deterioration of the structure [55]. SAR is calculated using Equation (2).

The SAR values of the sampled irrigation water in the study area varied between 0.01 and 10.34 meq L^{-1} (Table 3). The obtained SAR values showed that all tested samples of irrigation water, except one, were of the excellent class, and one sample of the good class (Table 4.). The spatial distribution of SAR values of the tested irrigation water samples is shown in Figure 7.

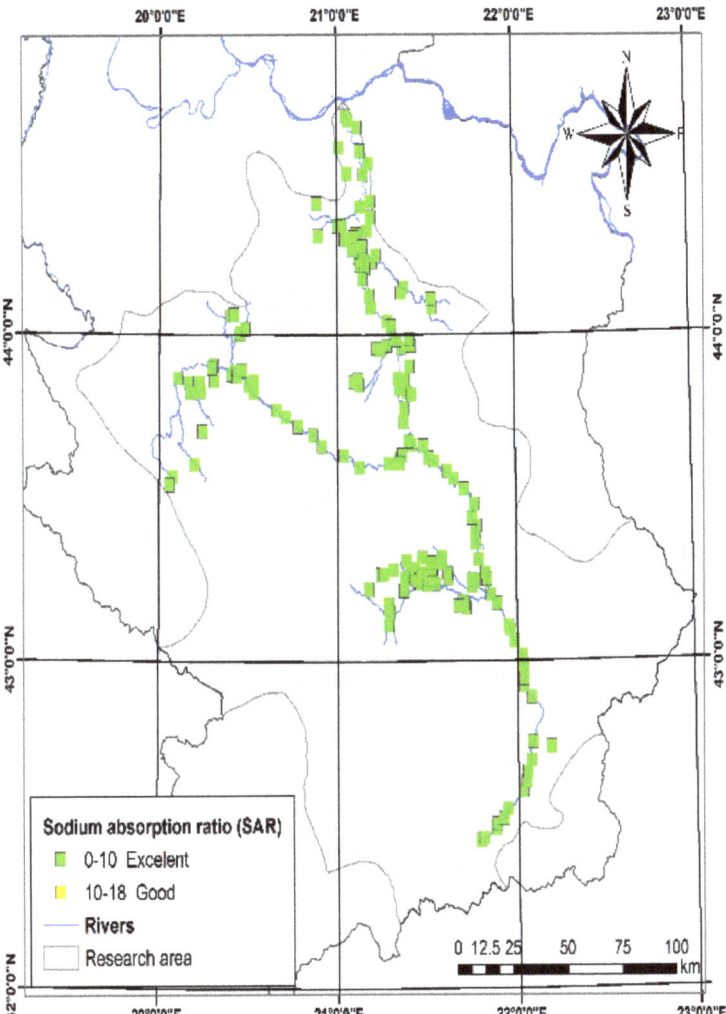

Figure 7. Spatial distribution of SAR.

3.1.4. Sodium Hazard (Na %)

For the assessment of the irrigation water quality, the percentage of sodium is one of the most important indicators. An excess of sodium with carbonate ions will help turn the soil into alkaline soil, in contrast, sodium mixed with chloride ions will accelerate the formation of saline soil, which ultimately worsens the infiltration capacity of the soil and reduces plant growth [56,57]. The percentage of sodium (%Na) is often used as a parameter to assess the suitability of irrigation water quality [28]. As a result of its reactivity with soil, sodium is considered an important ion for the classification of irrigation water and, if it occurs in excess, reduces the water permeability of the soil [58,59].

The Na % is determined by calculating the relative proportion of all *cations* available in water using the Equation (3) [28]:

The value of the specified parameter should not exceed 60% in irrigation waters. Table 3 shows that Na % in samples of water for irrigation in the study area ranged from 0.49 to 51.89%, with an average of 18.79%. One hundred and thirty-six samples, i.e., 63.26%, belonged to the class of excellent for irrigation, 72 samples (33.49%) to the class of good for irrigation, and seven samples (3.26%) to the class of permissible for irrigation. The analytical results were plotted on a Wilcox diagram (Figure 8). It shows the relationship between salinity hazards (expressed using EC values in μS cm^{-1}) and the sodium content in the water (expressed as % Na) [60] and is used to classify irrigation water samples. The spatial distribution of Na % content is shown in Figure 9.

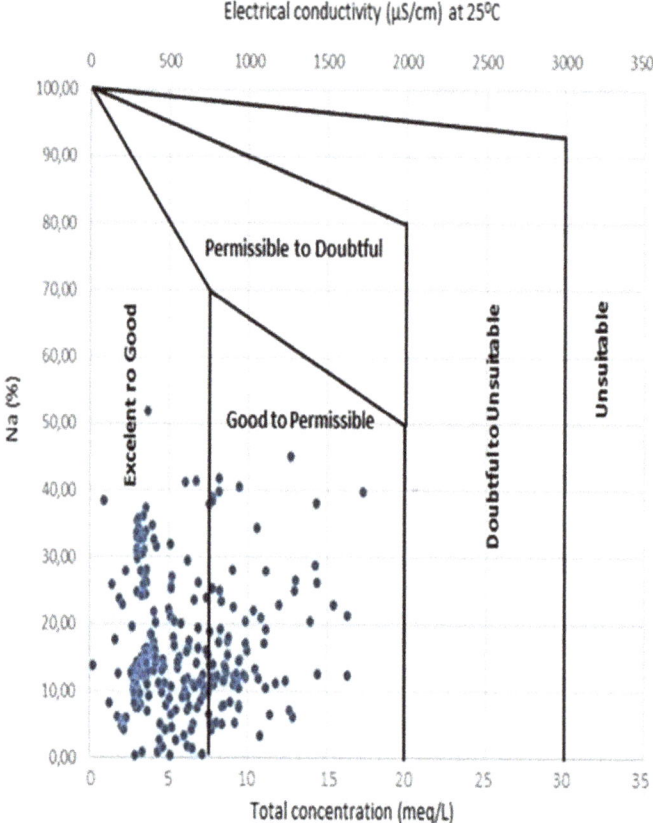

Figure 8. The plot of sodium percentage versus electrical conductivity [26].

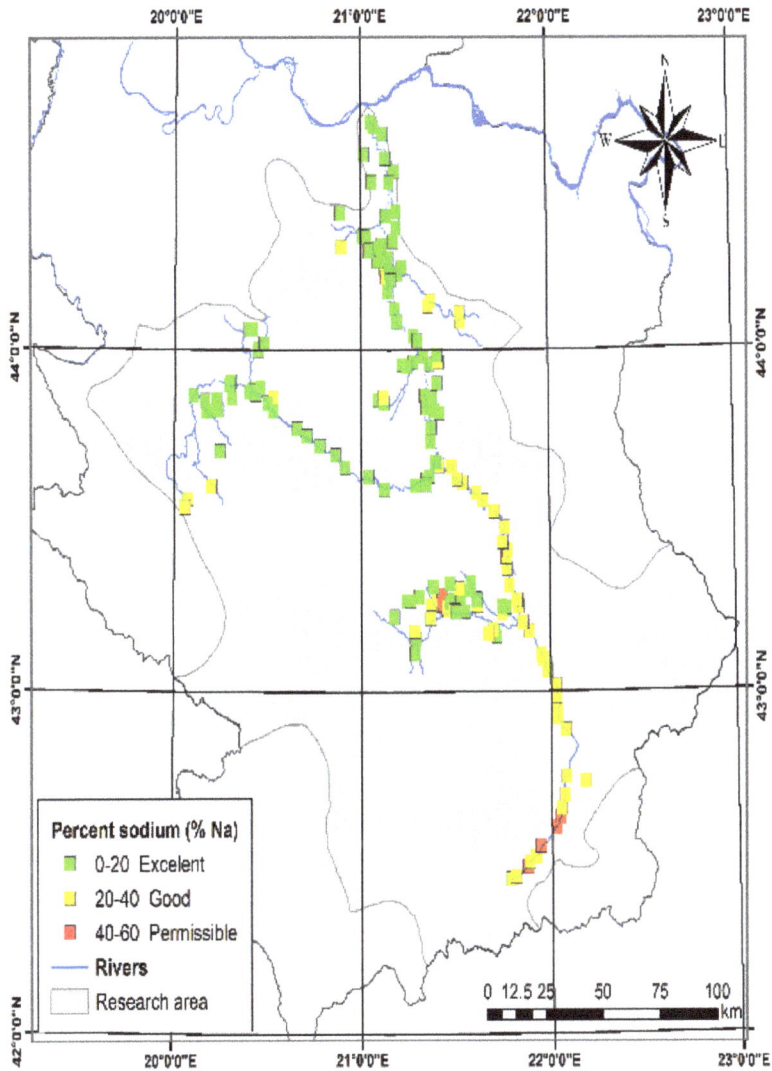

Figure 9. Spatial distribution of Na %.

3.1.5. Residual Sodium Carbonate (RSC)

Bicarbonate is an important component for assessing the quality of irrigation water [44]. By measuring the difference between the sum of carbonates and bicarbonates and the sum of calcium and magnesium, the residual sodium carbonate (RSC) value, proposed by Eaton [61], is determined to assess the impact of the danger of using irrigation water that is an alkaline reaction.

Soils irrigated with water of the stated quality, with high RSC, may lose their productive capacity because of structural deterioration due to the deposition of sodium carbonate [44]. Singraja [62] states that increased alkalinity can affect the decomposition of soil organic matter, which is also one of the negative consequences of using water for irrigation that is of inadequate quality.

The RSC index is often used to assess the suitability of water for irrigation in clayey soils that have a high cation exchange capacity. The presence of a higher amount of dissolved sodium compared to dissolved calcium and magnesium in irrigation water leads to the swelling of clayey soils or dispersion, which can lead to a drastic reduction in its infiltration capacity. The value of the specified parameter is determined by Equation (4).

According to RSC, groundwater is suitable for irrigation if RSC < 1.25, is marginal if it is higher than 1.25 but lower than 2.50, and is unsuitable if it exceeds 2.50 [20,63].

Residual sodium carbonate is classified into three categories: good, medium, and bad (Table 4). Of the examined samples, 91.63% belonged to the class of waters with an RSC value of less than 1.25. Those waters are generally safe for irrigation, and 8.37% belonged to the group whose value was higher than 2.5, i.e., generally not suitable for irrigation without improvement. The spatial distribution of a particular RSC is shown in Figure 10.

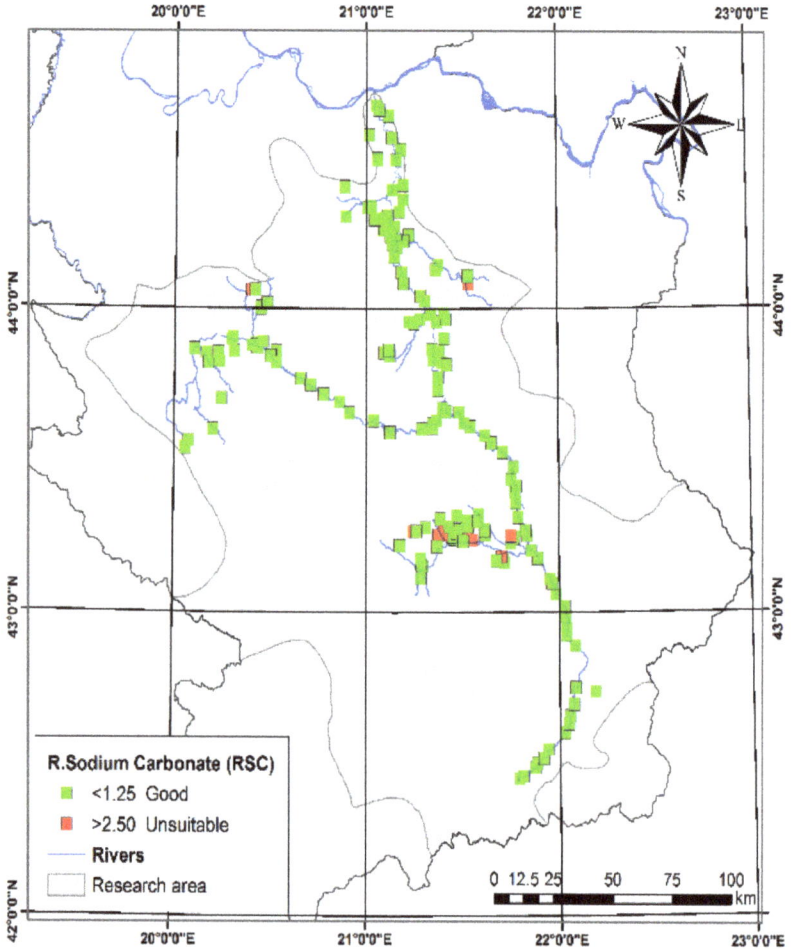

Figure 10. Spatial distribution of RSC.

3.1.6. Permeability Index (PI) and Doneen Diagram

The permeability index (PI) is one of the indicators of water suitability for irrigation and indicates problems of water permeability of soil that is flooded for a long time with water with high salt concentration [64]. It links the concentration of sodium, calcium, magnesium, and bicarbonate ions with the effect on soil permeability [65]. Prolonged irrigation with poor quality water can affect soil permeability [66]. To quantify the impact of long-term irrigation on soil quality, Doneen [29] proposed a criterion for assessing the suitability of water for irrigation based on the PI index determined by Equation (5)

Figure 11 presents a graph by which Doneen [29] presented the divisions of irrigation water quality based on the PI index. Class I and Class II waters are categorized as "good" and "suitable" with their higher maximum permeability [67].

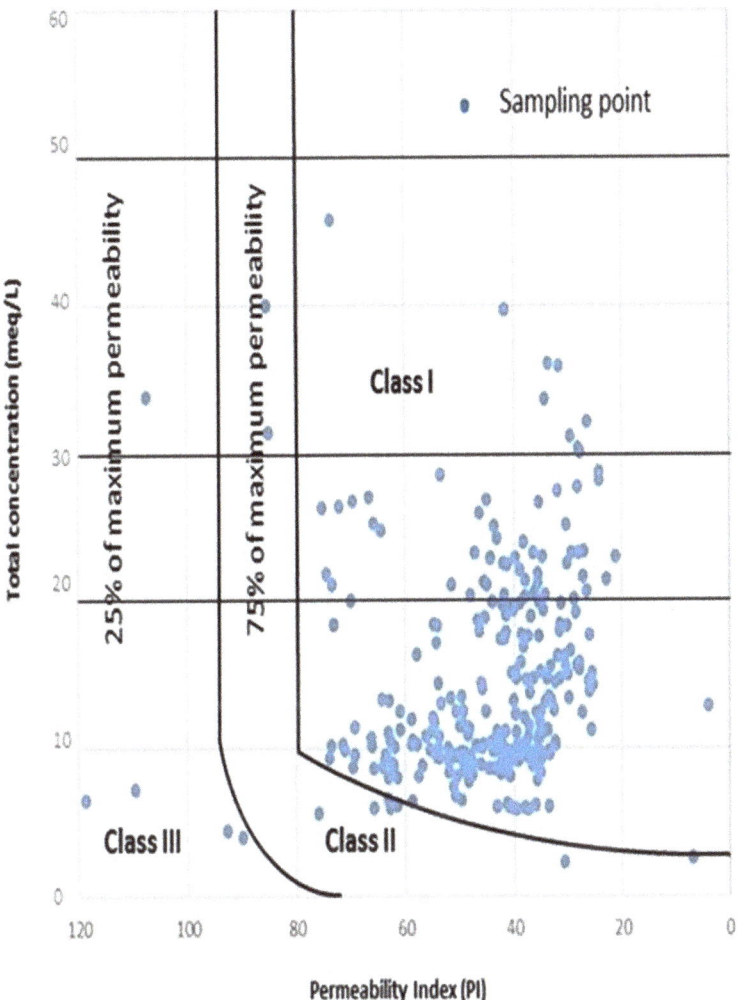

Figure 11. Doneen classification of irrigation water quality based on PI.

Based on the graphic classification of Doneen, this index determined the suitability of water for irrigation and categorized water as class I (PI > 75%), class II (25% < PI < 75%), and class III (PI < 25%) [49]. As shown in Table 4 and Figure 11, 201 samples (93.49%) were of excellent quality for PI-based irrigation and would not affect soil permeability; nine water samples were acceptable (4.19%); five water samples were unsuitable for irrigation, representing 2.33% of the total water samples.

The spatial distribution of the specific PI index of the tested irrigation water samples is shown in Figure 12.

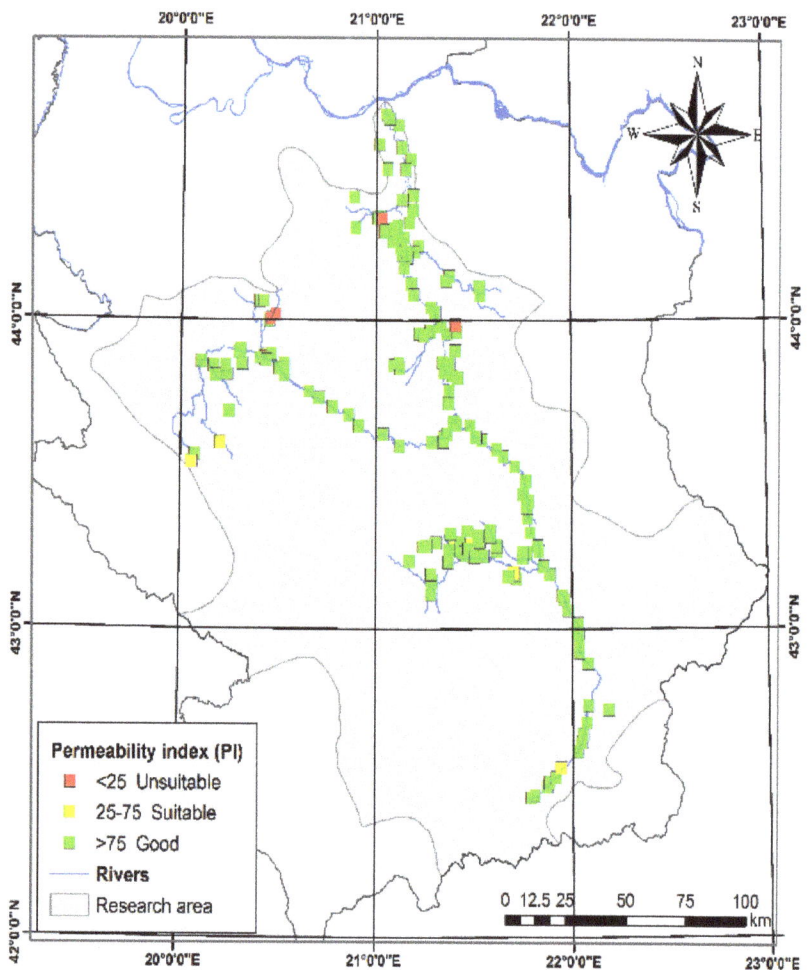

Figure 12. Spatial distribution of PI.

3.1.7. Magnesium Hazard (MH)

Szabolcs and Darab [68] proposed an assessment of the dangers of increased magnesium content as one of the indicators of the suitability of irrigation water.

For irrigation, the magnesium content also plays an important role, because the extreme magnesium content is a source of harmful effects on the soil. The risk of excess Mg^{2+} in water can be estimated by the ratio of magnesium (MH) and is calculated using Equation (6).

Water with a high content of magnesium often affects the properties of the soil on which it is applied, because with increased alkalinity there is a decrease in crop yield [46]. Carbonate and bicarbonate ions are responsible for high pH values since these anions are the main component affecting alkalinity [69]. Some 85.12% of the tested irrigation water samples corresponded to the suitable class, while 14.88% belonged to the non-compliant unsuitable class of irrigation water (Table 4). The spatial distribution of a certain MH index of tested irrigation water samples is shown in Figure 13.

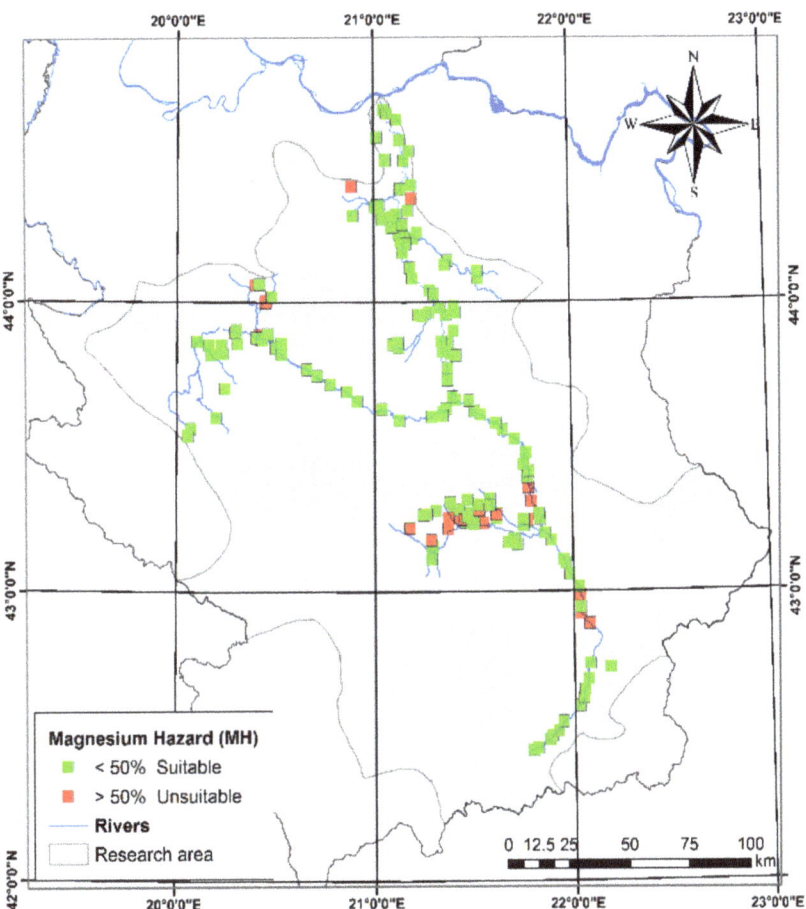

Figure 13. Spatial distribution of MH.

3.1.8. Kelley Index (KI)

The Kelley index (KI) [70] is used to determine the suitability of water for irrigation. The levels of Na, Ca, and Mg in water are used to calculate the value of KI, using Equation (2) [46,70]. Increased concentrations of Na, Ca, and Mg in water pose an alkaline hazard [71]. KI values lower than one (KI < 1) indicate that excess sodium has been found in water [47,70,72]. The obtained KI values of the tested samples of irrigation water varied between 0.004 and 4.416 meq L^{-1} (Table 4). According to the obtained KI values, 93.95% of the irrigation water samples belonged to the "suitable" class, and 6.05% are categorized as "unsuitable". The Kelley index was calculated using Equation (7).

The spatial distribution of a certain KI index of tested irrigation water samples is shown in Figure 14.

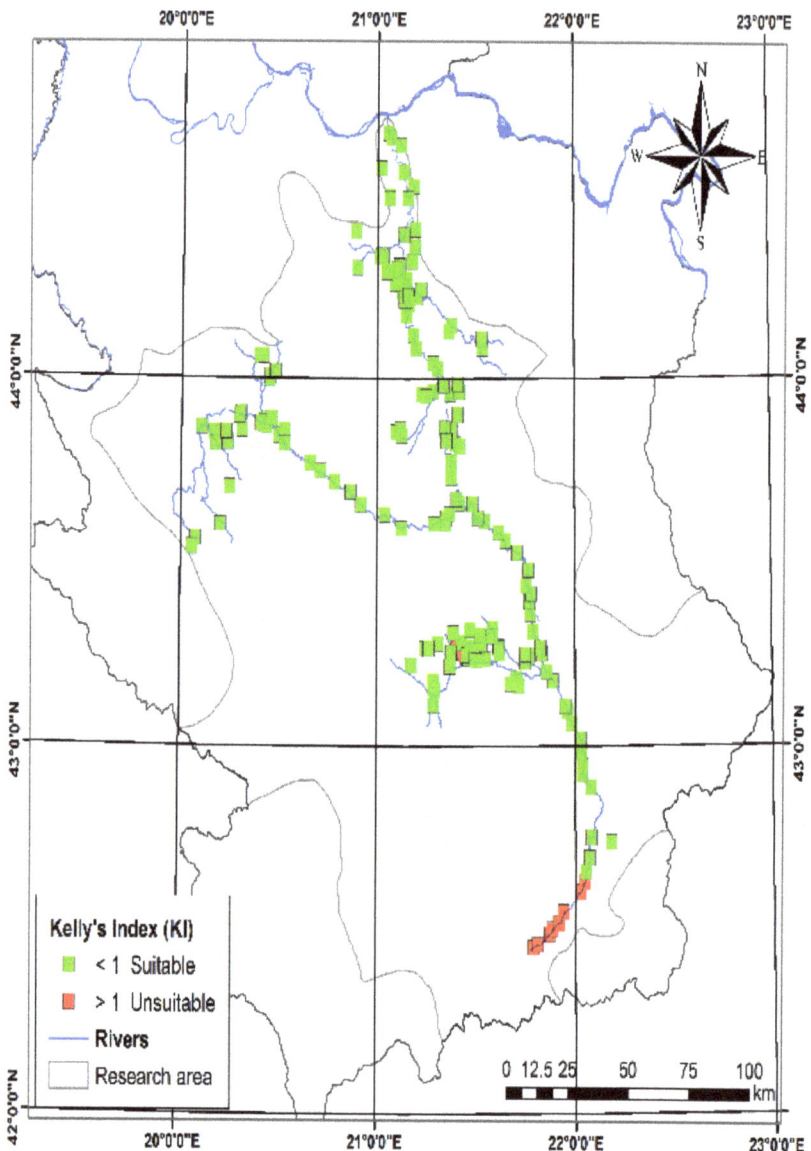

Figure 14. Spatial distribution of KI.

3.1.9. Potential Salinity (PS)

Potential salinity, determined as the sum of Cl^- and half-concentration of $SO_4{}^{2-}$ [29], is used as one of the classifications for assessing the suitability of water for irrigation. It is determined based on Equation (8).

In the examined irrigation water samples, 197 samples, i.e., 91.63% concerning the values of PS, were classified as excellent to good; 15 samples (6.98%) as good to injurious, and 3 samples (1.4%) as injurious to unsatisfactory (Table 4).

The spatial distribution of a particular PS of the tested irrigation water samples is shown in Figure 15.

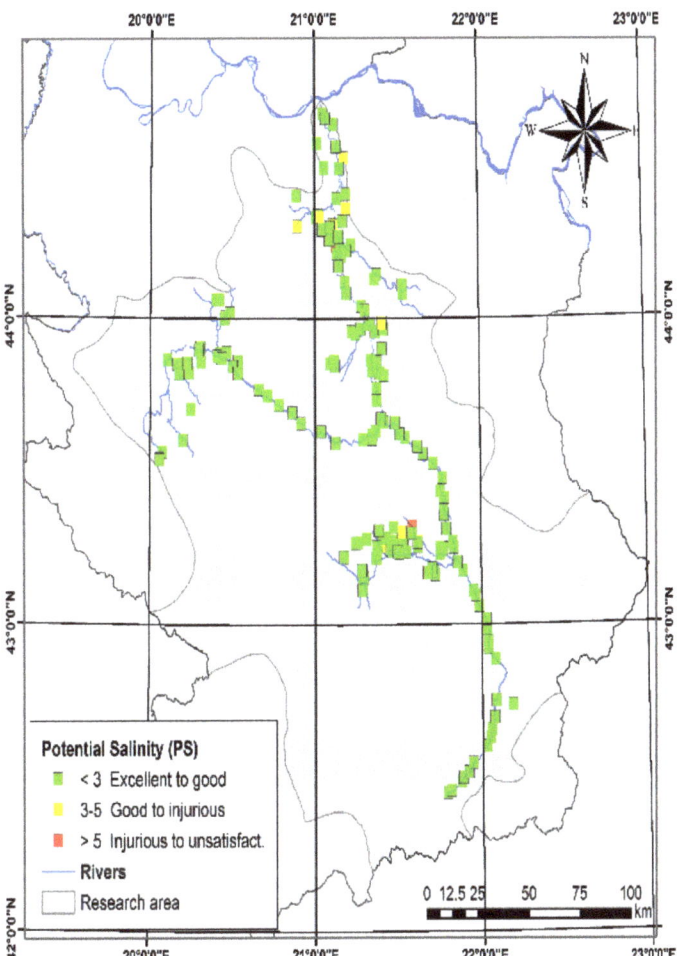

Figure 15. Spatial distribution of PS.

3.1.10. USSL Salinity Diagram

Salinity diagram—the USSL diagram [27], represents the relationship between salinity hazards (expressed in EC values) and sodium content in water (expressed in terms of sodium absorption coefficient, SAR; concentrations in meq L^{-1}). If the SAR value is in the range of 6 to 9, irrigation water will cause permeability problems in the types of clay soils that accumulate and swell [27]. He and Li [73] state that if the SAR values of irrigation water are less than 10 meq L^{-1}, they are classified as "excellent", for SAR values between 10 and 18 meq L^{-1}, they are classified as "good", and "suspicious" if the SAR values are between 18 and 26 meq L^{-1}. Waters with an SAR value higher than 26 meq L^{-1} are classified as "inappropriate" [60,74].

The study area was classified into six zones, based on USSL diagrams (Figure 16), as follows: (1) C1S1, (2) C2S1, (3) C3S1, (4) C4S1, (5) C4S2, (6) C3S2. According to this diagram, if the clusters of samples were located in the regions C1S1 and C2S2, they could be considered as very good, i.e., a good category of irrigation water. If the samples were in category C3S1, they belonged to moderately suitable irrigation waters, due to the high risk of salinity. Samples in the C3S2 and C4S1 categories were rated as irrigation water of medium to poor quality, due to the high risk of salinization, and they are not suitable for use on heavy soils and salt-sensitive plants.

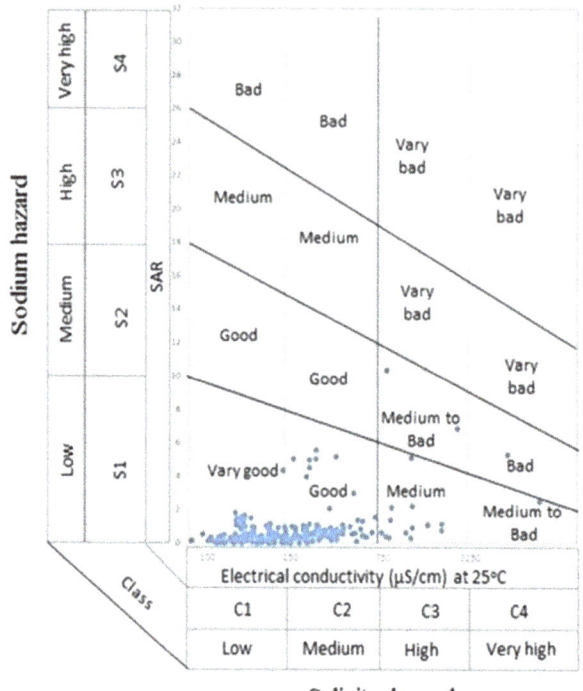

Figure 16. USSL diagram of water samples.

Samples from the C4S2 category belonged to the irrigation water of poor quality and could only be used on well-drained soils with caution to side effects. The USSL diagram indicates that the risk of salinization was expected in 8% of the tested irrigation water samples, and 82% of the samples could be considered as a very good or good category of irrigation water.

3.1.11. Irrigation Water Quality Index (IWQI)

The irrigation water quality index (IWQI) includes in the calculation only certain parameters of irrigation water quality based on the recommended limits for all soil types. Adimalla et al. [75] state that the IWQI method based on the analysis of irrigation water quality and its impact on soil and plants gives a clear categorization of the quality of applied water. It is based on the principle of comparing water quality parameters with specific standards, and defines irrigation water quality with a single value, thus avoiding water quality assessments that involve complex data intervals [11]. The irrigation water quality index is based on the recommended limits for continuous water use for all soil types [76,77] and indicates the following indicators of water quality for irrigation: salinity (which affects the availability of water to cultivated crops), permeability (which affects soil infiltration), toxicity (affects sensitive crops), and others. Based on two quality indicators, only certain irrigation water quality parameters were used for the IWQI calculation, as follows: electrical conductivity (EC), sodium adsorption ratio (SAR), and concentration of the ions such as sodium (Na^+), chloride (Cl^-), and bicarbonate (HCO_3^-) [24,78,79]. Based on the processed data of the estimated IWQI index, and the classification listed in Table 5, 164 samples of irrigation water, or 76.27%, were rated as excellent–no restriction water; 41 samples (19.07%) were estimated as good–low restriction water; 7 samples (3.26%) were rated as poor–moderate restriction water, and 3 samples (1.4%) as very poor–high restriction water. A graphical representation of the IWQI estimate is given in Figure 17.

Table 5. Classification of water quality for investigated sites based on IWQI.

IWQI	Exploitation Restrictions	Recommendation	
		Soil	Crops
(85–100)	No restriction (NR)	Water can be used for almost all types of soil. Soil is exposed to lower risks of salinity/sodicity problems	No toxicity risk for most plants
(70–85)	Low restriction (LR)	Irrigated soils with a light texture or moderate permeability can be adapted to this range. To avoid soil sodicity in heavy textures, soil leaching is recommended.	Elevated risks for salt sensitive plants
(55–70)	Moderate restriction (MR)	The water in this range would be better used for soils with moderate to high permeability values. Moderate leaching of salts is highly recommended to avoid soil degradation.	Plants with moderate tolerance to salts may be grow
(40–55)	High restriction (HR)	This range of water can be used in soils with high permeability without compact layers. High frquency irrigation schedule	Suitable for irrigation of plants with moderate to high tolerance to salts with special salinity control practices, except water with low Na, Cl and HCO_3 values
(0–40)	Severe restriction	Using this range of water for irrigaion under normal conditions should be avoided.	Only plants with high salt tolerance, except for waters with extremely low values of Na, Cl and HCO_3^-.

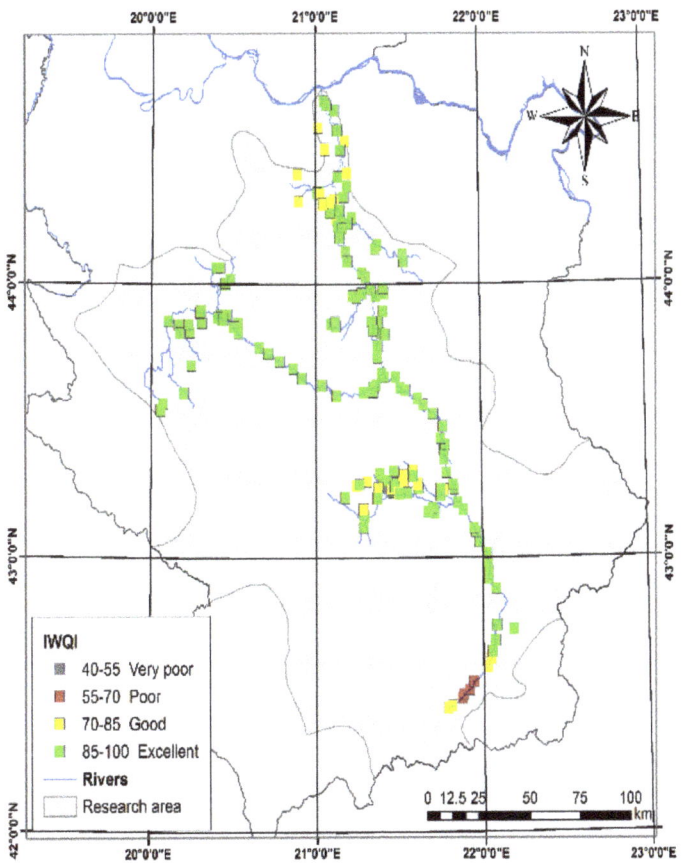

Figure 17. Spatial distribution of IWQI.

3.2. Geochemical Facies and Controlling Mechanisms

Most of the critical issues related to water hydrogeochemistry are often estimated based on the percentage concentrations of major *cations* and anions in meq L^{-1} in the [80] trilinear diagram. Piper's trilinear diagram is presented in Figure 18. The geochemical classification of water mainly depends on the concentration of cation and anion ions and their correlation. The Piper diagram is a frequently used and very efficient method for classifying water based on the basic geochemical characteristics of the major ions [16]. The chemical data of the analyzed samples collected from the research area were plotted in Piper's diagram (Figure 18). It was stated that there are three main hydrochemical types of tested samples of irrigation water, of which, type I: SO_4, Cl-Ca, Mg, belonged to 6.98% (15 water samples); type III: HCO_3, Na, belonged to 2.79% (6 water samples); type IV: HCO_3-Ca, Mg, belonged to 90.23% (194 samples). Based on the cationic triangle, it was noticeable that most of the samples belonged to the mixed zone, 54.43% of samples, followed by Ca^{2+} type with 23.25% of samples, then Mg^{2+} type with 19.53% of samples, and $Na^+ K^+$ type with 2.79% of samples. In the part of the anionic triangle, most of the samples, 93.02%, belonged to the type HCO_3^-, CO_3^-, followed by the mixed type with 6.52% of the samples, and only 0.46% of the samples belonged to the type Cl^-. The occurrence of individual examples of irrigation water samples with increased chloride content might be the result of pollution by sewage waste and leaching of salt residues in the soil [81], i.e., from household wastewater and untreated industrial waste [82]. The high concentration of Ca and Mg can be explained by the dissolution of dolomite limestones and Ca-Mg silicates (amphiboles, pyroxenes, olivine, biotite). Sodium and potassium in the aqueous system are obtained by atmospheric precipitation, dissolution of evaporites, and decomposition of silicates such as albite, anorthite, orthoclase, and microcline. The high concentration of K in some analyzed samples of irrigation water could be interpreted as a contribution of anthropogenic activities.

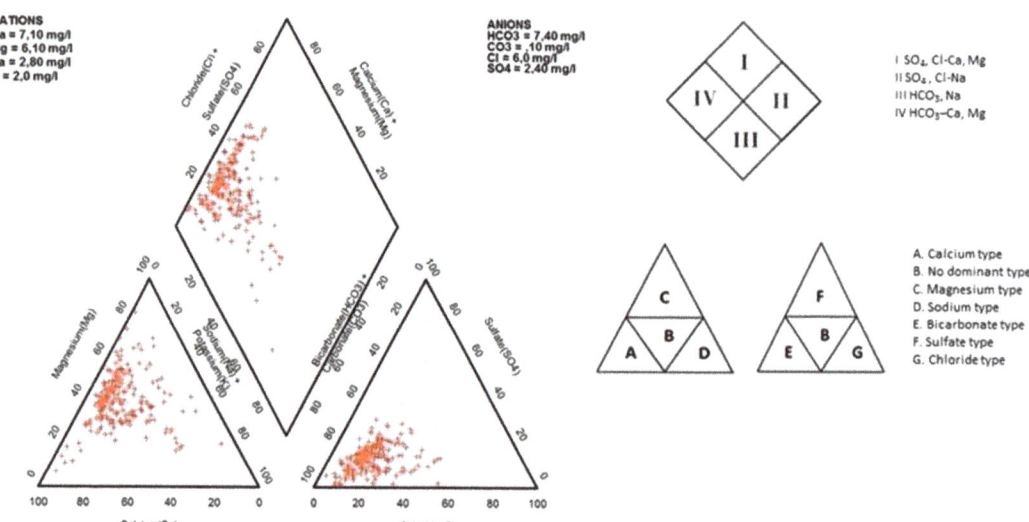

Figure 18. Piper's diagram for water samples.

In this paper, Chadha's diagram [83] was applied to interpret the hydrogeochemical properties of irrigation waters (Figure 19). It is formed by plotting the difference in meq L^{-1} between earth-alkaline ($Ca^{2+} + Mg^{2+}$) and alkali metals ($Na^+ + K^+$) on the X-axis, and the difference in meq L^{-1} between weakly acidic anions ($HCO_3^- + CO_3^{2-}$) and strong acid anions ($Cl^- + SO_4^{2-}$) on the Y-axis [84,85]. Each of the four fields formed by the above diagram has its hydrochemical significance. Field-1 (type $Ca^+–HCO_3^-$) indicates samples with recharging water filling capacity; Field-2 describes reverse ion exchange samples (Ca^+–Mg^+–Cl^- type); Field-3 indicates saltwater samples of outer members (type Na^+–Cl^-); Field-4 represents the description of the base ion exchange samples (type Na^+–HCO_3^-) [86]. The largest number of analyzed samples of irrigation water belonged to Field 1 (90.24%), followed by Field 4 (6.97%), and then Field 2 (2.79%), which was as per the findings from the Piper's diagram.

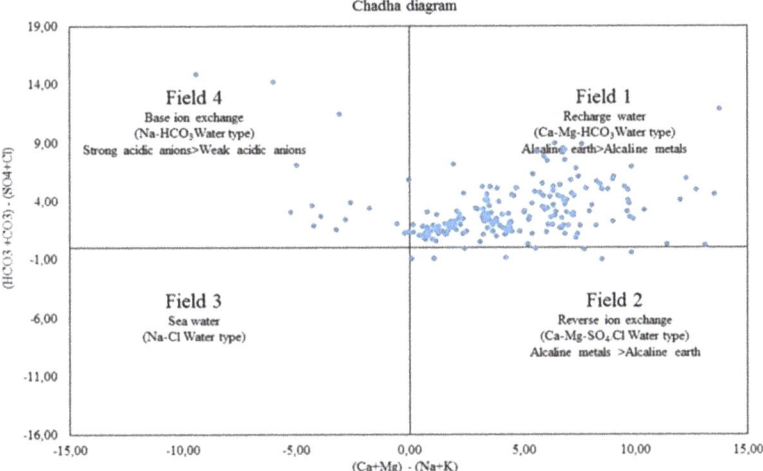

Figure 19. Geochemical classification based on Chadha's diagram.

Hydrochemical variability, among other things, can be shown by the Gibbs diagram, which shows the relationship between groundwater chemistry and aquifer lithology [87]. Based on this diagram, three main natural mechanisms can be found: the dominance of evaporation, the dominance of rocks, and the dominance of precipitation [30,88,89].

The Gibbs diagram can indicate the origin of solutes and hydrogeochemical processes [30]. It is a set of semi-logarithmic diagrams with the ratio of anions and *cations* ($Na^+/(Na^+ + Ca^{2+})$ and $Cl^-/(Cl^- + HCO_3^-)$), shown on the X-axis and TDS on the Y-axis (Figure 20). The diagram thus summarizes the most important natural mechanisms for controlling the chemical properties of water, i.e., precipitation control mechanisms, rock–geological substrate-control mechanisms, and evaporation-control mechanisms.

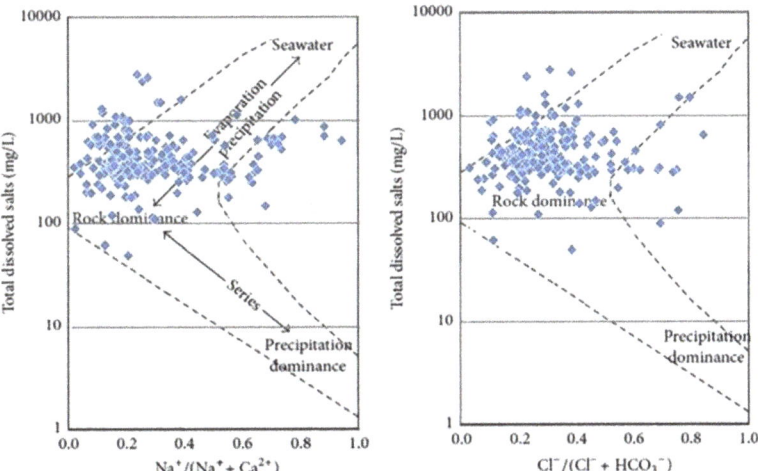

Figure 20. Gibbs diagram: (**a**) TDS vs. Na/(Na+Ca); (**b**) Cl/(Cl+HCO_3).

Although the Gibbs diagram can be used to determine the role of natural factors, it cannot exclude anthropogenic activities on the chemical properties of water. The Gibbs diagram also has some limitations. Water pollutants originating from anthropogenic activities such as mining, metallurgy and chemical industry, municipal communal services with their actions and contributions, can change the hydrochemical composition of water and increase the concentration of pollutants in water, such as Cl^-, SO_4^{2-} and TDS [90–92]. In addition, people change the hydrodynamic properties of water during the exploitation of water resources and thus affect the interactions of water and geological substrate (rocks) or the intensity of evaporation and change the concentration of individual elements.

Geochemical processes and their control mechanisms affect water quality and their suitability for irrigation. The similarities between the analyzed physicochemical components of the collected irrigation water samples were analyzed and shown graphically (Figure 21).

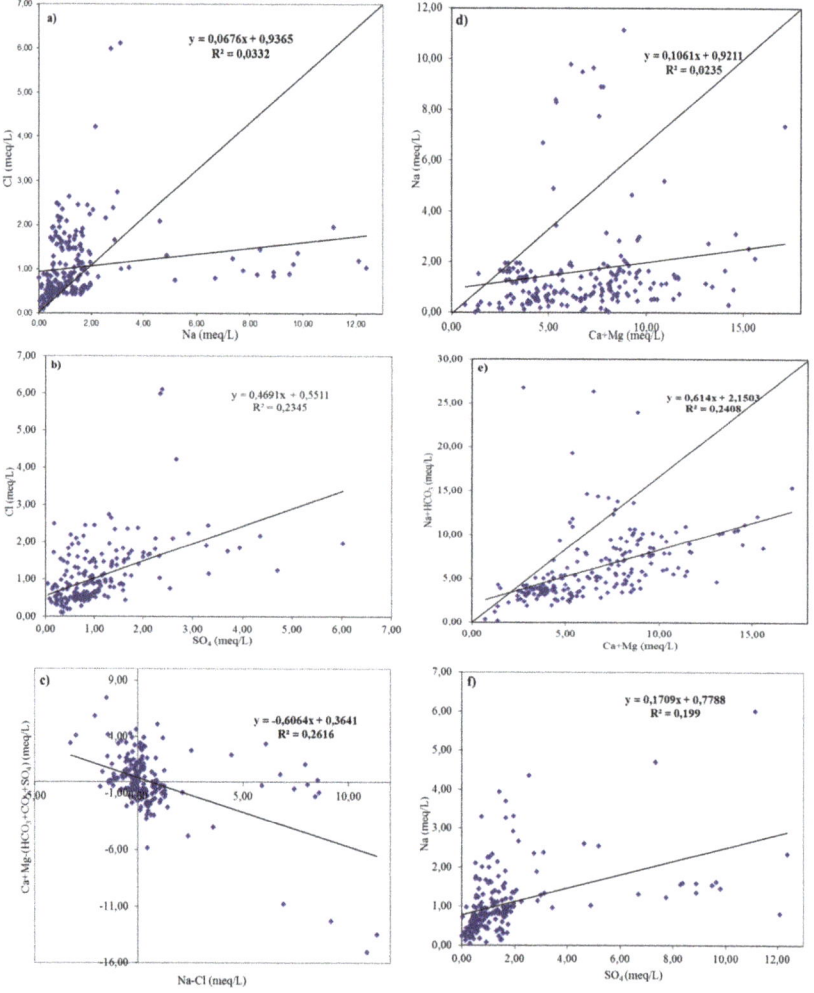

Figure 21. Major cation and anion relationships with ionic ratios (**a**) Na^+ vs. Cl^-, (**b**) SO_4^{2-} vs. Cl^-, (**c**) Na^+-Cl^- vs. Ca^{2+}+Mg^{2+}-(HCO_3^-+CO_3^{2-}+SO_4^{2-}), (**d**) Na^+ vs. Ca^{2+}+Mg^{2+}, (**e**) Ca^{2+}+Mg^{2+} vs. Na^++HCO_3^{3-}, (**f**) Na^+ vs. SO_4^{2-} on water samples.

The relationship between Na^+ and Cl^- is very important in identifying the potential occurrence of salinization (Figure 22). The correlation coefficient of the examined, listed parameters shows a low correlation ratio ($R^2 = 0.03$).

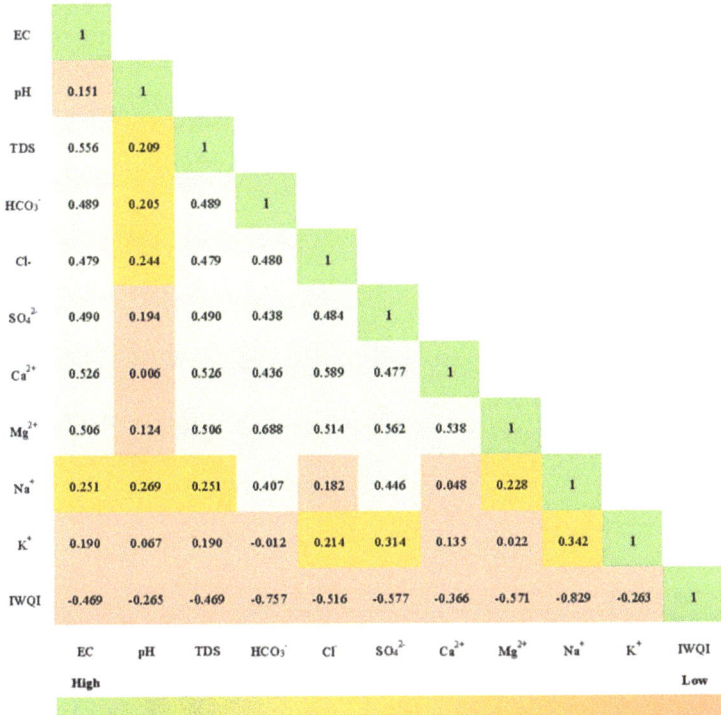

	EC	pH	TDS	HCO_3^-	Cl-	SO_4^{2-}	Ca^{2+}	Mg^{2+}	Na^+	K^+	IWQI
EC	1										
pH	0.151	1									
TDS	0.556	0.209	1								
HCO_3^-	0.489	0.205	0.489	1							
Cl-	0.479	0.244	0.479	0.480	1						
SO_4^{2-}	0.490	0.194	0.490	0.438	0.484	1					
Ca^{2+}	0.526	0.006	0.526	0.436	0.589	0.477	1				
Mg^{2+}	0.506	0.124	0.506	0.688	0.514	0.562	0.538	1			
Na^+	0.251	0.269	0.251	0.407	0.182	0.446	0.048	0.228	1		
K^+	0.190	0.067	0.190	-0.012	0.214	0.314	0.135	0.022	0.342	1	
IWQI	-0.469	-0.265	-0.469	-0.757	-0.516	-0.577	-0.366	-0.571	-0.829	-0.263	1

High   Low

Figure 22. Correlation matrix of hydrochemical parameters.

The same can be concluded from the relationship SO_4^{2-} vs. Cl^- (Figure 22), where the correlation ratio was determined ($R^2 = 0.23$), then from the ratio Na^+-Cl^- vs. Ca^{2+}+Mg^{2+}-(HCO_3^-+CO_3^{2-}+SO_4^{2-}), Figure 22, where the correlation coefficient was determined ($R^2 = 0.26$). Figure 22 shows the ratio Na^+ vs. Ca^{2+}+Mg^{2+}, where the correlation coefficient was determined ($R^2 = 0.02$). The correlation coefficient ($R^2 = 0.24$) was determined from the ratio Ca^{2+} + Mg^{2+} vs. Na^+ + HCO^{3-}, Figure 22, while from the ratio Na^+ vs. SO_4^{2-} the correlation coefficient ($R^2 = 0.19$) was determined, Figure 22.

Based on the conducted analysis, it can be concluded that there is no significant correlation between the analyzed parameters and that the probable disposal of untreated wastewater either from industry or anthropogenic origin in some samples leads to increased concentrations of Ca^{2+}, which is observed in some samples of irrigation water.

3.3. Multivariate Statistical Analysis

Groundwater quality is affected by various physicochemical variables and the degree of correlation between them can be assessed using a correlation matrix (Figure 22). The relationship between the two variables is established by estimating the correlation coefficient. Pearson was the first to develop this correlation analysis. A positive strong correlation represents the same sources of certain ions, while a weak correlation represents the sources of independent ions [93]. Analysis of the interdependence of variables was carried out by calculating linear Pearson correlation coefficients. It has been assumed that the regression modeling of the potential usefulness of the selected variable (explanatory) to model another variable (explained variable) determines the absolute value of the high correlation coefficient between these two variables. The statistical analysis usually assumes that if the correlation coefficient is >0.9, a very strong linear dependence exists; 0.7–0.9—significant linear dependence; 0.4–0.7—moderate linear dependence; 0.2–0.4—distinct linear dependence, but low; <0.2—no linear dependence [12].

The obtained results also imply moderate linear dependence for EC-TDS; EC-HCO$_3$; EC-Cl; EC-SO$_4$; EC-Ca; EC-Mg; TDS-HCO$_3$; TDS-Cl; TDS- SO$_4$; TDS-Ca; TDS-Mg; HCO^{3-}Cl; HCO$_3$-SO$_4$; HCO$_3$-Ca; HCO$_3$-Mg; HCO$_3$-Na; Cl-SO$_4$; Cl-Ca; Cl-Mg; SO$_4$-Ca; SO$_4$-Mg; SO$_4$-Na; Ca-Mg; end for EC-Na; EC-K; pH-TDS; pH- HCO$_3$; pH-Cl; pH-Na; TDS-Na; Cl-Na; SO$_4$-K; Mg-Na; Na-K distinct linear dependence, while for the rest of observed parameters there is no linear dependence (Figure 22).

The ion of Cl$^-$ shows a moderate correlation with Mg^{2+}, which indicates the possible leaching of secondary salts. The combination of SO$_4{}^{2-}$ with Ca^{2+} and Mg^{2+} can lead to the formation of insoluble salts such as CaSO$_4$ and MgSO$_4$, and the irrigation of arable soil with water containing these salts can cause their deposition on the surface and worsen its salinity, which affects the ecological environment of certain parts of the research area.

The correlation coefficients of TDS with HCO$_3{}^-$, Cl$^-$, SO$_4{}^{2-}$, Ca^{2+} and Mg^{2+} were higher than 0.4, which suggests that these five ions were dominant in the samples of tested irrigation water. Among these pairs, the correlation coefficient of HCO^{3-} and Mg^{2+} was the highest.

Several multivariate methods have been included in cluster analysis to identify the right groups of data sets, with similar groups belonging to the same class [94,95]. In cluster analysis, groups are divided based on similarity levels, and a dendrogram is formed where observations are combined. Cluster analysis (CA) was applied to determine the sources of changes in water resource quality by combining primary variables into a new set of variables. The results of the CA's basic physical and chemical parameters are presented (Figure 23). Two types of grouping were singled out, with the following parameters grouped in the same, Cluster I: CO$_3{}^2$, Na$^+$, K$^+$, pH, and the others in Cluster II, which was divided into two subclusters representing EC, Mg^{2+}, HCO^{3-} in one group and Cl$^-$, Ca^{2+}, TDS, SO$_4{}^{2-}$ in the other group.

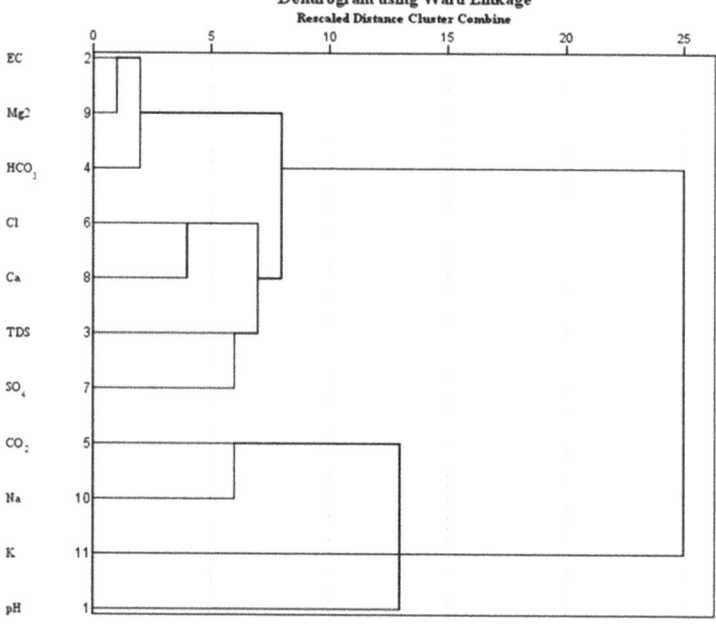

Figure 23. Cluster analysis (CA) for water physiochemical parameters for major ions.

The obtained results confirm that in the tested samples of water for irrigation the dominant *cations* were Ca^{2+}, Mg^{2+}, Na^{+}, and K^{+}, i.e., HCO_3^-, SO_4^{2-}, $C^-CO_3^{2-}$ were the dominant anions. These data are in agreement with the Piper diagram due to the effects of evaporation, weather, and anthropogenic influences. The interaction between geological substrate and water presented in Gibbs and Chadha's diagrams was also confirmed.

4. Conclusions

The assessment of the suitability of 215 tested irrigation water samples in the research area was performed based on the assessment of the hydrochemical results through the classifications SAR, Na%, RSC, PI, EC, TDS, MH, KI, PS, TH, IWQI. Most of the tested samples were suitable for irrigation and only a small number of samples were not suitable for the application.

Using the irrigation water quality index (IWQI), which was assessed based on comparing irrigation water quality parameters with specific standards with one value, it was determined that 95.34% of the tested samples were ranked as excellent and good, while poor and very poor were recorded in 4.66% of samples. By presenting the obtained results through Wilcox, Doneen, and USSL diagrams, the obtained results of observations were confirmed. The Piper's diagram showed that the dominant type of irrigation water in the study area was HCO3-Ca, Mg, which was found in 90.23% of the tested samples.

Based on Chadha's diagram, it was found that 90.21% of the tested irrigation water samples belong to the Ca^{2+}-HCO_3^- type, followed by 6.97% samples of the Na^+-HCO_3^- type, and 2.79% samples of the Ca^{2+}-$Mg^{2+}Cl^-$ type, which is as per Piper's diagram. The Gibbs diagram determined that there was no significant correlation between the analyzed parameters and that the probable disposal of untreated wastewater, either from industry or anthropogenic origin, led to increased concentrations of Ca^{2+} in some samples of irrigation water.

Industrial and intensive agricultural production, as well as anthropogenic pollution by the inflow from urban domestic sewage near the sampling site, were most likely the cause of inadequate quality irrigation water samples.

The graphical presentation of each of the examined parameters highlights the risk zones based on which it is possible to propose the application of some of the measures that will contribute to mitigating or eliminating identified deficiencies and problems. This will improve the current reporting approach and provide a basis for monitoring the quality of irrigation water in existing and planned irrigation systems.

This paper emphasizes the need to establish a real-time monitoring system for irrigation water quality at the research site. As the study area is characterized by intensive agricultural production, as such, it requires the establishment of continuous monitoring and risk management through tools for generating rapid reports, which would be primarily useful to policymakers and decision-makers on the use of irrigation water of the appropriate quality. It can be concluded that reporting can be carried out using the irrigation water quality index (IWQI).

Author Contributions: Conceptualization, R.P., A.S.-S., J.M. and Z.D.; fieldwork, R.P., A.S.-S. and J.M.; methodology, R.P., Z.D., H.M. and D.J.; software, D.J.; validation, A.S.-S., J.M. and Z.D.; formal analysis, J.M. and Z.D.; investigation, R.P., A.S.-S., J.M. and Z.D.; resources, R.P., A.S.-S., J.M., D.J., D.V. and Z.D.; data curation, R.P. and D.J.; writing—original draft preparation, R.P. and A.S.-S.; writing—review and editing, R.P., A.S.-S., J.M., D.J. and Z.D.; supervision, R.P.; project administration, R.P. All authors have read and agreed to the published version of the manuscript.

Funding: The study was financially supported by the Ministry of Education and Science of the Republic of Serbia, Project 451-03-68/2022-14/200011.

Institutional Review Board Statement: Not applicable.

Informed Consent Statement: Not applicable.

Data Availability Statement: All data are provided as tables and figures.

Acknowledgments: Ministry of Agriculture, Water Management and Forestry, Agricultural Land Administration, Republic of Serbia.

Conflicts of Interest: The authors declare that they have no known competing financial interest or personal relationships that could have appeared to influence the work reported in this paper.

References

1. Pivić, R.; Dinić, Z.; Stanojković Sebić, A. *Irrigation Water Quality of Drina River in Republic of Serbia*; LAP Lambert Academic Publishing, OmniScriptum GmbH & Co KG: Sunnyvale, CA, USA; Saarbrücken, Germany, 2017; p. 84.
2. Pivić, R.; Stanojković Sebić, A.; Maksimović, J.; Dinić, D. Soils of central Serbia areas and quality of available water for irrigation. Serbian Academy of Sciences and Arts. In *Rational Use of Land and Water in Serbia, Proceedings of the III Scientific Meeting of the Department of Chemical and Biological Sciences, Banja Koviljaca, Serbia, 25 September 2020*; Serbian Academy of Sciences and Arts: Belgrade, Serbia, 2020; pp. 71–87. (In Serbian)
3. Statistical Office of the Republic of Serbia. *Report of Irrigation in 2021*; Statistical Office Republic of Serbia: Belgrade, Serbia, No. 002-year LXXII; 2021. (In Serbian)
4. Ramakrishnaiah, C.R.; Adashiv, C.; Ranganna, G. Assessment of water quality index for the groundwater in Tumkur Taluk, Karnataka State, India. *E-J. Chem.* **2009**, *6*, 523–530. [CrossRef]
5. Ayers, R.S.; Westcot, D.W. *Water Quality for Agriculture*; FAO Irrigation and Drainage Paper: Rome, Italy, 1985; p. 29.
6. Mirabbasi, R.; Mazloumzadeh, S.M.; Rahnama, M.B. Evaluation of irrigation water quality using fuzzy logic. *Res. J. Environ. Sci.* **2008**, *2*, 340–352. [CrossRef]
7. Yıldız, S.; Karakuş, C.B. Estimation of irrigation water quality index with development of an optimum model: A case study. *Environ. Dev. Sustain.* **2020**, *22*, 4771–4786. [CrossRef]
8. Carpenter, S.R.; Caraco, N.F.; Correll, D.L.; Howarth, R.W.; Sharpley, A.N.; Smith, V.H. Nonpoint pollution of surface waters with phosphorus and nitrogen. *Ecol. Appl.* **1998**, *8*, 559–568. [CrossRef]
9. Jarvie, H.P.; Whitton, B.A.; Neal, C. Nitrogen and phosphorus in east coast British rivers: Speciation, sources and biological significance. *Sci. Total Environ.* **1998**, *210–211*, 79–109. [CrossRef]
10. Simeonova, V.; Stratisb, J.A.; Samarac, C.; Zachariadisb, G.; Voutsac, D.; Anthemidis, A.; Sofonioub, M.; Kouimtzisc, T. Assessment of the surface water quality in Northern Greece. *Water Res.* **2003**, *37*, 4119–4124. [CrossRef]
11. Ghazaryan, K.; Chen, Y. Hydrochemical assessment of surface water for irrigation purposes and its influence on soil salinity in Tikanlik oasis, China. *Environ. Earth Sci.* **2016**, *75*, 383. [CrossRef]
12. Amezketa, E. An integrated methodology for assessing soil salinization, a pre-condition for land desertification. *J. Arid. Environ.* **2006**, *67*, 594–606. [CrossRef]
13. Singh, S.; Ghosh, N.C.; Gurjar, S.; Krishan, G.; Kumar, S.; Berwal, P. Index-based assessment of suitability of water quality for irrigation purpose under Indian conditions. *Environ. Monit. Assess.* **2018**, *190*, 29. [CrossRef]
14. Singh, A.K.; Mondal, G.C.; Kumar, S.; Singh, T.B.; Tewary, B.K.; Sinha, A. Major ion chemistry, weathering processes and water quality assessment in upper catchment of Damodar River basin, India. *Environ Geol.* **2008**, *54*, 745–758. [CrossRef]
15. Etteieb, S.; Cherif, S.; Tarhouni, J. Hydrochemical assessment of water quality for irrigation: A case study of the Medjerda River in Tunisia. *Appl. Water Sci.* **2017**, *7*, 469–480. [CrossRef]
16. Zhang, W.; Ma, L.; Abuduwaili, J.; Ge, Y.; Issanova, G.; Saparov, G. Hydrochemical characteristics and irrigation suitability of surface water in the Syr Darya River, Kazakhstan. *Environ. Monit. Assess.* **2019**, *191*, 572. [CrossRef] [PubMed]
17. Gad, M.; El-Hattab, M. Integration of water pollution indices and DRASTIC model for assessment of groundwater quality in El Fayoum Depression, Western Desert, Egypt. *J. Afr. Earth Sci.* **2019**, *158*, 103554. [CrossRef]
18. Hem, J.D. *Study and Interpretation of the Chemical Characteristics of Natural Water*; Department of the Interior, US Geological: Alexandria, VA, USA, 1985; Volume 2254, pp. 117–120.
19. Cieszynska, M.; Wesolowski, M.; Bartoszewicz, M.; Michalska, M.; Nowacki, J. Application of physicochemical data for water-quality assessment of watercourses in the Gdansk Municipality (South Baltic coast). *Environ. Monit. Assess.* **2012**, *184*, 2017–2029. [CrossRef] [PubMed]
20. Li, P.; Wu, J.; Qian, H. Assessment of groundwater quality for irrigation purposes and identification of hydrogeochemical evolution mechanisms in Pengyang County, China. *Environ. Earth Sci.* **2013**, *69*, 2211–2225. [CrossRef]
21. Fakhre, A. Evaluation of hydrogeochemical parameters of groundwater for suitability of domestic and irrigational purposes: A case study from central Ganga Plain, India. *Arab. J. Geosci.* **2014**, *7*, 4121–4131. [CrossRef]
22. Saeedi, M.; Abessi, O.; Sharifi, F.; Meraji, H. Development of groundwater quality index. *Environ. Monit. Assess.* **2010**, *163*, 327–335. [CrossRef]
23. Meireles, A.; Andrade, E.M.; Chaves, L.; Frischkorn, H.; Crisostomo, L.A. A new proposal of the classification of irrigation water. *Rev. Cienc. Agron.* **2010**, *413*, 349–357. [CrossRef]
24. Abbasnia, A.; Radfard, M.; Mahvi, A.H.; Nabizadeh, R.; Yousefi, M.; Soleimani, H.; Alimohammadi, M. Groundwater quality assessment for irrigation purposes based on irrigation water quality index and its zoning with GIS in the villages of Chabahar, Sistan and Baluchistan, Iran. *Data Brief* **2018**, *19*, 623–631. [CrossRef]

25. Batarseh, M.; Imreizeeq, E.; Tilev, S.; Al Alaween, M.; Wael Suleiman, W.; Al Remeithi, A.M.; Al Tamimi, M.K.; Al Alawneh, M. Assessment of groundwater quality for irrigation in the arid regions using irrigation water quality index (IWQI) and GIS-Zoning maps: Case study from Abu Dhabi Emirate, UAE. *Groundw. Sustain. Dev.* **2021**, *14*, 100611. [CrossRef]
26. Hussein, H.A.; Ricka, A.; Kuchovsky, T.; El Osta, M.M. Groundwater hydrochemistry and origin in the south-eastern part of Wadi El Natrun, Egypt. *Arab. J. Geosci.* **2017**, *10*, 170–184. [CrossRef]
27. Richards, L.A. Diagnosis and Improvement of Saline Alkali Soils. In *Agriculture Handbook*; US Department of Agriculture: Washington, DC, USA, 1954; Volume 60, p. 160.
28. Wilcox, L.V. *Classification and Use of Irrigation Waters*; U.S. Department of Agriculture: Washington, DC, USA, 1955; p. 367.
29. Doneen, L.D. *Notes on Water Quality in Agriculture. Published as a Water Science and Engineering*; Department of Water Sciences and Engineering, University of California: Oakland, CA, USA, 1964; Volume 4001.
30. Gibbs, R.J. Mechanisms controlling world water chemistry. *Science* **1970**, *170*, 10881. [CrossRef] [PubMed]
31. Gad, M.; El Osta, M. Geochemical controlling mechanisms and quality of the groundwater resources in El Fayoum Depression, Egypt. *Arab. J. Geosci.* **2020**, *13*, 861. [CrossRef]
32. Kamtchueng, B.T.; Fantong, W.Y.; Wirmvem, M.J. Hydrogeochemistry and quality of surface water and groundwater in the vicinity of Lake Monoun, West Cameroon: Approach from multivariate statistical analysis and stable isotopic characterization. *Environ. Monit. Assess.* **2016**, *188*, 524. [CrossRef]
33. Rakotondrabe, F.; Ngoupayou, J.R.; Mfonka, Z. Water quality assessment in the Betare-Oya gold mining area (East-Cameroon): Multivariate statistical analysis approach. *Sci. Total Environ.* **2018**, *610*, 831–844. [CrossRef] [PubMed]
34. Li, P.; Tian, R.; Liu, R. Solute geochemistry and multivariate analysis of water quality in the Guohua Phosphorite Mine, Guizhou Province, China. *Expo. Health* **2019**, *11*, 81–94. [CrossRef]
35. Tanasijević, Đ.; Antonović, G.; Aleksić, Ž.; Pavićević, N.; Filipović, Đ.; Spasojević, M. *The Soils of the West and Northwestern Serbia*; Institute of Soil Science in Topčider: Belgrade, Serbia, 1966. (In Serbian)
36. Tanasijević, Đ.; Antonović, G.; Kovačević, R.; Aleksić, Ž.; Pavićević, N.; Filipović, Đ.; Jeremić, M.; Vojinović, Ž.; Spasojević, M. *The Soils of the Velika Morava and Mlava basin*; Institute of Soil Science in Topčider: Belgrade, Serbia, 1965. (In Serbian)
37. IUSS Working Group WRB. *World Reference Base for Soil Resources 2014. International Soil Classification System for Naming soils and Creating Legends for Soil Maps. Update 2015*; World Soil Resources Reports; FAO: Rome, Italy, 2015; Volume 106.
38. *Z1.111:1987*; Measurement of pH—Potentiometric Method. Institute for Standardization of Serbia SRPS: Belgrade, Serbia, 1987.
39. *SRPS EN 27888:1993*; Determination of Electrical Conductivity. Institute for Standardization of Serbia SRPS: Belgrade, Serbia, 1993.
40. APHA. *Standard Methods for the Examination of Water and Wastewater*, 18th ed.; Greenberg, A.E., Clesceri, L.S., Eaton, A.D., Eds.; American Public Health Association: Washington, DC, USA, 1992.
41. Rhoades, J.D.; Kandiah, A.; Mashali, M. *The Use of Saline Waters for Crop Production—FAO Irrigation & Drainage Paper*; FAO: Rome, Italy, 1992; Volume 48.
42. *ISO 11885:2007*; Water Quality—Determination of Selected Elements by Inductively Coupled Plasma Optical Emission Spectrometry (ICP-OES). ISO: Geneva, Switzerland, 2007.
43. APHA. *Standard Methods for the Examination of Water and Wastewater*, 21st ed.; American Public Health Association/American Water Works Association/Water Environment Federation: Washington, DC, USA, 2005.
44. Li, P.; Zhang, Y.; Yang, N.; Jing, L.; Yu, P. Major ion chemistry and quality assessment of groundwater in and around a mountainous tourist town of China. *Expo. Health* **2016**, *8*, 239–252. [CrossRef]
45. Deutsch, W.J. *Groundwater Geochemistry Fundamentals and Application to Contamination*; Lewis Publishers: Boca Raton, FL, USA, 1997.
46. Raghunath, H.M. *Groundwater*, 2nd ed.; Wiley Eastern Ltd.: New Delhi, India, 1987; p. 563.
47. Kelley, W.P. Use of saline irrigation water. *Soil Sci.* **1963**, *95*, 385–391. [CrossRef]
48. Durfor, C.N.; Becker, E. *Public Water Supplies of the 100 Largest Cities in the United States. Water Supply Paper*; Government Publishing Office: Washington, DC, USA, 1964; Volume 1812, pp. 343–346. [CrossRef]
49. Begum, A.; Harikrishna, R. Study on the Quality of Water in Some Streams of Cauvery River. *J. Chem.* **2008**, *2*, 377–384. [CrossRef]
50. Narsimha, A.; Sudarshan, V. Contamination of fluoride in groundwater and its effect on human health: A case study in hard rock aquifers of Siddipet, Telangana State, India. *Appl. Water Sci.* **2017**, *7*, 2501–2512. [CrossRef]
51. Adimalla, N.; Venkatayogi, S. Geochemical characterization and evaluation of groundwater suitability for domestic and agricultural utility in semi-arid region of Basara, Telangana State, South India. *Appl. Water Sci.* **2018**, *8*, 44. [CrossRef]
52. Sharma, D.A.; Rishi, M.S.; Keesari, T. Evaluation of groundwater quality and suitability for irrigation and drinking purposes in southwest Punjab, India using hydrochemical approach. *Appl. Water Sci.* **2017**, *7*, 3137–3150. [CrossRef]
53. Almeida, C.; Quintar, S.; González, P.; Mallea, M. Assessment of irrigation water quality. A proposal of a quality profile. *Environ. Monit. Assess.* **2008**, *142*, 149–152. [CrossRef] [PubMed]
54. Todd, D.K. *Groundwater Hydrology*, 1st ed.; John Wiley and Sons, Inc.: Hoboken, NJ, USA, 1960; p. 336.
55. González-Acevedo, Z.I.; Padilla-Reyes, D.A.; Ramos-Leal, J.A. Quality assessment of irrigation water related to soil salinization in Tierra Nueva, San Luis Potosí, Mexico. *Rev. Mex. De Cienc. Geológicas* **2016**, *33*, 271–285.
56. Rao, N.K.; Latha, P.S. Groundwater quality assessment using water quality index with a special focus on vulnerable tribal region of Eastern Ghats hard rock terrain, Southern India. *Arab. J. Geosci.* **2019**, *12*, 267. [CrossRef]

57. Sutradhar, S.; Mondal, P. Groundwater suitability assessment based on water quality index and hydrochemical characterization of Suri Sadar Sub-division, West Bengal. *Ecol. Inform.* **2021**, *64*, 101335. [CrossRef]
58. Vasanthavigar, M.; Srinivasamoorthy, K.; Prasanna, M.V. Evaluation of groundwater suitability for domestic, irrigational, and industrial purposes: A case study from Thirumanimuttar river basin, Tamilnadu, India. *Environ Monit Assess.* **2012**, *184*, 405–420. [CrossRef]
59. Taloor, A.K.; Pir, R.A.; Adimalla, N.; Ali, S.; Manhas, D.S.; Roy, S.; Singh, A.K. Spring water quality and discharge assessment in the Basantar watershed of Jammu Himalaya using geographic information system (GIS) and water quality Index (WQI). *Groundw. Sustain. Dev.* **2020**, *10*, 100364. [CrossRef]
60. Salifu, M.; Aidoo, F.; Hayford, M.S.; Adomako, D.; Asare, E. Evaluating the suitability of groundwater for irrigational purposes in some selected districts of the Upper West region of Ghana. *Appl. Water Sci.* **2017**, *7*, 653–662. [CrossRef]
61. Eaton, F.M. Significance of carbonates in irrigation waters. *Soil Sci.* **1950**, *69*, 123–134. [CrossRef]
62. Singaraja, C. Relevance of water quality index for groundwater quality evaluation: Thoothukudi District, Tamil Nadu, India. *Appl. Water Sci.* **2017**, *7*, 2157–2173. [CrossRef]
63. Ravikumar, P.; Mehmood, M.A.; Somasheka, R.K. Water quality index to determine the surface water quality of Sankey tank and Mallathahalli lake, Bangalore urban district, Karnataka, India. *Appl. Water Sci.* **2013**, *3*, 247–261. [CrossRef]
64. Rawat, K.S.; Singh, S.K.; Gautam, S.K. Assessment of groundwater quality for irrigation use: A peninsular case study. *Appl. Water Sci.* **2018**, *8*, 233. [CrossRef]
65. Singh, K.K.; Tewari, G.; Kumar, S. Evaluation of groundwater quality for suitability of irrigation purposes: A Case Study in the Udham Singh Nagar, Uttarakhand. *J. Chem.* **2020**, *2020*, 6924026. [CrossRef]
66. Li, P.; Wu, J.; Qian, H.; Zhang, Y.; Yang, N.; Jing, L.; Yu, P. Hydrogeochemical characterization of groundwater in and around a wastewater irrigated forest in the southeastern edge of the Tengger Desert, Northwest China. *Expo. Health* **2016**, *8*, 331–348. [CrossRef]
67. El-Amier, Y.A.; Kotb, W.K.; Bonanomi, G.; Fakhry, H.; Marraiki, N.A.; Abd-ElGawad, A.M. Hydrochemical Assessment of the Irrigation Water Quality of the El-Salam Canal, Egypt. *Water* **2021**, *13*, 2428. [CrossRef]
68. Szabolcs, I.; Darab, C. The Influence of Irrigation Water of High Sodium Carbonate Content of Soils. In Proceedings of the 8th International Congress of ISSS, Bucharest, Romania, 31 August–9 September 1964; Volume II, pp. 803–812.
69. Keesari, T.; Kulkarni, U.P.; Deodhar, A.; Ramanjaneyulu, P.S.; Sanjukta, A.K.; Kumar, U.S. Geochemical characterization of groundwater from an arid region in India. *Environ. Earth Sci.* **2014**, *71*, 4869–4888. [CrossRef]
70. Kelley, W.P. Permissible Composition and Concentration of Irrigated Waters. *Proc. Am. Soc. Civ. Eng.* **1940**, *66*, 607–613.
71. Dhembare, A.J. Assessment of water quality indices for irrigation of Dynaneshwar Dam Water, Ahmednagar, Maharashtra, India. *Arch. Appl. Sci. Res.* **2012**, *4*, 348–352.
72. Sundaray, S.K.; Nayak, B.B.; Bhatta, D. Environmental studies on river water quality with reference to suitability for agricultural purposes: Mahanadi river estuarine system, India—A case study. *Environ. Monit. Assess.* **2009**, *155*, 227–243. [CrossRef]
73. He, S.; Li, P. A MATLAB based graphical user interface (GUI) for quickly producing widely used hydrogeochemical diagrams. *Geochemistry* **2019**, *80*, 125550. [CrossRef]
74. Sadashivaiah, C.; Ramakrishnaiah, C.; Ranganna, G. Hydrochemical analysis and evaluation of groundwater quality in tumkur taluk, Karnataka state, India. *Int. J. Environ. Res. Publ. Health* **2008**, *5*, 158–164. [CrossRef] [PubMed]
75. Adimalla, N.; Taloor, A.K. Hydrogeochemical investigation of groundwater quality in the hard rock terrain of South India using Geographic Information System (GIS) and groundwater quality index (GWQI) techniques. *Groundw. Sustain. Dev.* **2020**, *10*, 100288. [CrossRef]
76. Stoner, J.D. *Water-Quality Indices for Specific Water Uses*; Department of the Interior, Geological Survey: Arlington, VA, USA, 1978; p. 12. [CrossRef]
77. Bortolini, L.; Maucieri, C.; Borin, M. A tool for the evaluation of irrigation water quality in the arid and semi-arid regions. *Agronomy* **2018**, *8*, 23. [CrossRef]
78. Spandana, M.P.; Suresh, K.R.; Prathima, B. Developing an irrigation water quality index for vrishabavathi command area. *Int. J. Eng. Res. Technol.* **2013**, *2*, 821–830.
79. Zaman, M.; Shahid, S.A.; Heng, L. Irrigation Water Quality. In *Guideline for Salinity Assessment, Mitigation and Adaptation Using Nuclear and Related Techniques*; Springer International Publishing: Cham, Switzerland, 2018; p. 113. [CrossRef]
80. Piper, A.M. A graphic procedure in the geochemical interpretation of water-analyses. *EOS Trans. Am. Geophys. Union.* **1944**, *25*, 914–928. [CrossRef]
81. Appelo, C.A.J.; Postma, D.; Balkema, A.A. *Geochemistry, Groundwater and Pollution*; Balkema, A.A., Ed.; Cambridge University Press: Rotterdam, The Netherlands, 1993; p. 536.
82. El Osta, M.; Milad Masoud, M.; Alqarawy, A.; Elsayed, S.; Gad, M. Groundwater Suitability for Drinking and Irrigation Using Water Quality Indices and Multivariate Modeling in Makkah Al-Mukarramah Province, Saudi Arabia. *Water* **2022**, *14*, 483. [CrossRef]
83. Chadha, D.K. A proposed new diagram for geochemical classification of natural waters and interpretation of chemical data. *Hydrogeol. J.* **1999**, *7*, 431–439. [CrossRef]
84. Varol, S.; Davraz, A. Assessment of geochemistry and hydrogeochemical processes in groundwater of the Tefenni plain (Burdur/Turkey). *Environ. Earth Sci.* **2014**, *71*, 4657–4673. [CrossRef]

85. Roy, A.; Keesari, T.; Mohokar, H.; Sinha, U.K.; Bitra, S. Assessment of groundwater quality in hard rock aquifer of central Telangana state for drinking and agriculture purposes. *Appl. Water Sci.* **2018**, *8*, 124. [CrossRef]
86. Chen, J.; Huang, Q.; Lin, Y.; Fang, Y.; Qian, H.; Liu, R.; Ma, H. Hydrogeochemical Characteristics and Quality Assessment of Groundwater in an Irrigated Region, Northwest China. *Water* **2019**, *11*, 96. [CrossRef]
87. Xu, P.; Feng, W.; Qian, H.; Zhang, Q. Hydrogeochemical Characterization and Irrigation Quality Assessment of Shallow Groundwater in the Central-Western Guanzhong Basin, China. *Int. J. Environ. Res. Public Health* **2019**, *16*, 1492. [CrossRef] [PubMed]
88. Wu, J.; Li, P.; Qian, H. Hydrochemical characterization of drinking groundwater with special reference to fluoride in an arid area of China and the control of aquifer leakage on its concentrations. *Environ. Earth Sci.* **2015**, *73*, 8575–8588. [CrossRef]
89. Nurlan, S.S. Current hydroecological state of the lower watercourse of Syrdarya and use of her resources of the drain. *Astrakhan Bull. Ecol. Educ.* **2017**, *2*, 50–55.
90. Issanova, G.; Jilili, R.; Abuduwaili, J.; Kaldybayev, A.; Saparov, G.; Yong Xiao, G. Water availability and state of water resources within water-economic basins in Kazakhstan. *Paddy Water Environ.* **2018**, *16*, 183–191. [CrossRef]
91. Karlykhanov, O.K.; Toktaganova, G.B. The assessment of irrigated land salinization in the Aral Sea Region. *Int. J. Environ. Sci. Educ.* **2016**, *11*, 7946–7960.
92. Pearson, K. Mathematical Contributions to the Theory of Evolution. III. Regression, Heredity, and Panmixia. *Philos. Trans. R. Soc. A Math. Phys. Eng. Sci.* **1896**, *187*, 253–318. [CrossRef]
93. Islam, A.R.M.T.; Mamun, A.A.; Rahman, M.M.; Zahid, A. Simultaneous comparison of modified-integrated water quality and entropy weighted indices: Implication for safe drinking water in the coastal region of Bangladesh. *Ecol. Indic.* **2020**, *113*, 106229. [CrossRef]
94. Danielsson, Å.; Cato, I.; Carman, R.; Rahm, L. Spatial clustering of metals in the sediments of the Skagerrak/Kattegat. *Appl. Geochem.* **1999**, *14*, 689–706. [CrossRef]
95. Lattin, J.M.; Carroll, J.D.; Green, P.E.; Green, P.E. *Analyzing Multivariate Data*; Thomson Brooks/Cole: Pacific Grove, CA, USA, 2003.

Article

Effects of Irrigation Schedules on Maize Yield and Water Use Efficiency under Future Climate Scenarios in Heilongjiang Province Based on the AquaCrop Model

Tangzhe Nie [1], Yi Tang [1], Yang Jiao [1], Na Li [1], Tianyi Wang [1,2], Chong Du [1,*], Zhongxue Zhang [2,3], Peng Chen [4], Tiecheng Li [2,3], Zhongyi Sun [5] and Shijiang Zhu [6]

1 School of Water Conservancy and Electric Power, Heilongjiang University, Harbin 150080, China; 2019036@hlju.edu.cn (T.N.); ty979794@163.com (Y.T.); jiaoyang0814@163.com (Y.J.); L17863523371@163.com (N.L.); wangtianyi7176@163.com (T.W.)
2 School of Water Conservancy and Civil Engineering, Northeast Agricultural University, Harbin 150030, China; zhangzhongxue@163.com (Z.Z.); litiecheng1212@126.com (T.L.)
3 Key Laboratory of Efficient Use of Agricultural Water Resources, Ministry of Agriculture and Rural Affairs, Northeast Agricultural University, Harbin 150030, China
4 College of Agricultural Science and Engineering, Hohai University, Nanjing 210098, China; chenpeng_isotope@163.com
5 College of Ecology and Environment, Hainan University, Haikou 570208, China; gis.rs@hainanu.edu.cn
6 College of Hydraulic and Environmental Engineering, China Three Gorges University, Yichang 443002, China; zhusjiang@aliyun.com
* Correspondence: duchong@hlju.edu.cn; Tel.: +86-136-3366-1083

Citation: Nie, T.; Tang, Y.; Jiao, Y.; Li, N.; Wang, T.; Du, C.; Zhang, Z.; Chen, P.; Li, T.; Sun, Z.; et al. Effects of Irrigation Schedules on Maize Yield and Water Use Efficiency under Future Climate Scenarios in Heilongjiang Province Based on the AquaCrop Model. *Agronomy* 2022, 12, 810. https://doi.org/10.3390/agronomy12040810

Academic Editors: Pantazis Georgiou and Dimitris Karpouzos

Received: 3 March 2022
Accepted: 26 March 2022
Published: 27 March 2022

Publisher's Note: MDPI stays neutral with regard to jurisdictional claims in published maps and institutional affiliations.

Copyright: © 2022 by the authors. Licensee MDPI, Basel, Switzerland. This article is an open access article distributed under the terms and conditions of the Creative Commons Attribution (CC BY) license (https://creativecommons.org/licenses/by/4.0/).

Abstract: Predicting the impact of future climate change on food security has important implications for sustainable food production. The 26 meteorological stations' future climate data in the study area are assembled from four global climate models under two representative concentration pathways (RCP4.5 and RCP8.5). The future maize yield, actual crop evapotranspiration (ET_a), and water use efficiency (*WUE*) were predicted by calibrated AquaCrop model under two deficit irrigation (the regulated deficit irrigation (RDI) at jointing stage(W1), filling stage(W2)), and full irrigation (W3) during the three periods (2021–2040, 2041–2060, and 2061–2080). The result showed that the maize yields under W1, W2, and W3 of RCP4.5 were 2.8%, 2.9%, and 2.5% lower than those in RCP8.5, respectively. In RCP8.5, the yield of W3 was 1.9% and 1.4% higher than W1 and W2, respectively. Under the RCP4.5, the ET_a of W1, W2, and W3 was 481.32 mm, 484.94 mm, and 489.12 mm, respectively. Moreover, the ET_a of W1 was significantly lower than W2 under the RCP4.5 and RCP8.5 ($p > 0.05$). In conclusion, regulated deficit irrigation at the maize jointing stage is recommended in the study area when considering *WUE*.

Keywords: maize; AquaCrop; actual crop evapotranspiration (ET_a); yield; water use efficiency (*WUE*); irrigation schedule; climate change

1. Introduction

Climate change has a great impact on agricultural systems. [1]. Climate change, characterized by temperature rise, the uncertain amount and patterns of precipitation (*Pe*), and elevated atmospheric CO_2 concentration [2], is a widely concerned issue in global development [3]. According to the Intergovernmental Panel on Climate Change (IPCC), the future temperature will increase by 1.5 °C or higher, particularly significant in the high latitudes and tropical regions [4]. In order to maintain the stable development of the agricultural economy and food security, it is necessary to predict the impact of future climate change on crop growth [5]. Due to sustaining the needs of the increasing population, the modern era model would replace the traditional model. Crop models would be one of the tools of future agricultural research. A well-efficient and validated model can be used to optimize resources and predict the yield [6].

With the temperature increasing, fluctuations in both amount and frequency of *Pe* and changes in CO_2 concentration would all directly impact crop evapotranspiration, actual evapotranspiration (ET_a), and irrigation water requirements [7]. With the crop growing, potential evapotranspiration and net irrigation water requirements would decrease if CO_2 concentrations increased [8]. Furthermore, the high temperature can actively promote the growth of most crops, advance crop phenology, shorten crop growth period, and reduce the cumulative biomass of crops, thus affecting the final yield [9,10]. Crop yields may vary due to climate change impacts in different regions, the region's latitude, and irrigation applications [11]. The maize's yield under normal, critical, and minimum irrigation in the North China Plain varied between 10,964–11,235 kg/ha [12]. In the Adana region, the highest maize yield (10,075 kg/ha) was obtained when irrigation limits were set between 25% ready water available (RAW) depletion and field capacity (FC), while the lowest yield (9837 kg/ha) was obtained when irrigation limits were set between 100% RAW depletion and FC [13]. However, many scholars proposed water-saving irrigation schedules to cope with the challenge of water shortage to ensure food security [14].

Water use efficiency (WUE) is an index for rational selection of irrigation schedule. Regulated deficit irrigation (RDI) is beneficial to yield increase by studying the effects of different water-deficit treatments on yield, ET_a, and *WUE* [15]. Appropriate water deficit treatment at the maize jointing stage can dramatically improve the utilization rate of irrigation water, and maize will not reduce production but improve maize yield traits [16]. When water deficit was affected in transpiration by reducing the irrigation and wet surface soil time, the ET_a would be decreased. RDI will generally have lower *WUE* than full irrigation [17]. Due to the uncertainty of the future climate, the crops may suffer water stress, and future yields would be unstable. Exploring the impact of different RDI schedules on maize for seed production under future climate conditions is very important for optimizing irrigation schedules and crop production selection. Volk et al. [18] forecasted the maize yield in Tanzania under the future climate, and they concluded that the establishment of different RDI schedules had no significant effect on maize yield. Jalil et al. [19] found that a reasonable selection of RDI system was of benefit to increasing the *WUE* and yield of winter wheat. However, there are relatively few studies on how to avoid yield loss by adjusting the deficit irrigation at each growth stage of crops under future climate scenarios.

Maize has played a significant role in meeting global food requirements and its yield accounts for nearly 30% of total global food production. According to the Food and Agriculture Organization of the United Nations (FAO) report, 23% of the world's maize yield comes from China, and China's maize harvested area accounts for more than one-fifth of the worldwide. Heilongjiang Province is an essential commercial grain base in China. Maize is the largest food crop in Heilongjiang Province, and its plant area and yield rank first in China. In 2019, the plant area of maize in the province was 5.87 million ha, and the yield was 39.4 billion kg [20]. Climate will seriously affect the growth of maize yield due to the uneven distribution of climate in Heilongjiang Province in the future [21]. It is necessary to combine rain-fed and RDI to ensure basic food security.

AquaCrop is an agricultural model that could reliably simulate maize yields under different irrigation schedules [22]. The AquaCrop model is sensitive to the water stress module. The simulation process has fewer steps and simple input parameters, which make the AquaCrop model widely used [23]. The AquaCrop model accurately simulated maize yield over some ranges. When assessing maize yield under different water stress irrigation schedules in semi-arid regions, compared to 50% field capacity irrigation, the AquaCrop model could more accurately simulate maize yield under 75% field capacity irrigation and full irrigation [24]. AquaCrop simulations have high accuracy in maize yield under full irrigation in the North China Plain. Under full irrigation, the error was 5% and 6% for grain yield and biomass, respectively. The error of full irrigation is smaller than that of rainfed. However, the model could be used to simulate maize yield and biomass under full irrigation [25]. Aquacrop was widely used in agricultural forecasting production around the world [26–28].

Studying the **optimal** selection of maize yield, ET_a, and *WUE* under different irrigation schedules in Heilongjiang under future climate changes will effectively provide optimal irrigation schedules for future maize planting systems. Our aims of this study were 1. to localize AquaCrop model parameters using observational data from irrigation experiments in the study region; and 2. to apply the calibrated model to evaluate the impacts of three irrigation schedules on maize yield and *WUE* in Heilongjiang Province under future climate scenarios.

2. Materials and Methods

2.1. Study Site and Field Data Sources

The research site is located in Heilongjiang Province in Northeast China, which is between 43°26′ and 53°33′ in latitude and 121°11′ and 135°05′ in longitude (Figure 1). The average temperature was between 3.1~4.6 °C, and the average annual *Pe* mainly was between 400 and 650 mm. *Pe* was concentrated primarily in June-August. The 6th accumulated temperature zone in Heilongjiang Province was unsuitable for maize planting. Thus, we didn't study the 6th accumulated temperature zone.

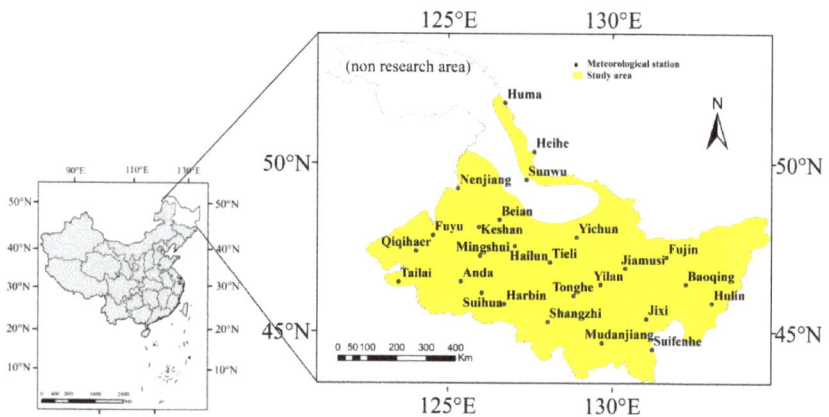

Figure 1. Locations of the Heilongjiang Province and 26 meteorological stations.

The field data of this study are from a four-year experiment established in the National Irrigation Experimental Center (45°43′09″ N, 126°36′35″ E, and altitude 140 m) in Harbin, Heilongjiang Province. The soil texture is loam, and the basic soil properties are as follows: N, 154.4 mg/kg; P_2O_5, 40.1 mg/kg; K_2O, 376.8 mg/kg; and pH, 7.27 [29]. We choose the maize growing four periods (emergence stage, jointing stage, tasseling stage, and filling stage) to study the effect of water stress at different growth stages on maize yield and ET_a (Table 1). A rain shelter was used to control precipitation. The size of each test pit used was 2.5 m × 2 m × 1.7 m.

Table 1. Irrigation treatments from 2014 to 2017.

Year	Treatments	Irrigation Upper and Lower Limit in Different Growth Stages of Maize (% of FC)			
		Emergence Stage	Jointing Stage	Tasseling Stage	Filling Stage
2014	T1	80–100%	50–100%	80–100%	80–100%
	T2	80–100%	80–100%	80–100%	50–100%
	T3	80–100%	80–100%	80–100%	80–100%
2015	T4	80–100%	45–100%	80–100%	80–100%
	T5	80–100%	80–100%	80–100%	45–100%
	T6	80–100%	80–100%	80–100%	80–100%
	T7	100%	100%	100%	100%
2016	T8	60–70%	70–80%	70–80%	70–80%
	T9	70–80%	50–60%	70–80%	70–80%
	T10	70–80%	70–80%	70–80%	70–80%
2017	T11	60–70%	70–80%	70–80%	70–80%
	T12	50–60%	70–80%	70–80%	70–80%
	T13	70–80%	70–80%	70–80%	70–80%

Note: Number before "–" in the table represents the lower limit of irrigation, and number after "–" in the table represents the upper limit of irrigation. "80–100%" in T1 treatment represents the irrigation starts when the soil moisture content reached 80% FC (the lower limit of irrigation), and the irrigation stops until the soil moisture content reached 100% FC (the upper limit of irrigation). Other explanations are the same as above. T8, T11, and T12 represent RDI treatments during the emergence stage; T1, T4, and T9 represent RDI treatments during the jointing stage; T2 and T5 represent RDI treatments during the filling stage; T3, T6, T7, T10, and T13 represent full irrigation treatments of maize growth. FC is the field capacity.

2.2. Future Climate Data

The Global Climate Model (GCM) is a tool used to study the earth's climate. In the Coupled Model Intercomparison Project phase 5, more than 60 GCMs have been proposed to contribute to future climate research [30]. In order to avoid the unreliability of a single GCM, multiple models were used to collect the predicted data of GCMs [31]. However, the future climate data were based on the ensemble datasets of four GCMs under the RCP4.5 and the RCP8.5, respectively. This study selected 26 meteorological stations distributed in different places in the Heilongjiang Province (Figure 1). Meteorological data consist of the daily maximum temperature (*Tmax*), daily minimum temperature (*Tmin*), *Pe*, and *Rad* from 2021 to 2080. We used the LARS-WG random weather generator downscaling method to generate future climate scenarios [32]. The calibration and verification data of the LARS-WG was derived from the historical meteorological data of daily *Tmax*, *Tmin*, *Pe*, and *Rad* of 26 meteorological stations from 1960 to 2015. The output meteorological data were the daily *Tmax*, *Tmin*, *Pe*, and *Rad* of four GCMs (Table 2) under RCP4.5 and RCP8.5, respectively, from 2021 to 2080. The period from 2021 to 2080 is divided into three research stages 2030s (2021–2040), 2050s (2041–2060), and 2070s (2061–2080). For details of the downscaling method, refer to [33]. The output meteorological data will be input into the AquaCrop. RCP8.5 represents radiation forcing values of more than 8.5 W/m^2 in 2100, and RCP4.5 means 4.5 W/m^2 when stable after 2100 [34]. We focus on the selection of RCP4.5 and RCP8.5 based on the socioeconomic conditions of radiative forcing currently faced by humans [35].

Table 2. 4 GCMs datasets in the LARS-WG model.

GCMs	Research Center	Countries and Regions	Grid Resolution
EC-EARTH	EC: Earth Consortium	Europe	1.125° × 1.125°
HadGEM2-ES	United Kingdom(UK) Meteorological Office	UK	1.25° × 1.88°
MIROC5	The University of Tokyo, National Institute for Environmental	Japan	1.39° × 1.41°
MPI-ESM-MR	Max Planck Institute for Meteorology	Germany	1.85° × 1.88°

2.3. AquaCrop Model Introduction and Settings

The AquaCrop model input data includes four modules, which are the climate module, crop module, management module, and soil module. Climate data includes daily *Tmax, Tmin, Pe, Rad,* and reference evapotranspiration (ET_0). The ET_a is calculated by ET_0-calculator software [36]. The crop data input section has some default values for crops growth parameters in this module. We need to adjust the corresponding plant parameters (including plant each growth period, sowing date, etc.) based on different climate and research sites. The crop growth was determined based on 14 agrometeorological observation stations in Heilongjiang Province. For the meteorological station without observation data, the data of the neighboring agrometeorological observation station in the same accumulated temperature area is selected as the calculation basis [37]. In order to improve the accuracy of output data during model simulation, parameters should be adjusted appropriately according to specific conditions. Irrigation management is specified by the irrigation method and the irrigation events. The irrigation schedule is formulated according to the irrigation time and depth of each stage of the crop growth period [14]. Soil data that describe soil properties in each layer are from The Soil Science Database (http://vdb3.soil.csdb.cn/, accessed on 30 March 2021).

To avoid the confounding effect of the non-productive consumptive water use (soil evaporation), the AquaCrop model calculates crop transpiration (*Tr*), soil evaporation (*E*), and ET_a using the following equation:

$$T_r = K_s K_{S_{Tr}} \left(K_{c_{Tr,x}} CC^* \right) ET_0 \tag{1}$$

$$E = K_r (1 - CC^*) K_{ex} ET_0 \tag{2}$$

$$ET_a = T_r + E \tag{3}$$

where, CC^* is the adjusted actual canopy coverage (%); $K_{c_{Tr,x}}$ is the maximum standard crop transpiration coefficient (dimensionless); K_S means the water stresses coefficient (dimensionless); $K_{S_{Tr}}$ is the temperature stresses coefficient (dimensionless); ET_0 is the reference evapotranspiration (mm); K_r is the evaporation reduction coefficient (dimensionless); K_{ex} is the maximum soil evaporation coefficient (dimensionless). ET_a was separated into *Tr* and *E* (mm).

In this study, the method recommended by FAO-66 was used to calculate the maize yield and *WUE*. As:

$$Y = f_{HI} HI_0 B \tag{4}$$

$$WUE = Y/ET_a \tag{5}$$

where *Y* is maize yield (kg/ha). f_{HI} is the harvest index adjustment factor (dimensionless). HI_0 is a reference harvest index (dimensionless), which means the yield ratio to biomass, *B* means the aboveground dry (kg/ha), *WUE* means water use efficiency (kg/m^3).

In this study, the inverse distance weighting (IDW) and Kriging methods of ArcGIS were used to interpolate the numerical values of each station output by AquaCrop into the study area, to analyze the spatial characteristics of the effects of different irrigation schedules on maize yield, ET_a, and *WUE* in Heilongjiang Province under future climate. The two-factor ANOVA of SPSS Statistics 17 was used to test the difference in yield, ET_a and *WUE* under different irrigation schedules.

In this study, we explored the effects of water stress at maize different growth stages on maize development. The generation of irrigation schedules in the AquaCrop model was used to evaluate or design a particular irrigation schedule. Irrigation practice was generated according to the specified time and a depth criterion when the model was running. In this study, RDI was set and generated at a specific time; the depth of irrigation depends on whether the soil moisture content reaches the set irrigation lower limit. When the soil moisture content falls to the set minimum limit, it will automatically irrigate to the fixed upper limit. The growth stage of maize is divided into four stages: emergence, jointing, tasseling, and filling. The jointing and filling stages of maize are essential stages of nutrient generation in crop growth [38]. Therefore, we set up three irrigation schedules, full irrigation and deficit irrigation in two crucial growth periods of maize. The three irrigation schedules sets are W1: water stress treatment in maize jointing stage. The lower and upper limits are 50% FC and 80% FC, respectively; W2: water stress treatment in maize filling stage. The lower and upper limits are 50% FC and 80% FC, respectively; W3: full irrigation schedule, which made maize suffer no water stress in the entire growth cycle.

2.4. AquaCrop Calibration and Verification

We selected the experimental data to calibrate the yield and ET_a in two years (2014–2015) main RDI (T1, T2, T3, T4, T5, T6, T7). Validation data were established by the six main RDI (T8, T9, T10, T11, T12, T13) experimental data from 2016 to 2017. The statistical parameters, including normalized RMSE ($CV(RMSE)$), determination coefficient (R^2), Willmott's agreement index (d), and model efficiency coefficient (EF) were determined for the performance evaluation of AquaCrop.

$$R^2 = 1 - \frac{\sum(yi - \hat{yi})^2}{\sum(yi - \hat{y})^2} \tag{6}$$

$$CV(RMSE) = \frac{1}{\bar{O}}\sqrt{\frac{\sum(Pi - Oi)^2}{n}} \tag{7}$$

$$d = 1 - \frac{\sum(Pi - Oi)^2}{\sum(|Pi - \bar{O}| + |Oi - \bar{O}|)^2} \tag{8}$$

$$EF = \frac{\sum_{i=1}^{n}(Oi - \bar{O})^2 - \sum_{i=1}^{n}(Si - Oi)^2}{\sum_{i=1}^{n}(Oi - \bar{O})^2} \tag{9}$$

where yi is the actual value, ŷi is the simulated value, and ŷ is the mean value. And $\bar{O}$ is the mean observations, pi is the simulated value, and Oi is the observed value. n means the research count. A simulation can be considered perfect if $CV(RMSE)$ is smaller than 10%, good if between 10 and 20%. d range is 0–1, with 0 indicating a bad fit and 1 indicating a good fit between the simulated and observed data. The EF value is smaller than 1; a positive value indicates that the simulated value better describes the measured data trend than the mean observations.

The flow chart showing the optimal selection of future irrigation schedules using the AquaCrop model is provided in Figure 2.

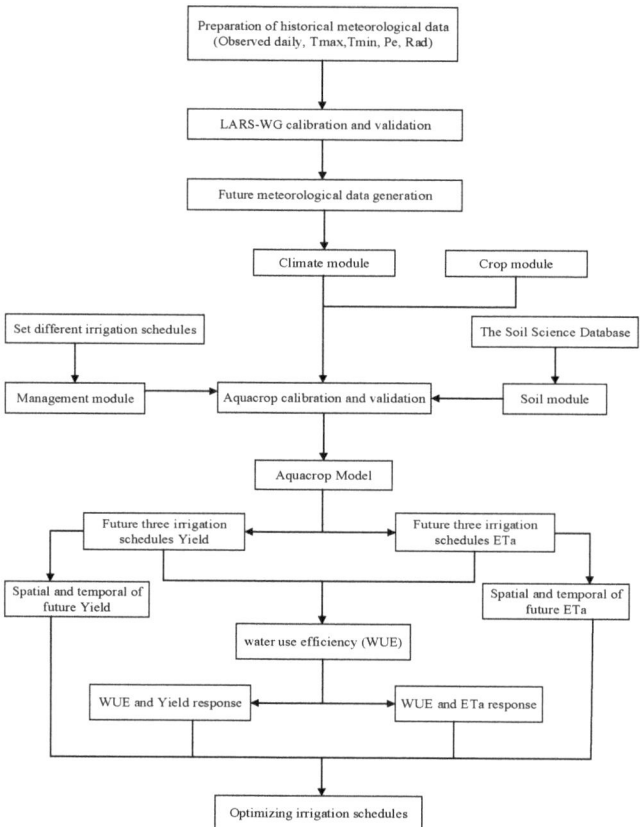

Figure 2. Flow chart for the steps involved in the estimation of future irrigation schedules using the AquaCrop model.

3. Results

3.1. Performance Evaluation of AquaCrop

Based on the AquaCrop model calibration and verification of the ET_a and yield, the model-simulated different irrigation schedules ET_a and yield agree well with the field-observed (Table 3). The R^2 of the simulated and observed maize yield reaches 0.72, and the average difference between simulated and observed yield under different treatments did not exceed 200 kg/ha (Figure 3). Each difference of the simulated and observed ET_a was less than 50 mm (Figures 4 and 5). When calibrating the AquaCrop model, the simulated and measured values of maize in different RDI were well fitted (Figure 4). From the six treatments in the validation, the model is reasonable in simulating the ET_a of the maize (Figure 5). Therefore, the AquaCrop model simulation results on maize ET_a and yield under different irrigation schedules during the whole growth stage were reliable and applicable for this study area.

Table 3. Fit indexes of AquaCrop model-simulated and measured ET_a and yield.

Parameter	CV (RMSE) (%)	d	R^2	EF
ET_a	8.21	0.99	0.97	0.97
Yield	4.44	0.91	0.72	0.68

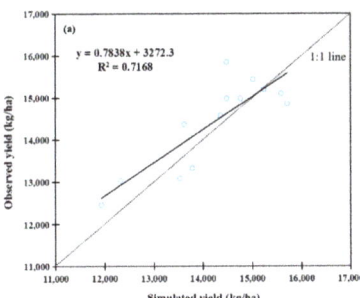

Figure 3. Calibration and validation of the AquaCrop model with yield (2014–2017).

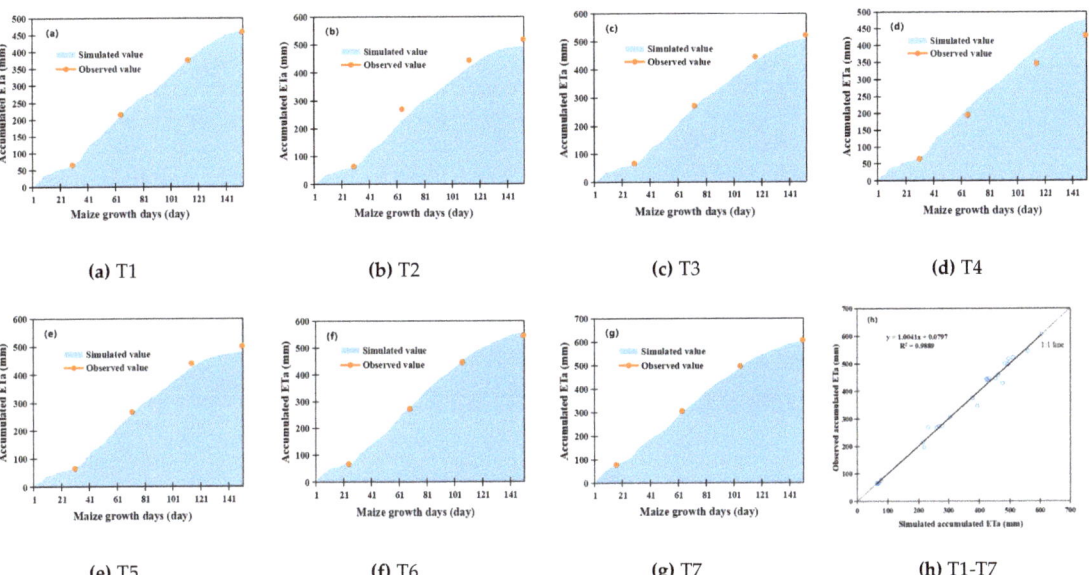

(a) T1 (b) T2 (c) T3 (d) T4

(e) T5 (f) T6 (g) T7 (h) T1-T7

Figure 4. Observational and simulated cumulative ET_a during the whole growth period of maize, (**a**) T1, (**b**) T2, (**c**) T3, (**d**) T4, (**e**) T5, (**f**) T6, (**g**) T7 and (**h**) calibration of ET_a for different stages (2014–2015).

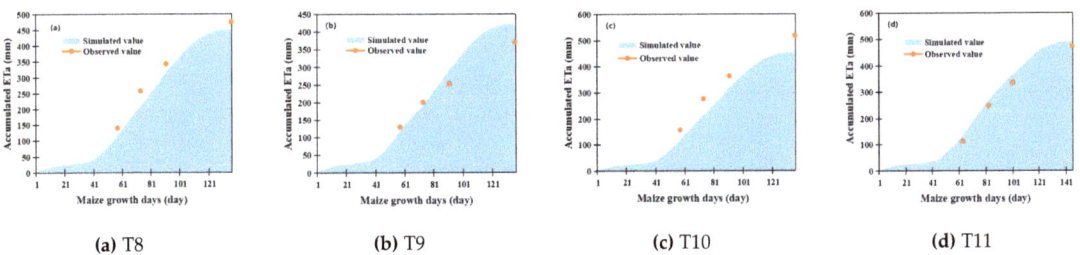

(a) T8 (b) T9 (c) T10 (d) T11

Figure 5. *Cont.*

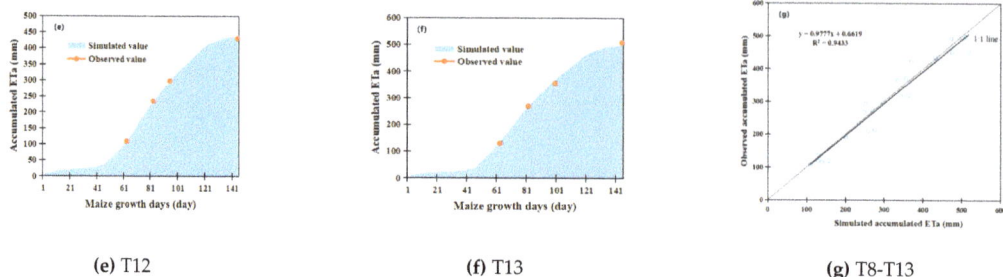

(e) T12   **(f)** T13   **(g)** T8-T13

Figure 5. Observational and simulated cumulative ET_a during the whole growth period of maize, (**a**) T8, (**b**) T9, (**c**) T10, (**d**) T11, (**e**) T12, (**f**) T13, and (**g**) validation of ET_a for different stages (2016–2017).

3.2. Projected Future Climate Change

Figure 6 presents the predicted future climate during the maize growing period. The highest *Pe*, *Tmax*, and *Tmin* appeared in the 2070s under the RCP4.5, they are 470.31 mm, 26.06 °C, and 14.69 °C, respectively, while the highest value of the *Rad* is 19.04 MJ/m² appeared at 2050s. In the RCP8.5, the average *Pe*, *Tmax* and *Tmin*, and *Rad* highest values appeared in the 2070s, they were 490.57 mm, 27.48 °C, 16.15 °C, and 19.08 MJ/m². RCP8.5's highest value of average *Pe*, *Tmax* and *Tmin*, and *Rad* was 4.31%, 5.45%, 9.94%, and 0.21% more than RCP4.5, respectively. Under the two RCPs, the maximum *Pe* appeared in the central and southern part of the study area; the highest *Rad* value mainly appears in the southwest. Moreover, the southern's *Tmin* has the highest value, and the *Tmax* high value was distributed in the southwestern and south.

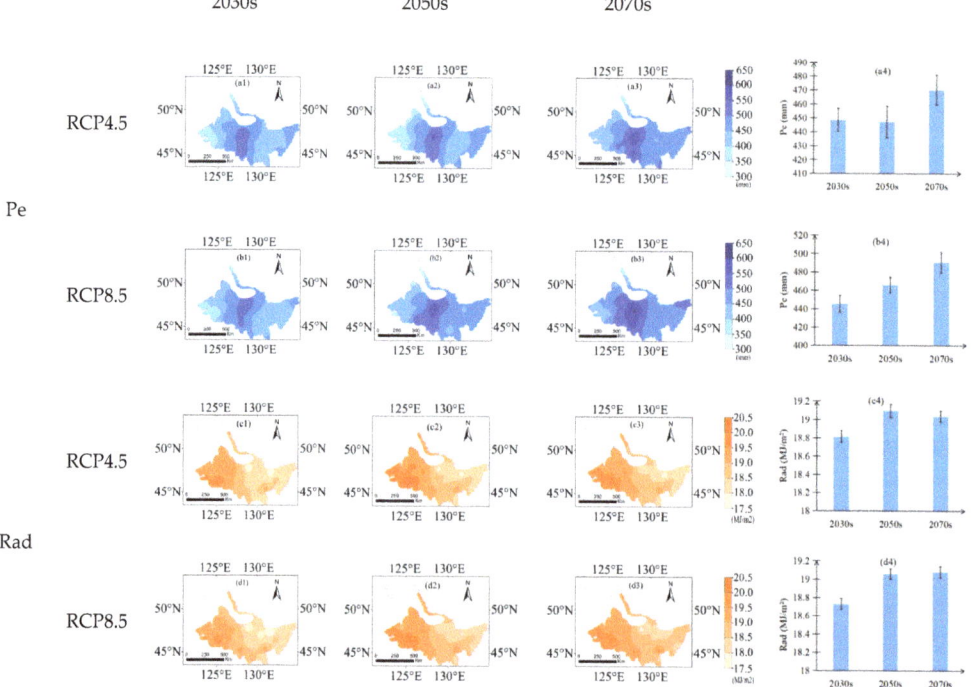

Figure 6. *Cont.*

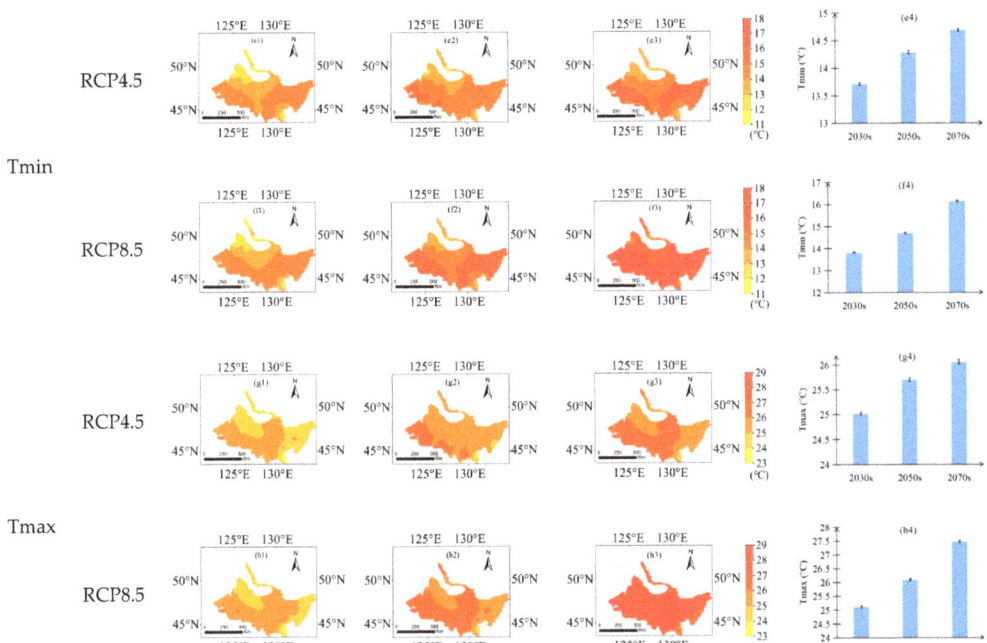

Figure 6. Spatial and temporal distribution of average precipitation (*Pe*) (**a1–a4,b1–b4**), solar radiation (*Rad*) (**c1–c4,d1–d4**), minimum temperature (*Tmin*) (**e1–e4,f1–f4**), and maximum temperature (*Tmax*) (**g1–g4,h1–h4**) for the multi-GCM ensemble during the maize growth stage under two RCPs from 2021–2080 in the study area.

3.3. ET_a Changes under Different Future Scenarios

The ET_a showed a declined trend from southwest to northeast in the study area (Figure 7). The ET_a had an upward trend under two RCPs with different magnitudes. The value of RCP8.5 is generally greater than the ET_a of RCP4.5. In the future, the maximum value of ET_a appears in the W3. Compared with the W3, W1, and W2 reduced by 1.5–1.6% and 0.4–0.6%.

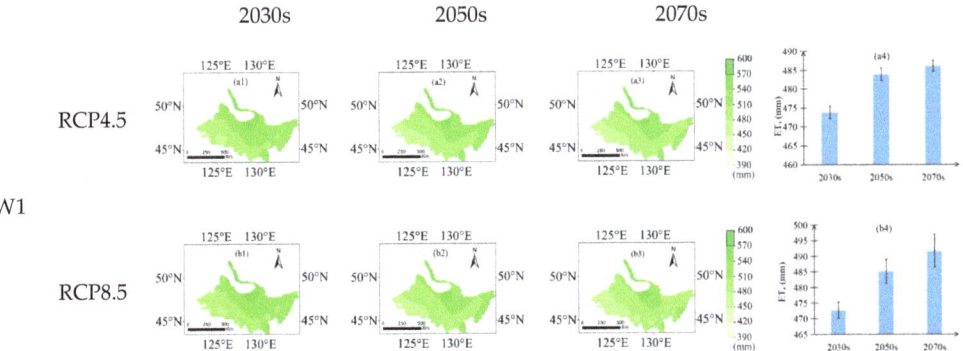

Figure 7. *Cont.*

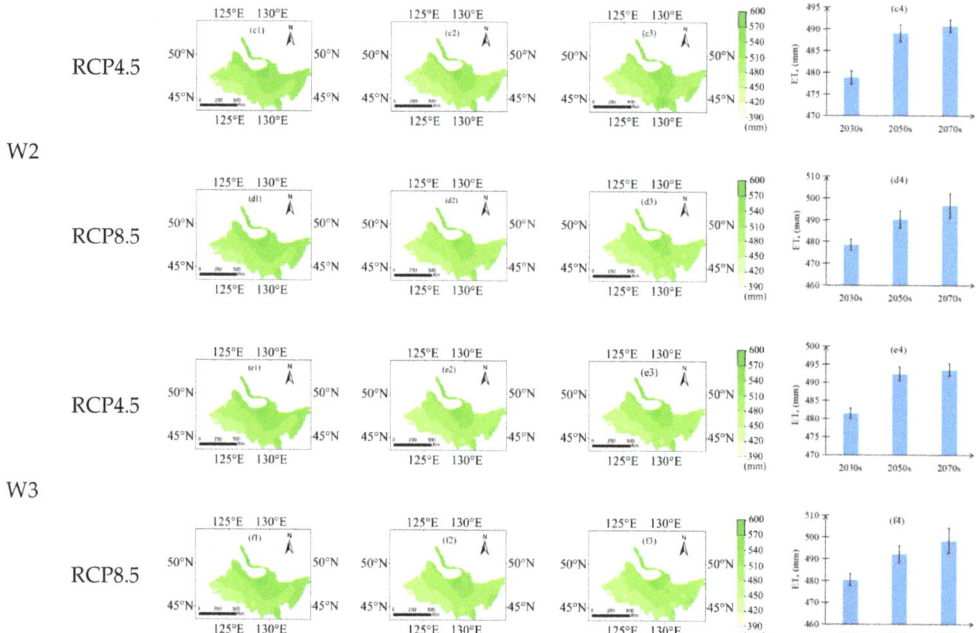

Figure 7. Spatial and temporal distribution of maize ET_a for the multi-GCM ensemble under different irrigation schedules ((**a1–a4,b1–b4**) W1, (**c1–c4,d1–d4**) W2,(**e1–e4,f1–f4**) W3) in the 2030s, 2050s, and 2070s with two RCPs.

3.4. Yield Changes under Different Future Scenarios

Maize yield under different irrigation schedules increased from the north part to the south area in the spatial distribution. The maximum value appeared in the southeast and southwest (Figure 8). Yield showed a growth trend from 2030s to 2070s. The maximum value for the three irrigation schedules appears in W3, which were 14,044 kg/ha (RCP4.5) and 14,402 kg/ha (RCP8.5).

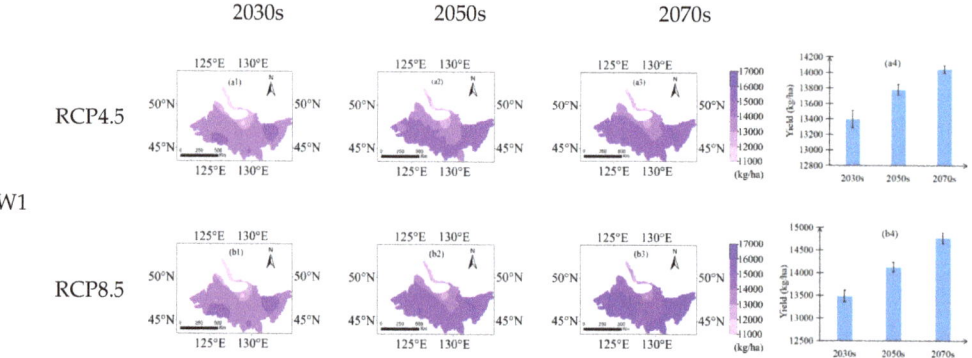

Figure 8. *Cont.*

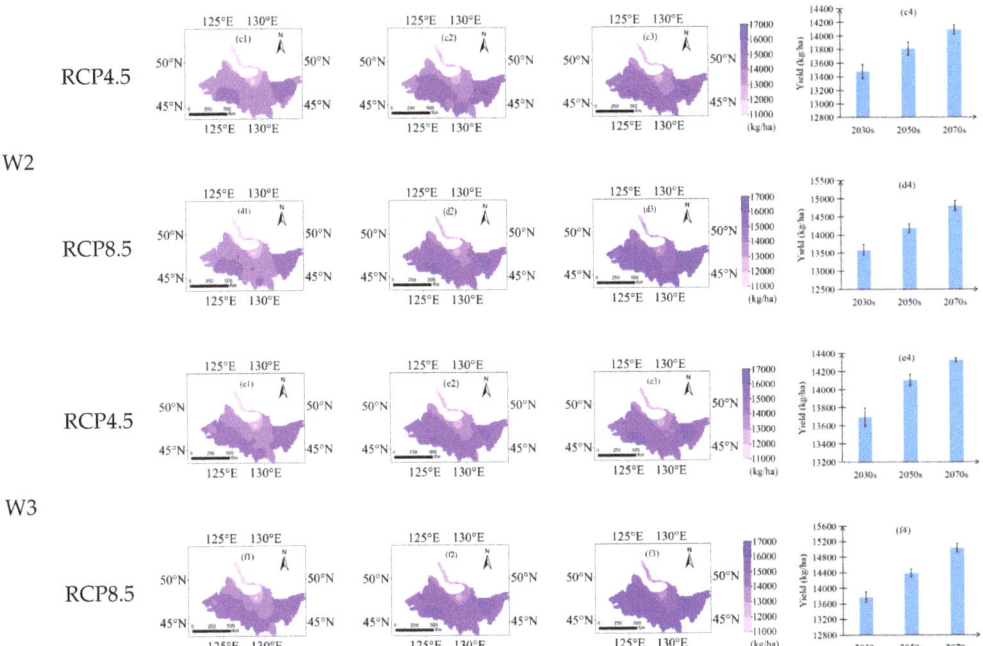

Figure 8. Spatial and temporal distribution of maize yield for the multi-GCM ensemble under different irrigation schedules ((**a1–a4,b1–b4**) W1,(**c1–c4,d1–d4**) W2,(**e1–e4,f1–f4**) W3) in the 2030s, 2050s, and 2070s with two RCPs.

3.5. WUE Changes under Different Future Scenarios

The *WUE* showed a growth trend from the west area to the east part (Figure 9). While the extent of increase in *WUE* under the two RCPs was different, and the RCP8.5's *WUE* was larger than RCP4.5's. Under both of two RCPs, the minimum value of *WUE* appears in W2. Compared with W2, the *WUE* of the W1 and W3 increased by 0.5–0.6% and 0.9–1.2%.

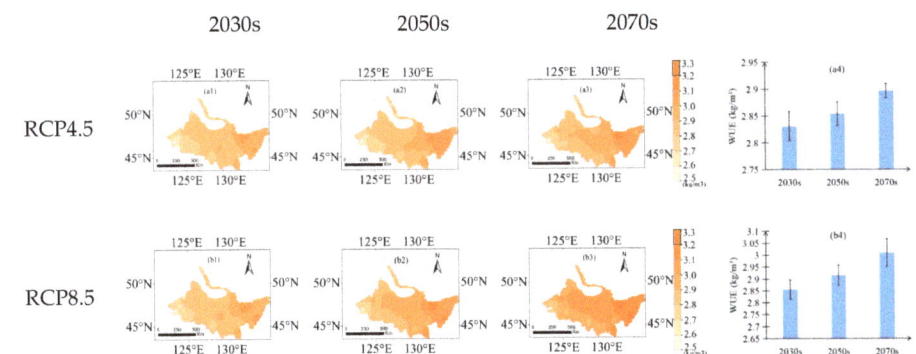

Figure 9. *Cont.*

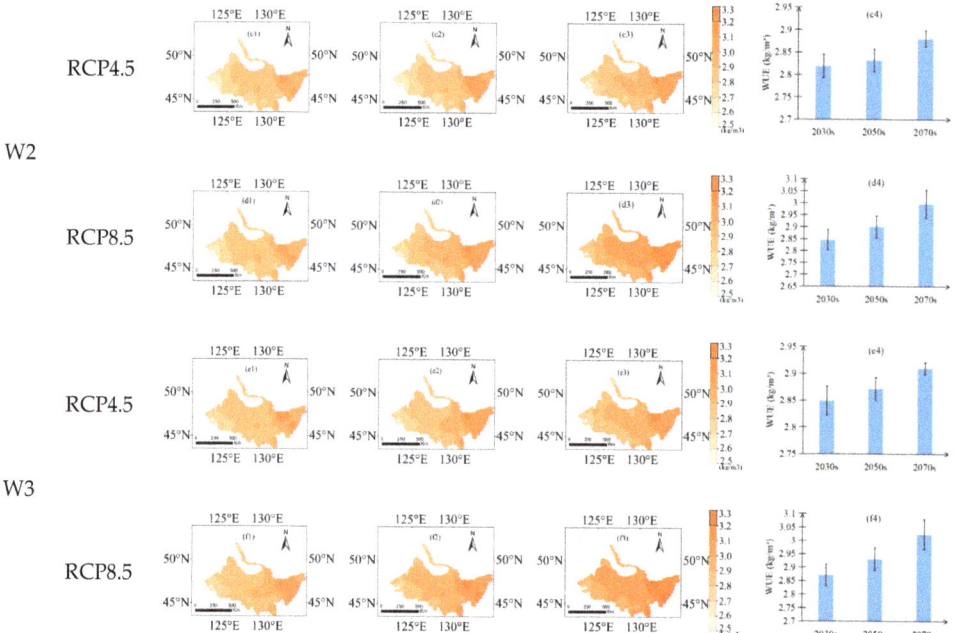

Figure 9. Spatial and temporal distribution of maize *WUE* for the multi-GCM ensemble under different irrigation schedules ((**a1–a4,b1–b4**) W1, (**c1–c4,d1–d4**) W2, (**e1–e4,f1–f4**) W3) in the 2030s, 2050s, and 2070s with two RCPs.

3.6. Assessment of Irrigation Optimization Scenarios and Corresponding Measures

Under the RCP4.5, the ET_a of W1, W2, and W3 were 481 mm, 486 mm, and 489 mm, respectively. The highest value of ET_a was in W3, and the lowest value was in W1. The difference between the three irrigation schedules is significant ($p \leq 0.05$) (Figure 10a). Yield under W3 was the highest. Although the W1's yield was less than W2's, the difference was not significant ($p > 0.05$) (Figure 11c). For *WUE*, W3's *WUE* was the highest among the three, and W1 and W2 were significantly lower than W3. Under the RCP8.5, the lowest value of ET_a appears in W1 ($p > 0.05$) (Figure 10b). Yields of maize under W1, W2, and W3 were 14,127kg/ha, 14,199 kg/ha, and 14,402 kg/ha, respectively. The results showed that the yield difference between W1 and W3 was significant ($p \leq 0.05$), and the yield difference between W2 and the other was not significant ($p > 0.05$) (Figure 11d). For *WUE*, W3 has the highest value. However, the difference between the three irrigation schedules is insignificant ($p > 0.05$). Overall, under the two RCPs, optimization selection W1 is recommended for farmers' reference as an optimal option under RDI, as W1 had a higher *WUE* without significantly decreasing yield.

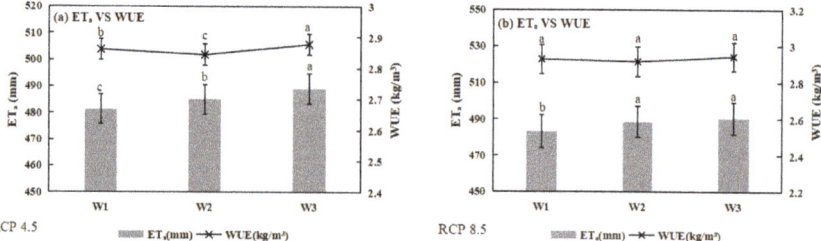

Figure 10. Relationship of *WUE* and ET_a under different irrigation optimization schedules. Different lowercase letters represent different levels of each column at $p \leq 0.05$.

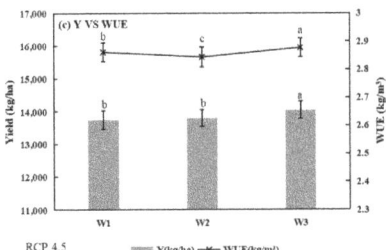 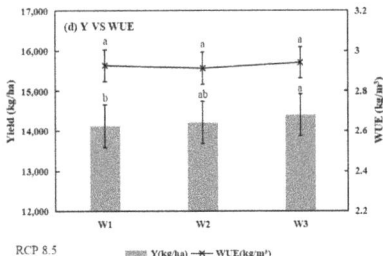

RCP 4.5   ▨ Y(kg/ha) —✱— WUE(kg/m³)   RCP 8.5   ▨ Y(kg/ha) —✱— WUE(kg/m³)

Figure 11. Relationship of Yield and *WUE* under different irrigation schedules. Different lowercase letters represent different levels of each column at $p \leq 0.05$.

4. Discussion

The four GCMs and two RCPs scenario models used in the study predicted an increase in temperature (*Tmin* and *Tmax*), *Pe*, and *Rad* in future study phases. Xiao et al. found that the crop growth stage would be shortened as the temperature increases. This condition usually reduces the transpiration of the crop during the growth stage [39]. The ET_a of most planting systems would somewhat decline with future climate change. Different ET_a would vary according to different cropping schedules under future climate scenarios [30]. Tao et al. [40] proposed that future temperature warming would significantly increase soil evaporation. Our results showed that ET_a increases as the temperature increases. The rising temperature may promote the photosynthesis and leaf expansion of crops, then accelerate the dry matter accumulation and crop growth, enhancing the transpiration, which is consistent with [41]. In this study area, the climate affected the southwestern Heilongjiang Province as a low-yield area. The relatively lower *Pe* and the higher ET_a, which may be due to the high temperature and *Rad*, leads to the lower *WUE* in this area. This study established analysis and comparison of irrigation schedules based on maize's different growth stages water stress. The results showed that maize's ET_a under treatments in which water stress appeared in the jointing stage was the least among all irrigation schedules. The reason may be that maize is more sensitive in the jointing stage with water stress. Jin et al. [42] indicated that maize's early growth stages were more vulnerable to water stress. The ET_a would reduce when water stress arises at filling stages. However, the results of this study were different from the result reported by Yu et al. [43]. They found that ET_a was the maximum when maize was exposed to water stress in the jointing stage. The results showed that photosynthesis in the early stage of maize growth was not only controlled by water stress, but the crop was still inhibited after rewatering, and it was difficult to recover. This result may be due to different research results obtained from other experimental locations. The degree of inhibition of crop growth is related to the duration of drought and the degree of stress. Future climate changes are complex, and the ET_a and yield of maize crops should be studied based on different future climate scenarios.

However, future climate changes would also affect maize yields. This climate change would impact crop yield and water supply and requirements. The increase in CO_2 concentration in the whole study area increases the possibility of photosynthesis and further promotes biomass accumulation in crops, thereby increasing yield [44,45]. This study showed that the RCP8.5's yield was higher than RCP4.5's, and the yield increased 386, 403, and 358 kg/ha under W1, W2, and W3, respectively. The results were similar to the results found by Qaisar et al. [46]. This study also showed that the yield under the RDI was lower than the yield under the full irrigation schedule, and the yield of W1 was lower than that of W2. This may be because the lack of development during the maize vegetative growth stage ultimately affects the reproductive development stage, and the high-quality output of ears and kernels reduces the yield of maize. Li et al. [47]'s research found that water decline leads to reduced leaf area, accelerated leaf senescence, and reduced photosynthesis, reducing leaf source activity and negatively affecting the seed setting rate, reducing maize yield. NeSmith et al. [48] proposed that the grain filling rate is not significantly affected by water deficit during grain filling. Therefore, our research results are consistent with theirs; the yield gap between W2 and W3 was not significant. Maize growth is a complicated process influenced by climatic, soil, and geographical factors, respectively [49]. Different irrigation schedules have less effect on the yield of maize, and the difference between them was not significant. Other researchers found that moderate water stress at the maize's jointing stage could increase crop *WUE* and control the growth redundancy of plants during the vegetative growth process, reduce the length of maize ears, and promote reproductive growth [50]. Cai et al. [51] and Wei et al. [52] found that appropriate water deficit treatments in the early stage of maize growth can improve crop *WUE* to varying degrees.

5. Conclusions

AquaCrop can simulate the maize ET_a and yield well, and the R^2 of the relationship between the model simulation and the observed were 0.99 and 0.71, respectively. Due to the increasing trend of future climate, the ET_a, yield, and *WUE* of maize in the two RCPs showed an increased trend during 2021–2080. The ET_a and yield showed an increasing trend from the north area to the south part in the study area, and *WUE* showed a downward trend from the east part to the west area. The ET_a, yield, and *WUE* of RCP8.5 were larger than these under RCP4.5. From the perspective of saving irrigation water without affecting the stability of maize yield, we recommend W1 for future maize planting. This study will supply useful knowledge with the impact of different irrigation schedules on crop growth under future climate, and help to optimize the selection of feasible irrigation schedules to balance the relationship between water scarcity and food security.

Author Contributions: T.N. collected the data; Y.T. and T.N. analyzed data; Y.T. wrote the paper; Y.T., N.L., Y.J. and T.W. drew the figures for this paper; Z.Z., P.C., T.L., C.D., S.Z. and Z.S. reviewed and edited this paper. All authors have read and agreed to the published version of the manuscript.

Funding: This work was fund by Basic Scientific Research Fund of Heilongjiang Provincial Universities [number: 2021-KYYWF-0019], Opening Project of Key Laboratory of Efficient Use of Agricultural Water Resources, Ministry of Agriculture and Rural Affairs of the People's Republic of China in Northeast Agricultural University [number: AWR2021002], National Natural Science Foundation Project of China [number: 51779046].

Data Availability Statement: Not applicable.

Acknowledgments: We thank the Chinese meteorological data sharing service (http://data.cma.cn, accessed on 30 March 2021) for providing the meteorological data. We thank all the members in Lab of Pumping, Hydraulic Teaching and Experimental Center of Heilongjiang University. Finally, we thank the anonymous reviewers and the editor for their suggestions, which substantially improved the manuscript.

Conflicts of Interest: The authors declare no conflict of interest.

References

1. Zhou, Y.; Li, N.; Dong, G.; Wu, W. Impact assessment of recent climate change on rice yields in the Heilongjiang Reclamation Area of north-east China. *J. Sci. Food Agric.* **2013**, *93*, 2698–2706. [CrossRef] [PubMed]
2. Piao, S.L.; Ciais, P.L.; Huang, Y.; Shen, Z.H.; Peng, S.S.; Li, J.S.; Zhou, L.P.; Liu, H.Y.; Ma, Y.C.; Ding, Y.H.; et al. The impacts of climate change on water resources and agriculture in China. *Nature* **2010**, *467*, 43–51. [CrossRef] [PubMed]
3. Ti, J.S.; Yang, Y.H.; Yin, X.G.; Liang, J.; Pu, L.L.; Jiang, Y.L.; Wen, X.Y.; Chen, F. Spatio-Temporal Analysis of Meteorological Elements in the North China District of China during 1960–2015. *Water* **2018**, *10*, 789. [CrossRef]
4. Intergovernmental Panel on Climate Change (IPCC). *Global warming of 1.5 °C: Impacts of 1.5 °C of Global Warming on Natural and Human Systems*; Contribution of Working Group I to the Sixth Assessment Report of the Intergovernmental Panel on Climate Change; IPCC Secretariat: Geneva, Switzerland, 2018.
5. Lin, Y.; Wenxiang, W.; Quansheng, G. CERES-Maize model-based simulation of climate change impacts on maize yields and potential adaptive measures in Heilongjiang Province, China. *J. Sci. Food Agric.* **2015**, *95*, 2838–2849. [CrossRef]
6. Fayaz, A.; Kumar, Y.R.; Lone, B.A.; Kumar, S.; Dar, Z.A.; Rasool, F.; Abidi, I.; Nisar, F.; Kumar, A. Crop Simulation Models: A Tool for Future Agricultural Research and Climate Change. *Asian J. Agric. Ext. Econ. Sociol.* **2021**, *39*, 146–154. [CrossRef]
7. Wang, W.; Peng, S.; Sun, F.; Xing, W.; Luo, F.; Xu, J. Spatialtemporal variations of rice irrigation water requirements in the mid-lower reaches of Yangtze River under changing climate. *Adv. Water Sci.* **2012**, *23*, 656–664. (In Chinese) [CrossRef]

8. Döll, P. Impact of climate change and variability on irrigation requirement: A global perspective. *Clim. Chang.* **2002**, *54*, 269–293. [CrossRef]
9. Tao, F.; Hayashi, Y.; Zhao, Z.; Sakamoto, T.; Yokozawa, M. Global warming, rice production, and water use in China: Developing a probabilistic assessment. *Agric. For. Meteorol.* **2008**, *148*, 94–110. [CrossRef]
10. Zheng, B.; Chenu, K.; Dreccer, M.; Chapman, S. Breeding for the future: What are the potential impacts of future frost and heat events on sowing and flowering time requirements for Australian bread wheat (Triticum aestivium) varieties? *Glob. Chang. Biol.* **2012**, *18*, 2899–2914. [CrossRef]
11. Kang, Y.; Khan, S.; Ma, X. Climate change impacts on crop yield, crop water productivity and food security—A review. *Prog. Nat. Sci.* **2009**, *19*, 1665–1674. [CrossRef]
12. Sun, H.; Zhang, X.; Liu, X.; Liu, X.; Shao, L.; Chen, S.; Wang, J.; Dong, X. Impact of different cropping systems and irrigation schedules on evapotranspiration, grain yield and groundwater level in the North China Plain. *Agric. Water Manag.* **2019**, *211*, 202–209. [CrossRef]
13. Aik, M.; Yetk, A.K.; Candoan, B.N.; Kuscu, H. Determining the yield responses of maize plant under different irrigation scenarios with AquaCrop model. *Int. J. Agric. Environ. Food Sci.* **2021**, *5*, 260–270. [CrossRef]
14. Sandhu, R.; Irmak, S. Performance of AquaCrop model in simulating maize growth, yield, and evapotranspiration under rainfed, limited and full irrigation. *Agric. Water Manag.* **2019**, *223*, 105687. [CrossRef]
15. Kang, S.Z.; Shi, W.J.; Hu, X.T.; Liang, Y.L. Effects of regulated deficit irrigation on physiological indexes and water use efficiency of maize. *Nongye Gongcheng Xuebao* **1998**, *4*, 83–87. (In Chinese)
16. Lv, P.P.; Lv, Z.D.; Bi, Y.J. Effects of regulated deficit irrigation with different irrigation amount on growth and yield of maize. *Glob. Water Resour. Dev. Manag.* **2020**, *2*, 9–13. (In Chinese) [CrossRef]
17. Trout, T.J.; Dejonge, K.C. Water productivity of maize in the US high plains. *Irrig. Sci.* **2017**, *35*, 251–266. [CrossRef]
18. Volk, J.; Gornott, C.; Sieber, S.; Lana, M.A. Can Tanzania's adaptation measures prevent future maize yield decline? A simulation study from Singida region. *Reg. Environ. Chang.* **2021**, *21*, 1–13. [CrossRef]
19. Jalil, A.; Akhtar, F.; Awan, U.K. Evaluation of the AquaCrop model for winter wheat under different irrigation optimization strategies at the downstream Kabul River Basin of Afghanistan. *Agric. Water Manag.* **2020**, *240*, 1–8. [CrossRef]
20. Heilongjiang Provincial People's Government. *Heilongjiang Yearbook*; Editorial Department: Harbin, China, 2020; pp. 244–245. ISSN 1008-0791.
21. Zhang, D.H.; Zhou, H.Q.; Lou, X. Problems and Countermeasures in Grain production in Heilongjiang Province. *Res. Agric. Mod.* **2012**, *33*, 411–414. (In Chinese)
22. Yang, C.Y.; Fraga, H.; Ieperen, W.V.; Santos, J.A. Assessment of irrigated maize yield response to climate change scenarios in Portugal. *Agric. Water Manag.* **2017**, *184*, 178–190. [CrossRef]
23. Cui, Y.; Lin, H.; Xie, Y.; Liu, S. Research on Application of AquaCrop Model in Crop Yield Prediction in Black Soil Region of Northeast China. *Sheng Tai Xue Bao* **2021**, *47*, 1–12. (In Chinese)
24. Abedinpour, M.; Sarangi, A.; Rajput, T.B.S.; Singh, M.; Pathak, H.; Ahmad, T. Performance evaluation of AquaCrop model for maize crop in a semi-arid environment. *Agric. Water Manag.* **2012**, *110*, 55–66. [CrossRef]
25. Shirazi, S.Z.; Mei, X.; Liu, B.; Liu, Y. Assessment of the AquaCrop Model under different irrigation scenarios in the North China Plain. *Agric. Water Manag.* **2021**, *257*, 107120. [CrossRef]
26. Abdalhi, M.A.M.; Jia, Z.H. Crop yield and water saving potential for AquaCrop model under full and deficit irrigation management. *Ital. J. Agron.* **2018**, *13*, 1288. [CrossRef]
27. Nie, T.Z.; Jiao, Y.; Tang, Y.; Li, N.; Wang, T.Y.; Du, C.; Zhang, Z.X.; Li, T.C.; Zhu, S.J.; Sun, Z.Y.; et al. Study on the Water Supply and the Requirements, Yield, and Water Use Efficiency of Maize in Heilongjiang Province Based on the AquaCrop Model. *Water* **2021**, *13*, 2665. [CrossRef]
28. Adeboye, O.B.; Schultz, B.; Adeboye, A.P.; Adekalu, K.O.; Osunbitan, J.A. Application of the AquaCrop model in decision support for optimization of nitrogen fertilizer and water productivity of soybeans. *Inf. Process. Agric.* **2020**, *8*, 528–534. [CrossRef]
29. Wang, B.; Li, F.H.; Huang, Y.; Sun, Y.L.; Zhang, Z.X. Experimental Study on High Efficient Regulated Deficit Irrigation System for Maize in Cold Area and Black Soil Area. *J. Irrig. Drain.* **2013**, *32*, 113–115. (In Chinese) [CrossRef]
30. Xiao, D.; Liu, D.L.; Feng, P.; Wang, B.; Tang, J. Future climate change impacts on grain yield and groundwater use under different cropping systems in the North China Plain. *Agric. Water Manag.* **2021**, *246*, 106685. [CrossRef]
31. Liu, D.L.; O'Leary, G.J.; Christy, B.; Macadam, I.; Wang, B.; Anwar, M.R.; Weeks, A. Effects of different climate downscaling methods on the assessment of climate change impacts on wheat cropping systems. *Clim. Chang.* **2017**, *144*, 687–701. [CrossRef]
32. Yang, L.H.; Zhong, P.A.; Zhu, F.L.; Ma, Y.F.; Wang, H.; Li, J.Y.; Xu, C.J. A comparison of the reproducibility of regional precipitation properties simulated respectively by weather generators and stochastic simulation methods. *Stoch. Environ. Res. Risk Assess.* **2021**, *36*, 495–509. [CrossRef]
33. Nie, T.Z.; Zhang, Z.X.; Qi, Z.J.; Chen, P.; Sun, Z.Y.; Liu, X.C. Characterizing Spatiotemporal Dynamics of CH4 Fluxes from Rice Paddies of Cold Region in Heilongjiang Province under Climate Change. *Int. J. Environ. Res. Public Health* **2019**, *16*, 692. [CrossRef] [PubMed]
34. Moss, R.H.; Edmonds, J.A.; Hibbard, K.A.; Manning, M.R.; Rose, S.K.; Vuuren Van, D.P.; Carter, T.R.; Emori, S.; Kainuma, M.; Kram, T. The next generation of scenarios for climate change research and assessment. *Nature* **2010**, *463*, 747–756. [CrossRef] [PubMed]

35. Xiao, D.; Liu, D.; Wang, B.; Feng, P.; Tang, J. Climate change impact on yields and water use of wheat and maize in the North China Plain under future climate change scenarios. *Agric. Water Manag.* **2020**, *238*, 106238. [CrossRef]
36. Tavakoli, A.R.; Moghadam, M.M.; Sepaskhah, A.R. Evaluation of the AquaCrop model for barley production under deficit irrigation and rainfed condition in Iran. *Agric. Water Manag.* **2015**, *161*, 136–146. [CrossRef]
37. Nie, T.Z.; Zhang, Z.X.; Lin, Y.Y.; Chen, P.; Sun, Z.Y. Temporal and spatial distribution characteristics of corn water demand in Heilongjiang Province from 1959 to 2015. *Nongye Jixie Xuebao* **2018**, *49*, 217–227. (In Chinese) [CrossRef]
38. Bai, R.; Yan, H.L.; Xue, Y.M.; Yu, S.Q. The influence of climatic conditions on the growth and development of maize. *J. Agric. Catastrophol.* **2021**, *11*, 89–90. (In Chinese)
39. Xiao, D.P.; Tao, F.L. Contributions of cultivars, management and climate change to winter wheat yield in the North China Plain in the past three decades. *Eur. J. Agron.* **2014**, *52*, 112–122. [CrossRef]
40. Tao, F.L.; Xiao, D.P.; Zhang, S.; Zhang, Z.; Rotter, R.P. Wheat yield benefited from increases in minimum temperature in the Huang-Huai-Hai Plain of China in the past three decades. *Agric. For. Meteorol.* **2017**, *239*, 1–14. [CrossRef]
41. Zhang, J.P.; Wang, C.Y.; Yang, X.G.; Zhao, Y.X.; Liu, Z.J.; Wang, J.; Chen, Y.Y. Forecast of the impact of future climate change on corn water demand in the three provinces of Northeast China. *Nongye Gongcheng Xuebao* **2009**, *25*, 50–55. (In Chinese) [CrossRef]
42. Jin, N.; He, J.Q.; Fang, Q.; Chen, C.; Yu, Q. The Responses of Maize Yield and Water Use to Growth Stage-Based Irrigation on the Loess Plateau in China. *Int. J. Plant Prod.* **2020**, *14*, 621–633. [CrossRef]
43. Yu, W.Y.; Ji, R.P.; Feng, R.; Zhao, X.L.; Zhang, Y.S. Responses of maize leaf photosynthetic characteristics and water use efficiency to water stress in different growth stages. *Acta Ecol. Sin.* **2015**, *9*, 2902–2909. (In Chinese) [CrossRef]
44. Araya, A.; Hoogenboom, G.; Luedeling, E.; Hadgu, K.M.; Kisekka, I.; Martorano, L.G. Assessment of maize growth and yield using crop models under present and future climate in southwestern Ethiopia. *Agric. For. Meteorol.* **2015**, *214*, 252–265. [CrossRef]
45. Dixit, P.N.; Telleria, R.; Khatibb, A.N.A.; Allouzi, S.F. Decadal analysis of impact of future climate on wheat production in dry Mediterranean environment: A case of Jordan. *Sci. Total Environ.* **2018**, *610*, 219–233. [CrossRef] [PubMed]
46. Saddique, Q.; Liu, D.L.; Wang, B.; Feng, P.Y.; Cai, H. Modelling future climate change impacts on winter wheat yield and water use: A case study in Guanzhong Plain, northwestern China. *Eur. J. Agron.* **2020**, *119*, 126113. [CrossRef]
47. Li, Y.; Tao, H.; Zhang, B.; Huang, S.; Wang, P. Timing of water deficit limits maize kernel setting in association with changes in the source-flow-sink relationship. *Front. Plant Sci.* **2018**, *9*, 01326. [CrossRef] [PubMed]
48. NeSmith, D.S.; Ritchie, J. Maize (Zea mays L.) response to a severe soil water-deficit during grain-filling. *Field Crops Res.* **1992**, *29*, 23–35. [CrossRef]
49. Cantore, V.; Lechkar, O.; Karabulut, E.; Sellami, M.H.; Albrizio, R.; Boari, F.; Stellacci, A.M.; Todorovic, M. Combined effect of deficit irrigation and strobilurin application on yield, fruit quality and water use efficiency of "cherry" tomato (*Solanum lycopersicum* L.). *Agric. Water Manag.* **2016**, *167*, 53–61. [CrossRef]
50. Sun, J.P.; Wei, Y.X.; Wang, Y.Y. Water-saving and Yield-increasing Effects of Regulated Deficit Irrigation of Corn in Western Heilongjiang. *J. Agric. Mech. Res.* **2016**, *38*, 186–190. (In Chinese) [CrossRef]
51. Cai, H.J.; Kang, S.Z.; Zhang, Z.H.; Chai, H.M.; Hu, X.T.; Wang, J. Research on the Optimal Time and the Degree of Regulated Deficit Irrigation for Crop. *Nongye Gongcheng Xuebao* **2000**, *3*, 24–27. (In Chinese) [CrossRef]
52. Wei, Y.X.; Ma, Y.Y.; Liu, H.; Zhang, Y.F.; Yang, J.M.; Zhang, Y. Drip Irrigation Maize Plant and Soil Moisture and Water-saving and Yield-increasing Effects under Regulated Deficit Irrigation. *Nongye Jixie Xuebao* **2018**, *49*, 253–260. (In Chinese) [CrossRef]

Article

AgroML: An Open-Source Repository to Forecast Reference Evapotranspiration in Different Geo-Climatic Conditions Using Machine Learning and Transformer-Based Models

Juan Antonio Bellido-Jiménez [1,2,*], Javier Estévez [1], Joaquin Vanschoren [2] and Amanda Penélope García-Marín [1]

[1]   Projects Engineering Area, Department of Rural Engineering, Civil Constructions and Engineering Projects, University of Córdoba, 14071 Córdoba, Spain; jestevez@uco.es (J.E.); amanda.garcia@uco.es (A.P.G.-M.)
[2]   Data Mining Group, Department of Mathematics and Computer Science, Eindhoven University of Technology, 5612 Eindhoven, The Netherlands; j.vanschoren@tue.nl
[*]   Correspondence: p22bejij@uco.es

Citation: Bellido-Jiménez, J.A.; Estévez, J.; Vanschoren, J.; García-Marín, A.P. AgroML: An Open-Source Repository to Forecast Reference Evapotranspiration in Different Geo-Climatic Conditions Using Machine Learning and Transformer-Based Models. *Agronomy* 2022, 12, 656. https://doi.org/10.3390/agronomy12030656

Academic Editors: Pantazis Georgiou and Dimitris Karpouzos

Received: 3 February 2022
Accepted: 5 March 2022
Published: 8 March 2022

Publisher's Note: MDPI stays neutral with regard to jurisdictional claims in published maps and institutional affiliations.

Abstract: Accurately forecasting reference evapotranspiration (ET_0) values is crucial to improve crop irrigation scheduling, allowing anticipated planning decisions and optimized water resource management and agricultural production. In this work, a recent state-of-the-art architecture has been adapted and deployed for multivariate input time series forecasting (transformers) using past values of ET_0 and temperature-based parameters (28 input configurations) to forecast daily ET_0 up to a week (1 to 7 days). Additionally, it has been compared to standard machine learning models such as multilayer perceptron (MLP), random forest (RF), support vector machine (SVM), extreme learning machine (ELM), convolutional neural network (CNN), long short-term memory (LSTM), and two baselines (historical monthly mean value and a moving average of the previous seven days) in five locations with different geo-climatic characteristics in the Andalusian region, Southern Spain. In general, machine learning models significantly outperformed the baselines. Furthermore, the accuracy dramatically dropped when forecasting ET_0 for any horizon longer than three days. SVM, ELM, and RF using configurations I, III, IV, and IX outperformed, on average, the rest of the configurations in most cases. The best NSE values ranged from 0.934 in Córdoba to 0.869 in Tabernas, using SVM. The best RMSE, on average, ranged from 0.704 mm/day for Málaga to 0.883 mm/day for Conil using RF. In terms of MBE, most models and cases performed very accurately, with a total average performance of 0.011 mm/day. We found a relationship in performance regarding the aridity index and the distance to the sea. The higher the aridity index at inland locations, the better results were obtained in forecasts. On the other hand, for coastal sites, the higher the aridity index, the higher the error. Due to the good performance and the availability as an open-source repository of these models, they can be used to accurately forecast ET_0 in different geo-climatic conditions, helping to increase efficiency in tasks of great agronomic importance, especially in areas with low rainfall or where water resources are limiting for the development of crops.

Keywords: machine learning; transformers; neural networks; support vector machine; reference evapotranspiration; forecasting; Bayesian optimization

1. Introduction

The worldwide population is increasing to alarming values that will require almost 50% more food to meet the demand in 2050 [1]. Therefore, research into new methodologies to outperform agroclimatic forecasts (solar radiation, precipitation, or evapotranspiration) is a relevant task that allows the optimization of water resource management, the improvement of irrigation scheduling, and, indeed, contributes to the great challenge of increasing food production. Furthermore, it is significantly impactful in arid and semiarid areas such as the Andalusian region (Southern Spain), where crop water uses are elevated and the scarce precipitation is limiting growth and agricultural yield.

Crop evapotranspiration measures the crops' water demand, being affected by atmospheric parameters (such as temperature, wind speed, or solar radiation), specific crop type, soil characteristics, as well as management and environmental conditions. The evapotranspiration rate from a reference surface with no shortage of water is named reference evapotranspiration (ET_0), which studies the evaporative demand of the atmosphere independently of the surface, the crop type, its development stage, and the management practices. Its calculation can be accurately determined using physics-based methods such as the FAO56-PM [2], which has been assessed globally in different climatic conditions and countries, including Korea [3], Argentina [4], and Tunisia [5], among others. However, measuring all the required parameters (air temperature, relative humidity, wind speed, and solar radiation) is very costly in installation and maintenance. Moreover, Automated Weather Stations (AWS) usually contain non-reliable long-term datasets, mainly for wind speed and solar radiation, due to a lack of maintenance or miscalibration [6]. These are the reasons why the geographical density of complete AWS is generally low, especially in rural areas and developing countries [7,8].

Therefore, developing new algorithms with fewer climatic input parameters is of high interest. In this context, Hargreaves and Samani [9] introduced an empirical equation (HS) that uses maximum and minimum daily air temperature values (Tx and Tn, respectively) and extraterrestrial solar radiation (Ra). Different studies have assessed HS in different aridity conditions and countries, such as Iran [10], Italy [11], Bolivia [11], China [12], and others. Nevertheless, advances in computation during the last several decades led to the application of new methodologies based on Artificial Intelligence (AI) with a very intensive computational cost. Thanks to the progress in CPU and GPU computation, the time spent training these models has dropped significantly, allowing scientists to apply them without needing a vast CPU/GPU farm and obtaining promising results in all sectors, especially agriculture. For example, Karimi et al. [13] evaluated the performance of random forest (RF) and other empirical methods to estimate ET_0 when several meteorological data were missing. RF surpassed the other models for temperature-based data availability when using Tx, Tn, Ra, and relative humidity (RH) as input features. Ferreira and da Cunha [14] assessed RF, extreme gradient boosting (XGB), multilayer perceptron (MLP), and convolutional neural network (CNN) to estimate daily ET_0 through different approaches, using hourly temperature and relative humidity as features in different AWS in Brazil. CNN outperformed the rest of the models for most statistics and locations in both local and regional approaches. However, no optimization algorithm was used during hyperparameter tuning. Yan et al. [15] evaluated XGB to estimate daily ET_0 in two different regions (an arid and humid region) from China and seven meteorological input combinations using maximum and minimum daily temperature (Tx and Tn, respectively), extraterrestrial solar radiation (Ra), relative humidity (RH), wind speed (U_2), and sunshine hours (n). In order to tune the different hyperparameters, the Whale Optimization Algorithm (WOA) was used. Their results showed that using local and external (neighbor stations) datasets obtained even better performance than using only local data in some cases. Therefore, this strategy is very promising when there is a lack of long-term records. Wu et al. [16] studied the performance of extreme learning machines (ELM) in different locations from China. They analyzed the use of the K-means clustering algorithm and the Firefly Algorithm (FFA) to estimate monthly mean daily ET_0 using Tx, Tn, Ra, and Tm (mean daily temperature). Nourani et al. [17] assessed support vector regression (SVR), Adaptive Fuzzy Inference System (ANFIS), MLP, and multiple linear regression (MLR) to forecast monthly ET_0 in Turkey, North Cyprus, and Iraq. Moreover, three ensemble methods were applied (simple averaging, weighted averaging, and neural ensemble) to outperform the performance and reliability of single modeling. The use of neural ensemble models highly outperformed single modeling in all cases, although simple and weighted averaging did not significantly perform better. Ferreira and da Cunha [18] evaluated the performance of daily ET_0 forecasts (up to 7 days) using CNN, long short-term memory (LSTM), CNN-LSTM, RF, and MLP using hourly data from different weather stations with heterogeneous aridity index characteristics in Brazil. In all cases, the use of the machine learning (ML) models outperformed the baselines, where CNN-LSTM performed the best in both local and regional scenarios using Tx, Tn, maximum and minimum relative humidity (RHx and RHn, respectively), wind speed, solar radiation (Rs), Ra, the day of the year (DOY), and ET_0 values from a lag window in the past (up to 30 days). In order to tune the different hyperparameters, a random search algorithm with 30 epochs was used.

In addition to these well-known and standard ML models, new architectures have been recently developed to deal with natural language programming (NLP) problems with outstanding results, called transformers [19]. The transformer model is an encoder–decoder architecture based on a self-attention mechanism that looks at an input sequence and decides which timesteps are valuable. The promising results of transformers have fostered their use on time series problems due to its apparent relationship. In both types of problems, words/parameter values are more or less meaningful based on their position. Therefore, several scientists have evaluated attention-based architectures in forecasting problems. For example, Wu et al. [20] proposed an Adversarial Sparse Transformer (AST) based on generative adversarial networks (GAN). They assessed it to forecast five different public datasets: (I) an hourly time series electricity consumption dataset, (II) an hourly traffic level from the San Francisco dataset, (III) an hourly solar power production dataset, and (IV) an hourly time series dataset from the M4 competition. Furthermore, [21] analyzed a transformed-based architecture to forecast influenza-like illness (ILI), obtaining promising results. Finally, Li et al. [22] evaluated the performance of transformers in time series forecasting using the same public datasets as Wu et al. [20] and obtained more accurate modeling with long-term dependencies.

This work is motivated by the need to minimize error in daily ET_0 forecasts, which is one of the main drawbacks in the reviewed literature, as well as the outstanding and promising performance of transformers and transformer-based models in different fields. Thereby, this work is the first one using a multivariate input transformer-based architecture in order to forecast daily ET_0 (from one to seven days ahead). The development and assessment have been carried out using past values of ET_0 and temperature-based measured variables as features in five sites of Andalusia (Córdoba, Málaga, Conil, Tabernas, and Aroche) with different geo-climatic characteristics. Moreover, standard ML models such as RF, MLP, SVR, ELM, CNN, and LSTM have been also evaluated in conjunction with Bayesian optimization to tune all their different hyperparameters. Thus, the main objectives of this work are (a) to assess the performance of the proposed transformer model to forecast ET_0 and to compare it to standard ML models and two simple baselines (historical monthly mean value and mean of previous seven days); (b) to evaluate different input feature configurations based on ET_0 past values and several temperature-based features to forecast ET_0, and (c) to analyze the forecast efficiency depending on the different geo-climatic characteristics of the sites.

2. Materials and Methods

2.1. Study Area and Dataset

Andalusia is located in the southwest of Europe, ranging from $37°$ to $39°$ N, from $1°$ to $7°$ W, and occupying an extension of 87,268 km^2. This work was carried out with data from five locations in Andalusia (Figure 1), with different geo-climatic characteristics and representing great variability in terms of UNEP aridity index [23] in this region (ranging from 0.555—dry subhumid—in Aroche, to 0.177—arid—in Tabernas). The coordinates and other characteristics of the AWS are reported in Table 1. In contrast, in Table 2, the minimum, mean, maximum, and standard deviation values of minimum, mean, and maximum daily air temperature (Tn, Tm, and Tx, respectively), relative humidity (RHn, RHm, RHx, respectively), wind speed (U_2), solar radiation (Rs), and reference evapotranspiration (ET_0) data are shown. The dataset used in this study belongs to the Agroclimatic Information Network of Andalusia (RIA), which can be downloaded at https://www.juntadeandalucia.es/agriculturaypesca/ifapa/ria/servlet/FrontController (accessed on 1 February 2022).

Figure 1. Spatial distribution of Aroche, Conil, Córdoba, Málaga, and Tabernas in the Andalusia region, south of Spain.

Table 1. Geo-climatic characteristics of the locations assessed in this work (ARO—Aroche, CON—Conil de la Frontera, COR—Córdoba, MAG—Málaga, and TAB—Tabernas). Time period from 2000 to 2018.

Site	Lon. (° W)	Lat. (° N)	Alt. (m)	Mean Annual Precipitation (mm)	UNEP Aridity Index	Total Available Days
Aroche (ARO)	6.94	37.95	293	632	0.555 (dry-subhumid)	6399
Conil de la Frontera (CON)	6.13	36.33	22	470	0.479 (semiarid)	5868
Córdoba (COR)	4.80	37.85	94	589	0.462 (semiarid)	6397
Málaga (MAG)	4.53	36.75	55	434	0.366 (semiarid)	6438
Tabernas (TAB)	2.30	37.09	502	237	0.178 (arid)	6694

In this work, because the accurate estimation of ET_0 using limited meteorological data has been improved in recent years [14,24] and due to the high availability of temperature records, only temperature-based and ET_0 values from the past have been used as input features to forecast ET_0. Specifically, two different windows have been evaluated, the use of 15 and 30 days from the past. Moreover, several temperature-based variables have been calculated, such as EnergyT (the area below the intraday temperature in a whole day), HourminTx (the time when Tx occurs), HourminTn (the time when Tn occurs), HourminSunset (the time when sunset occurs), HourminSunrise (the time when sunrise occurs), es (mean saturation vapor pressure), ea (actual vapor pressure) and VPD (vapor pressure deficit), Tx-Tn, HourminSunset-HourminTx, and HourminSunrise-HourminTn. All the configurations assessed in this work contained Tx, Tn, Tx-Tn, and Ra as features due to their very high Pearson correlation (Figure 2), and the rest of the configurations were selected based on their Pearson correlation values and the previous results on these same locations regarding ET_0 and solar radiation [24–26] estimations. The 27 different assessed configurations are shown in Table 3.

Table 2. Minimum (Min), mean, maximum (Max), and standard deviation (Std) values of all the daily parameters measured: maximum air temperature (Tx), mean air temperature (Tm), minimum air temperature (Tn), maximum relative humidity (RHx), mean relative humidity (RHm), minimum relative humidity (RHn), wind speed at 2 m height (U_2), solar radiation (Rs), reference evapotranspiration (ET_0) at each location (ARO—Aroche, CON—Conil de la Frontera, COR—Córdoba, MAG—Málaga, and TAB—Tabernas) and for the whole dataset (2000–2018).

		Tx (°C)	Tm (°C)	Tn (°C)	RHx (%)	RHm (%)	RHn (%)	U_2 (m/s)	Rs (MJ/m² day)	ET_0 (mm)
ARO	Min	2.5	−0.2	−8.0	32.5	17.2	5.0	0.3	1.0	0.3
	Mean	23.2	16.1	8.9	89.5	65.9	39.0	1.2	17.8	3.2
	Max	44.0	34.1	24.9	100.0	100.0	100.0	5.8	34.3	8.7
	Std	8.1	6.8	5.6	11.2	17.7	19.4	0.5	8.8	2.0
CON	Min	6.4	0.7	−5.3	39.9	24.3	6.9	0.0	0.5	0.4
	Mean	23.0	17.4	12.1	89.3	72.5	50.5	1.3	18.0	3.2
	Max	41.3	31.9	26.9	100.0	99.6	97.1	7.9	31.7	9.3
	Std	5.7	5.2	5.3	9.0	12.3	14.6	1.0	7.8	1.8
COR	Min	3.3	0.0	−8.3	38.9	21.8	4.3	0.0	0.5	0.3
	Mean	24.6	17.4	11.0	86.8	64.1	37.3	1.6	17.7	3.6
	Max	45.7	34.7	27.6	100.0	100.0	100.0	7.5	33.2	9.6
	Std	8.5	7.3	6.2	12.0	18.1	19.3	0.7	8.5	2.3
MAG	Min	6.2	3.3	−4.2	36.0	19.4	4.6	0.0	0.3	0.4
	Mean	23.9	18.2	12.6	85.1	63.4	39.1	1.3	18.2	3.4
	Max	42.7	33.7	26.8	100.0	99.7	98.3	4.6	32.4	10.3
	Std	6.3	5.8	5.5	10.5	14.2	15.1	0.5	8.2	1.9
TAB	Min	4.3	−1.2	−8.2	28.6	16.8	2.8	0.1	0.2	0.4
	Mean	23.2	16.4	9.8	85.7	59.9	32.9	1.9	18.4	3.8
	Max	42.5	32.1	26.0	100.0	97.5	95.0	9.9	32.8	10.6
	Std	7.2	6.6	6.2	11.9	15.1	14.8	0.9	7.8	2.0

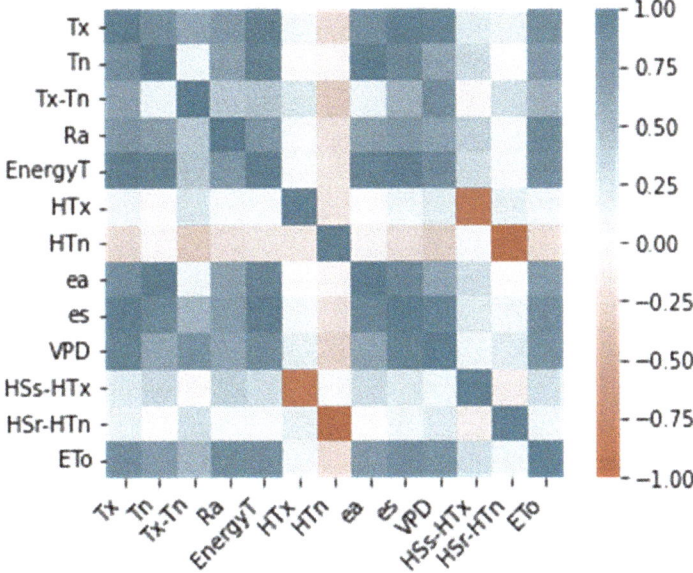

Figure 2. Pearson correlation values of the assessed features in all the stations.

Table 3. Configuration table with all configurations. HTx represents HourminTx, HTn represents HourminTn, HSs represents HourminSunset, and HSr represents HourminSunrise.

Conf.	Tx	Tn	Tx-Tn	Ra	EnergyT	ea	es	VPD	HTx	HTn	HSs-HTx	HSr-HTn	ET_0
I	X	X	X	X		X							X
II	X	X	X	X	X	X							X
III	X	X	X	X	X		X						X
IV	X	X	X	X	X								X
V	X	X	X	X	X						X		X
VI	X	X	X	X	X							X	X
VII	X	X	X	X	X					X			X
VIII	X	X	X	X	X	X			X				X
IX	X	X	X	X	X		X		X				X
X	X	X	X	X	X				X				X
XI	X	X	X	X	X				X		X		X
XII	X	X	X	X	X				X			X	X
XIII	X	X	X	X	X	X	X	X	X	X	X	X	X
XIV	X	X	X	X	X	X			X	X			X
XV	X	X	X	X	X		X		X	X			X
XVI	X	X	X	X	X				X	X			X
XVII	X	X	X	X	X				X	X	X		X
XVIII	X	X	X	X	X				X	X		X	X
XIX	X	X	X	X	X			X	X	X			X
XX	X	X	X	X	X			X	X				X
XXI	X	X	X	X	X			X					X
XXII	X	X	X	X			X						X
XXIII	X	X	X	X									X
XXIV	X	X	X	X							X		X
XXV	X	X	X	X								X	X
XXVI	X	X	X	X						X			X
XXVII	X	X	X	X					X				X

2.2. Preprocessing Methodology

In machine learning applications, a vital prerequisite to guarantee accurate modeling is the use of reliable datasets. In this work, the control guidelines reported by Estévez et al. [6] have been followed to identify erroneous and questionable data from sensor measurements by applying different tests (range, internal consistency, step, and persistence) and a spatial consistency test [27]. These quality assurance procedures have been successfully employed in different countries [4,28,29]. Afterward, the input and output matrices had to be built depending on the number of lag days from the past (15 or 30), the features to use (up to 27 input configurations), and the number of days to forecast (up to 7 days). In Figures 3 and 4, a mind map with all the possibilities is shown. It is worth noting that a MIMO (Multiple Input Multiple Output) approach was used in models that allowed it, whereas a direct approach was considered in the others according to the results of Ferreira and da Cunha [18].

Consequently, using configuration 1 and 15 lag days as an example, the values from day to day—14 of Tx, Tn, Tx-Tn, Ra, ea, and ET_0 are used as input features (a total of 90 values) for all the ML models (except for transformers—see Section 2.5.7), where Tx and Tn are directly given by AWS, and Ra and ea can be calculated using Tx, Tn, and the latitude, as stated by [2]. Finally, ET_0 is calculated using the well-known FAO56-PM method.

Later, in order to train, tune all the hyperparameters, and assess the final performance of the model, for each location, the dataset was split into training (70% of the entire dataset length), validation (20% of the training dataset length), and testing (30% of the entire dataset length) using a holdout technique. Next, the Bayesian optimization algorithm was used to tune all the hyperparameters (the hyperparameter space can be seen in Table S1, Supplementary Materials). Eventually, after the best hyperparameter set was found, the final model was trained using the entire training dataset (70% of the entire dataset length) and evaluated using the testing dataset. Figure 5 shows an overview of this methodology.

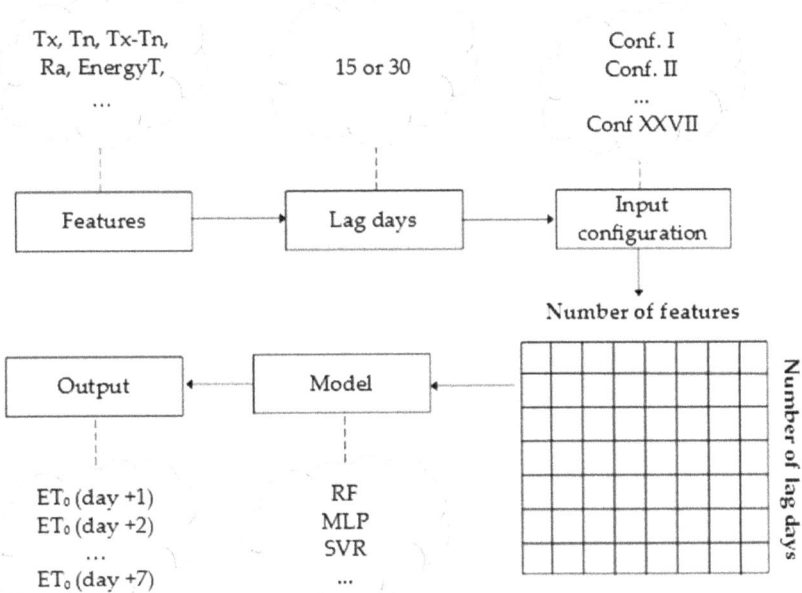

Figure 3. Mind map of the matrix data structure.

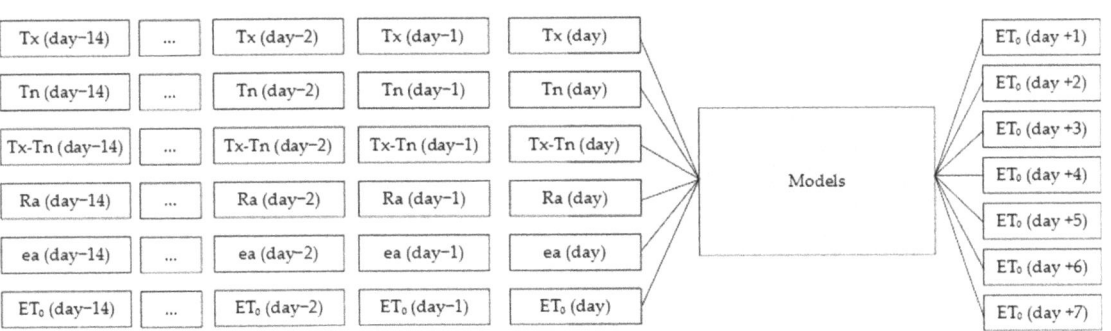

Figure 4. Forecasting approaches using configuration 1 as an example.

2.3. Reference Evapotranspiration Calculation

In this work, the ET_0 (FAO56-PM) values were used as input and target values. They were determined following the procedure of [2], and can be mathematically expressed as Equation (1):

$$ET_0 = \frac{0.408\Delta(Rn - G) + \gamma\frac{900}{T+273}U_2(es - ea)}{\Delta + \gamma(1 + 0.34U_2)} \tag{1}$$

where ET_0 is the reference evapotranspiration (mm day^{-1}), 0.408 corresponds to a coefficient (MJ^{-1} m^2 mm), Δ is the slope of the saturation vapor pressure versus temperature curve (kPa °C^{-1}), Rn is the net radiation calculated at the crop surface (MJ m^{-2} day^{-1}), G is the soil heat flux density at the soil surface (MJ m^{-2} day^{-1}), γ is the psychrometric constant (kPa °C^{-1}), T is the mean daily air temperature (°C), U_2 is the mean daily wind speed at 2 m height (m s^{-1}), and es and ea are the saturation vapor pressure and the mean actual vapor pressure, respectively (kPa).

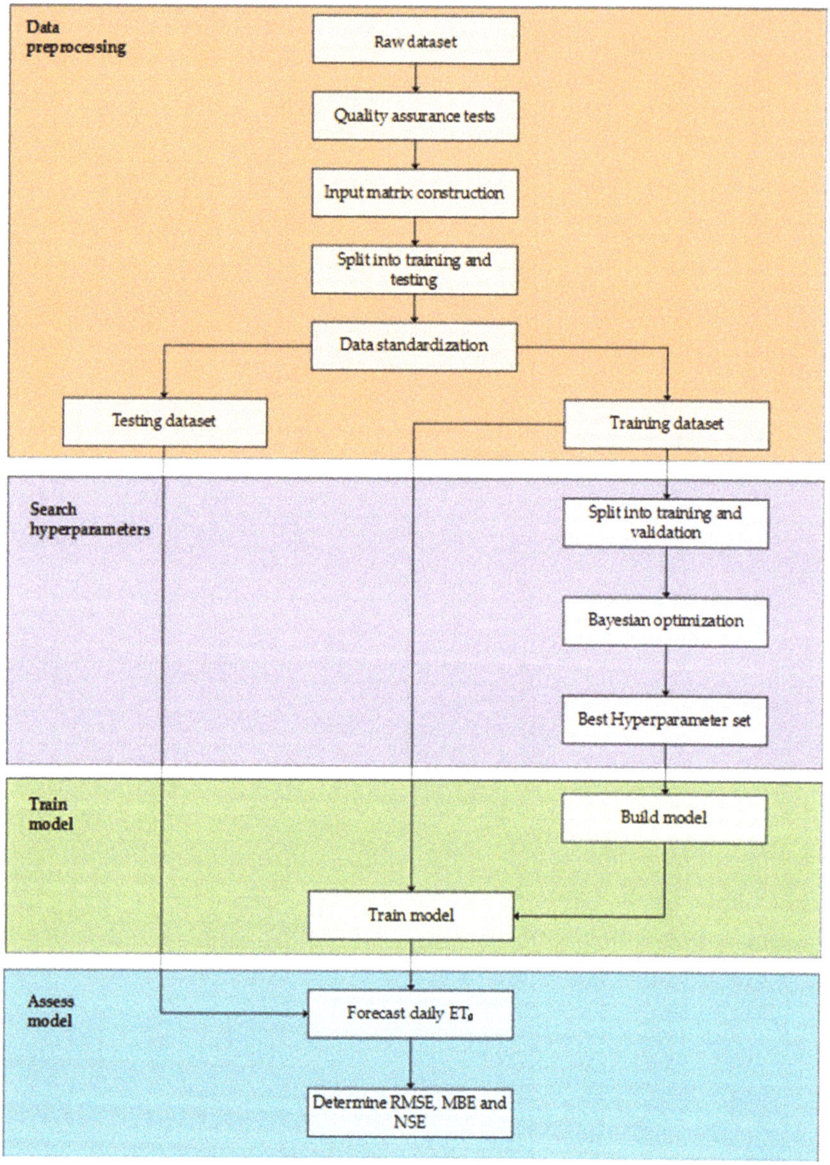

Figure 5. Methodology flowchart.

2.4. Baselines

In order to compare the performance of the developed models and configurations, it is crucial to have a baseline performance as a starting point. In this sense, two empirical baselines have been proposed in this work, following the methodology proposed by Ferreira and da Cunha [18]. In the first place, a moving average from the last 7 days was used. Secondly, the historical average monthly values from the training dataset were used for the corresponding forecast day.

2.5. Machine Learning Models

2.5.1. Multilayer Perceptron

The multilayer perceptron (MLP) is one of the most used agronomical and hydrological AI models [14,30,31]. Its popularity is based on its similarities to neurons in the biological nervous system, easy coding, and promising results in most cases. They are structured in three kinds of layers, the input and output layer, representing the inputs and outputs of the model, respectively, and the hidden layers, where all the neurons are located. The neurons work together to create stimuli (reference evapotranspiration forecast values) based on different inputs (the input matrix containing features from the past). A back-propagation algorithm makes the neurons learn (automatically update all weights and biases) and improve every mini batch every epoch. A single neuron architecture can be seen in Figure 6.

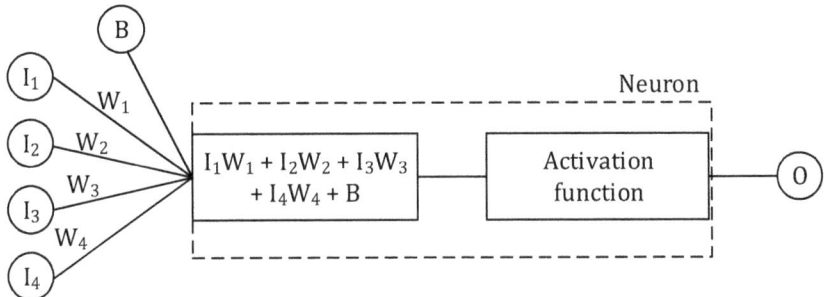

Figure 6. Single neuron architecture. I_1, I_2, I_3, and I_4 represent the inputs of the neuron; W_1, W_2, W_3, and W_4 correspond to the weights of every input; B is the bias, and O represents the output of the neuron after passing through an activation function.

2.5.2. Extreme Learning Machine

Extreme learning machine models (ELM) were first introduced by Huang et al. [32] as a single hidden layer feed-forward neural network with the following main characteristics: (I) the input weights and biases are randomly generated and (II) the output weights and biases are analytically determined. As a result, these models do not require any training process and have a meager computational cost, with promising results in ET_0 [24,33,34]. However, on the other hand, when the model is working with massive datasets, the amount of random access memory (RAM) required is enormous.

2.5.3. Support Vector Machine for Regression

Support vector machine (SVM) models for regression tasks, also known as support vector regression (SVR) models, are supervised AI models based on a different functionality than neuron-based architectures such as MLP and ELM. They search for the best hyperplane (and its margins) that contains all data points. Thus, it could be easily related to linear regression with the flexibility of defining how much error can be considered acceptable. Moreover, one of their most important features is the use of kernels to allow the model to operate on a high-dimensional feature space. SVMs can be mathematically expressed as a minimization problem of Equation (2) with the constraints in Equation (3).

$$MIN \left(\frac{1}{2} \|w\|^2 + C \sum_{i=1}^{n} |\xi_i| \right) \geq 0 \tag{2}$$

$$|y_i - w_i x_i| \leq \varepsilon + |\xi_i| \tag{3}$$

where w_i corresponds to the weight vector, x_i to the input vector, y_i to the output vector, ε represents the margins, ξ represents the deviation of values to the margins, C is a coefficient to penalize deviation to the margins, and n is the length of the training dataset. For further details, the work of [35] can be consulted.

2.5.4. Random Forest

A random forest (RF) is composed of the conjunction of multiple tree-based models in order to improve the overall result (ensemble model). The general idea is that different models are trained on different data samples (bootstrap) and feature sets. Instead of searching for the best features when splitting nodes, it searches among a random subset of the features. Thus, it results in greater diversity and better final performance.

2.5.5. Convolutional Neural Network

Convolutional neural network (CNN) models were first developed for image classification problems, where the convolution algorithm captures local patterns to learn a representation of figures to classify them. Moreover, this process can be extrapolated to 1D sequences of data such as time series datasets. One of the advantages of using convolutions is that they can obtain local features' relationships without the requirement of an extensive preprocessing method and can obtain outstanding results in ET_0 [14,36,37] and in other agro-climatic parameters [25,38,39].

Typically, such CNNs are composed of three layers: the convolutional layer, the pooling layer, and a fully connected layer. The convolutional layer is used to extract local relationships between the different features and timesteps. The pooling layer is added after the convolutional layer, and it gradually reduces the feature map. Finally, a fully connected layer is used to forecast the seven-day horizon ET_0 values (in this work). For further details, the work of Aloysius et al. [40] can be reviewed.

2.5.6. Long Short-Term Memory

Long short-term memory (LSTM) models were first introduced by Hochreiter et al. [41] as a recurrent neural network (RNN)-based model that could deal with long-term dependencies and address the vanishing gradient problem. In order to control the information flow, the LSTM block contains an input gate, an output gate, a forget gate, a cell state, and a hidden state. The gates are in charge of deciding which information is allowed on the cell state, i.e., whether a piece of information is relevant to keep or forget during training. The cell and hidden state can be seen as the memory of the network, used to carry relevant information throughout the sequence.

2.5.7. Transformers

A new state-of-the-art architecture has been recently presented for NLP problems, the transformers [19]; see Figure 7. One of the main motivations of transformers is to deal with the vanishing gradient problem of LSTM when working with long sequences. Although LSTMs can theoretically propagate crucial information over infinitely long sequences, due to the vanishing gradient problem, they pay more attention to recent tokens and eventually forget earlier tokens. In contrast, transformers use an attention mechanism, which learns the relevant subset of the sequences to accomplish the specific task. For a single head, the operation can be expressed as Equation (4),

$$Attention(Q,\ K,\ V) = Softmax(\frac{Q\ K^T}{\sqrt{d_k}})V \tag{4}$$

where Q, K, and V represent the query, key, and value, respectively, as an analogy to a database, and d_k corresponds to queries and keys' dimension. As stated by Yıldırım and Asgari-Chenaghlu (2021), the attention mechanism can be defined as follows: "*This can also be seen as a database where we use the query and keys in order to find out how much various items are related in terms of numeric evaluation. Multiplication of attention score and the V matrix produces the final result of this type of attention mechanism"*. In particular, transformers use a multi-head attention mechanism, which can be mathematically expressed as Equation (5).

$$MultiHead(Q, K, V) = [Head_1, \ldots, Head_h]W_0 \qquad (5)$$

where $Head_i$ is attention (QW_i, KW_i, VW_i) and W denotes all the learnable parameter matrices.

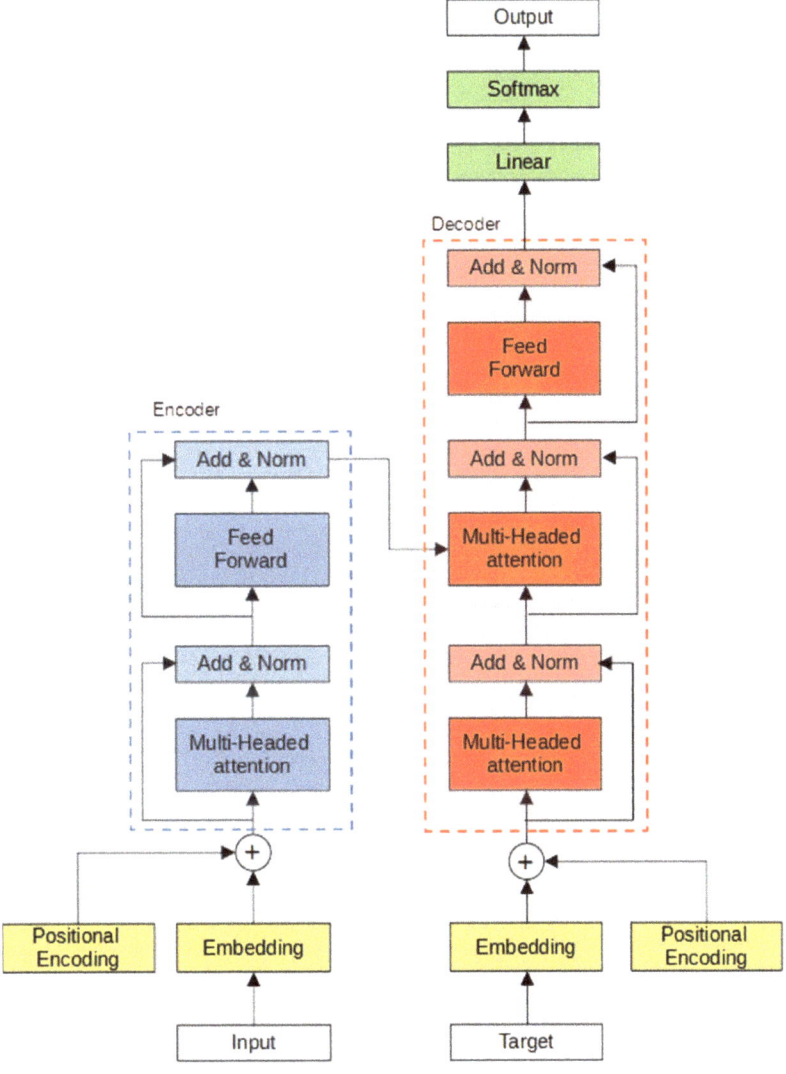

Figure 7. Original transformer architecture.

Generally, the transformer is an encoder–decoder architecture. Considering a translation task from English to Spanish, the encoder takes an input sequence ('I am from Spain') and maps it into a higher-dimensional space using a multi-headed attention, an adding, a normalization, and a fully connected feed-forward layer. The abstract vector obtained in the encoder module is fed into the decoder, which uses it to obtain the translated sentence ('Soy de España'). It is worth noting that both the encoder and decoder are composed of modules that can be stacked on top of each other multiple times. However, before carrying out any mathematical operation to the input data, it is required to convert words into numbers. The embedding layer is used for this purpose, transforming words into a vector of numbers that can be easily recognized by the model.

Another vital aspect to consider is the need for transformers to learn the temporal dependencies of the different timestamps through positional encoding because they do not inherently carry it out. In this work, the positional encoding was achieved using Equations (6) and (7) for monthly and daily values (Figure 8). In this way, 31 January and 2 February are close, but 5 May and 26 July are not.

$$PE_{(pos,\ 2i)} = \sin\left(\frac{pos}{10,000^{2i/d_{model}}}\right) \tag{6}$$

$$PE_{(pos,\ 2i+1)} = \cos\left(\frac{pos}{10,000^{2i/d_{model}}}\right) \tag{7}$$

where *pos* represents the position, d_{model} is the input dimension, and *i* represents the index in the vector. It is worth noting that this temporal dependency information is shared with the rest of the models in this work to make the comparison between models as fair as possible. Thus, new features are included in all configurations. For example, in configuration 1, the input features would be Tx, Tn, Tx-Tn, Ra, ea, ET0, Sin_day (sine of days over 31 days period), Cos_day (cosine of days over 31 days period), Sin_month (sine of days over 12 month period) and Cos_month (cosine of days over 12 month period).

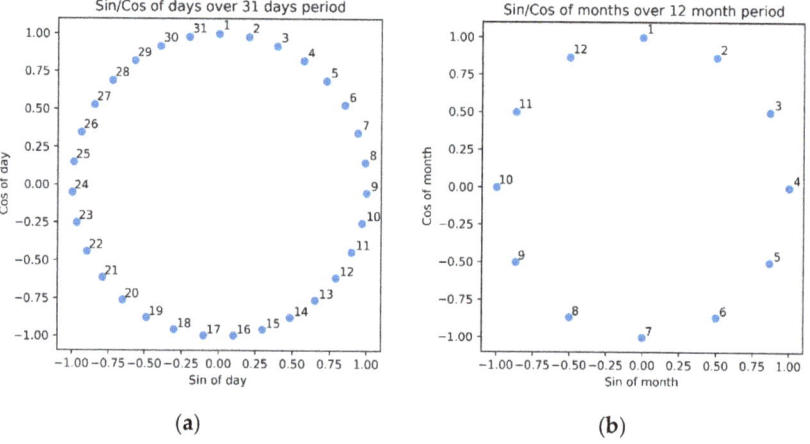

Figure 8. Sine/cosine positional encoding for 31 days in a month (**a**) and 12 months in a year (**b**).

The architecture used in this work can be seen in Figure 9. It is based on the original transformer architecture from Vaswani et al. [19] and the attention-based architecture of Song et al. [42]. Several aspects were modified. First, since the input data already have numerical values, the embedding layer was omitted. Then, the positional encoding included new features in the input matrix instead of adding their values to the "embedded vector". Consequently, four more features were used in this model (sine and cosine positional encoding for days in a month, and sine and cosine positional encoding for months in a year). Finally, the SoftMax layer was also deleted because we are dealing with a regression problem (forecasting ET0). Thus, the processing of data in the proposed transformer-based model can be described as follows. Firstly, the input matrix passes through a positional encoding mechanism. Then, the positional encoding features are added to the input matrix. Later, the data go to an attention-based block containing multi-head attention, dropout, normalization, addition, and feed-forward layers. Two different variations have been tested depending on the model used in the feed-forward layer: TransformerCNN, where a convolutional approach has been used, and TransformerLSTM, where an LSTM approach has been implemented. Eventually, the processed data go to an MLP model to carry out the regression task. The following works provide further details [19,21,43,44] and the code can be checked at the AgroML GitHub repository.

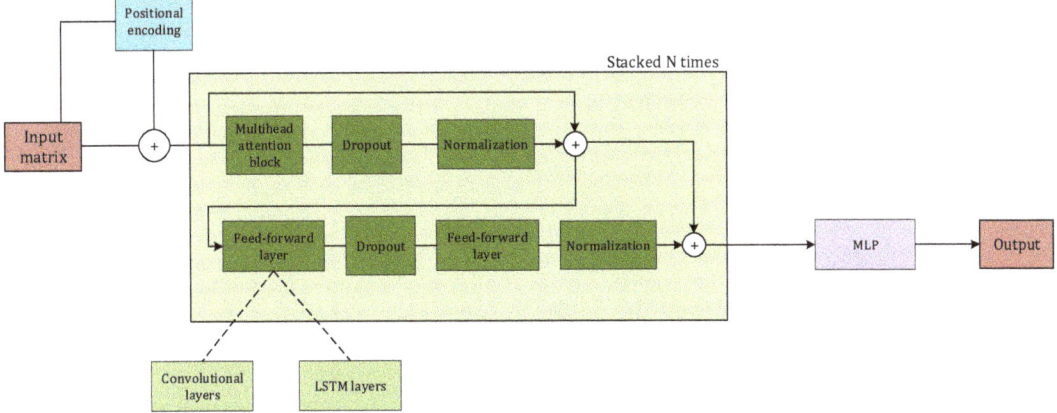

Figure 9. The architecture of the proposed multi attention-based model.

2.6. Bayesian Optimization

The most critical aspect to obtain accurate performance in machine learning models is choosing the fittest hyperparameter set. The results could dramatically change from outstanding to very poor. A prevalent practice among the scientific community in agronomy and hydrology is using a trial-and-error approach [14,18,36], evaluating from dozens to hundreds of sets. However, it is not an efficient approach because the process is too slow if the hyperparameter space is large, spending a significant amount of time on non-promising configurations. Otherwise, if the hyperparameter space is made to be small, one may obtain a suboptimal model. Several optimization algorithms have been assessed to solve this problem—for example, Particle Swarm Optimization (PSO), Grey Wolf Optimizer (GWO), Genetic Algorithms (GA), Bayesian Optimization (BO), and the Whale Optimization Algorithm (WOA), among others [31,45–47].

In this work, the BO algorithm has been proposed due to its high sample efficiency and popularity in automated machine learning libraries such as Auto-Weka 2.0 [48], Auto-Keras [49], and Auto-Sklearn [50] and they can be consulted in Hutter et al. [51]. Part of its popularity is related to the close relationship to human behavior when carrying out this same process [52,53], where prior results are considered to choose the following set. BO is based on Bayes' theorem, and it can be explained using the following four steps: (I) definition of the hyperparameter space; (II) the algorithm first tries several random sets; (III) the algorithm takes into account the previously assessed configuration sets when choosing the following one, balancing between exploitation (it exploits regions that are known to have good performance) and exploration (choosing region with higher uncertainty), and evaluating it; (IV) if the process has not finished yet, it goes to step 3.

In this work, BO has been implemented using Scikit-Optimize (gp_minimize) and Python 3.8. In all cases, this process was configured using 50 Bayesian epochs (80% of them were randomly chosen), selected after a trial-and-error algorithm among 50, 100, 150, and 200 Bayesian epochs, the mean absolute error (MAE) as the objective function, and the rest of parameters as default. The hyperparameter space can be found in Table S1, Supplementary Materials and their results in Table S2.

2.7. Evaluation Metrics

The models' performance has been evaluated by using the following parameters: mean bias error (MBE), root mean square error (RMSE), and the Nash–Sutcliffe model efficiency coefficient (NSE). The MBE, RMSE, and NSE are defined as Equations (8)–(10):

$$\text{MBE} = \frac{1}{n} \sum_{i=1}^{n} x_i - y_i \tag{8}$$

$$\text{RMSE} = \sqrt{\frac{1}{n} \sum_{i=1}^{n} (x_i - y_i)^2} \tag{9}$$

$$\text{NSE} = 1 - \frac{\sum_{i=1}^{n} (x_i - y_i)^2}{\sum_{i=1}^{n} (x_i - \bar{x})^2} \tag{10}$$

where x and y correspond to the observed and forecasted ET_0 values, respectively, n represents the number of records in the testing dataset, and the bar denotes the mean.

3. Results and Discussion

It is worth noting that the code developed in this work is available on GitHub in the public repository called AgroML, which can be found at https://github.com/Smarity/agroML (accessed on 1 February 2022). This new library focuses on helping scientists to research state-of-the-art machine learning models, mainly focused on agronomy estimations and forecasts but easily extrapolated to other sectors and problems. It lets new scientists test these models on their datasets, and experienced scientists commit new features and architectures. The code has been programmed in standard Python using Tensorflow, Scikit-Learn, Scikit-Optimize, Pandas, and Numpy.

3.1. Baseline Performance

Tables 4 and 5 show the RMSE and NSE performance for the baselines along the different forecast horizons (up to 1 week), where B1 refers to the moving average of the last seven ET_0 values and B2 the use of mean historical monthly ET_0 values (mean ET_0 values for each month of the year). Generally, B2 outperformed B1 for all the forecast horizons except for one day ahead, where B1 performed better in all sites. Moreover, B1 obtained the most accurate forecasts on the one day ahead horizon, and it gradually dropped when the forecast horizon increased. In Aroche, the most humid site, the best performance in both RMSE and NSE values was obtained (NSE = 0.9038 and RMSE = 0.6390), followed by Córdoba, Málaga, Conil, and Tabernas (the most arid site), in this order. This suggests a relationship between the aridity index, distance to the sea, and the performance of the models. In inland locations, the higher the aridity index, the fewer the forecasting errors. On the other hand, in coastal locations, the opposite occurs. The higher the aridity index and the farther from the sea, the more precise the ET_0 modeling. Finally, Table 6 shows the MBE values for the different stations and forecast horizons. In this case, B1 outperformed B2 in most of the cases.

Table 4. RMSE values for ET_0 forecast during seven forecast horizons and the two empirical baselines (B1—using the average value from the last seven days—and B2—using the mean monthly value from the training dataset).

Location	Baseline	Forecast Horizon						
		1	2	3	4	5	6	7
COR	B1	0.7551	0.8733	0.9365	0.9926	1.0172	1.0363	1.0644
	B2	0.8374	0.8374	0.8374	0.8374	0.8374	0.8374	0.8374
MAG	B1	0.7665	0.9084	0.9439	0.9632	0.9902	1.0140	1.0188
	B2	0.8143	0.8143	0.8143	0.8143	0.8143	0.8143	0.8143

Table 4. *Cont.*

Location	Baseline	Forecast Horizon						
		1	2	3	4	5	6	7
TAB	B1	0.8515	0.9961	1.0451	1.0938	1.1075	1.1568	1.1628
	B2	0.9176	0.9176	0.9176	0.9176	0.9176	0.9176	0.9176
CON	B1	0.7987	1.0675	1.1950	1.2474	1.2404	1.2444	1.2778
	B2	0.9567	0.9567	0.9567	0.9567	0.9567	0.9567	0.9567
ARO	B1	0.6390	0.7882	0.8840	0.9337	0.9820	0.9901	1.0032
	B2	0.8027	0.8027	0.8027	0.8027	0.8027	0.8027	0.8027
Mean	B1	0.7622	0.9277	1.0009	1.0461	1.0675	1.0883	1.1054
	B2	0.8667	0.8667	0.8667	0.8667	0.8667	0.8667	0.8667

Table 5. NSE values for ET_0 forecast during seven forecast horizons and the two empirical baselines (B1—using the average value from the last seven days—and B2—using the mean daily monthly value from the training dataset).

Location	Model	Forecast Horizon						
		1	2	3	4	5	6	7
COR	B1	0.8926	0.8564	0.8349	0.8145	0.8052	0.7978	0.7868
	B2	0.8680	0.8680	0.8680	0.8680	0.8680	0.8680	0.8680
MAG	B1	0.8376	0.7719	0.7538	0.7436	0.7290	0.7157	0.7129
	B2	0.8167	0.8167	0.8167	0.8167	0.8167	0.8167	0.8167
TAB	B1	0.8197	0.7531	0.7283	0.7023	0.6947	0.6671	0.6638
	B2	0.7906	0.7906	0.7906	0.7906	0.7906	0.7906	0.7906
CON	B1	0.8235	0.6844	0.6042	0.5684	0.5728	0.5695	0.5455
	B2	0.7465	0.7465	0.7465	0.7465	0.7465	0.7465	0.7465
ARO	B1	0.9038	0.8537	0.8160	0.7949	0.7732	0.7696	0.7636
	B2	0.8481	0.8481	0.8481	0.8481	0.8481	0.8481	0.8481
Mean	B1	0.8554	0.7849	0.7474	0.7247	0.7150	0.7039	0.6945
	B2	0.8140	0.8140	0.8140	0.8140	0.8140	0.8140	0.8140

Table 6. MBE values for ET_0 forecast during seven forecast horizons and the two empirical baselines (B1—using the average value from the last seven days—and B2—using the mean daily monthly value from the training dataset).

Location	Model	Forecast Horizon						
		1	2	3	4	5	6	7
COR	B1	−0.0002	−0.0001	−0.0001	0.0000	−0.0002	−0.0001	0.0007
	B2	0.1033	0.1033	0.1033	0.1033	0.1033	0.1033	0.1033
MAG	B1	0.0000	0.0002	0.0000	0.0000	−0.0008	−0.0016	−0.0015
	B2	0.0710	0.0710	0.0710	0.0710	0.0710	0.0710	0.0710
TAB	B1	0.0003	0.0003	0.0000	−0.0018	−0.0034	−0.0041	−0.0046
	B2	0.0972	0.0972	0.0972	0.0972	0.0972	0.0972	0.0972
CON	B1	0.0014	0.0047	0.0084	0.0117	0.0157	0.0198	0.0236
	B2	−0.0113	−0.0113	−0.0113	−0.0113	−0.0113	−0.0113	−0.0113
ARO	B1	0.0006	0.0011	0.0012	0.0021	0.0029	0.0036	0.0052
	B2	0.1787	0.1787	0.1787	0.1787	0.1787	0.1787	0.1787
Mean	B1	0.0004	0.0012	0.0019	0.0024	0.0028	0.0035	0.0047
	B2	0.0878	0.0878	0.0878	0.0878	0.0878	0.0878	0.0878

3.2. Analysis of ML Performance

Table 7 shows the minimum, mean, and maximum NSE, RMSE, and MBE values for all the sites and models using two different lag intervals (15 and 30 days). Generally, in terms of NSE and RMSE, the use of 15 days slightly outperformed all the models using 30 lag days for almost all the cases. On the other hand, the MBE performance for all models, locations, and lag days was very similar. Additionally, ML approaches highly outperformed the baselines, although the CNN and the transformer-based models gave the worst results in all sites. In Tabernas, the most arid site, in terms of NSE and RMSE, all the ML models surpassed the baseline performance. SVM obtained the best values (NSE = 0.869 and RMSE = 0.700 mm/day), followed very closely by RF (NSE = 0.867 and RMSE = 0.706 mm/day), which outperformed, on average, the rest of the models. On the other hand, the CNN model obtained the worst modeling for 30 lag days (NSE = 0.423 and RMSE = 1.438 mm/day). All the models obtained high mean MBE metrics, obtaining the highest MBE value (-0.974 mm/day) using CNN and 30 lag days. In Conil, the best values were obtained by SVM (RMSE = 0.684 mm/day), RF (RMSE = 0.703 mm/day), and ELM (RMSE = 0.717 mm/day), in this order and for 15 lag days. In terms of NSE, these three models also gave the best performance on mean values and for 15 lag days, whereas the worst were obtained by CNN (NSE = 0.520) for 30 lag days. In Córdoba, SVM and ELM using 15 lag days outperformed the rest of the models in both RMSE (0.605 and 0.614 mm/day) and NSE (0.934 and 0.932), respectively. Moreover, on average, the best results were obtained in Córdoba compared to the rest of the sites (NSE > 0.85, RMSE < 0.80 mm/day, and MBE ≈ 0.0 mm/day). In Aroche, the most humid site, the NSE values ranged from 0.737 (CNN model) to 0.922 (SVM model) and the RMSE values ranged from 0.597 mm/day (SVM model) to 1.097 mm/day (CNN model). Finally, in Málaga, the results using 30 lag days were slightly better for all models. SVM and RF outperformed the rest of the models in terms of NSE (0.894 and 0.892, respectively) and RMSE (0.631 mm/day and 0.640 mm/day, respectively), whereas the worst results were obtained using CNN (NSE = 0.409 and RMSE = 1.499 mm/day) and LSTM (NSE = 0.202 and RMSE = 1.739 mm/day).

In Figures 10–12, the RMSE and NSE values for all forecasting predictions in the different sites are shown in a boxplot, respectively. Firstly, no significant performance distinctions were observed from the two approaches depending on the number of lag days (15 and 30 days). However, the first approach (15 lag days) slightly outperformed the second (30 lag days) on mean values, and more precision was observed (a lower interquartile range). Moreover, the number of outliers having non-accurate modeling was much higher using the second approach. Then, as a general rule, using daily values from 15 days in the past is recommended over using 30 days. Furthermore, regarding the efficiency of different models, SVM, RF, and ELM were predominantly better than the rest of the models according to NSE and RMSE values, giving more precise results. In contrast, CNN and both transformer models were at the bottom in the ranking. Finally, the MBE results are plotted in a boxplot. The results were very accurate in both approaches and for all the models and sites, but CNN gave more outliers, especially using the 30 lag days approach.

To further analyze these results, Figures 13–15 show the best statistic values (NSE, RMSE, and MBE, respectively) of all the models and sites for the different forecast horizons used. In terms of NSE (Figure 13), all ML models highly outperformed B1 and B2 in all the forecast horizons and locations, except for Conil. In Conil, only SVM, RF, and ELM outperformed both B1 and B2 in all cases. On the other hand, the transformers, CNN, and MLP models underperformed B1 and B2 for a horizon higher than 3 days. Regarding RMSE, the results were similar to those shown in Figure 12. However, a more significant improvement in ML models is appreciated for most models and horizons. In terms of MBE (Figures 13–15), B2 obtained significantly worse results in Aroche, Córdoba, Málaga, and Tabernas, where ML performed very accurately in all cases. In Conil, there were no major differences in performance between all the models. Thereby, due to these results, it could be stated that the use of ML models to forecast ET_0 up to a week is highly recommended, especially SVM, RF, and ELM models. Generally, B1 highly outperformed B2 to forecast ET_0 values one day ahead, but its performance profoundly decreased for higher horizons, obtaining even worse results than B2. This denotes a low autocorrelation of daily ET_0 values but a higher relation with historical monthly values. Moreover, SVM generally showed the best performance in terms of NSE and RMSE, whereas, regarding MBE, all models performed very accurately. Finally, it is worth noting that in Conil (a coastal site with an aridity index close to being a dry sub-humid climate), the best ML models (SVM, RF, and ELM) could not highly outperform B2 as in the rest of the locations when forecasting more than two days ahead, due to the effect of the close distance to the sea and the higher aridity index.

Table 7. Minimum (Min.), mean, and maximum (Max.) of NSE, RMSE, and MBE values for all locations (TAB—Tabernas, CON—Conil, COR—Córdoba, ARO—Aroche, MAG—Málaga) and models using two different lag day windows (15 days and 30 days). T_CNN refers to transformer using CNN in the feed-forward layer, while T_LSTM refers to transformers using LSTM in this same layer.

Station	Model	Lag Days	NSE			RMSE			MBE		
			Min	Mean	Max	Min	Mean	Max	Min	Mean	Max
TAB	CNN	15	0.710	0.778	0.862	0.723	0.916	1.050	0.001	0.123	0.484
		30	0.423	0.752	0.848	0.734	0.939	1.438	0.000	−0.026	−0.974
	ELM	15	0.794	0.820	0.860	0.727	0.825	0.885	0.043	0.082	0.126
		30	0.778	0.807	0.853	0.722	0.830	0.892	−0.000	0.021	0.079
	LSTM	15	0.749	0.797	0.845	0.766	0.877	0.976	−0.003	0.088	0.236
		30	0.730	0.771	0.828	0.783	0.905	0.984	0.000	−0.009	−0.209
	MLP	15	0.769	0.810	0.854	0.743	0.848	0.936	0.000	0.046	0.265
		30	0.715	0.781	0.841	0.750	0.883	1.012	−0.000	−0.029	−0.210
	RF	15	0.802	0.821	0.867	0.710	0.823	0.866	0.057	0.094	0.117
		30	0.799	0.819	0.859	0.706	0.805	0.850	0.000	−0.011	−0.033
	SVM	15	0.779	0.817	0.869	0.704	0.831	0.915	0.000	0.074	0.183
		30	0.746	0.812	0.862	0.700	0.818	0.955	0.000	−0.018	0.121
	T_CNN	15	0.742	0.789	0.840	0.779	0.893	0.989	0.000	0.100	0.324
		30	0.705	0.770	0.841	0.750	0.905	1.029	−0.000	−0.017	−0.297
	T_LSTM	15	0.726	0.780	0.829	0.804	0.912	1.019	0.002	0.099	0.257
		30	0.699	0.765	0.831	0.775	0.916	1.040	0.000	−0.050	−0.312
CON	CNN	15	0.580	0.674	0.817	0.759	1.017	1.154	0.000	−0.037	−0.560
		30	0.303	0.520	0.724	0.889	1.164	1.409	0.002	−0.151	−0.706
	ELM	15	0.716	0.753	0.837	0.717	0.885	0.959	0.000	0.000	0.048
		30	0.635	0.697	0.779	0.796	0.927	1.021	−0.002	−0.057	−0.122
	LSTM	15	0.651	0.724	0.788	0.816	0.936	1.055	0.000	−0.029	−0.131
		30	0.378	0.552	0.706	0.919	1.126	1.326	0.000	−0.061	0.304
	MLP	15	0.579	0.709	0.808	0.778	0.959	1.160	0.000	−0.059	−0.260
		30	0.368	0.573	0.738	0.866	1.099	1.338	0.003	−0.153	−0.371
	RF	15	0.721	0.754	0.843	0.703	0.883	0.939	0.003	0.026	0.057
		30	0.667	0.704	0.799	0.759	0.915	0.967	−0.020	−0.054	−0.099
	SVM	15	0.640	0.752	0.851	0.684	0.885	1.065	0.000	−0.146	−0.250
		30	0.547	0.672	0.804	0.749	0.961	1.146	0.015	−0.235	−0.393
	T_CNN	15	0.561	0.679	0.800	0.794	1.008	1.184	0.000	−0.047	−0.225
		30	0.422	0.569	0.723	0.891	1.104	1.294	−0.001	−0.096	−0.451
	T_LSTM	15	0.570	0.674	0.746	0.895	1.018	1.177	0.000	−0.035	−0.166
		30	0.389	0.588	0.707	0.917	1.080	1.310	0.000	−0.082	−0.259

Agronomy **2022**, 12, 656

Table 7. *Cont.*

Station	Model	Lag Days	Min	NSE Mean	Max	Min	RMSE Mean	Max	Min	MBE Mean	Max
COR	CNN	15	0.818	0.882	0.929	0.630	0.808	1.011	0.000	0.056	−0.505
		30	0.522	0.853	0.913	0.670	0.873	1.592	0.000	0.035	1.003
	ELM	15	0.879	0.900	0.932	0.614	0.745	0.824	0.000	0.015	0.084
		30	0.848	0.874	0.909	0.686	0.813	0.896	−0.001	0.046	0.128
	LSTM	15	0.877	0.894	0.924	0.649	0.771	0.831	0.000	0.041	0.178
		30	0.835	0.865	0.902	0.713	0.841	0.932	0.000	0.027	0.193
	MLP	15	0.858	0.893	0.927	0.639	0.773	0.891	−0.000	0.038	0.211
		30	0.801	0.858	0.908	0.690	0.860	1.029	−0.001	0.011	0.172
	RF	15	0.892	0.903	0.928	0.633	0.734	0.776	0.011	0.029	0.045
		30	0.870	0.883	0.912	0.674	0.783	0.826	0.000	0.015	0.033
	SVM	15	0.869	0.900	0.934	0.605	0.744	0.855	−0.000	0.053	0.130
		30	0.832	0.875	0.914	0.667	0.809	0.942	0.000	0.064	0.167
	T_CNN	15	0.857	0.885	0.906	0.725	0.802	0.896	0.003	0.052	0.207
		30	0.815	0.855	0.892	0.749	0.870	0.988	0.000	0.023	−0.280
	T_LSTM	15	0.842	0.880	0.906	0.724	0.818	0.939	−0.000	0.048	0.204
		30	0.824	0.859	0.885	0.773	0.859	0.965	0.000	0.037	0.230
ARO	CNN	15	0.799	0.851	0.913	0.624	0.816	0.951	0.000	0.106	0.436
		30	0.737	0.840	0.916	0.620	0.851	1.097	0.001	0.056	0.256
	ELM	15	0.850	0.874	0.917	0.609	0.751	0.823	−0.001	0.056	0.113
		30	0.853	0.878	0.918	0.613	0.744	0.819	0.020	0.082	0.141
	LSTM	15	0.823	0.860	0.912	0.627	0.792	0.892	0.000	0.068	0.196
		30	0.798	0.850	0.908	0.647	0.827	0.960	−0.002	0.038	0.220
	MLP	15	0.803	0.861	0.911	0.632	0.789	0.943	−0.001	0.079	0.288
		30	0.793	0.853	0.913	0.630	0.815	0.972	0.000	0.020	0.164
	RF	15	0.860	0.877	0.914	0.620	0.742	0.794	0.022	0.098	0.139
		30	0.855	0.883	0.920	0.606	0.730	0.814	0.009	0.047	0.070
	SVM	15	0.817	0.869	0.918	0.607	0.764	0.908	−0.003	0.136	0.200
		30	0.810	0.868	0.922	0.597	0.772	0.931	0.006	0.091	0.201
	T_CNN	15	0.802	0.845	0.902	0.664	0.834	0.945	0.002	0.099	0.281
		30	0.794	0.845	0.901	0.674	0.840	0.970	0.000	0.018	0.210
	T_LSTM	15	0.800	0.843	0.885	0.719	0.840	0.950	0.000	0.089	0.278
		30	0.780	0.838	0.882	0.736	0.859	1.001	0.000	0.042	0.238
MAG	CNN	15	0.734	0.800	0.871	0.681	0.847	0.980	0.000	0.046	0.311
		30	0.409	0.819	0.880	0.672	0.823	1.499	0.000	−0.003	1.113
	ELM	15	0.821	0.841	0.878	0.662	0.756	0.804	0.000	0.031	0.071
		30	0.841	0.857	0.884	0.663	0.736	0.777	−0.001	−0.040	−0.084
	LSTM	15	0.810	0.830	0.862	0.705	0.782	0.828	0.000	0.036	0.132
		30	0.202	0.840	0.872	0.695	0.773	1.739	0.000	−0.069	−1.052
	MLP	15	0.773	0.823	0.872	0.678	0.798	0.904	0.000	0.036	0.195
		30	0.763	0.835	0.880	0.672	0.788	0.948	0.000	−0.048	−0.261
	RF	15	0.832	0.849	0.882	0.651	0.738	0.778	0.000	0.027	0.044
		30	0.859	0.869	0.892	0.640	0.704	0.732	−0.020	−0.039	−0.061
	SVM	15	0.797	0.843	0.885	0.643	0.750	0.855	0.000	0.049	−0.138
		30	0.814	0.858	0.894	0.631	0.731	0.839	0.000	−0.006	−0.094
	T_CNN	15	0.741	0.809	0.853	0.727	0.829	0.967	0.001	0.009	0.198
		30	0.773	0.825	0.864	0.716	0.812	0.928	0.002	−0.097	−0.371
	T_LSTM	15	0.768	0.801	0.835	0.771	0.846	0.916	0.000	0.001	−0.130
		30	0.787	0.827	0.852	0.749	0.808	0.897	0.000	−0.063	−0.247

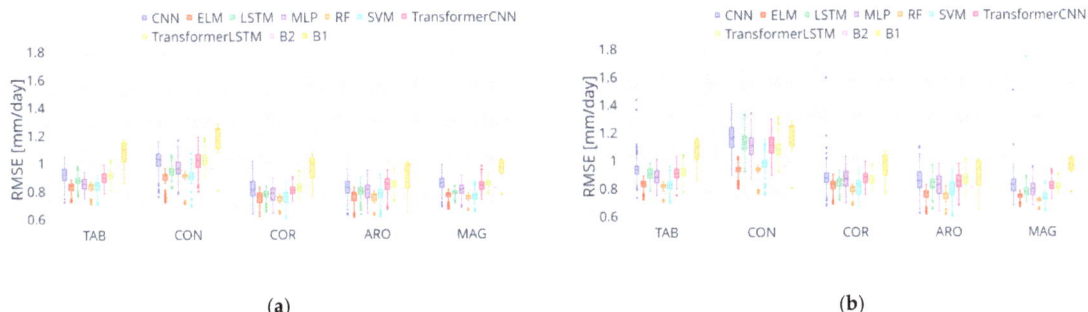

Figure 10. Boxplot with RMSE values from all models and configurations in the different AWS, using 15 lag days (**a**) and 30 lag days (**b**).

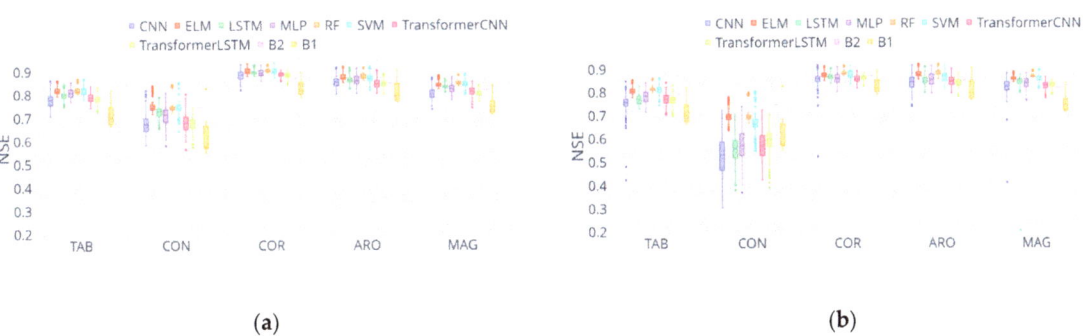

Figure 11. Boxplot with NSE values from all models and configurations in the different AWS, using 15 lag days (**a**) and 30 lag days (**b**).

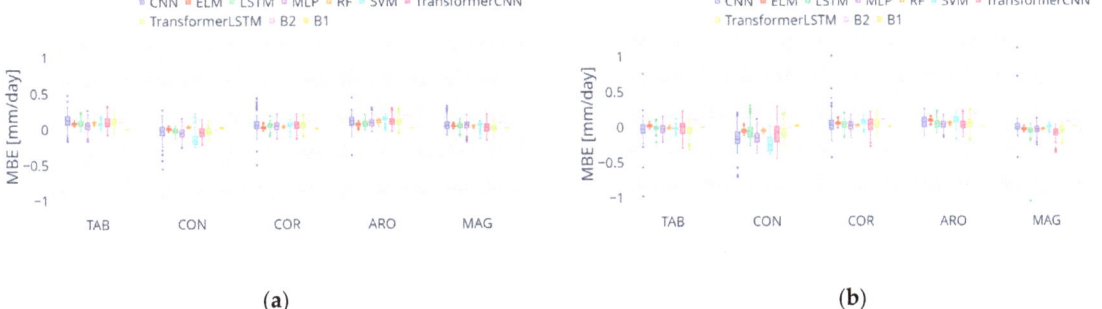

Figure 12. Boxplot with MBE values from all models and configurations in the different AWS, using 15 lag days (**a**) and 30 lag days (**b**).

3.3. Assessing the Different Configurations

In order to evaluate the performance of the different configurations at all locations, Table 8 shows the average and best RMSE values of each configuration in the different sites. In Tabernas, configurations III, XXII, IV, and IX obtained the most accurate results on mean, whereas configurations XVI, XII, and XXIV were the worst. In Conil, the best configurations in terms of mean RMSE were XXV, VI, and XX. Furthermore, configuration XXVI obtained the best value in absolute terms. On the other hand, configurations XIII, XI, and XII performed the worst on average. In Córdoba, regarding mean values, configurations XVII, XXIV, and V were at the bottom, whereas configurations III, XXVII, and II were at the top of the ranking. In Aroche, configuration V obtained the lowest RMSE value (RMSE = 0.598 mm/day). Moreover, considering the mean values, all configurations obtained very similar performance, beginning with RMSE = 0.764 mm/day (configuration I), followed closely by configurations IV (RMSE = 0.764 mm/day), III (RMSE = 0.767 mm/day), IX (RMSE = 0.767 mm/day), and XXII (RMSE = 0.768 mm/day), and finally RMSE = 0.788 mm/day (configurations XIII and XVII). Thus, it could be stated that in terms of the mean, although there were no significant differences in performance between the best and worst configurations, the use of configurations I, III, IV, and IX is recommended.

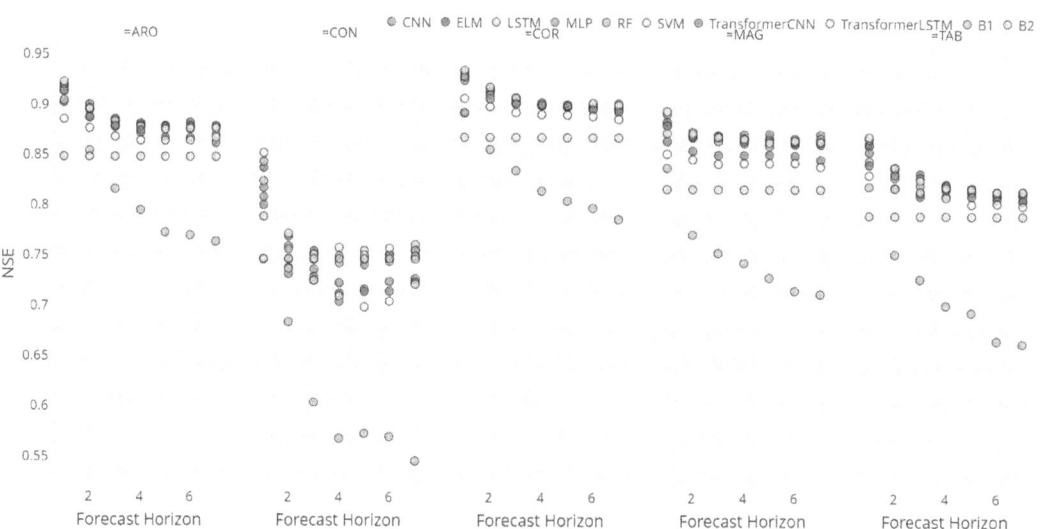

Figure 13. Scatter plot with the best NSE value for each model and location.

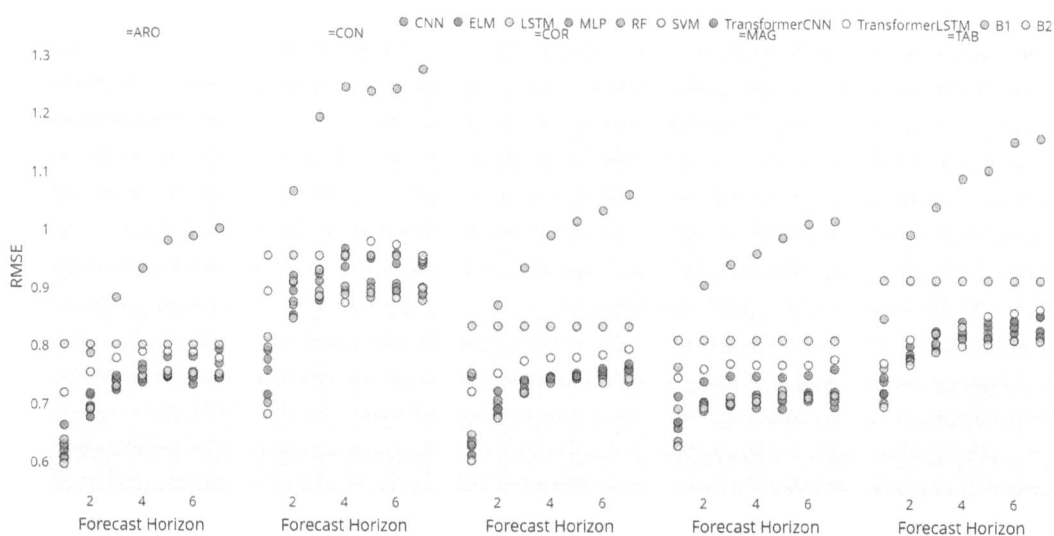

Figure 14. Scatter plot with the best RMSE value for each model and location.

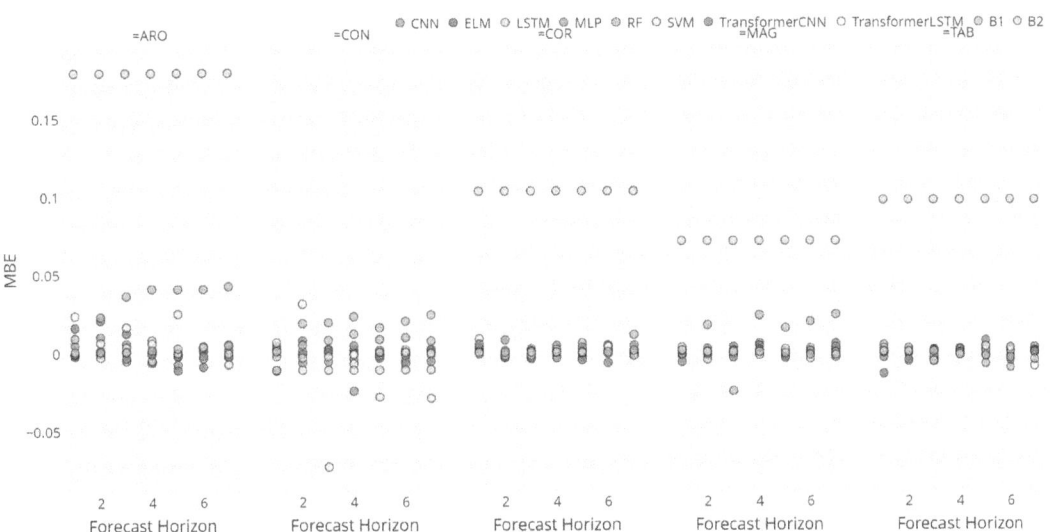

Figure 15. Scatterplot with the best MBE value for each model and location.

Table 8. Mean and minimum RMSE values (mm/day) for the different configurations at each location. The format is: mean (minimum). The best values are in bold.

Conf.	TAB	CON	COR	ARO	MAG	Mean
I	0.806 (0.704)	0.886 (0.695)	0.720 (0.614)	**0.686** (0.605)	0.724 (0.648)	**0.764**
II	0.801 (0.709)	0.909 (0.697)	0.718 (0.618)	0.703 (0.615)	0.732 (**0.631**)	0.772
III	**0.786** (0.701)	0.920 (0.694)	**0.710** (0.633)	0.693 (0.603)	0.730 (0.643)	0.767
IV	0.794 (0.703)	0.897 (0.694)	0.724 (0.630)	0.693 (0.604)	0.734 (0.646)	0.768
V	0.812 (0.706)	0.914 (0.700)	0.741 (0.621)	0.704 (0.598)	0.732 (0.632)	0.780
VI	0.812 (0.709)	0.870 (**0.687**)	0.720 (0.622)	0.725 (0.602)	0.743 (0.645)	0.774
VII	0.805 (0.703)	0.902 (0.689)	0.728 (0.621)	0.710 (0.601)	0.733 (0.648)	0.775
VIII	0.805 (0.709)	0.925 (0.693)	0.737 (0.617)	0.717 (0.606)	0.725 (0.642)	0.781
IX	0.799 (0.708)	0.883 (0.694)	0.735 (0.642)	0.693 (0.613)	0.726 (0.639)	0.767
X	0.803 (0.704)	0.897 (0.699)	0.734 (0.620)	0.687 (0.613)	0.730 (0.641)	0.770
XI	0.811 (0.709)	0.931 (0.698)	0.740 (0.617)	**0.686** (**0.597**)	**0.702** (0.640)	0.774
XII	0.823 (0.712)	0.926 (0.697)	0.732 (0.640)	0.706 (0.605)	0.722 (0.641)	0.781
XIII	0.814 (0.708)	0.933 (0.691)	0.734 (**0.605**)	0.726 (0.615)	0.737 (0.642)	0.788
XIV	0.809 (0.714)	0.892 (0.688)	0.737 (0.643)	0.721 (0.615)	0.741 (0.643)	0.780
XV	0.811 (0.708)	0.899 (0.715)	0.730 (0.614)	0.698 (0.612)	0.721 (0.645)	0.771
XVI	0.824 (0.709)	0.904 (0.693)	0.722 (0.619)	0.706 (0.599)	0.736 (0.633)	0.778

Table 8. *Cont.*

Conf.	TAB	CON	COR	ARO	MAG	Mean
XVII	0.810 (0.708)	0.921 (0.691)	0.753 (0.615)	0.726 (0.599)	0.734 (0.633)	0.788
XVIII	0.805 (0.707)	0.904 (0.718)	0.729 (0.622)	0.719 (0.606)	0.735 (0.647)	0.778
XIX	0.803 (0.707)	0.905 (0.688)	0.736 (0.616)	0.711 (0.605)	0.722 (0.633)	0.775
XX	0.816 (0.713)	0.879 (0.695)	0.733 (0.610)	0.719 (0.604)	0.747 (0.642)	0.778
XXI	0.801 (**0.700**)	0.920 (0.721)	0.725 (0.623)	0.696 (0.608)	0.738 (0.643)	0.776
XXII	0.792 (0.709)	0.893 (0.698)	0.728 (0.615)	0.709 (0.609)	0.722 (0.637)	0.768
XXIII	0.803 (0.713)	0.904 (0.696)	0.719 (0.627)	0.705 (0.604)	0.786 (0.643)	0.783
XXIV	0.823 (0.709)	0.917 (0.695)	0.741 (0.640)	0.696 (0.608)	0.731 (0.635)	0.781
XXV	0.821 (0.711)	**0.863** (0.691)	0.720 (0.618)	0.714 (0.613)	0.733 (0.655)	0.770
XXVI	0.822 (0.713)	0.894 (0.684)	0.736 (0.615)	0.711 (0.605)	0.730 (0.647)	0.778
XXVII	0.803 (0.710)	0.917 (0.699)	0.714 (0.627)	0.718 (0.612)	0.734 (0.636)	0.777

3.4. Overall Discussion

In this work, several aspects were evaluated in forecasting daily ET_0 at five locations in the Andalusia region (Southern Spain) with different geo-climatic conditions. Firstly, a new state-of-the-art architecture for NLP problems was assessed to forecast daily ET_0, the transformers. Specifically, two different approaches were evaluated, TransformerCNN and TransformerLSTM, and they were compared to standard machine learning models such as MLP, SVM, RF, or CNN, among others. In general, the results obtained using standard machine learning approaches such as RF, SVM, and ELM highly outperformed the rest of the models assessed in this work. Moreover, transformer-based models did not perform as expected in all cases when compared to standard ML models. However, their results were better than the baselines for most sites and cases (except for Conil). Secondly, another critical aspect to highlight in this work is that even using a self-attention mechanism (transformer-based models), the use of 30 lag days instead of 15 lag days was not beneficial to forecasting daily ET_0. On the contrary, slightly better results were obtained when 15 lag days were used, along with fewer serious outliers. Moreover, when comparing the different feature input configurations proposed in this study, none of them predominantly outperformed the rest, although configurations XIII, XIV, XX, and XXI were better on average. Figures 16–18 show a scatter plot of measured vs. predicted ET_0 values using the best ML model and configuration for 1 and 7 days ahead.

Furthermore, the results of the proposed models were significantly better than those reported by Ferreira and da Cunha [18] in terms of RMSE and NSE using different deep learning approaches in Brazil in AWS with an aridity index ranging from 0.3 to 1.6. The best NSE performances in Brazil ranged from 0.35 to 0.62 (approximately), whereas in this work, the best NSE values ranged from 0.60 to 0.95 (approximately). Moreover, this work also obtained slightly better NSE values than those reported by Nourani et al. [17] using ensemble modeling in different weather stations from Iran, Turkey, and Cyprus. These previous works used temperature, relative humidity, solar radiation, and wind speed values as input features, whereas all the configurations of this work were temperature-based variables. Additionally, comparing the results to those obtained by de Oliveira and Lucas et al. [54], the assessed models in the present work outperformed their CNN and ensemble CNN results in Brazil.

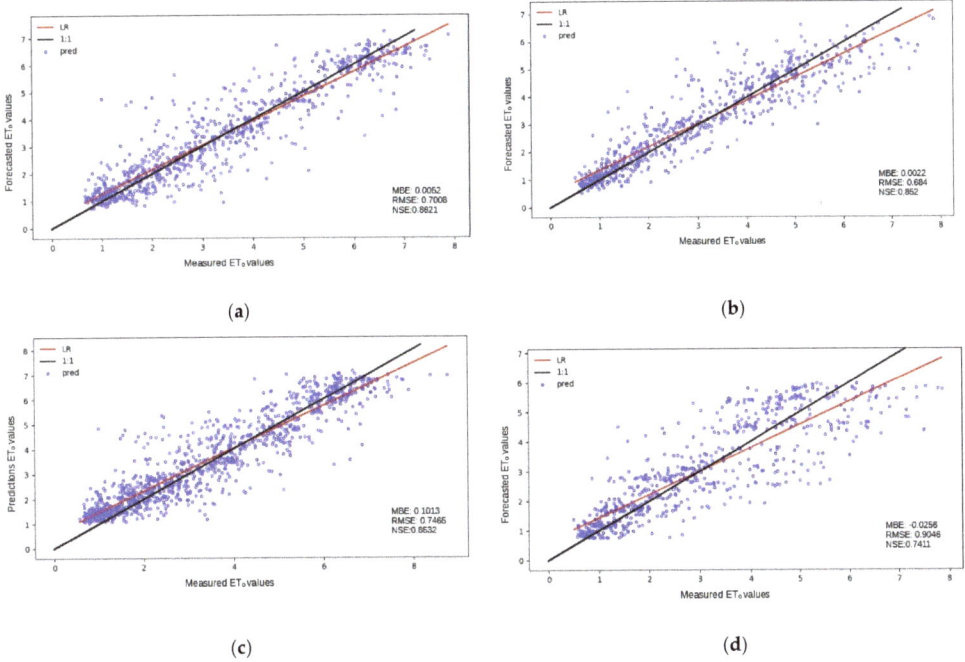

Figure 16. Scatter plot for measured vs. predicted values for (**a**) forecast horizon 1 in Tabernas, (**b**) forecast horizon 1 in Conil de la Frontera, (**c**) forecast horizon 7 in Tabernas, and (**d**) forecast horizon 7 in Conil de la Frontera.

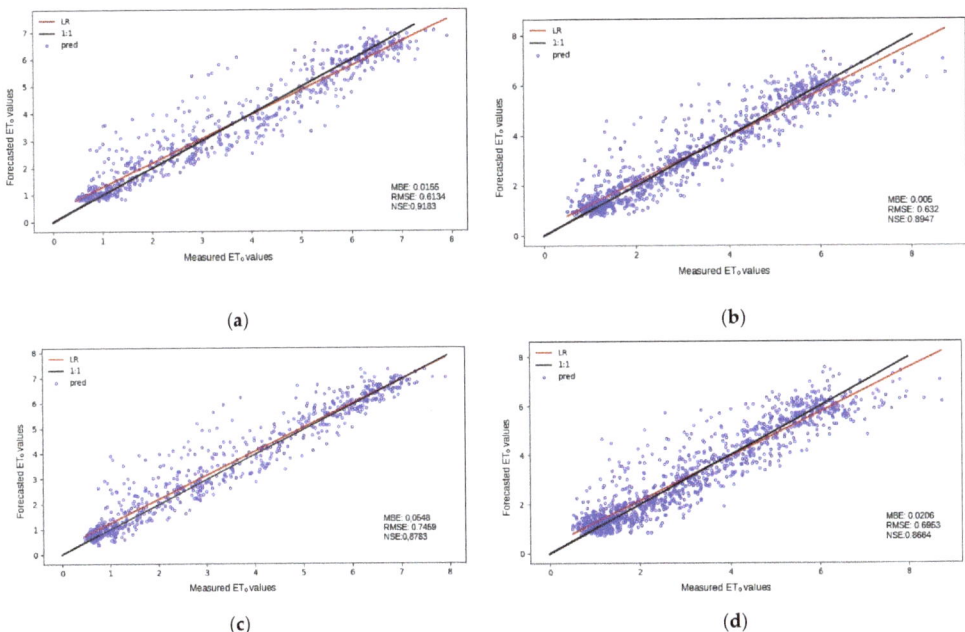

Figure 17. Scatter plot for measured vs. predicted values for (**a**) forecast horizon 1 in Aroche, (**b**) forecast horizon 1 in Málaga, (**c**) forecast horizon 7 in Aroche, and (**d**) forecast horizon 7 in Málaga.

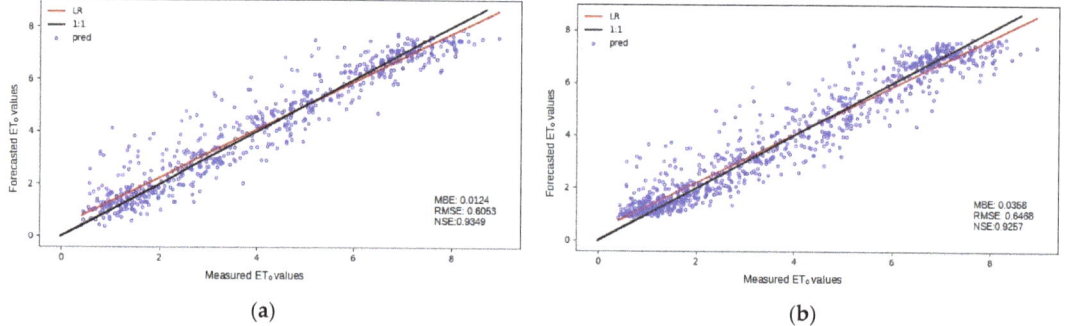

Figure 18. Scatter plot for measured vs. predicted values for (**a**) forecast horizon 1 in Córdoba, (**b**) forecast horizon 7 in Córdoba.

In all, the models developed in this work, especially SVM, ELM, and RF, are able to accurately forecast ET_0 for one week ahead using only temperature-based parameters and ET_0 past values. This issue is vital for improving crop irrigation scheduling, allowing adequate and anticipated planning, and contributing to agricultural production. Furthermore, providing reliable ET_0 future values positively impacts the current challenge of optimizing water resource management, especially in arid and semiarid locations.

4. Conclusions

In this work, several machine learning models have been developed and assessed for daily ET_0 forecasting from 1 to 7 days ahead using different input configurations, as well as different lag days. In general, all the ML approaches outperformed the baselines for all the forecast horizons and most locations, but SVM, RF, and ELM highly outperformed the rest of the models evaluated for most sites except for Conil de la Frontera, with unusually low wind speed values in this region. On the other hand, the transformers were, on average, at the bottom of the ranking. Moreover, all configurations obtained very similar results in terms of RMSE, but configurations I, III, IV, and IX slightly outperformed the rest. The NSE values were above 0.85 for Conil, Tabernas, and Málaga and above 0.9 for Córdoba and Aroche for their best modeling. In terms of RMSE, the average performance for Tabernas was 0.92 mm/day, 1.00 mm/day for Conil, 0.81 mm/day for Córdoba, 0.80 mm/day for Aroche, and 0.78 mm/day for Málaga. This denotes a relationship in performance regarding the aridity index and the distance to the sea. For inland locations, the higher the aridity index, the lower the error of forecasting ET_0 will be. On the other hand, for coastal sites, the higher the aridity index, the higher the error. Regarding MBE, most stations and models obtained very accurate values on average for most cases, with a mean performance value of 0.011 mm/day.

Further studies can deeply explore using these models in new regions with different geo-climatic conditions, different scenarios (a different time interval and a regional scenario), and for other parameters, such as solar radiation or precipitation. Moreover, accurate feature selection or reduction could be researched because, as could be stated based on the present results, the configurations containing the worst related features based on Pearson correlation (HTx, HTn, HSr-HTn) obtained very accurate minimum and mean RMSE (Table 8 and Figure 2). The approaches proposed in this work may result in greater efficiency for optimizing water resources, improving irrigation scheduling, and anticipating the decision-making for agricultural goals. Finally, the creation of an open-source repository will allow novel scientists to apply these models using their own datasets, as well as experienced scientists to commit improvements with new features and architectures. Overall, the ultimate aim is to democratize the use of machine learning to more efficiently solve today's agricultural problems.

Supplementary Materials: The following supporting information can be downloaded at: https://www. mdpi.com/article/10.3390/agronomy12030656/s1, Table S1. Hyperparameter space for all the models assessed in this work. MLP—Multilayer Perceptron, RF—Random Forest, SVR—Support Vector Regression, ELM—Extreme Learning Machine, CNN—Convolutional Neural Network, LSTM—Long Short-Term Memory, Transformer CNN—Transformer using CNN in the feed-forward layer, Transformer LSTM—Transformer using LSTM in the feed-forward layer. Table S2. Fittest hyperparameters for the best model and configuration at every location.

Author Contributions: Conceptualization, A.P.G.-M., J.A.B.-J., J.E. and J.V.; methodology, J.A.B.-J. and J.E.; software, J.A.B.-J. and J.E.; validation, A.P.G.-M., J.A.B.-J., J.E. and J.V.; formal analysis, J.A.B.-J., J.E. and J.V.; investigation, J.A.B.-J., J.E. and J.V.; resources, A.P.G.-M., J.A.B.-J. and J.E.; data curation, J.A.B.-J. and J.E.; writing—original draft preparation, J.A.B.-J. and J.E.; writing—review and editing, A.P.G.-M., J.A.B.-J., J.E. and J.V.; visualization, J.A.B.-J. and J.E.; supervision, A.P.G.-M., J.A.B.-J., J.E. and J.V.; project administration, A.P.G.-M. and J.E.; funding acquisition, A.P.G.-M., J.E. and J.A.B.-J. All authors have read and agreed to the published version of the manuscript.

Funding: This work was supported by the Spanish Ministry of Science, Innovation and Universities (grant number AGL2017-87658-R).

Acknowledgments: J.A. Bellido-Jiménez wishes to thank the University of Córdoba for providing a PIF scholarship funded by the research program and also funding part of this stay at Eindhoven, in collaboration with Banco Santander. To the Spanish Ministry of Science, Innovation, and Universities, grant number AGL2017-87658-R, for also funding this research. We also thank the Technological University of Eindhoven for its invitation to conduct research at its facilities.

Conflicts of Interest: The authors declare that they have no known competing financial interests or personal relationships that could appear to influence the work reported in this paper.

References

1. FAO. *The State of Food Security and Nutrition in the World 2021*; FAO: Rome, Italy, 2021. [CrossRef]
2. Allen, R.; Pereira, L.; Smith, M. *Crop Evapotranspiration-Guidelines for Computing Crop Water Requirements-FAO Irrigation and Drainage*; FAO: Rome, Italy, 1998; Volume 56.
3. Kwon, H.; Choi, M. Error Assessment of Climate Variables for FAO-56 Reference Evapotranspiration. *Meteorol. Atmos. Phys.* **2011**, *112*, 81–90. [CrossRef]
4. Estévez, J.; García-Marín, A.P.; Morábito, J.A.; Cavagnaro, M. Quality assurance procedures for validating meteorological input variables of reference evapotranspiration in mendoza province (Argentina). *Agric. Water Manag.* **2016**, *172*, 96–109. [CrossRef]
5. Jabloun, M.; Sahli, A. Evaluation of FAO-56 Methodology for Estimating Reference Evapotranspiration Using Limited Climatic Data. Application to Tunisia. *Agric. Water Manag.* **2008**, *95*, 707–715. [CrossRef]
6. Estévez, J.; Gavilán, P.; Giráldez, J.V. Guidelines on Validation Procedures for Meteorological Data from Automatic Weather Stations. *J. Hydrol.* **2011**, *402*, 144–154. [CrossRef]
7. Estévez, J.; Padilla, F.L.; Gavilán, P. Evaluation and Regional Calibration of Solar Radiation Prediction Models in Southern Spain. *J. Irrig. Drain. Eng.* **2012**, *138*, 868–879. [CrossRef]
8. WMO. *Guide to Instruments and Methods of Observations*; WMO: Geneva, Switzerland, 2018; Volume 8, ISBN 978-92-63-10008-5.
9. George, H.H.; Zohrab, A. Samani Reference Crop Evapotranspiration from Temperature. *Appl. Eng. Agric.* **1985**, *1*, 96–99. [CrossRef]
10. Raziei, T.; Pereira, L.S. Estimation of ETo with Hargreaves-Samani and FAO-PM Temperature Methods for a Wide Range of Climates in Iran. *Agric. Water Manag.* **2013**, *121*, 1–18. [CrossRef]
11. Ravazzani, G.; Corbari, C.; Morella, S.; Gianoli, P.; Mancini, M. Modified Hargreaves-Samani Equation for the Assessment of Reference Evapotranspiration in Alpine River Basins. *J. Irrig. Drain. Eng.* **2012**, *138*, 592–599. [CrossRef]
12. Luo, Y.; Chang, X.; Peng, S.; Khan, S.; Wang, W.; Zheng, Q.; Cai, X. Short-Term Forecasting of Daily Reference Evapotranspiration Using the Hargreaves-Samani Model and Temperature Forecasts. *Agric. Water Manag.* **2014**, *136*, 42–51. [CrossRef]
13. Karimi, S.; Shiri, J.; Marti, P. Supplanting Missing Climatic Inputs in Classical and Random Forest Models for Estimating Reference Evapotranspiration in Humid Coastal Areas of Iran. *Comput. Electron. Agric.* **2020**, *176*, 105633. [CrossRef]
14. Ferreira, L.B.; da Cunha, F.F. New Approach to Estimate Daily Reference Evapotranspiration Based on Hourly Temperature and Relative Humidity Using Machine Learning and Deep Learning. *Agric. Water Manag.* **2020**, *234*, 106113. [CrossRef]
15. Yan, S.; Wu, L.; Fan, J.; Zhang, F.; Zou, Y.; Wu, Y. A Novel Hybrid WOA-XGB Model for Estimating Daily Reference Evapotranspiration Using Local and External Meteorological Data: Applications in Arid and Humid Regions of China. *Agric. Water Manag.* **2021**, *244*, 106594. [CrossRef]

16. Wu, L.; Peng, Y.; Fan, J.; Wang, Y.; Huang, G. A Novel Kernel Extreme Learning Machine Model Coupled with K-Means Clustering and Firefly Algorithm for Estimating Monthly Reference Evapotranspiration in Parallel Computation. *Agric. Water Manag.* **2021**, *245*, 106624. [CrossRef]
17. Nourani, V.; Elkiran, G.; Abdullahi, J. Multi-Step Ahead Modeling of Reference Evapotranspiration Using a Multi-Model Approach. *J. Hydrol.* **2020**, *581*, 124434. [CrossRef]
18. Ferreira, L.B.; da Cunha, F.F. Multi-Step Ahead Forecasting of Daily Reference Evapotranspiration Using Deep Learning. *Comput. Electron. Agric.* **2020**, *234*, 106113. [CrossRef]
19. Vaswani, A.; Shazeer, N.; Parmar, N.; Uszkoreit, J.; Jones, L.; Gomez, A.N.; Kaiser, L.; Polosukhin, I. Attention Is All You Need. *Adv. Neural Inf. Process. Syst.* **2017**, *2017*, 5999–6009.
20. Wu, S.; Xiao, X.; Ding, Q.; Zhao, P.; Wei, Y.; Huang, J. Adversarial Sparse Transformer for Time Series Forecasting. *Adv. Neural Inf. Process. Syst.* **2020**, *33*, 17105–17115.
21. Wu, N.; Green, B.; Ben, X.; O'Banion, S. Deep Transformer Models for Time Series Forecasting: The Influenza Prevalence Case. *arXiv* **2020**, arXiv:2001.08317.
22. Li, S.; Jin, X.; Xuan, Y.; Zhou, X.; Chen, W.; Wang, Y.X.; Yan, X. Enhancing the Locality and Breaking the Memory Bottleneck of Transformer on Time Series Forecasting. In Proceedings of the Advances in Neural Information Processing Systems, Vancouver, BC, Canada, 8–14 December 2019; Volume 32.
23. Unep, N.M.; London, D.T. World Atlas of Desertification. *Land Degrad. Dev.* **1992**, *3*, 15–45.
24. Bellido-Jiménez, J.A.; Estévez, J.; García-Marín, A.P. New Machine Learning Approaches to Improve Reference Evapotranspiration Estimates Using Intra-Daily Temperature-Based Variables in a Semi-Arid Region of Spain. *Agric. Water Manag.* **2020**, *245*, 106558. [CrossRef]
25. Bellido-Jiménez, J.A.; Estévez, J.; García-Marín, A.P. Assessing Neural Network Approaches for Solar Radiation Estimates Using Limited Climatic Data in the Mediterranean Sea. In Proceedings of the 3rd International Electronic Conference on Atmospheric Sciences (ECAS 2020), online, 16–30 November 2020.
26. Bellido-Jiménez, J.A.; Estévez Gualda, J.; García-Marín, A.P. Assessing New Intra-Daily Temperature-Based Machine Learning Models to Outperform Solar Radiation Predictions in Different Conditions. *Appl. Energy* **2021**, *298*, 117211. [CrossRef]
27. Estévez, J.; Gavilán, P.; García-Marín, A.P. Spatial Regression Test for Ensuring Temperature Data Quality in Southern Spain. *Theor. Appl. Climatol.* **2018**, *131*, 309–318. [CrossRef]
28. Islam, A.R.M.T.; Shen, S.; Yang, S.; Hu, Z.; Chu, R. Assessing Recent Impacts of Climate Change on Design Water Requirement of Boro Rice Season in Bangladesh. *Theor. Appl. Climatol.* **2019**, *138*, 97–113. [CrossRef]
29. Yi, Z.; Zhao, H.; Jiang, Y. Continuous Daily Evapotranspiration Estimation at the Field-Scale over Heterogeneous Agricultural Areas by Fusing Aster and Modis Data. *Remote Sens.* **2018**, *10*, 1694. [CrossRef]
30. Sattari, M.T.; Apaydin, H.; Band, S.S.; Mosavi, A.; Prasad, R. Comparative Analysis of Kernel-Based versus ANN and Deep Learning Methods in Monthly Reference Evapotranspiration Estimation. *Hydrol. Earth Syst. Sci.* **2021**, *25*, 603–618. [CrossRef]
31. Tikhamarine, Y.; Malik, A.; Souag-Gamane, D.; Kisi, O. Artificial Intelligence Models versus Empirical Equations for Modeling Monthly Reference Evapotranspiration. *Environ. Sci. Pollut. Res.* **2020**, *27*, 30001–30019. [CrossRef]
32. Huang, G.B.; Zhu, Q.Y.; Siew, C.K. Extreme Learning Machine: Theory and Applications. *Neurocomputing* **2006**, *70*, 489–501. [CrossRef]
33. Zhu, B.; Feng, Y.; Gong, D.; Jiang, S.; Zhao, L.; Cui, N. Hybrid Particle Swarm Optimization with Extreme Learning Machine for Daily Reference Evapotranspiration Prediction from Limited Climatic Data. *Comput. Electron. Agric.* **2020**, *173*, 105430. [CrossRef]
34. Akusok, A.; Björk, K.-M.; Miche, Y.; Lendasse, A. High Performance Extreme Learning Machines: A Complete Toolbox for Big Data Applications. *IEEE Access* **2015**, *3*, 1011–1025. [CrossRef]
35. Smola, A.J.; Schölkopf, B. A Tutorial on Support Vector Regression. *Stat. Comput.* **2004**, *14*, 199–222. [CrossRef]
36. Chen, Z.; Zhu, Z.; Jiang, H.; Sun, S. Estimating Daily Reference Evapotranspiration Based on Limited Meteorological Data Using Deep Learning and Classical Machine Learning Methods. *J. Hydrol.* **2020**, *591*, 125286. [CrossRef]
37. de Oliveira, R.G.; Valle Júnior, L.C.G.; da Silva, J.B.; Espíndola, D.A.L.F.; Lopes, R.D.; Nogueira, J.S.; Curado, L.F.A.; Rodrigues, T.R. Temporal Trend Changes in Reference Evapotranspiration Contrasting Different Land Uses in Southern Amazon Basin. *Agric. Water Manag.* **2021**, *250*, 106815. [CrossRef]
38. Ghimire, S.; Deo, R.C.; Raj, N.; Mi, J. Deep Solar Radiation Forecasting with Convolutional Neural Network and Long Short-Term Memory Network Algorithms. *Appl. Energy* **2019**, *253*, 113541. [CrossRef]
39. Kim, S.; Hong, S.; Joh, M.; Song, S.K. DeepRain: ConvLSTM Network for Precipitation Prediction Using Multichannel Radar Data. *arXiv* **2017**, arXiv:1711.02316.
40. Aloysius, N.; Geetha, M. A Review on Deep Convolutional Neural Networks. In Proceedings of the 2017 IEEE International Conference on Communication and Signal Processing, ICCSP, Chenai, India, 6–8 April 2017; Institute of Electrical and Electronics Engineers Inc.: Piscataway, NJ, USA, 2018; Volume 2018, pp. 588–592.
41. Hochreiter, S.; Schmidhuber, J. Long Short-Term Memory. *Neural Comput.* **1997**, *9*, 1735–1780. [CrossRef]
42. Song, H.; Rajan, D.; Thiagarajan, J.J.; Spanias, A. Attend and Diagnose: Clinical Time Series Analysis Using Attention Models. In Proceedings of the 32th AAAI Conference on Artificial Intelligence, New Orleans, LA, USA, 2–7 February 2018.

43. Wolf, T.; Debut, L.; Sanh, V.; Chaumond, J.; Delangue, C.; Moi, A.; Cistac, P.; Rault, T.; Louf, R.; Funtowicz, M.; et al. *Transformers: State-of-the-Art Natural Language Processing*; Association for Computational Linguistics (ACL): Stroudsburg, PA, USA, 2020; pp. 38–45.

44. Mohammdi Farsani, R.; Pazouki, E. A Transformer Self-Attention Model for Time Series Forecasting. *J. Electr. Comput. Eng. Innov.* **2021**, *9*, 1–10. [CrossRef]

45. Alizamir, M.; Kisi, O.; Muhammad Adnan, R.; Kuriqi, A. Modelling Reference Evapotranspiration by Combining Neuro-Fuzzy and Evolutionary Strategies. *Acta Geophys.* **2020**, *68*, 1113–1126. [CrossRef]

46. Mohammadi, B.; Mehdizadeh, S. Modeling Daily Reference Evapotranspiration via a Novel Approach Based on Support Vector Regression Coupled with Whale Optimization Algorithm. *Agric. Water Manag.* **2020**, *237*, 106145. [CrossRef]

47. Gijsbers, P.; LeDell, E.; Thomas, J.; Poirier, S.; Bischl, B.; Vanschoren, J. An Open Source AutoML Benchmark. *arXiv* **2019**, arXiv:1907.00909.

48. Kotthoff, L.; Thornton, C.; Hoos, H.; Hutter, F.; Leyton-Brown, K. Auto-WEKA 2.0: Automatic Model Selection and Hyperparameter Optimization in WEKA. *J. Mach. Learn. Res.* **2017**, *18*, 826–830.

49. Jin, H.; Song, Q.; Hu, X. Auto-Keras: An Efficient Neural Architecture Search System. In Proceedings of the 25th ACM SIGKDD International Conference on Knowledge Discovery & Data Mining, Anchorage, AK, USA, 4–8 August 2019; pp. 1946–1956.

50. Feurer, M.; Klein, A.; Eggensperger, K.; Springenberg, J.T.; Blum, M.; Hutter, F. Auto-Sklearn:: Efficient and Robust Automated Machine Learning. In Proceedings of the Advances in Neural Information Processing Systems, Montreal, QC, Canada, 7–12 December 2015; Volume 2015, pp. 2962–2970.

51. Hutter, F.; Kotthoff, L.; Vanschoren, J. (Eds.) *Automated Machine Learning*; The Springer Series on Challenges in Machine Learning; Springer International Publishing: Cham, Switzerland, 2019; ISBN 978-3-030-05317-8.

52. Borji, A.; Itti, L. Bayesian Optimization Explains Human Active Search. In *Advances in Neural Information Processing Systems*; Curran Associates, Inc.: New York, NY, USA, 2013; Volume 26.

53. Shahriari, B.; Swersky, K.; Wang, Z.; Adams, R.P.; de Freitas, N. Taking the Human Out of the Loop: A Review of Bayesian Optimization. *Proc. IEEE* **2016**, *104*, 148–175. [CrossRef]

54. de Oliveira e Lucas, P.; Alves, M.A.; de Lima e Silva, P.C.; Guimarães, F.G. Reference Evapotranspiration Time Series Forecasting with Ensemble of Convolutional Neural Networks. *Comput. Electron. Agric.* **2020**, *177*, 105700. [CrossRef]

agronomy

Article

Daily Prediction and Multi-Step Forward Forecasting of Reference Evapotranspiration Using LSTM and Bi-LSTM Models

Dilip Kumar Roy [1,*], Tapash Kumar Sarkar [2], Sheikh Shamshul Alam Kamar [1], Torsha Goswami [3], Md Abdul Muktadir [4], Hussein M. Al-Ghobari [5], Abed Alataway [6], Ahmed Z. Dewidar [5,6], Ahmed A. El-Shafei [5,6,7] and Mohamed A. Mattar [5,6,8,*]

1   Irrigation and Water Management Division, Bangladesh Agricultural Research Institute, Joydebpur, Gazipur 1701, Bangladesh; alamkamar91@gmail.com
2   Grain Quality and Nutrition Division, Bangladesh Rice Research Institute, Joydebpur, Gazipur 1701, Bangladesh; tksarkar_engr@yahoo.com
3   Department of Veterinary Microbiology, Faculty of Veterinary and Animal Sciences, West Bengal University of Animal and Fishery Sciences, Kolkata 700056, West Bengal, India; toshami09@gmail.com
4   Centre for Carbon, Water and Food, The University of Sydney, Camperdown, NSW 2570, Australia; md.muktadir@sydney.edu.au
5   Department of Agricultural Engineering, College of Food and Agriculture Sciences, King Saud University, Riyadh 11451, Saudi Arabia; hghobari@ksu.edu.sa (H.M.A.-G.); adewidar@ksu.edu.sa (A.Z.D.); aelshafei1bn.c@ksu.edu.sa (A.A.-S.)
6   Prince Sultan Bin Abdulaziz International Prize for Water Chair, Prince Sultan Institute for Environmental, Water and Desert Research, King Saud University, Riyadh 11451, Saudi Arabia; aalataway@ksu.edu.sa
7   Department of Agricultural Engineering, Faculty of Agriculture (El-Shatby), Alexandria University, Alexandria 21545, Egypt
8   Agricultural Engineering Research Institute (AEnRI), Agricultural Research Centre, Giza 12618, Egypt
*   Correspondence: dilip.roy@my.jcu.edu.au (D.K.R.); mmattar@ksu.edu.sa (M.A.M.)

Citation: Roy, D.K.; Sarkar, T.K.; Kamar, S.S.A.; Goswami, T.; Muktadir, M.A.; Al-Ghobari, H.M.; Alataway, A.; Dewidar, A.Z.; El-Shafei, A.A.; Mattar, M.A. Daily Prediction and Multi-Step Forward Forecasting of Reference Evapotranspiration Using LSTM and Bi-LSTM Models. *Agronomy* **2022**, *12*, 594. https://doi.org/10.3390/agronomy12030594

Academic Editors: Pantazis Georgiou and Dimitris Karpouzos

Received: 29 December 2021
Accepted: 24 February 2022
Published: 27 February 2022

Publisher's Note: MDPI stays neutral with regard to jurisdictional claims in published maps and institutional affiliations.

Abstract: Precise forecasting of reference evapotranspiration (ET_0) is one of the critical initial steps in determining crop water requirements, which contributes to the reliable management and long-term planning of the world's scarce water sources. This study provides daily prediction and multi-step forward forecasting of ET_0 utilizing a long short-term memory network (LSTM) and a bi-directional LSTM (Bi-LSTM) model. For daily predictions, the LSTM model's accuracy was compared to that of other artificial intelligence-based models commonly used in ET0 forecasting, including support vector regression (SVR), M5 model tree (M5Tree), multivariate adaptive regression spline (MARS), probabilistic linear regression (PLR), adaptive neuro-fuzzy inference system (ANFIS), and Gaussian process regression (GPR). The LSTM model outperformed the other models in a comparison based on Shannon's entropy-based decision theory, while the M5 tree and PLR models proved to be the lowest performers. Prior to performing a multi-step-ahead forecasting, ANFIS, sequence-to-sequence regression LSTM network (SSR-LSTM), LSTM, and Bi-LSTM approaches were used for one-step-ahead forecasting utilizing the past values of the ET_0 time series. The results showed that the Bi-LSTM model outperformed other models and that the sequence of models in ascending order in terms of accuracies was Bi-LSTM > SSR-LSTM > ANFIS > LSTM. The Bi-LSTM model provided multi-step (5 day)-ahead ET_0 forecasting in the next step. According to the results, the Bi-LSTM provided reasonably accurate and acceptable forecasting of multi-step-forward ET_0 with relatively lower levels of forecasting errors. In the final step, the generalization capability of the proposed best models (LSTM for daily predictions and Bi-LSTM for multi-step-ahead forecasting) was evaluated on new unseen data obtained from a test station, Ishurdi. The model's performance was assessed on three distinct datasets (the entire dataset and the first and the second halves of the entire dataset) derived from the test dataset between 1 January 2015 and 31 December 2020. The results indicated that the deep learning techniques (LSTM and Bi-LSTM) achieved equally good performances as the training station dataset, for which the models were developed. The research outcomes demonstrated the

ability of the developed deep learning models to generalize the prediction capabilities outside the training station.

Keywords: deep learning; recurrent neural networks; machine learning algorithms; reference evapotranspiration

1. Introduction

Water conservation in irrigated agriculture has been a significant concern, as agriculture consumes the majority of the world's freshwater reserves. A considerable amount of water can be saved through accurate quantification of crop water requirements, which depends on the precise estimation of evapotranspiration (ET), one of the vital elements in computational frameworks of water balance equations. Being an essential element of the surface energy balances and water budgets, ET plays a central role in controlling interactions among soil, vegetation, and the atmosphere [1]. As such, proper design and efficient management of irrigation techniques and reliable planning for the allocation of scarce water resources largely depend on the accurate estimation of the ET [2]. The values of ET can be obtained through direct measurement techniques, including lysimeter methods, eddy covariance techniques, and the Bowen ratio–energy balance approach [3–5], which are expensive and deemed unavailable in many countries [6,7]. Alternatively, ET can be estimated indirectly utilizing a set of accessible climatological variables to determine reference evapotranspiration (ET_0). This indirect approach has been extensively used in many parts around the globe in which either unavailability or budgetary constraints prohibit direct estimation of ET. One of the most stable and well-established techniques of ET_0 computation is the FAO-56 Penman–Monteith (FAO-56 PM) equation [6]. It is also utilized to validate alternative ET_0 computation methods, as the equation was validated using lysimeter methods in different climates [8]. ET_0 computation using the FAO-56 PM equation requires a few climatological variables, including maximum and minimum air temperatures, wind speed, relative humidity, and solar radiation. Upon estimation of ET_0, crop evapotranspiration can be obtained by utilizing estimated ET_0 values and crop coefficient values for a particular crop.

Machine learning algorithms have recently been recognized as reliable tools in the prediction and future forecasting of ET_0. They have been used extensively in providing a reasonably accurate forecast of ET_0 in various hydrologic and climatic settings. The first implementation of ET_0 prediction modeling was based on the usage of artificial neural networks (ANN) [9–13]. Later, different variants of ANN and other machine learning algorithms have attained the researchers' interests. These include the usage of generalized regression neural networks [14,15], neural network with optimum time lags [16], adaptive neuro-fuzzy inference system (ANFIS) [17–23], random forests (RF) [14,24,25], CatBoost [26], hybrid extreme gradient boosting grey wolf optimizer (GWO) [27], extreme learning machine (ELM) [15,17,28–31], support vector regression (SVR) [23–25,31–33], multivariate relevance vector regression [34], genetic programming (GP) [35], Gaussian process regression (GPR) [36], multivariate adaptive regression splines (MARS) [2,9], M5 model tree (M5Tree) [2], radial basis M5Tree [37], gene-expression programming (GEP) [12,18,38–45], hierarchical fuzzy systems (HFS) [46], coupled extreme gradient boosting-whale optimization algorithm [47], coupled natural-extreme gradient boosting [48], hybrid model based on variational mode decomposition-GWO-SVM [49], and inter-model ensemble approaches [50]. Apart from machine learning approaches, there are other approaches of ET estimation, including the application of Sentinel-2 spectral information [51], comparison of different empirical methods [52], utilizing NASA POWER Reanalysis Products [53], and using lysimeter data [54]. Recently, Bellido-Jiménez et al. [55] examined several machine learning approaches to improve ET_0 estimations, considering only the temperature-based data (EnergyT and Hourmin) as inputs, and they determined that ELM outperformed the others. In another study, Vásquez et al. [56] proposed several methods based on maximum and minimum temperatures to enhance ET_0 computation under scarce data situations in the high tropical Andes. Nourani et al. [57] proposed one-, two-, and three-step-ahead predictions of ET_0 using ensembles of ANFIS, ANN, and MLR models in various climatic stations. This study evaluates deep learning algorithms' daily prediction and multi-step (5 steps)-ahead forecasting abilities.

The deep machine learning (DL) technique has attained substantial attention in recent years, being considered an advanced version of machine learning techniques. The DL technique has been successfully utilized in various research domains, including time series prediction [58–60], computer vision [61], classification of images [62], recognition of speech [63], language processing [64], forecasting of groundwater levels [65,66], and prediction of water quality parameters [67]. The DL techniques are primarily based on the recurrent neural networks (RNN), which, for their ability to preserve and utilize memory from the previous network states, are superior candidates for predicting and forecasting time series data [68–70]. Nevertheless, despite the ability to capture the trends of the time series data, the standard RNN model structures face difficulties in retaining the longer-term dependence among the variables and suffer from vanishing and exploding gradients-related issues [71]. Due to these two inherent problems of the standard RNN, network training becomes unrealistic as the network weights may either become zero or unnecessarily large during network training. The two most important criteria that ensure better network training are retaining necessary information and eluding redundant or unnecessary information among various network states. A long short-term memory (LSTM) network possesses these characteristics to overcome the training shortfall of RNNs. The LSTMs are the variants of standard RNNs and have widely been used in various research domains such as financial time series and language processing [72], traffic congestion, and traveling [73], including the application in the hydrologic time series prediction [74–77].

The application of DL-based models in predicting pan evaporation, reference evapotranspiration, and crop evapotranspiration in different climatic conditions have been found in recent literature. These include daily pan evaporation prediction using deep LSTM model [78], evapotranspiration computation estimation using deep neural network [79], daily reference evapotranspiration prediction using convolutional neural network (CNN) [80], one-step-ahead forecasting of reference evapotranspiration using LSTM [81], multi-step-ahead forecasting of daily reference evapotranspiration using LSTM and CNN-LSTM [82], multi-week-ahead forecasting of ET_0 using CNN-gated recurrent unit optimized with ant colony optimization [83], ET_0 estimation using deep learning-multilayer perceptrons [84], and short-term actual ET prediction using LSTM and NARX [85]. Despite the ET_0 prediction and forecasting application, the DL-based models, especially LSTM models, need to be evaluated for different combinations of input variables that provide better prediction accuracy. Recently, Zhang et al. [26] used only eight input combinations of different meteorological variables to estimate reference crop evapotranspiration using the CatBoost model. Another study by Maroufpoor et al. [86] used optimal input combinations to estimate reference evapotranspiration using a hybridized ANN model. Another study [87] used 29 different combinations of input variables from various meteorological variables to forecast daily reference evapotranspiration using ANN, SVR, and ELM. To the best of our knowledge, none of the previous studies evaluated all possible combinations of available input climatological variables to provide daily and multi-step forward ET_0 estimation using DL-based LSTM models. This is the first effort that has used various possible combinations of input variables using a deep learning model to predict daily and forecast multi-step-ahead reference evapotranspiration.

Another critical aspect of predictive modeling with the machine or deep learning approaches is evaluating the established models' ability to anticipate and forecast data from other meteorological stations. However, the generalization capabilities of the developed models for predicting and forecasting ET_0 in other meteorological stations have been given relatively little attention. For daily prediction of ET_0, Wang et al. [44] investigated the generalization capability of RF- and GEP-based machine learning tools, while Roy et al. [46] evaluated the potential of HFS models in generalizing the outputs using data from another meteorological station. For one-step-ahead forecasting of ET_0, Roy [81] utilized LSTM models; however, the study did not evaluate the generalization capability of the developed LSTM models for a new unseen test dataset. Nevertheless, model generalization has not been used for multi-step-ahead ET_0 forecasting using different combinations of input variables as well as using various machine and deep learning algorithms. To the best of our understanding, this study was the first attempt at providing daily prediction and multi-step-forward forecasting of ET_0 using LSTM and Bi-LSTM models.

Therefore, the prime objective and focus of this research were to (1) explore the capability of DL-based techniques, LSTM, and Bi-LSTM in predicting daily and forecasting multi-step (5 day)-ahead ET_0 estimates in the selected study areas in Bangladesh; (2) compare the prediction and forecasting skill of the proposed LSTM and Bi-LSTM models with that of the commonly used machine learning algorithms; and (3) assess the generalization capability of the proposed LSTM and Bi-LSTM models to predict and forecast ET_0 at a nearby station, at which the models were neither trained nor validated.

2. Material and Methods

2.1. Study Area and the Data

The study area consists of two upazillas (administrative units) in Gazipur and Pabna districts: Gazipur Sadar Upazilla and Ishurdi Upazilla (Figure 1). Meteorological data, including minimum and maximum daily temperatures, relative humidity, wind speed, and duration of sunshine, were acquired from two weather stations (Gazipur Sadar and Ishurdi). The climatic variables were gathered from different weather stations, as illustrated in Figure 1. A silicon photodiode type global solar radiation recorder (Licor-200SZ, LI-COR Biosciences, USA; accuracy = $\pm 5\%$; range = 0.3–4 µm; measurement height = 2 m) was used to measure the amount of sunshine along with length of the day. The maximum and minimum temperatures were measured employing the maximum and minimum thermometers (Zeal P1000, G. H. Zeal Ltd., London SW19 3UU, UK; accuracy = $\pm 0.2\,°C$; range and resolution = -50 to $+70\,°C$, $0.1\,°C$; measurement height = 2 m). Relative humidity was measured using a capacitive-type hygrometer (R. M. Young Company, Traverse City, MI 49686, USA; accuracy = $\pm 3\%$; range and resolution = 0–100%, 1%; measurement height = 2 m). The measurement of wind speed was performed using a rotating cup anemometer (Cup Anemometer 4.3018.10.000, Adolf Thies GmbH and Co. KG, Hauptstraße 76, 37083 Göttingen, Germany; accuracy = 1.2 m/s; range and resolution = 0.5–60 m/s, 0.1 m/s; measurement height = 10 m). It is noted that performing a thorough quality assurance procedure is often desirable to ensure the quality of climatic datasets, which enhances the reliability of ET_0 estimations using machine learning tools [88]. Although a detailed quality assurance procedure was not performed, the quality of the obtained climatic data was checked thoroughly for its correctness and completeness. The missing entries (less than 1%) were imputed using the 'movmedian' (Matlab MATLAB 2021a) approach of data imputation. Nevertheless, a few adjustments were performed to obtain the FAO-56 PM equation appropriate for local conditions following the recommendations provided in [89]. For instance, the wind speeds obtained at 10 m height (from the weather stations) were converted to wind speeds at the height of 2 m (keeping a lower limit of 0.5 m/s).

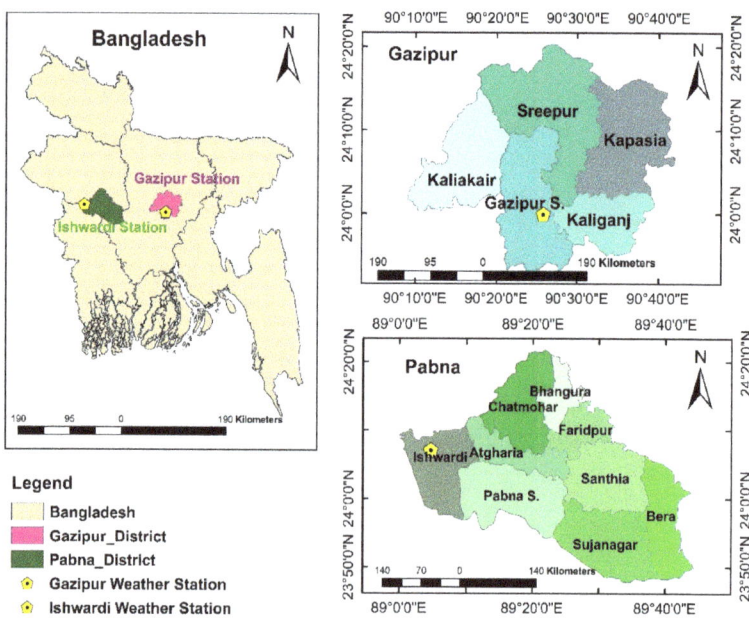

Figure 1. Weather station locations.

The weather station in Gazipur Sadar Upazilla was utilized as the training station for developing the proposed models, whereas data from the weather station in Ishurdi were used to evaluate the produced models' performance (testing station). The position of the weather station at Gazipur Sadar Upazilla is at 24.00° N latitude and 90.43° E longitude, being located 8.4 m above mean sea level (MSL). On the other hand, the test station is placed between 24.12° N latitude and 89.08° E longitude with an altitude of 18 m from the MSL. The weather data for the training station were obtained for a duration of 15.5 years (from 1 January 2004 to 30 June 2019). Descriptive statistics of the acquired weather data for the training station are presented in Table 1.

Table 1. Descriptive statistics of the weather data for the training station (Gazipur Sadar Upazilla).

Climatic Variables	Min	Max	Mean	Standard Deviation	Skewness	Kurtosis
Data Range: 1 January 2004 to 30 June 2019 (5660 Daily Entries)						
Minimum temperature, °C	4.40	34.50	21.17	5.64	−0.63	−0.88
Maximum temperature, °C	12.00	53.00	30.93	3.92	−1.10	2.11
Relative humidity, %	38.00	89.00	80.22	8.20	−0.63	0.75
Wind speed, m/s	0.68	5.06	2.79	1.05	−0.06	−1.32
Sunshine duration, h	0.00	11.40	5.54	3.09	−0.40	−1.04

The weather data for the test station were obtained for a duration of around 5.5 years (from 1 June 2015 to 31 December 2020). Descriptive statistics of the acquired weather data for the test station are presented in Table 2.

Table 2. Descriptive statistics of the entire, first half, and the second half of the weather data for the test station (Ishurdi Upazilla).

Climatic Variables	Mean	Standard Deviation	Skewness	Kurtosis
Entire dataset (1 June 2015 to 31 December 2020: 2041 daily entries)				
Minimum temperature, °C	21.37	5.98	−0.73	−0.76
Maximum temperature, °C	31.46	4.16	−0.83	0.28
Relative humidity, %	78.89	12.18	−1.23	1.93
Wind speed, m/s	1.43	0.23	0.07	0.22
Sunshine duration, h	5.90	3.19	−0.41	−0.71
First half data (1 June 2015 to 3 October 2018: 1221 daily entries)				
Minimum temperature, °C	21.06	6.08	−0.65	−0.92
Maximum temperature, °C	31.27	4.21	−0.71	0.26
Relative humidity, %	80.06	11.30	−1.24	2.25
Wind speed, m/s	1.43	0.23	0.06	0.35
Sunshine duration, h	5.75	3.18	−0.42	−0.98
Second half data (4 October 2018 to 31 December 2020: 820 daily entries)				
Minimum temperature, °C	21.69	5.87	−0.83	−0.56
Maximum temperature, °C	31.66	4.11	−0.95	0.35
Relative humidity, %	77.71	12.89	−1.18	1.54
Wind speed, m/s	1.44	0.23	0.09	0.08
Sunshine duration, h	6.05	3.19	−0.39	−0.44

Weather data acquired from the two weather stations for the specified duration were used to calculate the daily ET_0 values employing the FAO-56 PM equation (Equation (1)). The climatological variables (acquired weather data) and corresponding ET_0 values (computed using FAO-56 PM equation) were used as inputs and outputs from the proposed LSTM, Bi-LSTM, and other machine learning-based models. This approach of estimating ET_0 indirectly using the climatological variables has been a widely accepted method in situations where obtaining ET_0 directly becomes infeasible due to technical and budgetary constraints [6,15,90]. The FAO-56 PM equation is represented by

$$ET_0 = \frac{0.408\Delta(R_n - G) + \gamma\frac{900}{T_{mean}+273}u_2(e_s - e_a)}{\Delta + \gamma(1 + 0.34u_2)} \tag{1}$$

where ET_0 denotes reference evapotranspiration, mm/d; R_n represents the net radiation at the crop surface, $MJ/m^2/d$; G indicates heat flux density of soil, $MJ/m^2/d$; Δ represents the slope of the saturation vapor pressure curve, $kP_a/°C$; γ denotes psychometric constant, $kP_a/°C$; e_s represents the saturation vapor pressure, kP_a; e_a indicates the actual vapor pressure, kP_a; u_2 is the wind speed at the height of 2 m, m/s; and T_{mean} is the mean air temperature at 2.0 m height, °C.

The computed ET_0 values at the training station (Gazipur Sadar Upazilla) ranged between 0.92 and 8.02 mm/d, with the mean, standard deviation, skewness, and kurtosis values of 3.80 mm/d, 1.32 mm/d, 0.30, and −0.67, respectively. For the test station (Ishurdi), the computed ET_0 time series was divided into three sub-time series to test the generalization capability of the proposed modeling approach at different regions of the time series. The first time series considered was the entire dataset for which the ET_0 values had the mean, standard deviation, skewness, and kurtosis values of 3.67 mm/d, 1.24 mm/d, 0.28, and −0.62, respectively. The values of the mean, standard deviation, skewness, and kurtosis of the calculated ET_0 for the first half of the dataset were 3.57 mm/d, 1.25 mm/d, 0.35, and −0.62, respectively. The second half of the ET_0 time series contained the mean, standard deviation, skewness, and kurtosis values of 3.76 mm/d, 1.23 mm/d, 0.22, and −0.59, respectively.

For daily ET_0 prediction, meteorological variables and calculated ET_0 values using the FAO-56 PM equation were used as inputs and outputs. On the other hand, calculated ET_0 time series were used to develop the proposed models for one- and multi-step-ahead predictions by obtaining time-lagged characteristics from the time series data. For training the models, the entire dataset was divided into three parts: training data (40% of the entire dataset: 2264 daily entries—from 1 January 2004 to 13 March 2010), validation data (40% of the entire dataset: 2264 daily entries—from 14 March 2010 to 24 May 2016), and test data (remaining 20% of the total dataset: 1132 daily entries—from 25 May 2016 to 30 June 2019). To test the generalization capability of the proposed models, we partitioned the data from the test station as follows: entire dataset (2021 ET_0 values and associated meteorological variables ranging from 1 June 2015 to 31 December 2020), the first half of the entire dataset (1221 ET_0 values and associated meteorological variables ranging from 1 June 2015 to 3 October 2018), and the first half of the entire dataset (820 ET_0 values and associated meteorological variables ranging from 4 October 2018 to 31 December 2020).

2.2. Prediction Models

2.2.1. Long Short-Term Memory (LSTM) Networks

An LSTM is a variant of the neural network-based modeling approach, an upgraded version of RNNs capable of learning long-term dependence that exists at various steps in the sequential time series data. LSTMs safeguard against the vanishing and exploding gradient issues commonly observed in a standard RNN architecture, making an LSTM an ideal modeling tool to predict and forecast sequential time series data. To eliminate vanishing and exploding gradient problems, an LSTM integrates two important parameters called 'state dynamics' and 'gating functions' [91]. An LSTM network architecture is made up of several interconnected memory blocks that are connected to each other in a number of layers, each of which consists of many recurrently connected memory cells. The memory cells of LSTM architectures are comprised of three gates [92]: (a) input, (b) forget, and (c) output. For performing a regression task, an LSTM model employs four layers: a sequence input layer, an LSTM layer, a fully connected layer, and a regression layer. The input and fully connected layers correspond to the number of input and output variables, respectively. The LSTM layer accommodates the number of hidden units, whereas the regression layer performs the regression task. The sequence input and LSTM layers are the most important components of a fundamental LSTM network. The input layer is responsible for inputting the sequence data, e.g., time-series data to the network, whereas the LSTM layer facilitates learning long-term dependence among various time-steps of a sequential time series data. A comprehensive explanation of the LSTM model architecture is presented by Roy [81] and is not repeated in this effort. A bidirectional LSTM network (Bi-LSTM) architecture is similar to an LSTM network except that a Bi-LSTM network is associated with bidirectional long-term dependence among various time-steps of a sequential time series data.

In this study, both networks (LSTM and Bi-LSTM) have three hidden layers, each of which is followed by a dropout layer that is employed to prevent model overfitting. Each of the three hidden layers has a large number of hidden neurons. The first, second, and third hidden layers each had 100, 50, and 20 hidden neurons, respectively. In contrast, the dropout rates assigned for the associated dropout layers were chosen as 0.4, 0.3, and 0.2, respectively. The optimum numbers of hidden layers, hidden neurons, and dropout rates are determined by conducting a series of trials. Numerous combinations of varying numbers of these parameters are tested until a stable network is obtained. In addition, the best training options are selected upon conducting several trials, and similar training options are used for training both the LSTM and Bi-LSTM models for consistency. The training options used for training the LSTM and Bi-LSTM networks are provided in Table 3.

Table 3. Training options and the associated parameter values.

Training Options	Corresponding Parameter Values
Solver for optimization	'adam'
Maximum number of epochs	1000
Gradient threshold value	1
Preliminary learning rate	0.01
Minimum size of the batch	150
Length of sequence	1000

2.2.2. Adaptive Neuro-Fuzzy Inference System (ANFIS)

ANFIS, a variant of fuzzy inference systems (FIS), is adaptive in nature, incorporating fuzziness and ambiguity of input variables in developing input–output relationships of nonlinear systems [93]. An ANFIS grab holds the advantageous features of both the artificial neural networks and fuzzy set theory into an adaptive framework to model nonlinear and complex systems quite efficiently and effectively [94,95]. Due to less complexity and better learning ability [93], a Sugeno-type FIS is used to develop the ANFIS model utilizing a fuzzy c-means clustering (FCM) [96] algorithm to reduce the dimensionality of input variables. Detailed descriptions of ANFIS model structures are provided in Jang et al. [93] and are not repeated in this effort. Figure 2 presents an ANFIS model structure derived from a Sugeno-type FIS. The ANFIS models were developed in a MATLAB [97] environment.

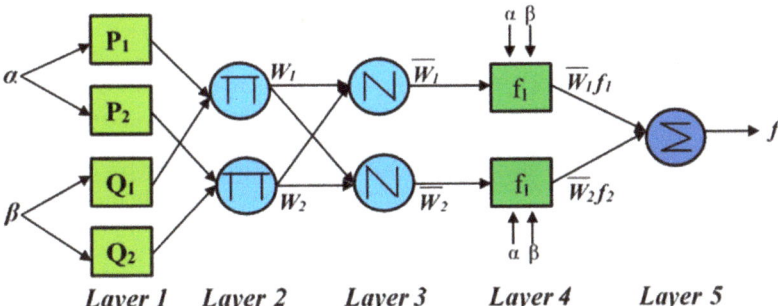

Figure 2. A schematic diagram of an ANFIS structure derived from a first-order Sugeno-type FIS. Reprinted with permission from Jang [98].

2.2.3. Gaussian Process Regression (GPR)

A GPR is a nonparametric modeling algorithm that is derived from the theories of probability and Gaussian process [99]. Following a Gaussian distribution, a GPR model provides the output, Y from the input variables, and $X(i)$ through developing a functional relationship, which can be mathematically represented as [100]

$$Y = f(X(i)) + \varepsilon \tag{2}$$

where ε is a Gaussian noise, the variance of which is denoted by σ_n^2.

The mean, $m(x_i)$, and covariance, $k(x_i, x_j)$, functions are the two important functional components of a typical GPR model. They can be mathematically expressed as [99]

$$m(x_i) = E[f(x_i)] \tag{3}$$

$$k(x_i, x_j) = E\big[(f(x_i) - m(x_i))(f(x_j) - m(x_j))\big] \tag{4}$$

On the basis of these two key functions, the functional relationship using Gaussian process theory is established by the following equation:

$$f(x) \sim gp\big(m(x_i), k(x_i, x_j)\big) \tag{5}$$

The prediction probability distribution of a GPR model is governed by the free parameters or hyperparameters, which are in essence the parameters of the mean and covariance functions. The values of free parameters or hyperparameters depend on the training dataset's log-likelihood function values. The GPR models were developed by utilizing the commands and functions of MATLAB [97].

2.2.4. M5 Model Trees (M5 Tree)

The development of the M5 tree is derived from the philosophies associated with the M5 technique [101,102] in building standalone trees. The prediction capabilities of M5 trees were demonstrated and well documented in various research domains [103,104]. In the M5 tree modeling approach, a complex modeling task is sub-divided into numerous sub-tasks via the divide-and-conquer technique, and the final result is the integration of solutions from all the sub-tasks [103]. This splitting technique results in a hierarchy of model trees in which non-terminal nodes are associated with splitting rules, whereas expert models are represented by the tree leaves [104]. Model development using the M5 tree technique is performed using three stepwise procedures: (1) development of an initial tree, (2) pruning of the tree, and (3) smoothing of the tree [105]. In the MATLAB environment, a toolbox "M5PrimeLab" [106] was used to develop M5 trees for predicting daily reference ET_0 values.

2.2.5. Multivariate Adaptive Regression Spline (MARS)

MARS [107] is a nonparametric modeling technique that is adaptive in nature and is believed to be a flexible and rapid approach to developing regression models. The MARS approach partitions the entire decision space into several input parameters on which standalone basis functions or splines are fitted to obtain the final MARS model [108]. Both a forward procedure and a backward procedure are utilized, i.e., MARS initially builds a comparatively complex model using the user-specified maximum number of basis functions in the forward step. In contrast, in the backward step, MARS parsimoniously selects the most significant input variables in predicting the output variable [109]. The backward step eliminates redundant input variables and assists in simplifying the final model while avoiding over-or under-fitting. The relationship between the input and output variables can be represented by the following equation [110]:

$$BF_i(x) = \max(0, x_j - \alpha) \text{ OR } BF_i(x) = \max(0, \alpha - x_j) \tag{6}$$

$$y = f(x) = \beta \pm \gamma_k \times BF_i(x) \tag{7}$$

where i represents the index of Basis functions, j denotes the index of input variables, BF_i symbolizes the ith Basis function, x_j is the jth input variable, α is a threshold value used by the MARS model during model building, β is a constant, γ_k indicates the respective coefficient of $BF_i(x)$, and y denotes the model prediction (output variable).

A MATLAB toolbox 'ARESLab' [106] was employed to build MARS-based ET_0 prediction models. This study used both piecewise-linear and piecewise-cubic modeling approaches to predict daily ET_0 values.

2.2.6. Probabilistic Linear Regression (PLR)

PLR utilizes Bayesian inference techniques to develop prediction models through probabilistically performing linear regression. The PLR approach is often referred to as empirical Bayesian linear regression, using either an expectation-maximization (EM) algorithm [111] or a Mackay fixpoint iteration method [112]. The EM algorithm is generally utilized to formulate the PLR models. As such, the present study used the EM algorithm in developing PLR-based ET_0 prediction models. Mo Chen [113] developed a MATLAB toolbox in this research to develop PLR models.

2.2.7. Support Vector Regression (SVR)

SVRs are derived from the principles of the support vector machine (SVM) algorithm [114], which has been attaining significant attention in recent years for its capability to solve a diversified range of regression and classification problems [115]. SVRs are developed via a nonlinear mapping technique that utilizes required data from the input space to a high-dimensional feature space on which linear regressions are executed [116]. An elaborated explanation of the theory of the SVR approach has been provided in Chevalier et al. [117], and only a brief account of the SVR theorem is presented in this effort. The following equation symbolizes the training dataset in developing a linear SVR model:

$$\{(\overline{x_1}, y_1), (\overline{x_2}, y_2), \ldots, (\overline{x_l}, y_l)\} \tag{8}$$

$\overline{x_i} \in R^d$, $y_i \in R$, and $l =$ number of data entries

In this case, the solution function can be expressed as

$$f(\overline{x}) = \sum_{i=1}^{l}(\alpha_i - \alpha_i^*) < \overline{x_i}, \overline{x} > +b \tag{9}$$

where $< ., . >$ denotes dot product, and α_i, α_i^*, and b represent coefficients computed by the SVR model.

A data transformation step is performed to build nonlinear SVR models, including a nonlinear mapping function $\varnothing$ [118] that transforms low-dimensional input space into a high-dimensional feature space. The computation $\varnothing$ becomes challenging during progressive mapping of the input–output data into higher dimensions. This limitation is handled using the Mercers theorem, which can be represented by the following equation:

$$< \varnothing(\overline{u}), \varnothing(\overline{v}) >= k(\overline{u}, \overline{v}) \tag{10}$$

For a particular mapping $\varnothing$, the Mercers theorem introduces the concept of using a kernel function k, which is used to calculate the dot product of any two points $(\overline{u}, \overline{v})$, and the computation of dot products in this approach bypasses the explicit calculation of high-dimensional and nonlinear mapping. The prediction performance of nonlinear SVR models depends on the kernel function, which is regarded as one of the most important parameters in SVR modeling.

2.3. Ranking of the ET$_0$ Prediction Models: Shannon's Entropy

ET$_0$ prediction models were ranked using performance-based weights assigned to standalone models using Shannon's entropy principle. For this, a decision matrix of prediction models (m) and performance indices (PI) is formulated, which can be represented in the form of the following equation [119]:

$$ET_{ij} = \begin{bmatrix} ET_{11} & ET_{21} & \cdots & ET_{m1} \\ ET_{12} & ET_{22} & \cdots & ET_{m2} \\ \vdots & \vdots & \vdots & \vdots \\ ET_{1PI} & ET_{2PI} & \cdots & ET_{mPI} \end{bmatrix} \tag{11}$$

To reduce the adverse impacts of index dimensionality, we standardized the performance index values between 0 and 1 $\{S_{ij} \in [0,1], i = 1, 2, \ldots, m; j = 1, 2, \ldots, PI\}$. The standardization component S_{ij} was performed using the following equation [119]:

$$S_{ij} = \begin{cases} \frac{ET_{ij}}{\max(ET_{i1}, ET_{i2}, \ldots, ET_{iPI})}, & \text{for benefit indexes} \\ \frac{X_{ij}}{\min(ET_{i1}, ET_{i2}, \ldots, ET_{iPI})}, & \text{for cos t indexes} \end{cases} \tag{12}$$

Shannon's entropy-based ranking was performed using a five-step stepwise procedure described in Roy et al. [21], which was not repeated here.

2.4. Selection of Input Variables for Daily Predictions

All possible combinations of the five input variables (minimum temperatures, maximum temperatures, relative humidity, wind speed, and sunshine hours) were used. A total of 31 models were developed on the basis of the 31 combinations (single, two-input combinations, three-input combinations, four-input combinations, and all five inputs) of input variables. Two-, three-, and four-input combinations are presented in Table 4.

Table 4. Different combinations of two-, three-, and four-input combinations.

Two-Input Combinations	Three-Input Combinations	Four-Input Combinations
Min temp, max temp	Min temp, max temp, humidity	Min temp, max temp, humidity, wind speed
Min temp, humidity	Min temp, max temp, wind speed	Min temp, max temp, humidity, sunshine hours
Min temp, wind speed	Min temp, max temp, sunshine hours	Min temp, max temp, wind speed, sunshine hours
Min temp, sunshine hours	Min temp, humidity, wind speed	Min temp, humidity, wind speed, sunshine hours
Max temp, humidity	Min temp, humidity, sunshine hours	Max temp, humidity, wind speed, sunshine hours
Max temp, wind speed	Min temp, wind speed, sunshine hours	
Max temp, sunshine hours	Max temp, humidity, wind speed	
Humidity, wind speed	Max temp, humidity, sunshine hours	
Humidity, sunshine hours	Max temp, wind speed, sunshine hours	
Wind speed, sunshine hours	Humidity, wind speed, sunshine hours	

These combinations of input variables were evaluated for two deep learning algorithms (LSTM and Bi-LSTM). The 62 models (31 LSTM + 31 Bi-LSTM) developed were ranked on the basis of their prediction accuracies using Shannon's entropy by incorporating a number of benefit (correlation coefficient, Nash–Sutcliffe efficiency coefficient, Willmott's index of agreement) and cost (normalized or relative root mean squared error, maximum absolute error, median absolute deviation) performance evaluation indices. The best-input combinations thus obtained were used to develop the other shallow machine learning algorithms.

2.5. Model Performance Evaluation

The performances of the proposed models were evaluated using various statistical evaluation indices as follows:

- Correlation coefficient, R

$$R = \frac{\sum_{i=1}^{n}\left(ET_{i,a} - \overline{ET_a}\right)\left(ET_{i,a} - \overline{ET_p}\right)}{\sqrt{\sum_{i=1}^{n}\left(ET_{i,a} - \overline{ET_a}\right)^2}\sqrt{\sum_{i=1}^{n}\left(ET_{i,p} - \overline{ET_p}\right)^2}} \tag{13}$$

- Nash–Sutcliffe efficiency coefficient, NS [120]

$$NS = 1 - \frac{\sum_{i=1}^{n}\left(ET_{i,a} - ET_{i,p}\right)^2}{\sum_{i=1}^{n}\left(ET_{i,a} - \overline{ET_a}\right)^2} \tag{14}$$

- Index of agreement, IOA [121]

$$IOA = 1 - \frac{\sum_{i=1}^{n}\left(ET_{i,a} - ET_{i,p}\right)^2}{\sum_{i=1}^{n}\left(\left|ET_{i,p} - \overline{ET_a}\right| + \left|ET_{i,a} - \overline{ET_a}\right|\right)^2} \tag{15}$$

- Root mean square error, RMSE [122]

$$RMSE = \sqrt{\frac{1}{n}\sum_{i=1}^{n}\left(ET_{i,a} - ET_{i,p}\right)^2} \tag{16}$$

- Normalized RMSE, NRMSE

$$NRMSE = \frac{RMSE}{\overline{ET_a}} \tag{17}$$

- Maximum absolute error, MAE

$$MAE = \max\left[\left|ET_{i,a} - ET_{i,p}\right|\right] \tag{18}$$

- Median absolute deviation, MAD

$$MAD(ET_a, ET_p) = \operatorname{median}\left(\left|ET_{1,a} - ET_{1,p}\right|, \left|ET_{2,a} - ET_{2,p}\right|, \ldots, \left|ET_{n,a} - ET_{n,p}\right|\right) \atop \text{for } i = 1, 2, \ldots, n \tag{19}$$

where $ET_{i,a}$ and $ET_{i,p}$ are ET_0 quantities at the ith data points acquired from the FAO-56 PM computed and model predicted values, respectively; $\overline{ET_a}$ represents the arithmetic mean of the FAO-56 PM computed ET_0 values; and n is the amount of input–output data.

3. Results and Discussion

3.1. Daily Prediction of ET_0 Using Various Machine Learning Algorithms at the Training Station (Gazipur Sadar)

To determine the optimum numbers of input variables combinations, we used 31 possible combinations of five input variables to develop 31 LSTM and 31 Bi-LSTM models. Learning (training) and testing of the ET_0 models were performed simultaneously. Prediction errors on the test dataset in terms of RMSE criterion for the 31 developed models are presented in Table 5. As evidenced by the numerical values presented in Table 5, the LSTM model predictions were slightly better than those of the Bi-LSTM models when the RMSE criterion was used as a deciding factor. It was also observed that both the LSTM- and Bi-LSTM-based ET_0 prediction models produced the lowest RMSE values (best daily ET_0 predictions) when all five variables were used. The performance of LSTM (RMSE = 0.081 mm/d) was slightly better than that of the Bi-LSTM (RMSE = 0.087 mm/d) model. However, in situations where adequate data are not available, the use of fewer input variables may be employed to achieve a realistically precise prediction of ET_0 values. For instance, four climatological variables (a combination of maximum temperature, relative humidity, wind speed, and sunshine hours) could be used to obtain sufficiently accurate daily ET_0 predictions using LSTM (test error in terms of RMSE value equals 0.107 mm/d) and Bi-LSTM (test error in terms of RMSE value equals 0.116 mm/d) models. Other combinations of four meteorological variables, e.g., (minimum temperature, maximum temperature, relative humidity, sunshine hours) and (minimum temperature, relative humidity, wind speed, sunshine hours) provided reasonably accurate daily ET_0 predictions (Table 5). In addition, combinations of three meteorological variables (relative humidity, wind speed, sunshine hours) and (minimum temperature, relative humidity, sunshine hours) produced reasonable accurate predictions, with test RMSE values ranging between 0.333 and 0.377 mm/d.

Table 5. Prediction errors of deep learning-based ET_0 models (LSTM and Bi-LSTM) with different input combinations on the test dataset.

Model No.	Different Input Combinations	Test RMSE, mm/d	
		LSTM	Bi-LSTM
	Single Input Combinations		
M1	Min temp	0.880	0.964
M2	Max temp	0.775	0.781
M3	Humidity	1.124	*1.211*
M4	Wind speed	*1.177*	1.105
M5	Sunshine hours	0.732	0.807
	Two Inputs combinations		
M6	Min temp, max temp	0.765	0.779
M7	Min temp, humidity	0.729	0.751
M8	Min temp, wind speed	1.004	1.049
M9	Min temp, sunshine hours	0.527	0.514
M10	Max temp, humidity	0.634	0.602
M11	Max temp, wind speed	0.734	0.743
M12	Max temp, sunshine hours	0.501	0.430
M13	Humidity, wind speed	0.727	0.760
M14	Humidity, sunshine hours	0.531	0.983
M15	Wind speed, sunshine hours	0.527	0.627
	Three Inputs Combinations		
M16	Min temp, max temp, humidity	0.570	0.574
M17	Min temp, max temp, wind speed	0.729	0.722
M18	Min temp, max temp, sunshine hours	0.512	0.447
M19	Min temp, humidity, wind speed	0.726	0.723
M20	Min temp, humidity, sunshine hours	0.337	0.377
M21	Min temp, wind speed, sunshine hours	0.470	0.501
M22	Max temp, humidity, wind speed	0.567	0.566
M23	Max temp, humidity, sunshine hours	0.300	0.239
M24	Max temp, wind speed, sunshine hours	0.409	0.394
M25	Humidity, wind speed, sunshine hours	0.337	0.333
	Four Inputs Combinations		
M26	Min temp, max temp, humidity, wind speed	0.577	0.561
M27	Min temp, max temp, humidity, sunshine hours	0.262	0.229
M28	Min temp, max temp, wind speed, sunshine hours	0.382	0.404
M29	Min temp, humidity, wind speed, sunshine hours	0.271	0.238
M30	Max temp, humidity, wind speed, sunshine hours	0.107	0.116
	All Inputs		
M31	Min temp, max temp, humidity, wind speed, sunshine hours	**0.081**	**0.087**

RMSE = root mean squared error, LSTM = long short-term memory networks, Bi-LSTM = bi-directional long-short term memory networks. The numbers in boldface indicate the best performance, whereas the numbers in boldface and italicized represent the worst performance.

Nonetheless, decision making in such situations is challenging, as the RMSE criterion alone is insufficient as a decision-making tool. To assist in the decision-making process, we used three benefit (the higher numeric values indicate better model performances: R, NS, IOA) and three cost (the lower the numeric values, the better the model performance: NRMSE, MAE, MAD) performance evaluation indices in the decision-making process with the aid of Shannon's entropy. On the testing dataset, we computed the R, NS, IOA, NRMSE, MAE, and MAD criteria for all 31 LSTM and 31 Bi-LSTM models. These evaluation indices were used to rank proposed models using Shannon's entropy-based decision theory. Table 6 shows the ranking results together with the corresponding ranking values.

Table 6. Ranking of the LSTM and Bi-LSTM models using Shannon's entropy.

Sl. No.	LSTM		Bi-LSTM	
	Model	Ranking Value	Model	Ranking Value
1	M31	0.996	M31	0.966
2	M30	0.906	M30	0.913
3	M27	0.702	M27	0.704
4	M23	0.687	M23	0.696
5	M20	0.657	M29	0.688
6	M29	0.652	M25	0.642
7	M25	0.640	M20	0.621
8	M28	0.604	M24	0.600
9	M24	0.600	M28	0.594
10	M21	0.584	M12	0.581
11	M12	0.563	M18	0.576
12	M18	0.561	M21	0.563
13	M14	0.560	M26	0.557
14	M22	0.558	M9	0.555
15	M26	0.556	M22	0.551
16	M15	0.555	M16	0.551
17	M9	0.555	M10	0.535
18	M16	0.554	M15	0.522
19	M10	0.535	M17	0.488
20	M11	0.496	M19	0.485
21	M17	0.493	M11	0.482
22	M19	0.491	M7	0.478
23	M13	0.491	M13	0.475
24	M7	0.483	M6	0.462
25	M5	0.482	M2	0.460
26	M6	0.470	M5	0.451
27	M2	0.470	M14	0.384
28	M1	0.415	M1	0.376
29	M8	0.364	M8	0.336
30	M3	0.306	M4	0.311
31	M4	0.209	M3	0.256

It is perceived from the results presented in Table 6 that models that used all five input variables (M31) were the top-ranked predictors, followed by M30, M27, and M23 for both LSTM and Bi-LSTM algorithms. Models M3 and M4 appeared to be the worst performers when using LSTM or Bi-LSTM algorithms for model development. The findings are in accordance with the work of Kisi et al. [37], who indicated that considering all input variables greatly increased the accuracy of the prediction model (radial basis M5Tree) for the data acquired from the three weather stations. Therefore, the results suggest that all input variables would be employed to better predict the daily ET_0 for the meteorological data and the corresponding ET_0 values presented in this study. Consequently, to arrange for an impartial comparison, we developed other prediction modeling algorithms (ANFIS, GPR, M5Tree, MARS, PLR, and SVR) using all five input variables available for the study area. Similar evaluation indices were computed for all the other prediction modeling algorithms proposed in this research. The prediction results are presented in Table 7.

The prediction results in Table 7 indicated that all ET_0 prediction models are reasonably accurate at predicting daily ET_0 values, as evidenced by the different performance indices computed on the testing dataset. While no standalone model exhibited the best performance for all evaluation indices, the individual prediction models provided the estimates of daily ET_0 values superior to others. All ET_0 models had satisfactory prediction accuracy as all models had better (higher) values R, NS, and IOA and lower NRMSE, MAE, and MAD values. LSTM and Bi-LSTM models had superior performance in comparison with others according to all performance evaluation indices. PLR was found to be the worst-performing model.

Table 7. Performance indices of the developed ET_0 prediction models for the testing dataset.

Model	Performance Evaluation Indices					
	R	NS	IOA	NRMSE	MAE, mm/d	MAD, mm/d
LSTM	0.998	0.995	0.999	0.021	0.666	0.025
Bi-LSTM	0.998	0.995	0.999	0.023	0.582	0.027
ANFIS	0.991	0.981	0.995	0.043	0.706	0.061
GPR	0.993	0.985	0.996	0.038	0.650	0.052
M5 Tree	0.985	0.970	0.993	0.054	1.153	0.062
MARS_C	0.992	0.983	0.996	0.041	0.869	0.054
MARS_L	0.992	0.983	0.996	0.040	0.760	0.054
PLR	0.973	0.943	0.985	0.075	1.489	0.114
SVR	0.993	0.985	0.996	0.038	0.676	0.050

MARS_C = piecewise cubic, MARS_L = piecewise linear.

To provide an additional evaluation regarding the prediction capabilities of the proposed machine learning algorithms (ET_0 prediction models), we presented and compared the absolute error boxplots. Figure 3 illustrates the absolute error boxplots for all the developed models. Absolute error boxplots represent a relative assessment of the statistical distributions of the absolute errors between the FAO-56 PM-computed and model-predicted ET_0 values and supports the evaluation of the degree of general distributions of the inaccuracies provided by the models. The horizontal lines inside the boxplots represent the median values of the absolute errors, whereas the black circles mark the mean (average) of the absolute errors. Absolute error boxplots also demonstrated the superior performance of the LSTM- and Bi-LSTM-based models.

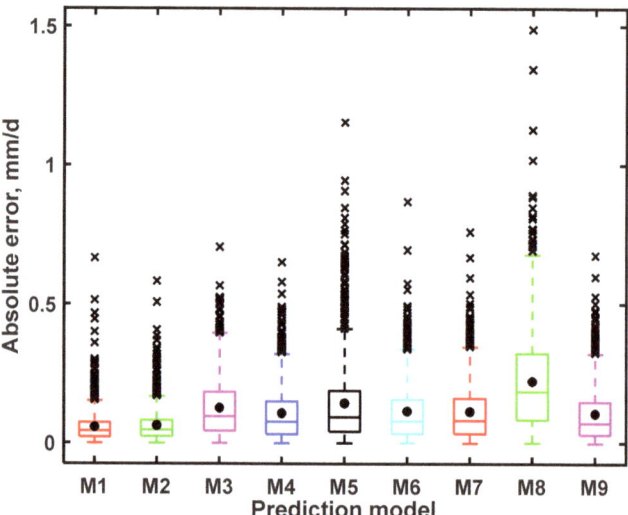

Figure 3. Absolute error boxplots. M1–M9 represent LSTM, Bi-LSTM, ANFIS, GPR, M5 tree, MARS_C, MARS_L, PLR, and SVR models, respectively.

As far as the two best models are considered, the LSTM model performed better than Bi-LSTM when NRMSE and MAD criteria were considered. In contrast, Bi-LSTM outperformed the LSTM model according to the MAE criterion. On the other hand, both LSTM and Bi-LSTM performed equally well with respect to R, NS, and IOA criteria. Therefore, it is concluded that ET_0 prediction models showed differing precisions depending on the model evaluation indices calculated on the FAO-56 PM and model predicted ET_0 values, which indicated an inconsistency in the model performance when divergent or non-identical evaluation indices were employed. Decision making in this situation is extremely arduous and can be smoothed by employing a decision theory that integrates a number of different model evaluation indices in decision making. This study employed Shannon's entropy as a decision-making tool.

The ranking of the proposed ET_0 models computed using Shannon's entropy is presented in Table 8. The greater the values of Shannon's entropy, the better the model's performance. Table 8 suggests that LSTM was the top-performing model followed by Bi-LSTM, although the difference between the ranking values of these two models was negligible.

Table 8. Shannon's entropy values for different models and their corresponding ranks.

Model	Shannon's Entropy Value	Rank
LSTM	0.979	1
Bi-LSTM	0.978	2
ANFIS	0.807	6
GPR	0.839	3
M5 tree	0.734	8
MARS_C	0.794	7
MARS_L	0.810	5
PLR	0.665	9
SVR	0.836	4

The performance index values for the best model (LSTM) are as follows (Table 7): R = 0.998, NS = 0.995, IOA = 0.999, NRMSE = 0.021, MAE = 0.666 mm/d, and MAD = 0.025 mm/d. Although an explicit comparison between the findings of this research and other studies is not possible due to variations in study conditions (modeling tools and geographical locations), the numeric values of various performance indices were observed as being comparable to or even better than those found in the recent literature on ET_0 modeling. For instance, the present study's findings are superior to those obtained by Tao et al. [123], who obtained NRMSE and R^2 values of 0.043 and 0.97, respectively, using an optimization algorithm-tuned ANFIS model to predict ET_0 in the Bur Dedougou, Burkina Faso. The LSTM model proposed in this study also shows better performance than the optimization algorithm tuned SVR model developed in Ahmadi et al. [32], who obtained the following performance indices at various stations: RMSE = 0.540 mm/d and R = 0.983 at Mashhad station; RMSE = 0.404 mm/d and R = 0.980 at Arak station; RMSE = 0.299 mm/d and R = 0.989 at Shiraz station; RMSE = 0.559 mm/d and R = 0.978 at Tehran station; RMSE = 0.457 mm/d and R = 0.962 at Bandar Abbas station; and RMSE = 0.399 mm/d and R = 0.986 at Yazd station. The present study's findings are also in good agreement with the findings presented in Chia et al. [124], who obtained RMSE and R^2 values of 0.001–0.197 mm/d and 1.000–0.949, respectively, at three stations using an optimization algorithm-tuned ELM model. The findings are also compared with those presented in Mohammadi and Mehdizadeh [125] that are based on RMSE and R^2 criteria. Our proposed LSTM model shows superior performance over the best models developed with the daily data in Ferreira and da Cunha [80], who reported NS values of 0.69 to 0.84 and R^2 values of 0.79 to 0.88. The present study's findings are superior to the optimization algorithm-tuned ELM model developed by Wu et al. [30] that reported R^2 and NRMSE values of 0.993 and 0.0554, respectively. Elbeltagi et al. [126] reported R values of 0.94, 0.95, and 0.95 at the Ad Daqahliyah, Kafr ash Shaykh, and Ash Sharqiyah regions, respectively, using the DNN model. These R values were lower than the R-value obtained using the proposed LSTM model in the present study (R = 0.998). The NS value of the present study (NS = 0.995) is also superior to the NS value (NS = 0.959) presented in Gao et al. [127], indicating the better performance of the proposed LSTM model. The findings of our study are also comparable to those presented in Chia et al. [50], who reported minimum MAE and RMSE values of 0.444 mm/d and 0.543 mm/d, respectively.

Nevertheless, an apple-to-apple comparison can be performed between the findings obtained from the LSTM model presented in this effort with the models investigated in

Roy et al. [21] (an ensemble of ANFIS models) and in Roy et al. [20] (optimization algorithm tuned ANFIS model). With the optimization algorithm-tuned ANFIS model for the same study area, Roy et al. [20] obtained the following performance indices: R = 0.993, NS = 0.986, IOA = 0.996, MAD = 0.054 mm/d, NRMSE = 0.038. Our proposed LSTM model performed better than the ANFIS model presented by Roy et al. [20] with respect to all of these performance indices (R = 0.998, NS = 0.995, IOA = 0.999, NRMSE = 0.021, and MAD = 0.025 mm/d in the present study). Statistical indices provided by the LSTM model (R = 0.998, NS = 0.995, IOA = 0.999, and MAD = 0.025 mm/d) proposed in this research also appeared to be superior than those presented by Roy et al. [21] using ensemble of ANFIS models (R = 0.993, NS = 0.985, IOA = 0.996, and MAD = 0.054 mm/d). Furthermore, the proposed LSTM model's performance is superior to the performance of the optimization algorithm tune hierarchical fuzzy systems (HFS) presented by Roy et al. [46] with respect to R (LSTM = 0.998, HFS = 0.987), NRMSE (LSTM = 0.021, HFS = 0.052), and MAD (LSTM = 0.025 mm/d, HFS = 0.068 mm/d) criteria.

3.2. One-Step-Ahead Prediction of ET_0 Using Different Modeling Approaches at the Training Station (Gazipur Sadar)

3.2.1. One-Step-Ahead Forecast Using Sequence to Sequence Regression LSTM (SSR-LSTM) Network

An SSR-LSTM network-based model was trained by employing the historical ET_0 dataset (time series) computed using the FAO-56 PM equation from the meteorological variables. In an SSR-LSTM model, the outputs from the model correspond to the training sequences (ET_0 time series) with ET_0 values moved to a one-time step ahead. At every time step of the ET_0 sequence, an SSR-LSTM network learns how to predict ET_0 values for the next time step. For training the proposed SSR-LSTM model, the historical ET_0 time series was partitioned into training and test sets (90% of the entire data was used for training, whereas the remaining 10% was used for testing the model). Model parameters including the number of hidden layers and neurons were decided upon by conducting several trials. An SSR-LSTM model with one hidden layer having 200 hidden neurons in the hidden layer provided the best results for both the model training and testing phases. The optimal values of other model parameters were solver = 'adam', number of epochs = 250, gradient threshold = 1, initial learning rate = 0.005, and multiplying factor for the learn rate dropping = 0.2. Model performance is presented in Figure 4.

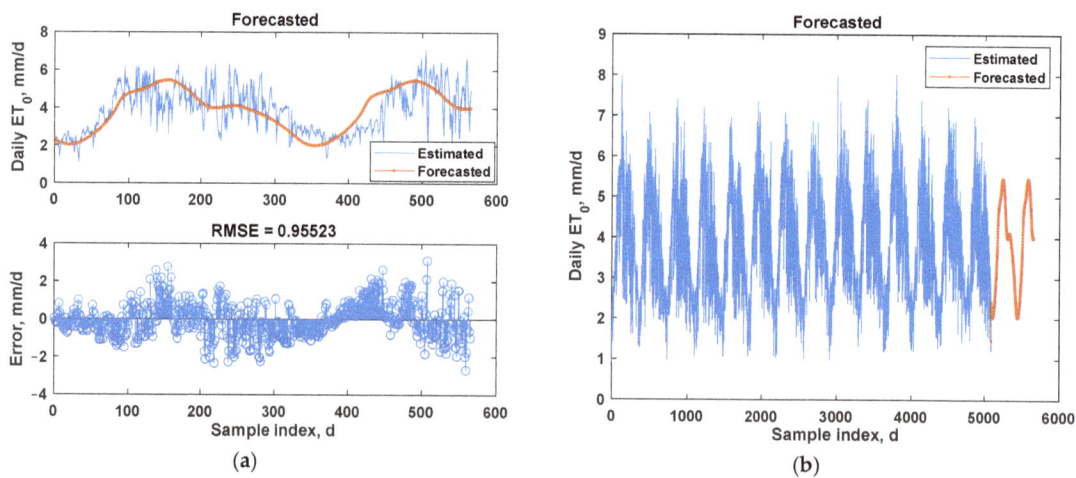

Figure 4. SSR-LSTM performance: (**a**) estimated (FAO-56 PM-computed) and SSR-LSTM-forecasted ET_0 values for the test dataset; (**b**) future projections beyond the observed ET_0 values.

It is observed from Figure 4 that even though the SSR-LSTM model adequately apprehended the trends of the ET_0 time series for the test set of the data (Figure 4b), the SSR-LSTM forecasts were comparatively flat compared to the original ET_0 time-series data (Figure 4a). This necessitates the improvement in the forecasting performance of the initial SSR-LSTM model. One way of improving performance is to update the SSR-LSTM network state using the observed ET_0 values instead of the predicted ET_0 values. Resetting the network's state is used in this study to prevent previous predictions from impacting the results.

This was performed by resetting the network state in order to prevent previous predictions from affecting the predictions on the new dataset. The forecasting results obtained from the updated network state of the SSR-LSTM model are presented in Figure 5.

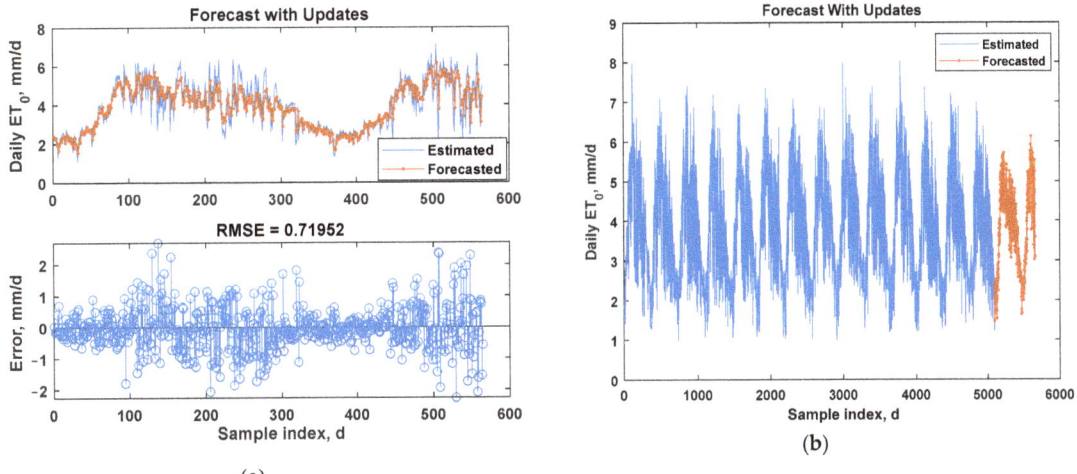

Figure 5. SSR-LSTM performance after network updating: (**a**) estimated (FAO-56 PM-computed) and SSR-LSTM-forecasted ET_0 values for the test dataset; (**b**) future projections beyond the observed ET_0 values.

3.2.2. One-Step-Ahead Forecast Using ANFIS, LSTM, and Bi-LSTM Models

For developing ANFIS, LSTM, and Bi-LSTM models to provide one-step-ahead forecasts, we computed PACF functions to obtain time-lagged information from the daily ET_0 time series. This information obtained from the PACF functions was employed to assess the time-based dependences between ET_0 for a present day (ET_t) and the ET_0 values at a particular day in a prior period (e.g., at a lag time of ET_{t-1}, ET_{t-2}, ET_{t-3}, ET_{t-4}, and ET_{t-5}). These time-based dependences in the ET_0 time series were assessed for 50 time lags (e.g., ET_{t-1} to ET_{t-50}), as shown in Figure 6. In Figure 6, the blue lines indicate the 95% confidence band, whereas the red vertical lines represent the corresponding values of ACF and PACF. Time-lagged ET_0 values serve as the inputs to the ANFIS, LSTM, and Bi-LSTM models to forecast one-day-ahead ET_0 values (outputs from the models). The optimal sets of time-lagged ET_0 inputs for model development were selected carefully after observing the PACF functions.

(a) (b)

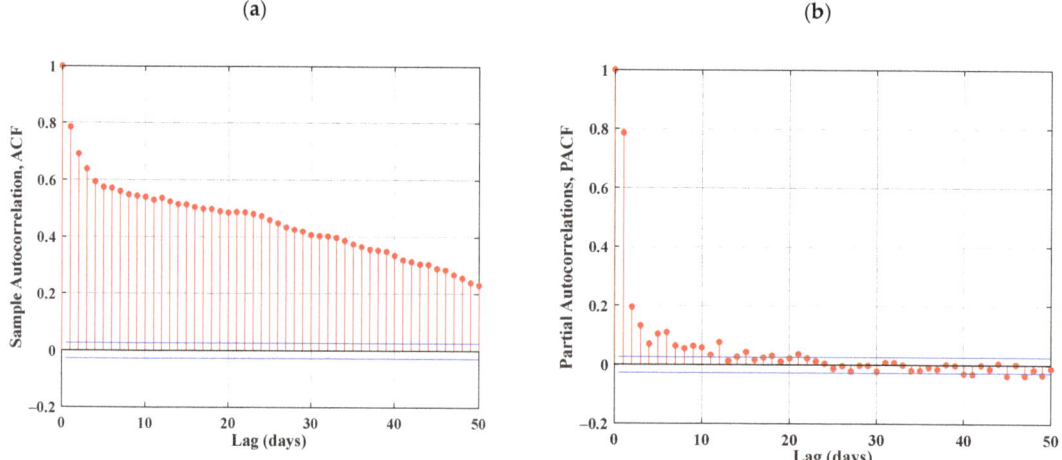

Figure 6. ACF (**a**) and PACF (**b**) plots of the ET_0 time series for 50 lags at Gazipur station.

A careful observation of the PACF plot shown in Figure 6 determines the following time-lagged ET_0 values as inputs to the developed models:

ET_t, ET_{t-1}, ET_{t-2}, ET_{t-3}, ET_{t-4}, ET_{t-5}, ET_{t-6}, ET_{t-7}, ET_{t-8}, ET_{t-9}, ET_{t-10}, ET_{t-11}

The outputs from the developed models were ET_{t+1}(one-day-ahead ET_0 values).

ANFIS outputs: The results of the one-step-ahead forecast using the ANFIS model are presented in Figure 7 and Table 9. Figure 7 presents ANFIS forecasts through scatter plots and hydrographs, whereas Table 8 shows model prediction capabilities based on several statistical performance evaluation indices. Hydrographs and scatterplots presented in Figure 7 demonstrate the reasonable precision of the one-day-ahead ET_0 forecasts by the ANFIS model. It is observed from Figure 7 that the training and test RMSE (0.759 and 0.789 mm/d, respectively, for the training and testing phases) did not vary considerably, which indicates a better model fit without model over- or under-fitting. Figure 7 also indicates acceptable values of training and test R-values (0.825 and 0.755, respectively, for the training and testing phases). As far as other performance evaluation indices are considered, the ANFIS model produced the following values of performance measures computed on the test dataset: NS = 0.567, IOA = 0.858, NRMSE = 0.207 mm/d, MAE = 2.710 mm/d, and MAD = 0.308 mm/d.

LSTM and Bi-LSTM outputs: Comparison of FAO-56 PM-calculated and model-predicted ET_0 values, error plots, and projected (one-step-ahead) ET_0 values produced by the LSTM and Bi-LSTM models are presented in Figures 8 and 9, respectively. It is noticed from Figures 8 and 9 that both LSTM and Bi-LSTM models captured the trend of the ET_0 time series precisely and that Bi-LSTM model forecasts were superior to those of the LSTM model. The performance evaluation results based on several statistical performance evaluation indices are presented in Table 9. The LSTM model produced the following values of performance measures computed on the test dataset: R = 0.698, NS = 0.698, IOA = 0.429, NRMSE = 0.237 mm/d, MAE = 3.047 mm/d, and MAD = 0.334 mm/d. On the other hand, the Bi-LSTM model produced the following values of performance measures computed on the test dataset: R = 0.999, NS = 0.998, IOA = 0.999, NRMSE = 0.014 mm/d, MAE = 0.491 mm/d, and MAD = 0.017 mm/d.

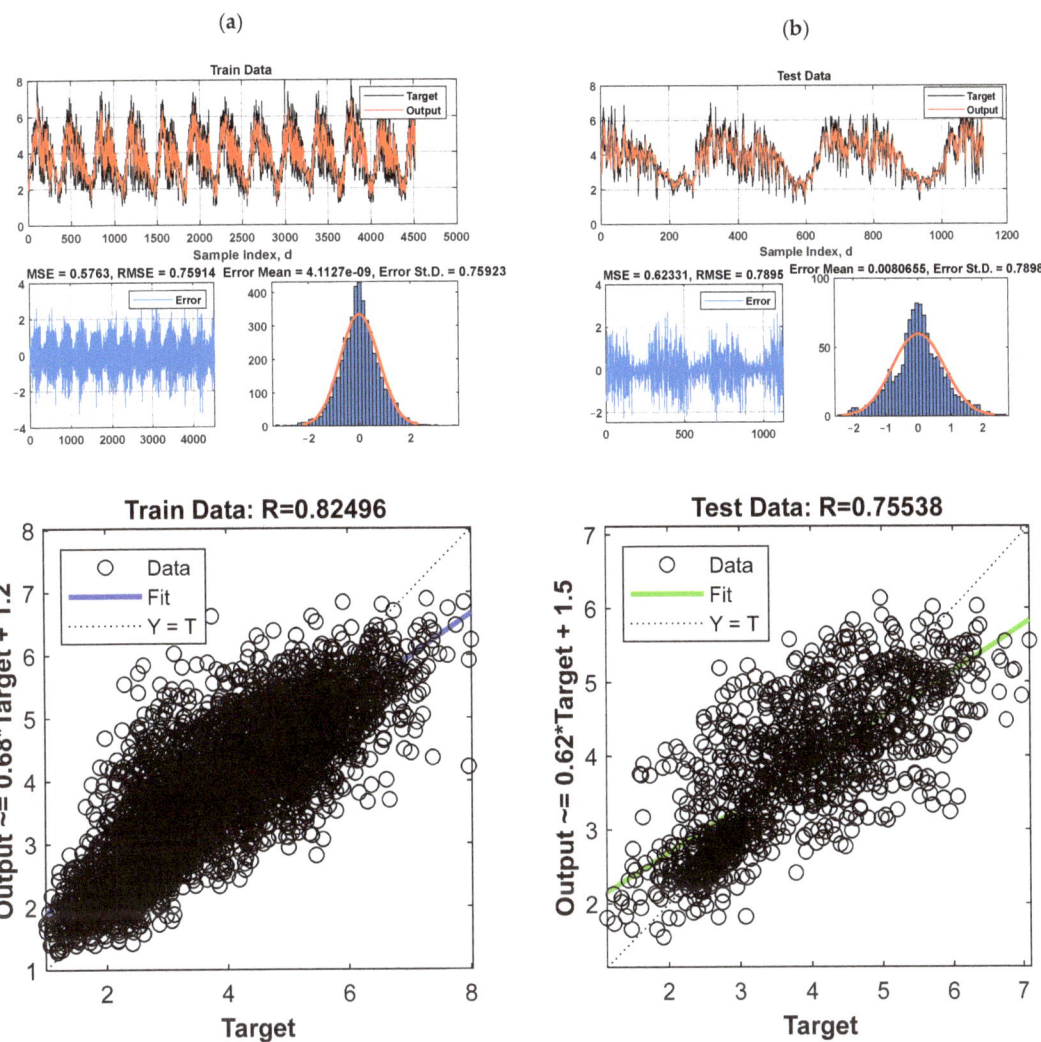

Figure 7. Scatter plots and regression plots for the values of FAO-56 PM-calculated ET_0 and ANFIS-forecasted ET_0 for the training (**a**) and testing (**b**) phases.

Table 9. Performance indices of the one-day-ahead ET_0 prediction models for the testing dataset.

Model	Performance Evaluation Indices					
	R	NS	IOA	NRMSE	MAE, mm/d	MAD, mm/d
ANFIS	0.755	0.567	0.858	0.207	2.710	0.308
Bi-LSTM	0.999	0.998	0.999	0.014	0.491	0.017
LSTM	0.698	0.429	0.833	0.237	3.047	0.334
SSR-LSTM	0.818	0.666	0.898	0.184	2.687	0.279

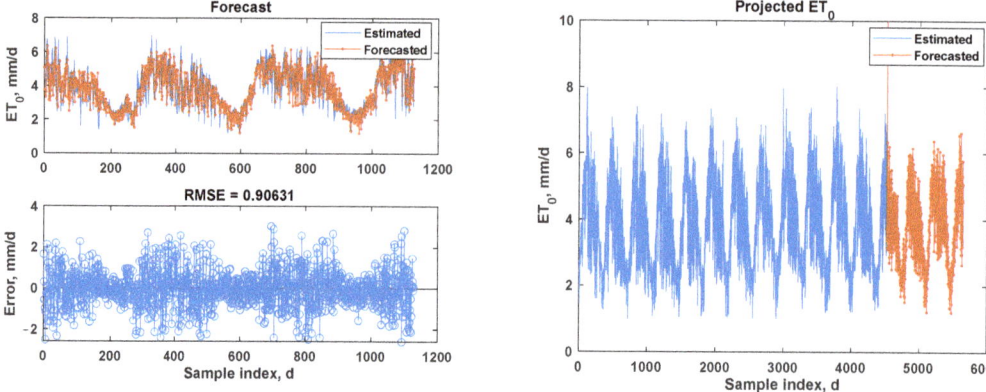

Figure 8. FAO-56 PM-calculated and LSTM-projected ET_0 values with error plots computed on the test dataset.

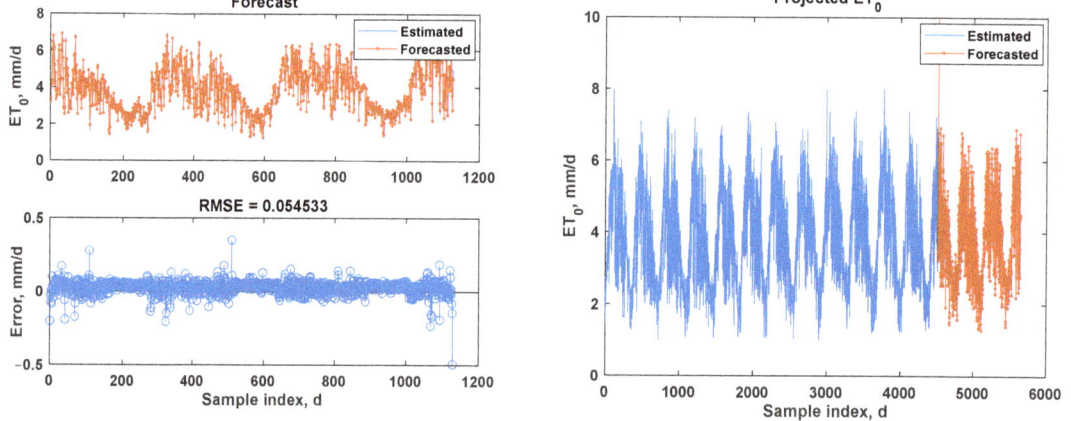

Figure 9. FAO-56 PM-calculated and Bi-LSTM-projected ET_0 values with error plots computed on the test dataset.

It is observed from Table 9 that the Bi-LSTM model provided a superior performance compared to the other models (SSR-LSTM, ANFIS, and LSTM) according to the statistical indices computed on the test dataset. It is noted that the prediction results with respect to the calculated performance indices did not demonstrate a considerable inconsistency. However, to reach a solid conclusion regarding the best-performing model, we applied the concept of Shannon's entropy to provide a performance ranking (Table 10). It is observed from Table 10 that Bi-LSTM appeared to be the best performer, while SSR-LSTM, ANFIS, and LSTM held the second, third, and fourth positions, respectively. Therefore, according to the performance results for one-step-ahead forecasting, the best-performing Bi-LSTM model was employed to provide multi-step (5 day)-ahead forecasting.

Table 10. Shannon's entropy-based model ranking for one-day-ahead ET_0 forecasts.

Model	Shannon's Entropy Value	Rank
Bi-LSTM	1.00	1
SSR-LSTM	0.30	2
ANFIS	0.27	3
LSTM	0.24	4

3.3. Multi-Step (5 Day-Ahead) Forecasting Using the Bi-LSTM Model

The forecasting performance of the developed ET_0 prediction model using the Bi-LSTM algorithm was evaluated using several statistical performance indices on the test dataset. However, to ascertain that no model over- or under-fitting occurred, we quantitatively evaluated the results obtained from both the training and validation phases. Five Bi-LSTM models were developed to forecast 1, 2, 3, 4, and 5 day-ahead ET_0 values. For all models, the selected time-lagged variables were served as inputs to the Bi-LSTM models. Table 11 presents the performances of the developed Bi-LSTM models on the training and validation datasets. It is evident from Table 11 that the absolute variances between the training and validation performances increased with the increase in the forecasting horizon. Overall, the training performances were satisfactory for all forecasting horizons.

Table 11. Training and validation performances of the developed Bi-LSTM models at Gazipur station.

Forecasting Horizon	Training RMSE, mm/d	Validation RMSE, mm/d
1 day	0.08	0.11
2 days	0.12	0.17
3 days	0.09	0.18
4 days	0.10	0.22
5 days	0.10	0.28

The trained and validated Bi-LSTM models were then used to forecast ET_0 values on the test dataset, which were selected from the entire dataset. Testing performances were assessed using several evaluation indices, as shown in Table 12. It is perceived from Table 12 that the forecasting horizon greatly influenced the forecasting accuracies. The accuracy decreased with the increase in the forecasting horizon as in the case of the training and validation performances. However, the overall performances of the Bi-LSTM model for all forecasting horizons showed particularly good performance, as indicated by the computed statistical performance evaluation indices. The performance of the developed models was also assessed using line graphs and error plots as shown in Figure 10.

Table 12. Multi-day-ahead forecasting performance of the Bi-LSTM model on the test dataset at Gazipur station.

Indices	Forecasting Horizon				
	1 Day	2 Days	3 Days	4 Days	5 Days
RMSE, mm/d	0.11	0.17	0.18	0.22	0.28
NRMSE	0.03	0.04	0.05	0.06	0.07
R	1.00	0.99	0.99	0.98	0.97
MAD, mm/d	0.03	0.04	0.04	0.06	0.08
MAE, mm/d	0.07	0.08	0.10	0.13	0.17
NS	0.99	0.98	0.98	0.97	0.95
IOA	1.00	0.99	0.99	0.99	0.99

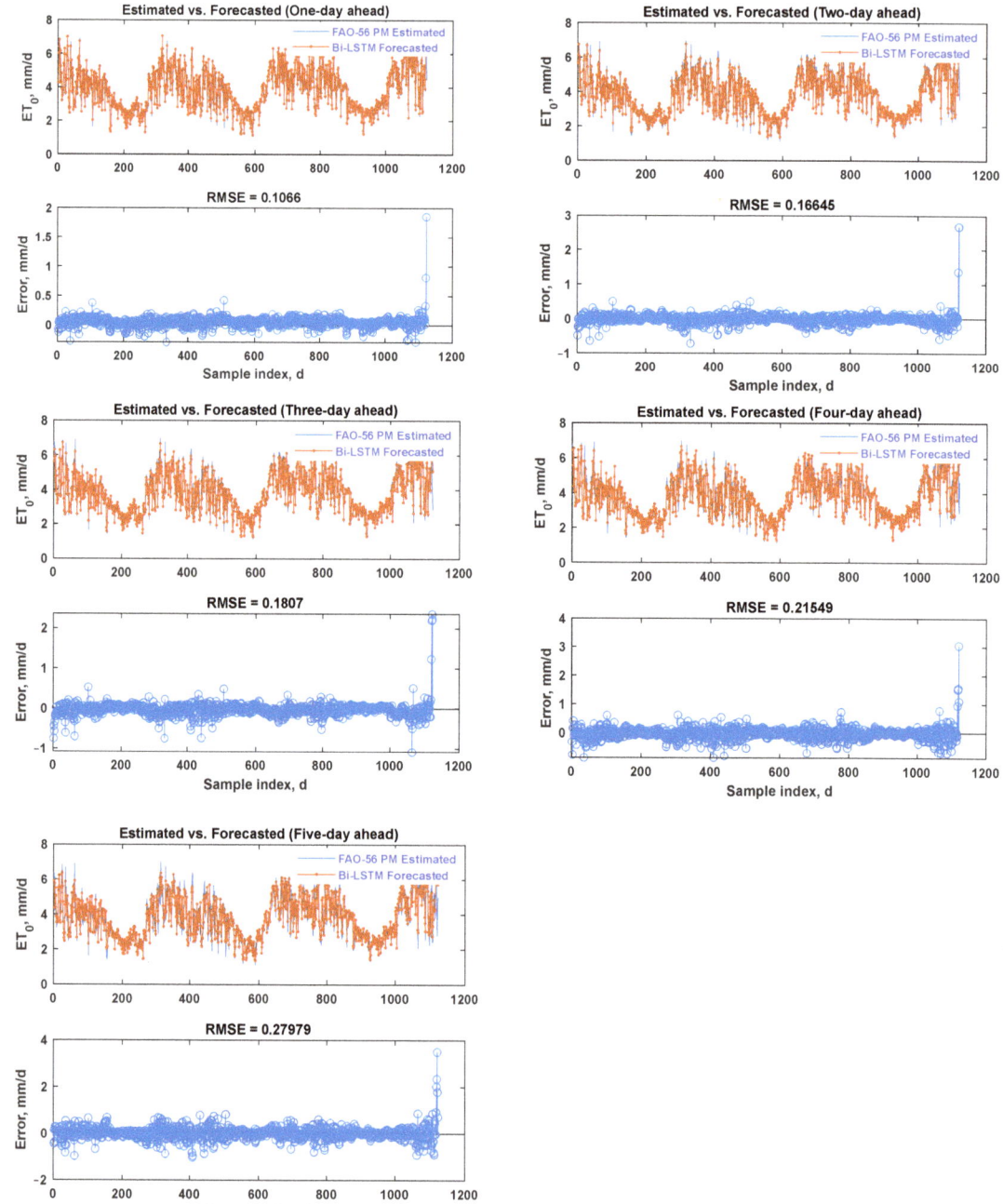

Figure 10. Line graph and error plots for 1, 2, 3, 4, and 5 day-ahead forecasting at Gazipur station.

It is observed from Table 12 that the Bi-LSTM model showed reasonably good performance, as evidenced by the computed performance indices. It produced lower values of cost indices (RMSE, NRMSE, MAD, and MAE) as well as higher values of benefit indices (R, NS, IOA). However, it is noted that the forecasting accuracy largely depended on the forecasting horizon, i.e., the sequence of forecasting accuracies are as follows:

1 day > 2 days > 3 days > 4 days > 5 days. This finding is in good agreement with the work of Yin et al. [128], who also stated that forecasting accuracy decreased with the increased forecasting horizon. Nevertheless, the forecasting accuracy of the Bi-LSTM model at 5 days ahead was also found acceptable for deep learning-based modeling of ET_0. Ferreira and da Cunha [82] also reported better deep learning model performance (CNN-LSTM) on the first and second forecasting days. Our findings using the Bi-LSTM model (RMSE = 0.11–0.28 mm/d) outperformed the CNN-LSTM model proposed by Ferreira and da Cunha [82] (mean RMSE values of 0.87 to 0.88 mm/d) with respect to RMSE criterion. Our proposed Bi-LSTM model performed better than the Bi-LSTM model proposed by Yin et al. [128] with respect to RMSE, R, and NS criteria. For instance, for 1 day-ahead forecasting, Yin et al. [128] obtained RMSE, R, and NS values of 0.159 mm/d, 0.992, and 0.988, respectively, whereas the values of RMSE, R, and NS in our study were found to be 0.11 mm/d, 1.00, and 0.99, respectively. Similarly, our proposed Bi-LSTM model outperformed the Bi-LSTM model presented by Yin et al. [128] for 4 day-ahead ET_0 forecasting. Moreover, our results also showed superior performance than the Bi-LSTM model results presented by Roy [81] in terms of R and IOA criteria for 1 day-ahead ET_0 forecasting. Roy [81] reported R and IOA values of 0.698 to 0.999 and 0.833 to 0.999, respectively, while the present study provided R and IOA values of 1.00 and 1.00, respectively. Therefore, it can be inferred that our proposed Bi-LSTM model is suitable for forecasting multi-step-ahead ET_0 values quite efficiently and precisely. It is noted that the Bi-LSTM model produced a slightly higher forecast error, especially at the end of the ET_0 time series. This comparatively big error at the end of the dataset may have arisen from higher values of ET_0 (outliers), which was not smoothed in order to evaluate the performance of the proposed modeling approaches for datasets containing outliers. Nevertheless, these values are still acceptable in the context of modeling ET_0 using machine learning approaches.

3.4. Generalization Capability of the Proposed Best ET_0 Prediction Models

The validation of the proposed best models (LSTM for daily predictions and Bi-LSTM for multi-step-ahead forecasts) was performed using data obtained from a new test station at which the models were not developed. The entire dataset of the test station (Ishurdi station) was split into three separate sets, each of which was employed to validate the models developed at the training station (Gazipur Sadar station). These three standalone datasets were fed into the LSTM and Bi-LSTM models to predict daily ET_0 values and forecast multi-day-ahead ET_0, respectively. The outputs from the models were weighed against the FAO-56 PM-computed ET_0 values using numerous statistical performance evaluation indices.

3.4.1. Generalization Capability of Proposed Best LSTM Model: Daily Prediction of ET_0

Table 13 summarizes the evaluation results for a variety of performance indices. The LSTM model exhibited a reasonably good performance at the test station data's three different sets (entire, first half, and second half). The computed performance indices indicated a satisfactory performance of the proposed LSTM model. It produced reasonably higher values of benefit indices (R, NS, and IOA) and lower values of the cost indices (RMSE, NRMSE, MAD, and MAE) for the entire, the first half, and the second half of the test station data. It is also observed that the first half of the dataset produced relatively better performance when compared to that of the second half and the entire dataset. Overall, the performance is satisfactory. On this basis, it is arguably concluded that the proposed LSTM model at Gazipur Sadar station can predict daily ET_0 values at Ishurdi station without developing a model at Ishurdi station. Additionally, performance data were presented using scatter and error plots, as illustrated in Figure 11, which depict the distribution of errors at individual data points.

Table 13. Performance of the LSTM model for predicting daily ET_0 values on the Ishurdi dataset.

Performance Indices	Entire Dataset	First Half Data	Second Half Data
RMSE, mm/d	0.65	0.49	0.84
NRMSE	0.18	0.13	0.23
R	0.87	0.92	0.83
MAD, mm/d	0.18	0.18	0.20
MAE, mm/d	0.44	0.39	0.52
NS	0.72	0.84	0.57
IOA	0.97	0.98	0.96

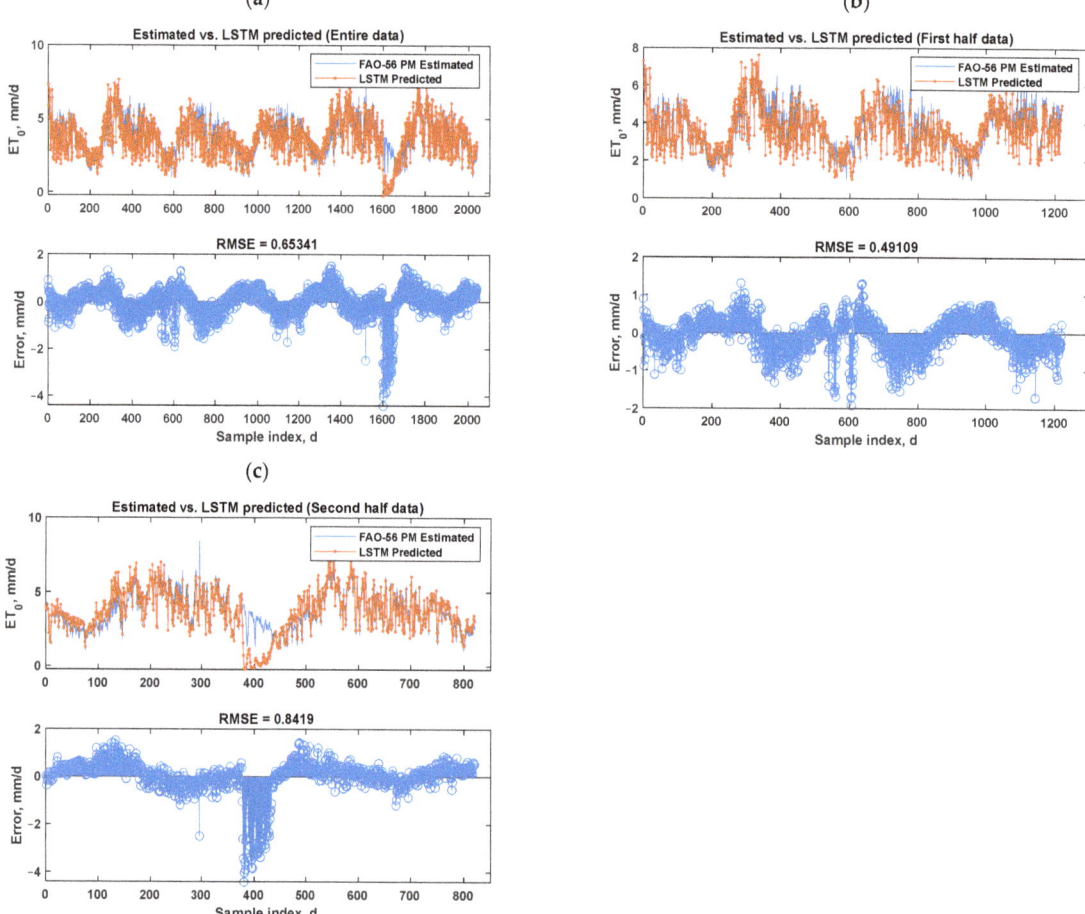

Figure 11. Line graph and error plots of FAO-56 PM-computed and LSTM-predicted daily ET_0 at Ishurdi station: (**a**) entire dataset, (**b**) first half of the dataset, and (**c**) second half of the dataset.

3.4.2. Generalization Capability of Proposed Best Bi-LSTM Model: Multi-Step (Multi-Day)-Ahead ET_0 Forecasting

For multi-step (multi-day)-ahead ET_0 forecasting, new Bi-LSTM models were developed because the nature of data was different. However, a similar model structure and parameters as in the case of Gazipur station were used. As a Bi-LSTM model performed better for one-step-ahead prediction at Gazipur station, the Bi-LSTM model was used to develop models for forecasting 1, 2, 3, and 5 day-ahead ET_0 values at the Ishurdi station.

For this, time-lagged information from the ET_0 time series was collected for 50 lags. The most significant input variables were determined by observing partial autocorrelation functions of the lagged time series, as shown in Figure 12.

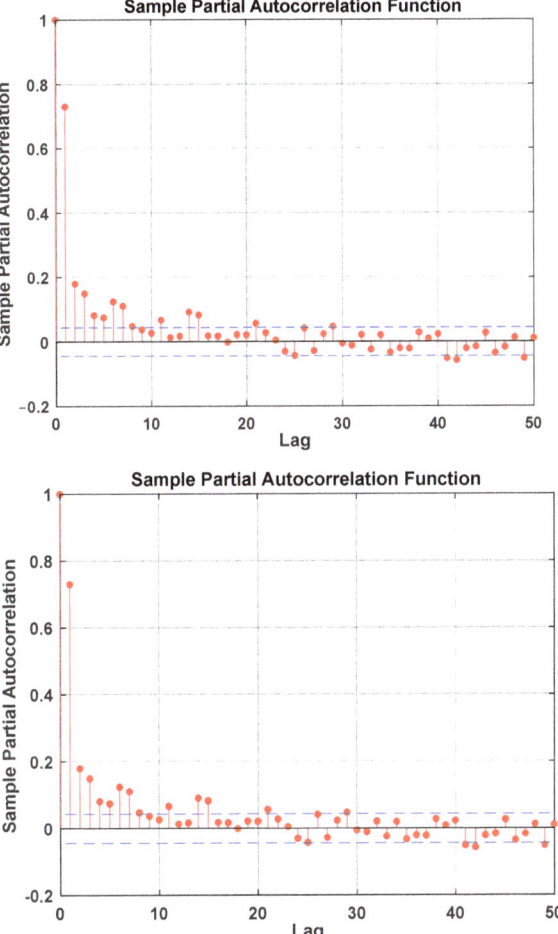

Figure 12. Sample partial autocorrelation functions of the lagged ET_0 time series at Ishurdi station.

Five Bi-LSTM models were developed to forecast 1, 2, 3, 4, and 5 day-ahead ET_0 forecasting. For all models, the selected time-lagged variables were served as inputs to the Bi-LSTM models. Table 14 presents the performances of the proposed Bi-LSTM models at the training and validation datasets. The absolute variances between the training and validation performances increased with the increase in the forecasting horizon. Overall, the training performances were satisfactory for all forecasting horizons.

Table 14. Training and validation performances of the developed Bi-LSTM models at Ishurdi station.

Forecasting Horizon	Training RMSE, mm/d	Validation RMSE, mm/d
1 day	0.09	0.12
2 days	0.10	0.17
3 days	0.11	0.29
4 days	0.12	0.56
5 days	0.10	0.73

The trained and validated Bi-LSTM models were then used to forecast ET_0 values on the test dataset, which were selected from the entire dataset. Testing performances were assessed using several statistical index values, as shown in Table 15. The forecasting horizon greatly influenced the forecasting accuracies. The accuracy decreased with the increase in the forecasting horizon, as in the case of the training and validation performances. However, the overall performances of the Bi-LSTM model for all forecasting horizons showed particularly good performance, as indicated by the computed statistical performance evaluation indices. Performance evaluation results of the developed models were also assessed with the aid of line graphs and error plots, as shown in Figure 13. The performance results illustrated in Figure 13 were in good agreement with the statistical index values presented in Table 15. As observed in the line graphs and error plots, forecasting accuracy largely depended on the forecasting horizon: forecasting accuracy decreased with increases in the forecasting horizon.

Table 15. Multi-day-ahead forecasting performance of the Bi-LSTM model on the test dataset at Ishurdi station.

Indices	Forecasting Horizon				
	1 Day	2 Days	3 Days	4 Days	5 Days
RMSE, mm/d	0.12	0.17	0.29	0.56	0.73
NRMSE	0.03	0.05	0.08	0.16	0.20
R	1.00	0.99	0.98	0.90	0.86
MAD, mm/d	0.04	0.05	0.08	0.14	0.24
MAE, mm/d	0.09	0.12	0.19	0.37	0.56
NS	0.99	0.98	0.95	0.81	0.69
IOA	1.00	1.00	0.99	0.95	0.91

It is observed from Figure that 1 day- and 2 day-ahead forecasting results were relatively better when compared to the results produced in three, four, and five day-ahead forecasts with respect to the RMSE criterion. A closer look at the line graphs also revealed the superiority of one day- and two day-ahead forecasts over the other three forecasting horizons and that Bi-LSTM models captured the lower values of the ET_0 time series quite accurately in comparison with the higher values for one day-, two day-, and three day-ahead forecasts. While producing acceptable results, the Bi-LSTM models followed similar trends for both the lower and higher values in the ET_0 time series in the case of the four day- and five day-ahead forecasts. It is also perceived from the line graphs that errors were relatively smaller at the end of the time series for the one day- and two day-ahead forecasts, while the Bi-LSTM models produced relatively higher errors at the end of the dataset for the three, four, and five day-ahead forecasts. Although performed differently at different forecast horizons, the Bi-LSTM model forecasts were quite accurate and closer to the FAO-56 PM-estimated ET_0 values. This is also evident from the statistical performance evaluation indices presented in Table 15. In particular, the NRMSE values of 0.03, 0.05, 0.08, 0.16, and 0.20 for the one, two, three, four, and five day-ahead forecasts, respectively, revealed the reasonable accurate forecasts of the proposed Bi-LSTM model. A model's performance is said to be excellent when the NRMSE value is lower than 0.1, good when the NRMSE value is between 0.1 and 0.2, fair when the NRMSE value is between 0.2 and 0.3, poor when the NRMSE is greater than 0.3 [129,130].

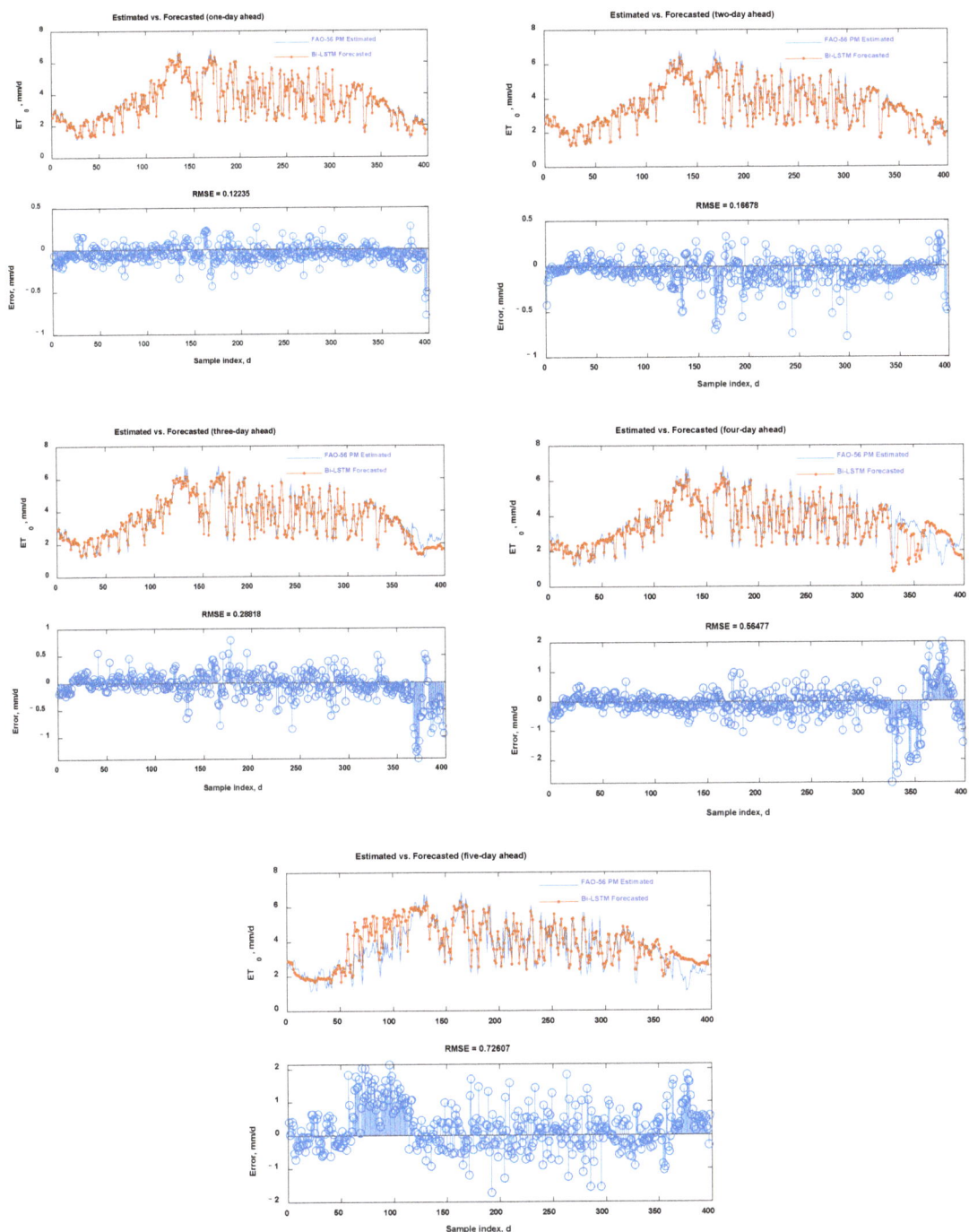

Figure 13. Line graph and error plots for 1, 2, 3, 4, and 5 day-ahead forecasting at Ishurdi station.

Agronomy **2022**, *12*, 594

4. Conclusions

Precise prediction and forecasting of ET_0 have been a critical and emerging first step for developing a justifiable and effective irrigation scheduling plan. This research provided a selection of the best machine and deep learning algorithms to develop robust prediction and forecasting tools for daily and multi-step (5 day)-ahead ET_0 prediction and forecasting, respectively. The selection results indicated the superiority of the LSTM model for daily ET_0 predictions, whereas for multi-step-ahead forecasting, the Bi-LSTM model provided superior performance. For daily ET_0 prediction, a number of meteorological variables were used as inputs to the model, whereas the computed ET_0 values were used as outputs from the model. For multi-step (5 day)-ahead forecasting, the appropriate daily time-lagged ET_0 values were used as inputs to the Bi-LSTM model, and the outputs from the Bi-LSTM model were the one, two, three, four, and five step-ahead ET_0 values. On the basis of the results of the one-step-ahead prediction performed previously for model selection, we found that the Bi-LSTM model was further employed to provide multi-step (5 day)-ahead forecasting. Results revealed the suitability of the Bi-LSTM model in predicting multi-step-ahead ET_0 values.

In a further step, best models for daily prediction (LSTM) and multi-step-ahead forecasting (Bi-LSTM) were used to generalize the ET_0 values for the data obtained from a different weather station, for which the models were neither trained nor validated. More specifically, the LSTM network was used to generalize the daily ET_0 predictions in a nearby meteorological station without developing a model for that station. On the other hand, the Bi-LSTM model was developed for the Ishurdi station to forecast 1, 2, 3, 4, and 5 day-ahead ET_0 forecasting. The relatively low errors obtained by the LSTM and Bi-LSTM approaches led to a good fit of the models in predicting daily ET_0 values and forecasting multi-step-ahead ET_0 values. This can be expected to be very useful in the practice of irrigation water management, for which ET_0 is an important parameter.

Author Contributions: Conceptualization, supervision, methodology, formal analysis, writing—original draft preparation, writing—review and editing, D.K.R., T.K.S., S.S.A.K.; data curation, project administration, investigation, T.G., M.A.M. (Md Abdul Muktadir); conceptualization, funding acquisition, writing—review and editing, writing—original draft preparation, H.M.A.-G., A.A., A.Z.D., A.A.E.-S., M.A.M. (Mohamed A. Mattar). All authors have read and agreed to the published version of the manuscript.

Funding: This research was financially supported by the Vice Deanship of Research Chairs at King Saud University.

Institutional Review Board Statement: Not applicable.

Informed Consent Statement: Not applicable.

Data Availability Statement: The data presented in this study are available on request from the corresponding author.

Acknowledgments: This work was financially supported by the Vice Deanship of Research Chairs at King Saud University.

Conflicts of Interest: The authors declare no conflict of interest.

References

1. Liu, S.M.; Xu, Z.W.; Zhu, Z.L.; Jia, Z.Z.; Zhu, M.J. Measurements of evapotranspiration from eddy-covariance systems and large aperture scintillometers in the Hai River Basin, China. *J. Hydrol.* **2013**, *487*, 24–38. [CrossRef]
2. Kisi, O. Modeling reference evapotranspiration using three different heuristic regression approaches. *Agric. Water Manag.* **2016**, *169*, 162–172.
3. Kool, D.; Agam, N.; Lazarovitch, N.; Heitman, J.L.; Sauer, T.J.; Ben-Gal, A. A review of approaches for evapotranspiration partitioning. *Agric. For. Meteorol.* **2014**, *184*, 56–70. [CrossRef]
4. Martí, P.; González-Altozano, P.; López-Urrea, R.; Mancha, L.A.; Shiri, J. Modeling reference evapotranspiration with calculated targets. Assessment and implications. *Agric. Water Manag.* **2015**, *149*, 81–90. [CrossRef]

5. Zhang, B.; Liu, Y.; Xu, D.; Zhao, N.; Lei, B.; Rosa, R.D.; Paredes, P.; Paço, T.A.; Pereira, L.S. The dual crop coefficient approach to estimate and partitioning evapotranspiration of the winter wheat–summer maize crop sequence in North China Plain. *Irrig. Sci.* **2013**, *31*, 1303–1316. [CrossRef]
6. Allen, R.G.; Pereira, L.S.; Raes, D.; Smith, M. *Crop Evapotranspiration-Guidelines for Computing crop Water Requirements*; FAO: Rome, Italy, 1998.
7. Ding, R.; Kang, S.; Zhang, Y.; Hao, X.; Tong, L.; Du, T. Partitioning evapotranspiration into soil evaporation and transpiration using a modified dual crop coefficient model in irrigated maize field with ground-mulching. *Agric. Water Manag.* **2013**, *127*, 85–96. [CrossRef]
8. Landeras, G.; Ortiz-Barredo, A.; López, J.J. Comparison of artificial neural network models and empirical and semi-empirical equations for daily reference evapotranspiration estimation in the Basque Country (Northern Spain). *Agric. Water Manag.* **2008**, *95*, 553–565. [CrossRef]
9. Ferreira, L.B.; da Cunha, F.F.; Fernandes Filho, E.I. Exploring machine learning and multi-task learning to estimate meteorological data and reference evapotranspiration across Brazil. *Agric. Water Manag.* **2022**, *259*, 107281. [CrossRef]
10. Kelley, J.; Pardyjak, E.R. Using neural networks to estimate site-specific crop evapotranspiration with low-cost sensors. *Agronomy* **2019**, *9*, 108. [CrossRef]
11. Yassin, M.A.; Alazba, A.A.; Mattar, M.A. Modelling daily evapotranspiration using artificial neural networks under hyper arid conditions. *Pak. J. Agric. Sci.* **2016**, *53*, 695–712.
12. Yassin, M.A.; Alazba, A.A.; Mattar, M.A. Artificial neural networks versus gene expression programming forestimating reference evapotranspiration in arid climate. *Agric. Water Manag.* **2016**, *163*, 110–124. [CrossRef]
13. Kumar, M.; Raghuwanshi, N.S.; Singh, R.; Wallender, W.W.; Pruitt, W.O. Estimating evapotranspiration using artificial neural network. *J. Irrig. Drain. Eng.* **2002**, *128*, 224–233. [CrossRef]
14. Feng, Y.; Cui, N.; Gong, D.; Zhang, Q.; Zhao, L. Evaluation of random forests and generalized regression neural networks for daily reference evapotranspiration modelling. *Agric. Water Manag.* **2017**, *193*, 163–173. [CrossRef]
15. Feng, Y.; Peng, Y.; Cui, N.; Gong, D.; Zhang, K. Modeling reference evapotranspiration using extreme learning machine and generalized regression neural network only with temperature data. *Comput. Electron. Agric.* **2017**, *136*, 71–78. [CrossRef]
16. Gocić, M.; Arab Amiri, M. Reference evapotranspiration prediction using neural networks and optimum time lags. *Water Resour. Manag.* **2021**, *35*, 1913–1926. [CrossRef]
17. Doğan, E. Reference evapotranspiration estimation using adaptive neuro-fuzzy inference systems. *Irrig. Drain.* **2009**, *58*, 617–628. [CrossRef]
18. Dou, X.; Yang, Y. Evapotranspiration estimation using four different machine learning approaches in different terrestrial ecosystems. *Comput. Electron. Agric.* **2018**, *148*, 95–106. [CrossRef]
19. Gavili, S.; Sanikhani, H.; Kisi, O.; Mahmoudi, M.H. Evaluation of several soft computing methods in monthly evapotranspiration modelling. *Meteorol. Appl.* **2018**, *25*, 128–138. [CrossRef]
20. Roy, D.K.; Lal, A.; Sarker, K.K.; Saha, K.K.; Datta, B. Optimization algorithms as training approaches for prediction of reference evapotranspiration using adaptive neuro fuzzy inference system. *Agric. Water Manag.* **2021**, *255*, 107003. [CrossRef]
21. Roy, D.K.; Barzegar, R.; Quilty, J.; Adamowski, J. Using ensembles of adaptive neuro-fuzzy inference system and optimization algorithms to predict reference evapotranspiration in subtropical climatic zones. *J. Hydrol.* **2020**, *591*, 125509. [CrossRef]
22. Shiri, J.; Marti, P.; Nazemi, A.H.; Sadraddini, A.A.; Kisi, O.; Landeras, G.; Fakheri Fard, A. Local vs. external training of neuro-fuzzy and neural networks models for estimating reference evapotranspiration assessed through k-fold testing. *Hydrol. Res.* **2013**, *46*, 72–88. [CrossRef]
23. Tabari, H.; Kisi, O.; Ezani, A.; Hosseinzadeh, T.P. SVM, ANFIS, regression and climate based models for reference evapotranspiration modeling using limited climatic data in a semi-arid highland environment. *J. Hydrol.* **2012**, *444–445*, 78–89. [CrossRef]
24. Huang, G.; Wu, L.; Ma, X.; Zhang, W.; Fan, J.; Yu, X.; Zeng, W.; Zhou, H. Evaluation of CatBoost method for prediction of reference evapotranspiration in humid regions. *J. Hydrol.* **2019**, *574*, 1029–1041. [CrossRef]
25. Wu, T.; Zhang, W.; Jiao, X.; Guo, W.; Alhaj Hamoud, Y. Evaluation of stacking and blending ensemble learning methods for estimating daily reference evapotranspiration. *Comput. Electron. Agric.* **2021**, *184*, 106039. [CrossRef]
26. Zhang, Y.; Zhao, Z.; Zheng, J. CatBoost: A new approach for estimating daily reference crop evapotranspiration in arid and semi-arid regions of Northern China. *J. Hydrol.* **2020**, *588*, 125087. [CrossRef]
27. Lu, X.; Fan, J.; Wu, L.; Dong, J. Forecasting multi-step ahead monthly reference evapotranspiration using hybrid extreme gradient boosting with grey wolf optimization algorithm. *Comput. Model. Eng. Sci.* **2020**, *125*, 699–723.
28. Abdullah, S.S.; Malek, M.A.; Abdullah, N.S.; Kisi, O.; Yap, K.S. Extreme Learning Machines: A new approach for prediction of reference evapotranspiration. *J. Hydrol.* **2015**, *527*, 184–195. [CrossRef]
29. Feng, Y.; Cui, N.; Zhao, L.; Hu, X.; Gong, D. Comparison of ELM, GANN, WNN and empirical models for estimating reference evapotranspiration in humid region of Southwest China. *J. Hydrol.* **2016**, *536*, 376–383. [CrossRef]
30. Wu, L.; Zhou, H.; Ma, X.; Fan, J.; Zhang, F. Daily reference evapotranspiration prediction based on hybridized extreme learning machine model with bio-inspired optimization algorithms: Application in contrasting climates of China. *J. Hydrol.* **2019**, *577*, 123960. [CrossRef]

31. Yin, Z.; Feng, Q.; Yang, L.; Deo, R.C.; Wen, X.; Si, J.; Xiao, S. Future projection with an extreme-learning machine and support vector regression of reference evapotranspiration in a mountainous inland watershed in north-west China. *Water* **2017**, *9*, 880. [CrossRef]

32. Ahmadi, F.; Mehdizadeh, S.; Mohammadi, B.; Pham, Q.B.; Doan, T.N.C.; Vo, N.D. Application of an artificial intelligence technique enhanced with intelligent water drops for monthly reference evapotranspiration estimation. *Agric. Water Manag.* **2021**, *244*, 106622. [CrossRef]

33. Ferreira, L.B.; da Cunha, F.F.; de Oliveira, R.A.; Fernandes Filho, E.I. Estimation of reference evapotranspiration in Brazil with limited meteorological data using ANN and SVM—A new approach. *J. Hydrol.* **2019**, *572*, 556–570. [CrossRef]

34. Torres, A.F.; Walker, W.R.; McKee, M. Forecasting daily potential evapotranspiration using machine learning and limited climatic data . *Agric. Water Manag.* **2011**, *98*, 553–562. [CrossRef]

35. Gocić, M.; Motamedi, S.; Shamshirband, S.; Petković, D.; Ch, S.; Hashim, R.; Arif, M. Soft computing approaches for forecasting reference evapotranspiration. *Comput. Electron. Agric.* **2015**, *113*, 164–173. [CrossRef]

36. Karbasi, M. Forecasting of Multi-Step Ahead Reference Evapotranspiration Using Wavelet- Gaussian Process Regression Model. *Water Resour. Manag.* **2018**, *32*, 1035–1052. [CrossRef]

37. Kisi, O.; Keshtegar, B.; Zounemat-Kermani, M.; Heddam, S.; Trung, N.T. Modeling reference evapotranspiration using a novel regression-based method: Radial basis M5 model tree. *Theor. Appl. Climatol.* **2021**, *145*, 639–659. [CrossRef]

38. Mattar, M.A. Using gene expression programming in monthly reference evapotranspiration modeling: A case study in Egypt. *Agric. Water Manag.* **2018**, *198*, 28–38. [CrossRef]

39. Mattar, M.A.; Alazba, A.A. GEP and MLR approaches for the prediction of reference evapotranspiration. *Neural Comput. Appl.* **2019**, *31*, 5843–5855. [CrossRef]

40. Alazba, A.A.; Yassin, M.A.; Mattar, M.A. Modeling daily evapotranspiration in hyper-arid environment using gene expression programming. *Arab J. Geosci.* **2016**, *9*, 202. [CrossRef]

41. Yassin, M.A.; Alazba, A.A.; Mattar, M.A. Comparison Between Gene Expression Programming and Traditional Models for Estimating Evapotranspiration under Hyper Arid Conditions. *Water Resour.* **2016**, *43*, 412–427. [CrossRef]

42. Shiri, J.; Sadraddini, A.A.; Nazemi, A.H.; Kişi, Ö.; Landeras, G.; Fakheri Fard, A.; Marti, P. Generalizability of Gene Expression Programming-based approaches for estimating daily reference evapotranspiration in coastal stations of Iran. *J. Hydrol.* **2014**, *508*, 1–11. [CrossRef]

43. Shiri, J.; Kişi, Ö.; Landeras, G.; López, J.J.; Nazemi, A.H.; Stuyt, L.C.P.M. Daily reference evapotranspiration modeling by using genetic programming approach in the Basque Country (Northern Spain). *J. Hydrol.* **2012**, *414–415*, 302–316. [CrossRef]

44. Wang, S.; Lian, J.; Peng, Y.; Hu, B.; Chen, H. Generalized reference evapotranspiration models with limited climatic data based on random forest and gene expression programming in Guangxi, China. *Agric. Water Manag.* **2019**, *221*, 220–230. [CrossRef]

45. Wang, S.; Fu, Z.-Y.; Chen, H.-S.; Nie, Y.-P.; Wang, K.L. Modeling daily reference ET in the karst area of northwest Guangxi (China) using gene expression programming (GEP) and artificial neural network (ANN). *Theor. Appl. Climatol.* **2016**, *126*, 493–504. [CrossRef]

46. Roy, D.K.; Saha, K.K.; Kamruzzaman, M.; Biswas, S.K.; Hossain, M.A. Hierarchical fuzzy systems integrated with particle swarm optimization for daily reference evapotranspiration prediction: A novel approach. *Water Resour. Manag.* **2021**, *35*, 5383–5407. [CrossRef]

47. Yan, S.; Wu, L.; Fan, J.; Zhang, F.; Zou, Y.; Wu, Y. A novel hybrid WOA-XGB model for estimating daily reference evapotranspiration using local and external meteorological data: Applications in arid and humid regions of China. *Agric. Water Manag.* **2021**, *244*, 106594. [CrossRef]

48. Başağaoğlu, H.; Chakraborty, D.; Winterle, J. Reliable evapotranspiration predictions with a probabilistic machine learning framework. *Water* **2021**, *13*, 557. [CrossRef]

49. Fu, T.; Li, X.; Jia, R.; Feng, L. A novel integrated method based on a machine learning model for estimating evapotranspiration in dryland. *J. Hydrol.* **2021**, *603*, 126881. [CrossRef]

50. Chia, M.Y.; Huang, Y.F.; Koo, C.H. Resolving data-hungry nature of machine learning reference evapotranspiration estimating models using inter-model ensembles with various data management schemes. *Agric. Water Manag.* **2022**, *261*, 107343. [CrossRef]

51. Pasqualotto, N.; D'Urso, G.; Bolognesi, S.F.; Belfiore, O.R.; Van Wittenberghe, S.; Delegido, J.; Pezzola, A.; Winschel, C.; Moreno, J. Retrieval of evapotranspiration from sentinel-2: Comparison of vegetation indices, semi-empirical models and SNAP biophysical processor approach. *Agronomy* **2019**, *9*, 663. [CrossRef]

52. Rodrigues, G.C.; Braga, R.P. A Simple application for computing reference evapotranspiration with various levels of data availability—ETo tool. *Agronomy* **2021**, *11*, 2203. [CrossRef]

53. Rodrigues, G.C.; Braga, R.P. Estimation of daily reference evapotranspiration from NASA POWER reanalysis products in a hot summer mediterranean climate. *Agronomy* **2021**, *11*, 2077. [CrossRef]

54. Zheng, S.; Ni, K.; Ji, L.; Zhao, C.; Chai, H.; Yi, X.; He, W.; Ruan, J. Estimation of evapotranspiration and crop coefficient of rain-fed tea plants under a subtropical climate. *Agronomy* **2021**, *11*, 2332. [CrossRef]

55. Bellido-Jiménez, J.A.; Estévez, J.; García-Marín, A.P. New machine learning approaches to improve reference evapotranspiration estimates using intra-daily temperature-based variables in a semi-arid region of Spain. *Agric. Water Manag.* **2021**, *245*, 106558. [CrossRef]

56. Vásquez, C.; Célleri, R.; Córdova, M.; Carrillo-Rojas, G. Improving reference evapotranspiration (ETo) calculation under limited data conditions in the high Tropical Andes. *Agric. Water Manag.* **2022**, *262*, 107439. [CrossRef]
57. Nourani, V.; Elkiran, G.; Abdullahi, J. Multi-step ahead modeling of reference evapotranspiration using a multi-model approach. *J. Hydrol.* **2020**, *581*, 124434. [CrossRef]
58. Tien Bui, D.; Hoang, N.-D.; Martínez-Álvarez, F.; Ngo, P.-T.T.; Hoa, P.V.; Pham, T.D.; Samui, P.; Costache, R. A novel deep learning neural network approach for predicting flash flood susceptibility: A case study at a high frequency tropical storm area. *Sci. Total Environ.* **2020**, *701*, 134413. [CrossRef]
59. Xu, H.; Zhou, J.; Asteris, P.G.; Jahed Armaghani, D.; Tahir, M.M. Supervised machine learning techniques to the prediction of tunnel boring machine penetration rate. *Appl. Sci.* **2019**, *9*, 3715. [CrossRef]
60. Yang, H.-F.; Chen, Y.-P.P. Hybrid deep learning and empirical mode decomposition model for time series applications. *Expert Syst. Appl.* **2019**, *120*, 128–138. [CrossRef]
61. Fang, W.; Zhong, B.; Zhao, N.; Love, P.E.D.; Luo, H.; Xue, J.; Xu, S. A deep learning-based approach for mitigating falls from height with computer vision: Convolutional neural network. *Adv. Eng. Inform.* **2019**, *39*, 170–177. [CrossRef]
62. Fan, L.; Zhang, T.; Zhao, X.; Wang, H.; Zheng, M. Deep topology network: A framework based on feedback adjustment learning rate for image classification. *Adv. Eng. Inform.* **2019**, *42*, 100935. [CrossRef]
63. Cummins, N.; Baird, A.; Schuller, B.W. Speech analysis for health: Current state-of-the-art and the increasing impact of deep learning. *Methods* **2018**, *151*, 41–54. [CrossRef] [PubMed]
64. Plappert, M.; Mandery, C.; Asfour, T. Learning a bidirectional mapping between human whole-body motion and natural language using deep recurrent neural networks. *Rob. Auton. Syst.* **2018**, *109*, 13–26. [CrossRef]
65. Bowes, B.D.; Sadler, J.M.; Morsy, M.M.; Behl, M.; Goodal, J.L. Forecasting groundwater table in a flood prone coastal city with long short-term memory and recurrent neural networks. *Water* **2019**, *11*, 1–38. [CrossRef]
66. Supreetha, B.S.; Shenoy, N.; Nayak, P. Lion algorithm-optimized long short-term memory network for groundwater level forecasting in Udupi District, India. *Appl. Comput. Intell. Soft Comput.* **2020**, *2020*, 8685724. [CrossRef]
67. Barzegar, R.; Aalami, M.T.; Adamowski, J. Short-term water quality variable prediction using a hybrid CNN–LSTM deep learning model. *Stoch. Environ. Res. Risk Assess.* **2020**, *34*, 415–433. [CrossRef]
68. Chang, F.-J.; Chang, L.-C.; Huang, C.-W.; Kao, I.-F. Prediction of monthly regional groundwater levels through hybrid soft-computing techniques. *J. Hydrol.* **2016**, *541*, 965–976. [CrossRef]
69. Daliakopoulos, I.N.; Coulibaly, P.; Tsanis, I.K. Groundwater level forecasting using artificial neural networks. *J. Hydrol.* **2005**, *309*, 229–240. [CrossRef]
70. Guzman, S.M.; Paz, J.O.; Tagert, M.L.M. The use of NARX neural networks to forecast daily groundwater levels. *Water Resour. Manag.* **2017**, *31*, 1591–1603. [CrossRef]
71. Bengio, Y.; Simard, P.; Frasconi, P. Learning long-term dependencies with gradient descent is difficult. *IEEE Trans. Neural Netw. Learn. Syst.* **1994**, *5*, 157–166. [CrossRef]
72. Fischer, T.; Krauss, C. Deep learning with long short-term memory networks for financial market predictions. *Eur. J. Oper. Res.* **2018**, *270*, 654–669. [CrossRef]
73. Zhao, Z.; Chen, W.; Wu, X.; Chen, P.C.Y.; Liu, J. LSTM network: A deep learning approach for short-term traffic forecast. *IET Intell. Transp. Syst.* **2017**, *11*, 68–75. [CrossRef]
74. Hu, C.; Wu, Q.; Li, H.; Jian, S.; Li, N.; Lou, Z. Deep learning with a long short-term memory networks approach for rainfall-runoff simulation. *Water* **2018**, *10*, 1543. [CrossRef]
75. Liang, C.; Li, H.; Lei, M.; Du, Q. Dongting lake water level forecast and its relationship with the three Gorges dam based on a long short-term memory network. *Water* **2018**, *10*, 1389. [CrossRef]
76. Tian, Y.; Xu, Y.-P.; Yang, Z.; Wang, G.; Zhu, Q. Integration of a parsimonious hydrological model with recurrent neural networks for improved streamflow forecasting. *Water* **2018**, *10*, 1655. [CrossRef]
77. Zhang, J.; Zhu, Y.; Zhang, X.; Ye, M.; Yang, J. Developing a Long Short-Term Memory (LSTM) based model for predicting water table depth in agricultural areas. *J. Hydrol.* **2018**, *561*, 918–929. [CrossRef]
78. Majhi, B.; Naidu, D.; Mishra, A.P.; Satapathy, S.C. Improved prediction of daily pan evaporation using Deep-LSTM model. *Neural Comput. Appl.* **2020**, *32*, 7823–7838. [CrossRef]
79. Hu, X.; Shi, L.; Lin, G.; Lin, L. Comparison of physical-based, data-driven and hybrid modeling approaches for evapotranspiration estimation. *J. Hydrol.* **2021**, *601*, 126592. [CrossRef]
80. Ferreira, L.B.; da Cunha, F.F. New approach to estimate daily reference evapotranspiration based on hourly temperature and relative humidity using machine learning and deep learning. *Agric. Water Manag.* **2020**, *234*, 106113. [CrossRef]
81. Roy, D.K. Long short-term memory networks to predict one-step ahead reference evapotranspiration in a subtropical climatic zone. *Environ. Process.* **2021**, *8*, 911–941. [CrossRef]
82. Ferreira, L.B.; da Cunha, F.F. Multi-step ahead forecasting of daily reference evapotranspiration using deep learning. *Comput. Electron. Agric.* **2020**, *178*, 105728. [CrossRef]
83. Ahmed, A.A.; Deo, R.C.; Feng, Q.; Ghahramani, A.; Raj, N.; Yin, Z.; Yang, L. Hybrid deep learning method for a week-ahead evapotranspiration forecasting. *Stoch. Environ. Res. Risk Assess.* **2021**, *36*, 831–849. [CrossRef]
84. Saggi, M.K.; Jain, S. Reference evapotranspiration estimation and modeling of the Punjab Northern India using deep learning. *Comput. Electron. Agric.* **2019**, *156*, 387–398. [CrossRef]

85. Granata, F.; Di Nunno, F. Forecasting evapotranspiration in different climates using ensembles of recurrent neural networks. *Agric. Water Manag.* **2021**, *255*, 107040. [CrossRef]
86. Maroufpoor, S.; Bozorg-Haddad, O.; Maroufpoor, E. Reference evapotranspiration estimating based on optimal input combination and hybrid artificial intelligent model: Hybridization of artificial neural network with grey wolf optimizer algorithm. *J. Hydrol.* **2020**, *588*, 125060. [CrossRef]
87. Yu, H.; Wen, X.; Li, B.; Yang, Z.; Wu, M.; Ma, Y. Uncertainty analysis of artificial intelligence modeling daily reference evapotranspiration in the northwest end of China. *Comput. Electron. Agric.* **2020**, *176*, 105653. [CrossRef]
88. Estévez, J.; García-Marín, A.P.; Morábito, J.A.; Cavagnaro, M. Quality assurance procedures for validating meteorological input variables of reference evapotranspiration in mendoza province (Argentina). *Agric. Water Manag.* **2016**, *172*, 96–109. [CrossRef]
89. Allen, R.G.; Pereira, L.S.; Raes, D. *Evapotranspiracion ´Del Cultivo. Guías Para la Determinacion ´ de Los Requerimientos de Agua de Los Cultivos (Technical Report)*; FAO: Roma, Italia, 2006.
90. Shiri, J.; Nazemi, A.H.; Sadraddini, A.A.; Landeras, G.; Kişi, O.; Fakheri Fard, A.; Marti, P. Comparison of heuristic and empirical approaches for estimating reference evapotranspiration from limited inputs in Iran. *Comput. Electron. Agric.* **2014**, *108*, 230–241. [CrossRef]
91. Hochreiter, S.; Schmidhuber, J. Long short-term memory. *Neural Comput.* **1997**, *9*, 1735–1780. [CrossRef]
92. Yuan, X.; Chen, C.; Lei, X.; Yuan, Y.; Muhammad Adnan, R. Monthly runoff forecasting based on LSTM–ALO model. *Stoch. Environ. Res. Risk Assess.* **2018**, *32*, 2199–2212. [CrossRef]
93. Jang, J.-S.R.; Sun, C.T.; Mizutani, E. *Neuro-Fuzzy and Soft Computing: A Computational Approach to Learning and Machine Intelligence*; Prentice Hall: Upper Saddle River, NJ, USA, 1997.
94. Sugeno, M.; Yasukawa, T. A fuzzy-logic-based approach to qualitative modeling. *IEEE Trans. Fuzzy Syst.* **1993**, *1*, 7. [CrossRef]
95. Takagi, T.; Sugeno, M. Fuzzy identification of systems and its applications to modeling and control. *IEEE Trans. Syst. Man. Cybern.* **1985**, *SMC-15*, 116–132. [CrossRef]
96. Bezdek, J.C.; Ehrlich, R.; Full, W. FCM: The fuzzy c-means clustering algorithm. *Comput. Geosci.* **1984**, *10*, 191–203. [CrossRef]
97. *MATLAB Version R2019b*; The MathWorks, Inc.: Natick, MA, USA, 2019.
98. Jang, J.-S.R. ANFIS: Adaptive-network-based fuzzy inference system. *IEEE Trans. Syst. Man. Cybern.* **1993**, *23*, 665–685. [CrossRef]
99. Rasmussen, C.E.; Williams, C.K. *Gaussian Process for Machine Learning*; The MIT Press: Cambridge, MA, USA, 2006.
100. Bishop, C. *Pattern Recognition and Machine Learning*; Springer: New York, NY, USA, 2006.
101. Quinlan, J.R. Learning with continuous classes. In Proceedings of the Australian Joint Conference on Artificial Intelligence, Hobart, Australia, 16–18 November 1992; pp. 343–348.
102. Wang, Y.; Witten, I. Induction of model trees for predicting continuous classes. *Work. Pap.* **1996**, *96*, 23.
103. Bhattacharya, B.; Solomatine, D.P. Neural networks and M5 model trees in modelling water level–discharge relationship. *Neurocomputing* **2005**, *63*, 381–396. [CrossRef]
104. Solomatine, D.P.; Dulal, K. Model trees as an alternative to neural networks in rainfall-runoff modelling. *Hydrol. Sci. J.* **2003**, *48*, 399–411. [CrossRef]
105. Solomatine, D.P.; Yunpeng, X. M5 model trees and neural networks: Application to flood forecasting in the upper reach of the Huai River in China. *J. Hydrol. Eng.* **2004**, *9*, 491–501. [CrossRef]
106. Jekabsons, G. *M5PrimeLab: M5' Regression Tree, Model Tree, and Tree Ensemble Toolbox for Matlab/Octave*; The MathWorks, Inc.: Natick, MA, USA, 2020. Available online: http://www.cs.rtu.lv/jekabsons/regression.html (accessed on 23 December 2021).
107. Friedman, J.H. Multivariate adaptive regression splines (with discussion). *Ann. Stat.* **1991**, *19*, 1–67.
108. Bera, P.; Prasher, S.O.; Patel, R.M.; Madani, A.; Lacroix, R.; Gaynor, J.D.; Tan, C.S.; Kim, S.H. Application of MARS in simulating pesticide concentrations in soil. *Trans. Asabe* **2006**, *49*, 297–307.
109. Salford-Systems. *SPM Users Guide: Introducing MARS*; Minitab, LLC.: State College, PA, USA, 2019. Available online: https://www.minitab.com/content/dam/www/en/uploadedfiles/content/products/spm/IntroMARS.pdf (accessed on 23 December 2021).
110. Roy, D.K.; Datta, B. Multivariate adaptive regression spline ensembles for management of multilayered coastal aquifers. *J. Hydrol. Eng.* **2017**, *22*, 4017031. [CrossRef]
111. Dempster, A.P.; Laird, N.M.; Rubin, D.B. Maximum likelihood from incomplete data via the EM algorithm. *J. R. Stat. Soc. Ser.* **1977**, *39*, 763–768.
112. MacKay, D.J.C. The evidence framework applied to classification networks. *Neural Comput.* **1992**, *4*, 720–736. [CrossRef]
113. Chen, M. Probabilistic Linear Regression. 2021. Available online: https://www.mathworks.com/matlabcentral/fileexchange/55832-probabilistic-linear-regression (accessed on 23 December 2021).
114. Yu, P.-S.; Chen, S.-T.; Chang, I.-F. Support vector regression for real-time flood stage forecasting. *J. Hydrol.* **2006**, *328*, 704–716. [CrossRef]
115. Yoon, H.; Jun, S.-C.; Hyun, Y.; Bae, G.-O.; Lee, K.-K. A comparative study of artificial neural networks and support vector machines for predicting groundwater levels in a coastal aquifer. *J. Hydrol.* **2011**, *396*, 128–138. [CrossRef]
116. Basak, D.; Pal, S.; Patranabis, D.C. Support vector regression. *Neural Inf. Process.* **2007**, *11*, 203–224.
117. Chevalier, R.F.; Hoogenboom, G.; McClendon, R.W.; Paz, J.A. Support vector regression with reduced training sets for air temperature prediction: A comparison with artificial neural networks. *Neural Comput. Appl.* **2011**, *20*, 151–159. [CrossRef]

118. Zhang, G.; Ge, H. Prediction of xylanase optimal temperature by support vector regression. *Electron. J. Biotechnol.* **2012**, *15*, 7. [CrossRef]
119. Wu, J.; Sun, J.; Liang, L.; Zha, Y. Determination of weights for ultimate cross efficiency using Shannon entropy. *Expert Syst. Appl.* **2011**, *38*, 5162–5165. [CrossRef]
120. Nash, J.E.; Sutcliffe, J.V. River flow forecasting through conceptual models part I—A discussion of principles. *J. Hydrol.* **1970**, *10*, 282–290. [CrossRef]
121. Willmot, C.J. On the validation of models. *Phys. Geogr.* **1981**, *2*, 184–194. [CrossRef]
122. Legates, D.R.; McCabe Jr, G.J. Evaluating the use of "goodness-of fit" measuresin hydrologic and hydroclimatic model validation. *Water Resour. Res.* **1999**, *35*, 233–241. [CrossRef]
123. Tao, H.; Diop, L.; Bodian, A.; Djaman, K.; Ndiaye, P.M.; Yaseen, Z.M. Reference evapotranspiration prediction using hybridized fuzzy model with firefly algorithm: Regional case study in Burkina Faso. *Agric. Water Manag.* **2018**, *208*, 140–151. [CrossRef]
124. Chia, M.Y.; Huang, Y.F.; Koo, C.H. Swarm-based optimization as stochastic training strategy for estimation of reference evapotranspiration using extreme learning machine. *Agric. Water Manag.* **2021**, *243*, 106447. [CrossRef]
125. Mohammadi, B.; Mehdizadeh, S. Modeling daily reference evapotranspiration via a novel approach based on support vector regression coupled with whale optimization algorithm. *Agric. Water Manag.* **2020**, *237*, 106145. [CrossRef]
126. Elbeltagi, A.; Deng, J.; Wang, K.; Malik, A.; Maroufpoor, S. Modeling long-term dynamics of crop evapotranspiration using deep learning in a semi-arid environment. *Agric. Water Manag.* **2020**, *241*, 106334. [CrossRef]
127. Gao, L.; Gong, D.; Cui, N.; Lv, M.; Feng, Y. Evaluation of bio-inspired optimization algorithms hybrid with artificial neural network for reference crop evapotranspiration estimation. *Comput. Electron. Agric.* **2021**, *190*, 106466. [CrossRef]
128. Yin, J.; Deng, Z.; Ines, A.V.; Wu, J.; Rasu, E. Forecast of short-term daily reference evapotranspiration under limited meteorological variables using a hybrid bi-directional long short-term memory model (Bi-LSTM). *Agric. Water Manag.* **2020**, *242*, 106386. [CrossRef]
129. Heinemann, A.B.; Oort, P.A.V.; Fernandes, D.S.; Maia, A. Sensitivity of APSIM/ORYZA model due to estimation errors in solar radiation. *Bragantia* **2012**, *71*, 572–582. [CrossRef]
130. Li, M.-F.; Tang, X.-P.; Wu, W.; Liu, H.-B. General models for estimating daily global solar radiation for different solar radiation zones in mainland China. *Energy Convers. Manag.* **2013**, *70*, 139–148. [CrossRef]

 *agronomy*

Article

Estimate Cotton Water Consumption from Shallow Groundwater under Different Irrigation Schedules

Guohua Zhang [1] and Xinhu Li [2,3,4,*]

1   China Irrigation and Drainage Development Center, Beijing 210054, China; zgh311133@163.com
2   State Key Laboratory of Desert and Oasis Ecology, Xinjiang Institute of Ecology and Geography, Chinese Academy of Sciences, Urumqi 830011, China
3   Akesu National Station of Observation and Research for Oasis Agro-Ecosystem, Akesu 843017, China
4   University of Chinese Academy of Sciences, Beijing 100049, China
*   Correspondence: lixinhu@ms.xjb.ac.cn

Abstract: Shallow groundwater is considered an important water resource to meet crop irrigation demands. However, limited information is available on the application of models to investigate the impact of irrigation schedules on shallow groundwater depth and estimate evaporation while considering the interaction between meteorological factors and the surface soil water content (SWC). Based on the Richards equation, we develop a model to simultaneously estimate crop water consumption of shallow groundwater and determine the optimal irrigation schedule in association with a shallow groundwater depth. A new soil evaporation function was established, and the control factors were calculated by using only the days after sowing. In this study, two irrigation scheduling methods were considered. In Method A, irrigation was managed based on the soil water content; in Method B, irrigation was based on the crop water demand. In comparison with Method B, Method A was more rational because it could use more groundwater, and the ratio of soil evaporation to total evapotranspiration was low. In this model, the interaction between meteorological factors and the SWC was considered to better reflect the real condition; therefore, the model provided a better way to estimate the crop water consumption.

Keywords: arid region; evapotranspiration partitioning; soil evaporation; soil water content

Citation: Zhang, G.; Li, X. Estimate Cotton Water Consumption from Shallow Groundwater under Different Irrigation Schedules. *Agronomy* **2022**, *12*, 213. https://doi.org/10.3390/agronomy12010213

Academic Editors: Pantazis Georgiou and Dimitris Karpouzos

Received: 16 December 2021
Accepted: 14 January 2022
Published: 16 January 2022

Publisher's Note: MDPI stays neutral with regard to jurisdictional claims in published maps and institutional affiliations.

1. Introduction

Water scarcity is a great concern for irrigation agriculture worldwide. More than half of the farmland in the world exists in arid and semiarid regions [1]. Irrigation is essential to crop production in arid regions and plays an important role in crop water demands throughout the world. The site-specific application of irrigation water within a field improves water use efficiency and reduces water usage for sustainable crop production, especially in arid and semiarid regions [2].

Rational irrigation scheduling is essential to irrigation management, and many irrigation scheduling studies have been performed in arid regions [3–5]. In general, one type of irrigation scheduling is based on evapotranspiration demand, and another is based on the soil water content of the root zone [5]. Shallow groundwater exists in many areas of the world [6], and some farmlands are irrigated with shallow groundwater in arid regions. Shallow groundwater exists in many areas of the world and plays a vital role in sustaining agricultural productivity in many irrigated areas [3,7,8]. Irrational or intensive irrigation leads to a decline in the shallow groundwater table. The variation in a shallow groundwater table strongly influences the water balance [9]. Water cycling in soils with shallow groundwater is complex due to the deep percolation and groundwater evapotranspiration (ET_g) that occur in arid regions [10]. An improved irrigation schedule could reduce the amount of deep percolation [8].

Several studies have used lysimeters to measure crop water consumption from a shallow groundwater table [11–15]. The lysimeters were accurate, but their use is limited because of their high cost [3]. Therefore, some studies have attempted to quantify the groundwater as a part of the SPAC (soil-plant-atmosphere continuum) at the field and regional scales using models such as SWAP [3], EPIC [16], and DSSAT [17]; however, the models mentioned above usually oversimplified the influence of groundwater [18]. Han et al. [18] used the Hydrus-1D model coupled with a simplified crop growth model from SWAT to estimate the effect of groundwater on the water balance in the cotton root zone, but their study did not include a method for irrigation scheduling and was not implemented or systematically used by the majority of growers. Huo et al. [3] simulated the various amounts of irrigation applied to soil with different water tables, but the irrigation schedule was fixed in their study. Therefore, information relative to the irrigation scheduling method is still limited.

Accurate estimations of evapotranspiration (ET) are urgently needed. ET includes soil evaporation (E) and crop transpiration (T); transpiration is considered a physiological process, whereas soil evaporation is a physical process. Soil evaporation is an important component of the total crop water consumption, and the E ratio is higher in the earlier growing season due to wet surface soils and low crop cover (CC) [19]; however, evaporation may be lower in the later growing season when the surface soils are drier, and the CC is higher [20]. Unkovich et al. [20] reviewed the published field measurements from Australia and found an average of 38% crop water consumption due to soil evaporation. Kool et al. [21] noted that the E/ET ratio exceeded 30% in 32 of the studies and noted that E usually constituted a large fraction of ET and should be independently considered. Numerous measurement methods and analytical models have been developed to estimate T and E separately, but large variability exists in ET partitioning, suggesting that obtaining accurate ET partitioning is relatively challenging [21]. In general, T is controlled by atmospheric evaporative demand, leaf area index (LAI) or crop cover (CC), surface soil water content (SWC), and reference crop evapotranspiration (ET_0) [22–25]. These factors influence T and E, as well as ET partitioning. Therefore, most previous studies focused on the influence of crops on ET partitioning using regression functions between CC and T/ET or E/ET [26–30]. Zhao et al. [25] established a function between ET partitioning and control factors (CC and SWC), but the metrological factors were neglected, and the CC and SWC had to be acquired by field experiment observations. The establishment of simple functions that are correlated with the main control factors without field experiment observations was necessary. It will provide a better way to interpret evapotranspiration partitioning and to model water consumption in cotton fields.

Existing models include Hydrus-1D; however, E or SWC must be acquired by field experimental observations. For SWC, E is just an input parameter, leading to the interaction between SWC and E being neglected. However, this interaction effect is present. For example, a high soil water content could cause high soil evaporation, but meteorological factors also influence the SWC [20,21]. If the SWC or E are included as only input parameters, then the interaction effect will be neglected, and the real condition will not be reflected. Although the HYDRUS model can simulate irrigation and water consumption at a given groundwater table, the model can only set a fixed or initial groundwater table. In practice, changes in groundwater table are not entirely dependent on evaporation consumption from cropland (but lysimeter with impermeable bottom) but involve lateral recharge and groundwater consumption from surrounding areas. Our model takes into account an actual groundwater change (groundwater variation with time as an input parameter), which better reflects a real situation. Therefore, in this study, we developed a model to estimate crop water consumption and considered an actual groundwater table fluctuation and the interaction between SWC and E.

The objective of our research was to (1) develop a model based on the Richards equation to simultaneously estimate crop water consumption with shallower groundwater, (2) develop a new function for estimating soil evaporation considering the interaction between SWC and E, (3) validate this model using cotton field experiment observations, and (4) apply the new tool to estimate a suitable irrigation schedule.

2. Materials and Methods

2.1. Mathematic Model

2.1.1. Water Flow Equations

The Richards model was used to simulate the soil water movement as follows [31]:

$$\begin{cases} \frac{\partial \theta}{\partial t} = \frac{\partial}{\partial z}\left[K\left(\frac{\partial h}{\partial z} + 1\right)\right] - S, 0 \le z \le L_r \\ \frac{\partial \theta}{\partial t} = \frac{\partial}{\partial z}\left[K\left(\frac{\partial h}{\partial z} + 1\right)\right], L_r \le z \le L \end{cases} \tag{1}$$

where S is the root water uptake (cm^{-1}), h is the pressure water head (cm), θ is the soil water content ($cm^3\ cm^{-3}$), L_r is the crop root zone depth (cm), L is the soil depth (cm), and K is the unsaturated hydraulic conductivity (cm day^{-1}). The van Genuchten–Mualem model was used to estimate the soil hydraulic properties [32].

2.1.2. Root Water Uptake

The root water uptake is described as follows [33]:

$$S = \alpha(h) \times b(z) \times T_p(t) \tag{2}$$

where $\alpha(h)$ is the soil water stress response function [34]. The parameters in the Feddes et al. [34] model for cotton are $h_1 = -10$ cm, $h_2 = -25$ cm, $h_{3max} = -200$ cm, $h_{3min} = -600$ cm, and $h_4 = -14{,}000$ cm [35,36], $b(z)$ is the root water uptake distribution function [37] and $T_p(t)$ is the cotton actual transpiration during one day (cm), which was calculated as follows [4]:

$$T_p(t) = ET_p(t) \times \left(1 - exp\left(-\gamma\left(1 + \delta\left|sin\frac{\pi(t_d - S_N)}{12}\right|\right)LAI\right)\right) \tag{3}$$

where $ET_p(t)$ is the potential evapotranspiration of cotton (cm day^{-1}), which was calculated using the FAO 56 Penman–Monteith equation [38], LAI is the leaf area index and is a function of the time after sowing, δ and γ are the empirical coefficients, t_d is time (h), and S_N is time at noon (h).

2.2. Field Experiment and Model Parameterization

2.2.1. Study Area

This experiment was conducted at the Aksu National Field Research Station of Agroecosystems (40°37′ N, 80°51′ E) in the northwestern Tarim Basin, Xinjiang Province, China. Cotton was the main crop, which is largely dependent on irrigation. The precipitation was variable with a mean value of 45.7 mm, the annual pan evaporation was approximately 2500 mm, and the mean temperature was approximately 8 °C. The groundwater table is shallow and approximately 2 m. The soil physical properties at the experimental site are shown in Table 1.

Table 1. Soil bulk density and particle size distribution from 0 to 80 cm in experiment sites.

Depth (cm)	Bulk Density (g/cm³)	Soil Particle Size Distribution (%)		
		<0.002 mm	0.002–0.05 mm	0.05–2.0 mm
0–10	1.33	8.03	78.8	13.17
10–20	1.37	8.36	80.32	11.32
20–40	1.5	11.56	82.32	6.12
40–60	1.44	13.7	83.21	3.09
60–80	1.41	9.42	75.96	14.62

2.2.2. Field Experimental Design

The experiment was conducted in 2010 and 2011, and the plot area was 150 m × 90 m. The cotton was sown in April and harvested in November. The phenological phases of cotton growth are shown in Table 2, and the irrigation schedule is shown in Table 3. Additional spring and winter irrigation was applied before seeding and after harvest, providing approximately 200 mm of water to leach the salt.

Table 2. Phenological phases of cotton growth in 2010 and 2011.

Phenological Phases	2010	2011
Seeding	30 April 2010	5 May 2011
Emergence stage	2 May 2010	5 May 2011
Squaring stage	10 June 2010	12 June 2011
Flowering stage	3 July 2010	28 June 2011
Boll opening stage	25 August 2010	19 September 2011
Harvest	1 November 2010	2 November 2011

Table 3. The rainfall date and irrigation schedules in 2010 and 2011 in cotton field.

Irrigation Date	Irrigation Amount (mm)	Rainfall Date	Rainfall Amount (mm)	Rainfall Date	Rainfall Amount (mm)
25 June 2010	180	5 June 2010	1.6	25 October 2010	15.7
12 June 2010	100	12 June 2010	0.2	31 October 2010	0.6
2 August 2010	150	16 June 2010	0.4	5 May 2011	4.8
29 August 2010	130	20 June 2010	0.3	10 May 2011	3.2
30 June 2011	199	25 June 2010	0.3	15 May 2011	4.4
15 July 2011	240	31 June 2010	3.0	20 May 2011	0.6
9 August 2011	148	20 August 2011	1.8	5 June 2011	2.6
25 August 2011	157	25 August 2010	0.5	20 June 2011	8.2
		20 September 2010	2.3	15 August 2011	3.1
		26 September 2010	11.2	19 August 2011	1.5
		29 September 2010	7.8	5 September 2011	2.3

2.2.3. Measurements

The soil water content and groundwater table were observed every 5 days (Figure 1). The soil water content was measured using a neutron probe (CNC503DR, Beijing Hean Nuclear Instrument Company, Shenzhen, China) with three replications at depths of 10, 20, 30, 40, 50, 70, 90, 110, 130, and 150 cm. This neutron probe consists of two main components: a probe and a rate meter. The density of slow neutrons formed around the probe is nearly proportional to the concentration of hydrogen in the medium of the probe. Equation (4) was used to calculate soil water content:

$$\theta = 0.615 \frac{R}{R_w} + 0.0289, \ r = 0.952 \tag{4}$$

where R was the slow neutron count in soil, and R_w was standard neutron count.

The groundwater table was measured using a water level scale every 5 days. An automatic meteorological station was installed in the field to measure wind speed (Model 010c, Met One, Grants Pass, OR, USA), solar radiation (Model LI200X, Campbell Scientific, Logan, UT, USA), precipitation (Model 52202, RM Young, Traverse City, MI, USA), air temperature, and relative humidity (Model Hmp45c, Campbell Scientific, Logan, UT, USA), and a datalogger (Model CR1000, Campbell Scientific, Logan, UT, USA) was used to monitor the sensors. Leaf samples were collected with five replications for each phenological phase, and the LAI was calculated by the leaf width (W) and length (L) [39], the mean LAI was 0.16 (std, 0.01), 0.92 (std, 0.03), 2.91 (std, 0.16), 3.23 (std, 0.33), and 1.76 (std, 0.15) from emergence to harvest stage, respectively.

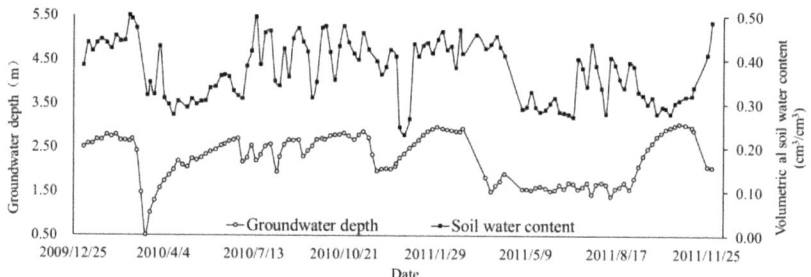

Figure 1. Volumetric soil water content at a depth of 20 cm and groundwater depths in experiment sites in 2010 and 2011.

2.2.4. Numerical Modeling

The initial conditions were set based on the soil moisture. The bottom boundary was assigned at the pressure head boundary condition using the observed groundwater table depths, and the upper boundary was set according to the atmospheric boundary condition as follows [33]:

$$\frac{\partial \theta}{\partial t} = -\frac{\partial q}{\partial x} \tag{5}$$

where q is the soil surface water flux. If the soil surface is flooded, $q = I(t)$, $I(t)$ is infiltration rate, and if the soil surface water has begun to evaporate, the $q = E(\theta)$, where $E(\theta)$ is the evaporation intensity.

We defined the five soil layers (0–10, 10–20, 20–40, 40–60, and 60–80 cm) in the soil profile (Table 1) based on previous studies [18]. The soil hydraulic parameters of the van Genuchten–Mualem model are shown in Table 4, and the parameters remained constant for both model calibration and validation. The pore connectivity parameter was constant (0.5). The parameter estimation tool (PEST) [40] was used to optimize the parameters θ_r, α, n, γ and δ. The time from 30 April to 31 October in 2010 and the time from 2 May to 2 November in 2011 were used to calibrate t and validate the model, respectively. The optimized parameters of the van Genuchten–Mualem analytical functions [32] are given in Table 4, with $\gamma = 0.7937$ and $\delta = 0.10364$.

Table 4. Soil hydraulic parameters of the van Genuchten–Mualem model at the experimental site.

Depth (cm)	θ_r (cm³/cm³)	θ_s (cm³/cm³)	α (cm⁻¹)	n	K_s (cm·day⁻¹)
0–10	0.0563	0.43	0.0050	1.70	25
10–20	0.057	0.42	0.0053	1.68	15
20–40	0.0062	0.41	0.0062	1.63	9
40–60	0.0671	0.44	0.0059	1.64	6
60–270	0.055	0.41	0.0053	1.68	5

Soil evaporation is also controlled by meteorological factors and the CC. The CC was adequately described by the logistic growth curve with the days after sowing (t), and ET_0 reflected the meteorological factor; thus, we were able to establish the function by combining the ET_0 and CC to calculate the soil evaporation as follows:

$$E(\theta_{top}) = (a_1\theta_{top} + a_2)^{a_3} \times (ET_0)^{a_4} \times (1 + exp(-a_5 t)) \tag{6}$$

The constants a_1, a_2, a_3, a_4, and a_5 were obtained by fitting the parameters with observed data, and they were 1.3181, 0.3400, 8.4880, 0.1282, and 1.5734, respectively. The R² was 0.77, and the RMSE was 0.049. θ_{top} represents the mean soil water content (0–5 cm). This equation does not include a measured parameter "CC"; the CC was reflected by the last term (power function).

The relationship between the LAI and the days after sowing is expressed in Equation (6), with $R^2 = 0.99$ and RMSE = 1.15:

$$LAI(t) = -4.64 \times 10^{-3} + 2.89 \times 10^{-2}t - 1.48 \times 10^{-3}t^2 + 3.27 \times 10^{-5}t^3 - 1.70 \times 10^{-7}t^4 \quad (7)$$

2.3. Model Solutions

The soil to a depth of 2.7 m was simulated, and the grid size was 0.01 m. Equation (1) was solved using the finite-element method, and the solutions were the same as those of the Hydrus-1D model [33]. A software package (MATLAB 7.0 for Windows) was used to edit the code.

2.3.1. Model Evaluation

The correlation coefficient (R^2) and the root mean square error (RMSE) were used to evaluate the agreement between the simulated results and the observed results:

$$\text{RMSE} = \left[\frac{1}{n} \sum_{i=1}^{n} (C_{si} - C_{ob})^2 \right]^{1/2} \quad (8)$$

$$R^2 = \left[\frac{\sum_{i=1}^{n} (C_{si} - \overline{C_{si}})(C_{ob} - \overline{C_{ob}})}{\sum_{i=1}^{n} (C_{si} - \overline{C_{si}}) \sum_{i=1}^{n} (C_{ob} - \overline{C_{ob}})} \right]^2 \quad (9)$$

where C_{si} is the model prediction at time interval i, C_{ob} is the field observation, n is the total number of observed values, $\overline{C_{si}}$ is the mean of the simulated values, and $\overline{C_{ob}}$ is the mean of the observed values.

2.3.2. Sensitivity Experiments

The numeric simulation experiments were mainly conducted to determine the influence of the irrigation schedule and groundwater depth on the soil root zone water balance. The effects of the irrigation schedule and groundwater table depth on the water capacity, capillary rise, actual evaporation, and actual transpiration were estimated. The root zone was from the surface soil (0 cm) to a depth of 60 cm. Fourteen different modeling scenarios were considered (Table 5). The groundwater table fluctuation had a range from 1.4 to 3.1 m during the growing season in 2010 and 2011 (Figure 1). The modeling scenarios cover this range to reflect the real situation; thus, we conduct seven groundwater tables from 1.0 to 4.0 m.

Table 5. The 14 different modeling scenarios included the seven groundwater depths and two irrigation schedules for the numerically simulated experiments.

Irrigation Schedule		Groundwater Depth (m)
Method A	**Method B**	
Irrigation Managed Based on the Soil Water Content	**Irrigation Was Based on the Crop Water Demand**	
A1	B1	1.0
A2	B2	1.5
A3	B3	2.0
A4	B4	2.5
A5	B5	3.0
A6	B6	3.5
A7	B7	4.0

We assumed that the groundwater table was a natural fluctuation, which was not entirely influenced by water consumption from cropland. It can be seen in Figure 1, which was the observed variation of the groundwater table with time. The variation trend does not show a strong relation with irrigation or water consumption. For each scenario, the groundwater table time series were generated by adding a constant value (ranging from −1.5 to 1.5 m) to the groundwater table depth measured in 2010. It was an input parameter to run the model (Figure 2). The mean groundwater table was 2.5 m (measured value during growing season) in 2010, the seven groundwater table (mean value) were conducted as follow: 1.0 m (2.5 m + (−1.5 m)), 1.5 m (2.5 m + (−1.0 m)), 2.0 m (2.5 m + (−0.5 m)), 2.5 m (2.5 m + 0 m), 3.0 m (2.5 m + 0.5 m), 3.5 m (2.5 m + 1.0 m), and 4.0 m (2.5 m + 1.5 m) (Table 5), respectively, all the change trend of groundwater tale (14 treatments) were consistent with measured data in 2010 during growing season (Figure 2). Two common irrigation scheduling methods were selected. In Method A, the irrigation was managed based on the soil water content, and the lower limits of the controlled soil water contents at field capacity (θ_f) were $0.55\theta_f$, $0.6\theta_f$, and $0.7\theta_f$ for the planned soil moisture layers of 0.2 m, 0.4 m, 0.6 m and 0.8 m in the emergence stage and squaring stage, respectively. The higher limit of the soil water contents at field capacity was 0.95. With Method B, the irrigation amount and timing were designed based on the crop water demand (local farmer custom), depending on the potential ET in the next N (or 20) days.

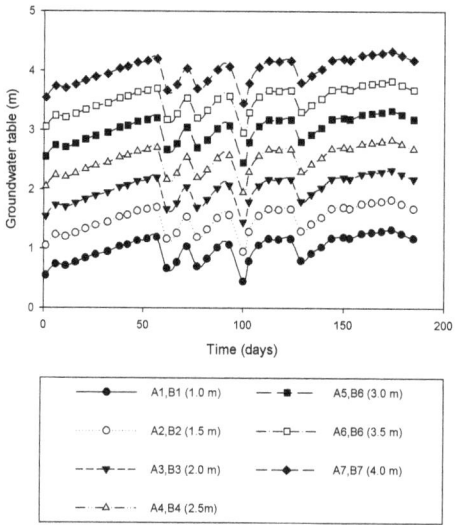

Figure 2. The variation of groundwater table in 14 different modeling scenarios.

3. Results and Discussion

3.1. Model Calibration and Validation

The relationship between the measured and simulated soil water storage in the root zone (0–60 cm) is shown in Figure 3. Suitable agreement between the measured and simulated values (soil water storage) was found. The model slightly overpredicted the water content during the calibration process; on the contrary, the validation process showed slight underprediction. More statistical tests were conducted to evaluate the performance of the model. The simulated and measured soil water contents from the soil depths of 20 and 150 cm are shown in Figure 4, which indicates that the predicted values are in agreement with the observed values. The R^2 and RMSE values for the soil water contents and soil water storage (0–60 cm) are shown in Table 6. The R^2 values ranged from 0.92 to 0.94, the RMSE values of the soil water storage at 60 cm, and the soil water content were 20–24 mm and 0.04, respectively. These statistical tests indicated that the model had suitable performance.

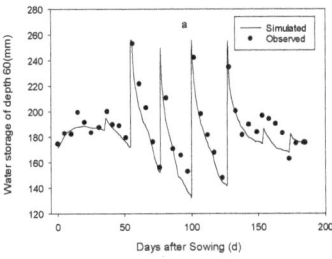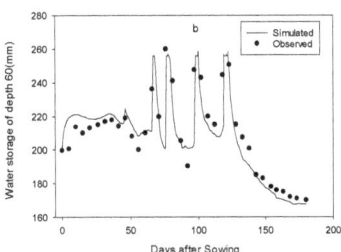

Figure 3. The observed and simulated soil water storage in the root zone (0 to 60 cm) in 2010 (**a**) and 2011 (**b**).

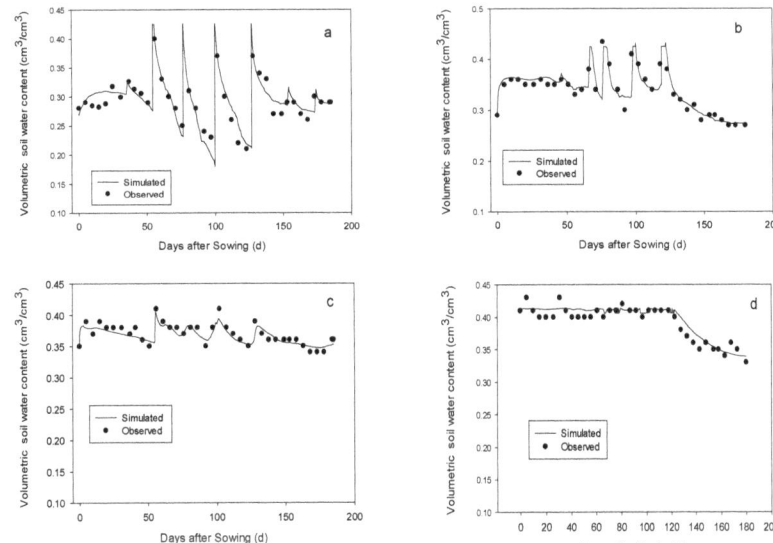

Figure 4. The variations in soil water content at (**a**) soil depths of 20 cm in 2010, (**b**) soil depths of 20 cm in 2011, (**c**) soil depths of 150 cm in 2010, and (**d**) soil depths of 150 cm in 2011.

Table 6. Statistical tests for the modeling results in 2011 and 2012.

Period	Year	Item	Soil Water Content ($-$)	Soil Water Storage (mm)
Calibration process	2010	R^2	0.92	0.93
		RMSE	0.04	20.83
Validation process	2011	R^2	0.94	0.93
		RMSE	0.04	24.06

3.2. Scenario Simulation

3.2.1. Soil Water Storage Variation in the Root Zone

The soil water storage in the root zone (0–60 cm) showed variations under the different treatments (Figure 5). The soil water storage showed sharp increases after irrigation and then decreased gradually due to percolation and evaporation. When the groundwater table was shallow, the soil water storage showed small variations with high values because the groundwater could supply water into the soil root zone. In contrast, when the groundwater table was deep, the water supply from the groundwater was weaker and could not satisfy the water consumption by the cotton, which caused low soil water content in the root zone, deepened the groundwater table, and caused greater fluctuations in the soil water content.

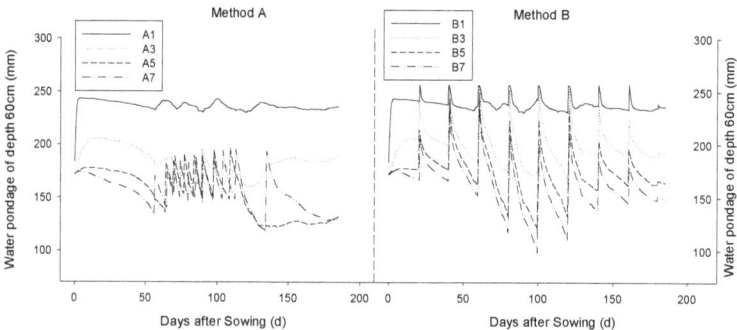

Figure 5. The variations in soil water storage in the soil root zone under different treatments.

In general, shallow groundwater is considered a water resource for crop irrigation. In our research, the irrigation times increased from treatment A1 (1.0 m) to A7 (4.0 m) with the increase in groundwater depth. The groundwater table was maintained at depths of 1.0 (treatment A1) and 1.5 m (treatment A2), and the groundwater was able to supply enough water to the root zone for evaporation (Figure 6 and Table 7). The groundwater table does not show a significant decrease (Figure 2); therefore, irrigation was not necessary with a shallower groundwater table (1.0 and 1.5 m). Kahlown et al. [14] investigated the water tables that were maintained at shallower depths (e.g., 0.5 m for wheat) and noted that the groundwater could meet the crop water requirements, but differences were observed for different crops [14] and soil textures [3]. For treatment A7, the first irrigation occurred on the 56th day. This is because the soil evaporation was small due to low surface soil water content, which led to low soil evaporation. Usually, the surface soil water content showed a strong influence on soil evaporation, which has been reported in many studies [41–43]. The dry surface soils lead to vapor transport through the surface [44], which will be greatly inhibited the soil evaporation, thus retained the water in below the surface. The cotton was seeding around at depth of 5 cm, thus it could obtain the water by root at the emergence stage.

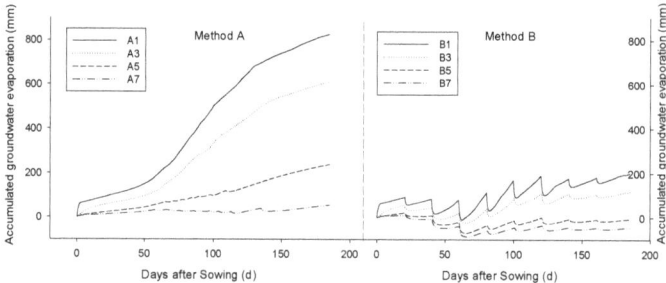

Figure 6. The dynamic variation of accumulated groundwater evaporation under different treatments.

For Method B (Treatments B1 to B7), the irrigation amounts and times were similar among the treatments (Table 7). Compared with Method A, the soil water storage (0–60 cm) in Method B showed more significant fluctuations (Figure 5). Method B had fixed irrigation times and amounts, and the soil water potential was higher in the root zone than in the lower layer after irrigation and resulted in downward soil water movement. In contrast, irrigation began when the soil water content reached the prescribed lower limit; therefore, the soil water potential was lower in the root zone than in the low layer, causing the groundwater to move upward into the root zone (Figure 6).

Table 7. The simulated irrigation schedule for 14 treatments.

Treatments	Irrigation Schedule	Irrigation Time (d)/Irrigation Amount (mm)											Total
A1	Irrigation time	0											
	Irrigation amount	0											0
A2	Irrigation time	0											
	Irrigation amount	0											0
A3	Irrigation time	89											
	Irrigation amount	42											42
A4	Irrigation time/(d)	65	74	82	89	105							
	Irrigation amount	35.00	36.75	40.25	42.00	47.25							201.25
A5	Irrigation time	64	70	77	83	89	97	109					
	Irrigation amount	46.55	35.00	38.50	40.25	42.00	45.50	49.00					296.8
A6	Irrigation time	60	66	71	77	83	88	96	103	112	141		
	Irrigation amount	41.65	35.00	36.75	38.50	40.25	42.00	43.75	47.25	49.00	72.80		446.95
A7	Irrigation time	56	65	69	74	80	85	90	97	103	112	134	
	Irrigation amount	39.20	35.00	35.00	36.75	38.50	40.25	42.00	45.50	47.25	49.00	75.60	484.05
B1–B7	Irrigation time/(d)	20	40	60	80	100	120	140	160	180			
	Irrigation amount	38.31	73.79	103.75	97.21	90.89	106.88	45.85	34.44	3.72			594.84

3.2.2. Evapotranspiration Dynamics

Figure 6 shows the accumulated soil evaporation, transpiration, evapotranspiration, ET_0, and irrigation amounts under the different treatments. The amount of soil evaporation was larger when the groundwater table was shallower, with the soil evaporation reaching 252.6 and 244.1 mm at a depth of 1.0 m and reaching only 68.5 mm and 81.5 mm at a depth of 4.0 m for Method A and Method B, respectively (Figure 7a). However, the cotton transpiration was not significantly different for the different groundwater tables and irrigation methods, showing a range from 485.0 to 505.7 mm for all treatments, and the amounts of T and E were similar in the two irrigation methods (Figure 7a,b). Except for the seeding and emergence stages, cotton transpiration was less than soil evaporation because of the small CC. However, after the emergence stage, the increased, causing larger cotton transpiration and smaller soil evaporation. As the irrigation amount and time increased, the soil evaporation and transpiration showed increasing trends (Figure 7d), but the variation in transpiration was small. This trend occurred because the soil evaporation was higher after irrigation and resulted in increased soil evaporation. The soil evaporation variation trend was similar to the variation pattern.

As mentioned above, cotton transpiration was strongly correlated with meteorological factors and was also correlated with the irrigation amount and soil water content. Cotton transpiration was described by the logistic curve (Figure 7b), with cotton transpiration greater than soil evaporation, but the soil evaporation increased significantly or was higher than transpiration after irrigation, indicating that the irrigation had a great influence on the soil evaporation. The ET_0 was calculated using the meteorological factor by the FAO 56 Penman–Monteith equation [38]. The ET_0 directly reflects the evaporation demand by ambient conditions. The high ET_0 in the early stage is due to the increase in evaporation demand (temperature and radiation), while the actual evaporation (evapotranspiration) is controlled by the soil surface water content, and the crop transpiration is small at this stage. Therefore, the high ET_0 in the early stage but the evapotranspiration smaller than ET_0 (Figure 7c). With the cotton growing, the cotton transpiration increased significantly (Figure 7b), the evapotranspiration is gradually close to ET_0 and has the same trend, indicating that the evapotranspiration was dominated by cotton transpiration after the flowering stage.

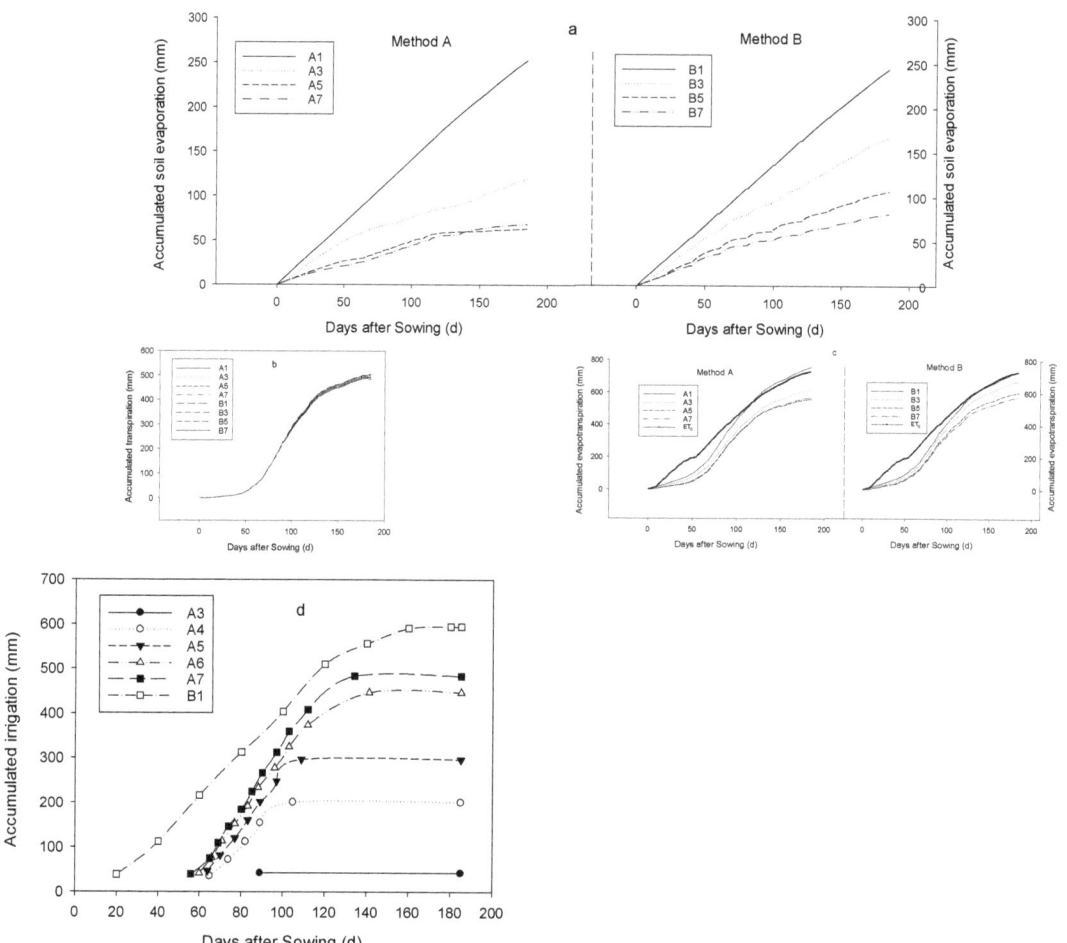

Figure 7. The dynamic variations in accumulated soil evaporation (**a**), crop transpiration (**b**), crop evapotranspiration (**c**), and irrigation amount (**d**).

Ding et al. [45] found that higher soil evaporation occurred in early growing seasons due to the small canopy in the maize field, particularly after rainfall or irrigation events. The wet soil surface may have caused increased soil evaporation, and the soil evaporation gradually decreased as the soil dried [25]. Our research results were similar to the results of these studies. The soil evaporation was approximately 11% to 34% of the total evapotranspiration, and the soil evaporation was noticeably reduced with a shallower groundwater table and less irrigation. The soil evaporation was influenced by the groundwater table and showed some differences with the irrigation times and different groundwater table depths. The surface soil was relatively wet when the groundwater table was shallow (e.g., 1.0 m), and the soil evaporation was approximately 33% of the total evapotranspiration. For the deeper groundwater table depth in Method B, the ratio (E/ET) showed a decreasing trend from 33% to 14%, whereas for Method A, the soil evaporation was only 11% of the total ET, with treatment 5 showing the smallest ratio of all treatments (Figure 8).

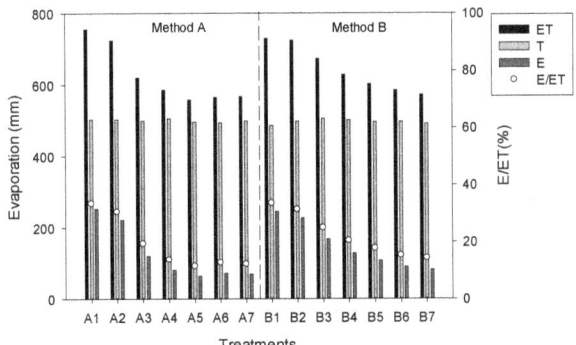

Figure 8. The variation of soil evaporation (E) and transpiration (T) evaporation in different ground-water tables and irrigation schedules.

As noted above, previous studies have presented some regression functions between CC or LAI and T/ET or E/ET to obtain the ET partition [26–30]; however, some functions neglected metrological factors. As mentioned above, the SWC or E was the input parameter, and some parameters, including CC, must be acquired by field observations, while the interaction effect was neglected. Our functions (Equation (5)) were established in correlation with the SWC, CC, and meteorological factors, and the parameter was calculated by days after sowing (Equation (6)), considering the interaction effect. Therefore, we provided a better way to reflect evapotranspiration partitioning and to model water consumption in a cotton field.

3.2.3. Water Balance in the Root Zone

Table 8 shows the water balance in the root zone, including evapotranspiration, soil evaporation, cotton transpiration, and soil water and groundwater evaporation. The relative error was less than 3%, which indicates that the model performs well when quantifying the water balance in the cotton field. The cotton water consumption showed great differences under the different treatments. For example, the source of cotton water consumption was mainly derived from the groundwater when the groundwater table was shallower in Method A, but the source gradually transformed to irrigation when the groundwater table was deeper. The cotton water consumption was 826 mm when the groundwater table remained at a depth of 1 m, almost meeting the cotton water demands. When the groundwater table was deeper, the groundwater evaporation decreased gradually. The groundwater evaporation sharply decreased to below 2.5 m. The groundwater evaporation was 383 mm at 2.5 m, which only partially met the cotton water consumption demand.

The simulated evapotranspiration ranged from 558 to 755 mm, but the cotton transpiration ranged from 485 to 505 mm. The irrigation of cotton was not necessary when the groundwater table was less than 1.5 m deep because the water source for cotton consumption mainly came from the groundwater. The irrigation amount was greater than 300 mm when the groundwater table was deeper than 3.0 m.

For Method B, as the groundwater table decreased, the groundwater evaporation decreased because the irrigation times and amounts were consistent. The groundwater evaporation reached 199 mm at the groundwater table depth of 1.0 m, which was 27% of the total evapotranspiration. However, the groundwater evaporation became negative when the groundwater table was deeper than 3.0 m, indicating that the irrigation caused the soil water to drain deeply, recharging the groundwater. The deep soil water drainage showed an increasing trend with increasing groundwater table depth. Therefore, we conclude that Method A used more groundwater than Method B throughout the growth period. Method A was more rational because it allowed the use of more groundwater. Although irrigating the cotton was not necessary when the groundwater table was less than 1 m, a shallower groundwater table increased the soil evaporation, thus leading to more groundwater consumption, which may be harmful to root growth.

Table 8. The water balance component under 14 treatments.

Simulated Treatments	ET/mm	T/mm	E/mm	ET_g/mm	Soil Water Consumption/mm	Irrigation Amount/mm	Irrigation Times	Error/%
A1	755.23	502.66	252.57	826.36	−52.90	0.00	0	−2.41
A2	724.00	503.09	220.91	780.66	−41.65	0.00	0	−2.07
A3	620.01	500.01	120.00	610.80	−17.98	42.00	1	−2.39
A4	585.86	505.65	80.20	383.96	8.93	201.25	5	−1.41
A5	558.09	494.75	63.34	238.12	39.53	296.80	7	−2.93
A6	564.76	493.05	71.72	100.54	31.21	446.95	10	−2.47
A7	566.80	498.33	68.48	55.71	42.10	484.05	11	−2.66
B1	729.11	484.99	244.12	199.63	−52.94	594.84	9	−1.70
B2	723.59	497.14	226.46	179.97	−42.43	595.85	9	−1.35
B3	672.99	505.43	167.56	118.33	−20.86	593.22	9	−2.63
B4	628.12	500.22	127.90	48.33	−4.58	595.85	9	−1.83
B5	602.00	495.26	106.74	−5.99	7.44	593.98	9	1.09
B6	585.00	495.89	89.11	−26.54	16.35	595.85	9	−0.11
B7	572.61	491.09	81.52	−46.12	23.17	594.84	9	0.13

4. Conclusions

A model was developed to quantify cotton water consumption and to estimate a suitable irrigation schedule according to the groundwater depth (1.0–4.0 m). In this study, two irrigation scheduling methods were considered. In Method A, irrigation was managed based on the soil water content; in Method B, irrigation was based on the crop water demand. The simulation results were verified with soil water storage measurements in the root zone (0–60 cm) and soil water contents at depths of 20 and 150 cm from the cotton field. Suitable agreement was presented between the simulation results and experimental data obtained from cotton field experiments, indicating that the model showed suitable performance.

A new function was established in correlation with the surface soil water content, crop cover, and reference crop evapotranspiration to calculate the soil evaporation and determine the evaporation partition. These factors were calculated using the days after sowing and did not require observations. The interaction effect between meteorological factors and surface soil water content was also considered and better reflected the real condition, thus providing a better way to interpret evapotranspiration partitioning.

With a deeper groundwater table depth from 1.0 to 4.0 m, the ratio between the soil evaporation and evapotranspiration showed a decreasing trend, but Method A showed a smaller ratio than Method B. In addition, for Method A, the groundwater could supply enough water to the soil root zone, and irrigation was not necessary when the groundwater table was less than at a depth of 1.5 m; however, the groundwater could not supply enough water to meet the crop water consumption demands even if the groundwater table was at a depth of 1.0 m, and the crops could not use the groundwater when the groundwater was deeper than 3.0 m in Method B. Therefore, we suggest that the irrigation schedule should be managed based on the soil water content because this irrigation schedule could use more groundwater and have a smaller soil evaporation to evapotranspiration ratio.

Author Contributions: Data curation; writing—original draft: G.Z.; Conceptualization, Formal analysis, Methodology; Writing—review and editing, X.L. All authors have read and agreed to the published version of the manuscript.

Funding: This work was supported by the West Light Foundation of the Chinese Academy of Sciences (grant No. 2020-XBQNXZ-012), National Natural Science Foundation of China (grant No. 41977013).

Institutional Review Board Statement: Not applicable.

Informed Consent Statement: Not applicable.

Data Availability Statement: These data sets were derived from the following public domain resources (Chinese Ecosystem Research Network): (http://www.cnern.org.cn/data/initDRsearch?classcode=STA) (accessed on 25 April 2016).

Acknowledgments: We thank Ouyang Ying from the USDA Forest Service, who revised this article.

Conflicts of Interest: The authors declare no conflict of interest.

References

1. Rijsberman, F.R. Water scarcity: Fact or fiction? *Agric. Water Manag.* **2006**, *80*, 5–22. [CrossRef]
2. Neupane, J.; Guo, W. Agronomic Basis and Strategies for Precision Water Management: A Review. *Agronomy* **2019**, *9*, 87. [CrossRef]
3. Huo, Z.; Feng, S.; Dai, X.; Zheng, Y.; Wang, Y. Simulation of hydrology following various volumes of irrigation to soil with different depths to the water table. *Soil Use Manag.* **2012**, *28*, 229–239. [CrossRef]
4. Kang, Y.; Wang, R.; Wan, S.; Hu, W.; Jiang, S.; Liu, S. Effects of different water levels on cotton growth and water use through drip irrigation in an arid region with saline ground water of Northwest China. *Agric. Water Manag.* **2012**, *109*, 117–126. [CrossRef]
5. Steele, D.D.; Stegman, E.C.; Knighton, R.E. Irrigation management for corn in the northern Great Plains, USA. *Irrig. Sci.* **2000**, *19*, 107–114. [CrossRef]
6. Babajimopouios, C.; Panoras, A.; Georgoussis, H.; Arampatzis, G.; Hatzigiannakis, E.; Papamichail, D. Contribution to irrigation from shallow water table under field conditions. *Agric. Water Manag.* **2007**, *92*, 205–210. [CrossRef]
7. Mejia, M.N.; Madramootoo, C.A.; Broughton, R.S. Influence of water table management on corn and soybean yields. *Agric. Water Manag.* **2000**, *46*, 73–89. [CrossRef]
8. Ayars, J.E.; Christen, E.W.; Soppe, R.W.; Meyer, W.S. The resource potential of in-situ shallow ground water use in irrigated agriculture: A review. *Irrig. Sci.* **2005**, *24*, 147–160. [CrossRef]
9. Cui, Y.L.; Shao, J.L. The role of ground water in arid/semiarid ecosystems, Northwest China. *Groundwater* **2005**, *43*, 471–477. [CrossRef]
10. Xue, J.; Guan, H.; Huo, Z.; Wang, F.; Huang, G.; Boll, J. Water saving practices enhance regional efficiency of water consumption and water productivity in an arid agricultural area with shallow groundwater. *Agric. Water Manag.* **2017**, *194*, 78–89. [CrossRef]
11. Wallender, W.W.; Grimes, D.W.; Henderson, D.W.; Stromberg, L.K. Estimating the Contribution of a Perched Water-Table to the Seasonal Evapotranspiration of Cotton. *Agron. J.* **1979**, *71*, 1056–1060. [CrossRef]
12. Zhang, L.; Dawes, W.R.; Slavich, P.G.; Meyer, W.S.; Thorburn, P.J.; Smith, D.J.; Walker, G.R. Growth and ground water uptake responses of lucerne to changes in groundwater levels and salinity: Lysimeter, isotope and modelling studies. *Agric. Water Manag.* **1999**, *39*, 265–282. [CrossRef]
13. Soppe, R.W.O.; Ayars, J.E. Characterizing ground water use by safflower using weighing lysimeters. *Agric. Water Manag.* **2003**, *60*, 59–71. [CrossRef]
14. Kahlown, M.A.; Ashraf, M. Effect of shallow groundwater table on crop water requirements and crop yields. *Agric. Water Manag.* **2005**, *76*, 24–35. [CrossRef]
15. Yang, J.; Wan, S.; Deng, W.; Zhang, G. Water fluxes at a fluctuating water table and groundwater contributions to wheat water use in the lower Yellow River flood plain, China. *Hydrol. Process.* **2007**, *21*, 717–724. [CrossRef]
16. Williams, J.R.; Jones, C.A.; Kiniry, J.R.; Spanel, D.A. The EPIC Crop Growth Model. *Trans. ASAE* **1989**, *32*, 0497–0511. [CrossRef]
17. Jones, J.W.; Hoogenboom, G.; Porter, C.H.; Boote, K.J.; Batchelor, W.D.; Hunt, L.A.; Wilkens, P.W.; Singh, U.; Gijsman, A.J.; Ritchie, J.T. The DSSAT cropping system model. *Eur. J. Agron.* **2003**, *18*, 235–265. [CrossRef]
18. Han, M.; Zhao, C.; Šimůnek, J.; Feng, G. Evaluating the impact of groundwater on cotton growth and root zone water balance using Hydrus-1D coupled with a crop growth model. *Agric. Water Manag.* **2015**, *160*, 64–75. [CrossRef]
19. Eastham, J.; Gregory, P.J. The influence of crop management on the water balance of lupin and wheat crops on a layered soil in a Mediterranean climate. *Plant Soil* **2000**, *221*, 239–251. [CrossRef]
20. Unkovich, M.; Baldock, J.; Farquharson, R. Field measurements of bare soil evaporation and crop transpiration, and transpiration efficiency, for rainfed grain crops in Australia—A review. *Agric. Water Manag.* **2018**, *205*, 72–80. [CrossRef]
21. Kool, D.; Agam, N.; Lazarovitch, N.; Heitman, J.L.; Sauer, T.J.; Ben-Gal, A. A review of approaches for evapotranspiration partitioning. *Agric. For. Meteorol.* **2014**, *184*, 56–70. [CrossRef]
22. Mitchell, P.J.; Veneklaas, E.; Lambers, H.; Burgess, S.S.O. Partitioning of evapotranspiration in a semi-arid eucalypt woodland in south-western Australia. *Agric. For. Meteorol.* **2009**, *149*, 25–37. [CrossRef]
23. Jiang, X.; Kang, S.; Li, F.; Du, T.; Tong, L.; Comas, L. Evapotranspiration partitioning and variation of sap flow in female and male parents of maize for hybrid seed production in arid region. *Agric. Water Manag.* **2016**, *176*, 132–141. [CrossRef]
24. Raz-Yaseef, N.; Yakir, D.; Schiller, G.; Cohen, S. Dynamics of evapotranspiration partitioning in a semi-arid forest as affected by temporal rainfall patterns. *Agric. For. Meteorol.* **2012**, *157*, 77–85. [CrossRef]
25. Zhao, P.; Kang, S.; Li, S.; Ding, R.; Tong, L.; Du, T. Seasonal variations in vineyard ET partitioning and dual crop coefficients correlate with canopy development and surface soil moisture. *Agric. Water Manag.* **2018**, *197*, 19–33. [CrossRef]

26. Wu, Y.; Du, T.; Ding, R.; Tong, L.; Li, S.; Wang, L. Multiple Methods to Partition Evapotranspiration in a Maize Field. *J. Hydrometeorol.* **2017**, *18*, 139–149. [CrossRef]
27. Kato, T.; Kimura, R.; Kamichika, M. Estimation of evapotranspiration, transpiration ratio and water-use efficiency from a sparse canopy using a compartment model. *Agric. Water Manag.* **2004**, *65*, 173–191. [CrossRef]
28. Zhang, B.Z.; Kang, S.Z.; Zhang, L.; Du, T.S.; Li, S.E.; Yang, X.Y. Estimation of seasonal crop water consumption in a vineyard using Bowen ratio-energy balance method. *Hydrol. Process.* **2007**, *21*, 3635–3641. [CrossRef]
29. Yan, H.; Zhang, C.; Oue, H.; Wang, G.; He, B. Study of evapotranspiration and evaporation beneath the canopy in a buckwheat field. *Theor. Appl. Climatol.* **2014**, *122*, 721–728. [CrossRef]
30. Wang, L.; Good, S.P.; Caylor, K.K. Global synthesis of vegetation control on evapotranspiration partitioning. *Geophys. Res. Lett.* **2014**, *41*, 6753–6757. [CrossRef]
31. Richards, L.A. Capillary conduction of liquids through porous mediums. *J. Appl. Phys.* **1931**, *1*, 318–333. [CrossRef]
32. Van Genuchten, M.T. A closed-form equation for predicting the hydraulicconductivity of unsaturated soils. *Soil Sci. Soc. Am. J.* **1980**, *44*, 892–898. [CrossRef]
33. Simunek, J.; Van Genuchten, M.T.; Sejna, M. *The HYDRUS-1D Softwarepackage for Simulating the One-Dimensional Movement of Water, Heat, Andmultiple Solutes in Variably-Saturated Media*; Department of Environmental Sciences: Riverside, CA, USA, 2005.
34. Feddes, R.A.; Kowalik, P.J.; Zaradny, H. *Simulation of Field Water Use and Crop. Yield*; John Wiley & Sons: New York, NY, USA, 1978.
35. Forkutsa, I.; Sommer, R.; Shirokova, Y.I.; Lamers, J.P.A.; Kienzler, K.; Tischbein, B.; Martius, C.; Vlek, P.L.G. Modeling irrigated cotton with shallow groundwater in the Aral Sea Basin of Uzbekistan: II. Soil salinity dynamics. *Irrig. Sci.* **2009**, *27*, 319–330. [CrossRef]
36. Wang, Z.; Jin, M.; Šimůnek, J.; van Genuchten, M.T. Evaluation of mulched drip irrigation for cotton in arid Northwest China. *Irrig. Sci.* **2013**, *32*, 15–27. [CrossRef]
37. Hoffman, G.J.; Van Genuchten, M.T. *Soil Properties and Efficient Water Use: Water Management for Salinity Control*; American Society of Agronomy: Madison, WI, USA, 1983.
38. Allen, R.G.; Pereira, L.S.; Raes, D.; Smith, M. *Crop. Evapotranspiration—Guidelines for Computing Crop Water Requirements*; FAO: Rome, Italy, 1998.
39. Watson, D.J. Comparative physiological studies in the growth of field crops: I. Variation in net assimilation rate and leaf area between species and varieties, and within and between years. *Ann. Bot.* **1947**, *47*, 41–76. [CrossRef]
40. Computing, W. *PEST—Model.-Independent Parameter Estimation*, 5th ed.; Watermark Computing: Brisbane, Australia, 2005.
41. Camillo, P.J.; Gurney, R.J. A Resistance Parameter for Bare-Soil Evaporation Models. *Soil Sci.* **1986**, *141*, 95–105. [CrossRef]
42. Daamen, C.C.; Simmonds, L.P. Measurement of Evaporation from Bare Soil and its Estimation Using Surface Resistance. *Water Resour. Res.* **1996**, *32*, 1393–1402. [CrossRef]
43. Merlin, O.; Stefan, V.G.; Amazirh, A.; Chanzy, A.; Ceschia, E.; Er-Raki, S.; Gentine, P.; Tallec, T.; Ezzahar, J.; Bircher, S.; et al. Modeling soil evaporation efficiency in a range of soil and atmospheric conditions using a meta-analysis approach. *Water Resour. Res.* **2016**, *52*, 3663–3684. [CrossRef]
44. van de Griend, A.A.; Owe, M. Bare soil surface resistance to evaporation by vapor diffusion under semiarid conditions. *Water Resour. Res.* **1994**, *30*, 181–188. [CrossRef]
45. Ding, R.; Kang, S.; Zhang, Y.; Hao, X.; Tong, L.; Du, T. Partitioning evapotranspiration into soil evaporation and transpiration using a modified dual crop coefficient model in irrigated maize field with ground-mulching. *Agric. Water Manag.* **2013**, *127*, 85–96. [CrossRef]

Article

Analysis of the Acceptance of Sustainable Practices in Water Management for the Intensive Agriculture of the Costa de Hermosillo (Mexico)

Claudia Ochoa-Noriega, Juan F. Velasco-Muñoz, José A. Aznar-Sánchez * and Belén López-Felices

Department of Economy and Business, Research Centre on Mediterranean Intensive Agrosystems and Agrifood Biotechnology (CIAIMBITAL), University of Almería, 04120 Almería, Spain; claudia08a@hotmail.com (C.O.-N.); jfvelasco@ual.es (J.F.V.-M.); blopezfelices@ual.es (B.L.-F.)
* Correspondence: jaznar@ual.es

Abstract: Mexico, as many countries, relies on its aquifers to provide at least 60% of all irrigation water to produce crops every year. Often, the water withdrawal goes beyond what the aquifer can be replenished by the little rainfall. Mexico is a country that has experienced a successful process of regional development based on the adoption of intensive agricultural systems. However, this development has occurred in an unplanned way and displays shortcomings in terms of sustainability, particularly in the management of water resources. This study analysed the case of Costa de Hermosillo, which is one of the Mexican regions in which this model of intensive agriculture has been developed and where there is a high level of overexploitation of its groundwater resources. Based on the application of a qualitative methodology involving different stakeholders (farmers, policymakers, and researchers), the main barriers and facilitators for achieving sustainability in water resources management have been identified. A series of consensus-based measures were contemplated, which may lead to the adoption of sustainable practices in water management. Useful lessons can be drawn from this analysis and be applied to other agricultural areas where ground and surface water resources are overexploited, alternative water sources are overlooked, and where stakeholders have conflicting interests in water management.

Keywords: intensive agriculture; water management; participatory assessment; stakeholders; sustainable development

Citation: Ochoa-Noriega, C.; Velasco-Muñoz, J.F.; Aznar-Sánchez, J.A.; López-Felices, B. Analysis of the Acceptance of Sustainable Practices in Water Management for the Intensive Agriculture of the Costa de Hermosillo (Mexico). *Agronomy* **2022**, *12*, 154. https://doi.org/10.3390/agronomy12010154

Academic Editors: Jorge F. S. Ferreira

Received: 10 November 2021
Accepted: 5 January 2022
Published: 8 January 2022

Publisher's Note: MDPI stays neutral with regard to jurisdictional claims in published maps and institutional affiliations.

1. Introduction

Of the objectives included in the 2030 Agenda of the United Nations, the eradication of poverty and hunger and access to drinking water are the most urgent for the survival of a large part of the population [1]. These objectives are closely related and their fulfilment is threatened by different factors. First, the population is growing much faster than food producers' capacity to respond [2]. It is estimated that the population will increase from 7.7 to 9.7 billion people by 2050 [3]. Furthermore, global economic development has given rise to the expansion of the population segment classified as middle class, which has a higher level of income, generating a modification in consumption patterns due to the evolution of global lifestyles [2]. Consequently, consumer preferences require a greater use of resources, which threatens the sustainability of the production system. It is estimated that in order to satisfy global demand for the year 2050, based on current consumption patterns, the resources equivalent to those of three planet earths would be necessary [4]. In food production alone, it has been estimated that by the year 2050, an increase in production of between 25 and 110% will be required, depending on the different possible scenarios [5,6].

As a principal supplier, not only of food but also a wide range of raw materials, agriculture plays a prominent role in ensuring food security [7]. In addition to satisfying the growing demand, agricultural production systems must adapt to the consequences of global climate change [8]. These consequences include the alteration of rainfall cycles, long periods of drought and imbalances in the supply of water; more frequent and more unpredictable and extreme weather phenomena; and changes in soil humidity, evapotranspiration flows and surface run-off [9,10]. The agricultural expansion and intensification taking place over the last few decades has enabled unprecedented growth in food production. However, it has had a severe impact on forest and aquatic systems [11]. Deforestation practices related to agriculture are the world's second largest threat in terms of conservation of biodiversity [12,13], given that approximately three quarters of the world's forests have been lost due to this activity [14].

The main limiting factor for the expansion and intensification of agriculture is the availability of water [15]. Furthermore, as the leading consumer of water resources on a global level, agriculture has reduced the quality and quantity of available water on a global level in recent decades [11,16]. Agriculture uses between 60 and 90% of the available water, depending on the climate and economic development of the region [17,18]. An increase in irrigation to satisfy the growing demand for food will severely affect the availability of water for the natural ecosystems and even human supply [19,20]. According to the 2020 United Nations report on water resources, there are currently 2.2 billion people across the world who have limited access to drinking water [21].

Mexico has become an agricultural power in terms of cultivated area, production and volume of exports [22]. It is also one of the world's principal suppliers of food [23]. The country has an area of 198 million hectares, of which approximately 73% is used for agricultural activities [24,25]. Agriculture accounts for approximately 4% of Gross National Product (GNP) [26]. In recent years, the share of Mexican agricultural products in foreign markets has increased, thanks to their quality and variety and the tariff advantages derived from the North American Free Trade Agreement (NAFTA) [27]. Furthermore, the agricultural activity has played a fundamental role in the regional development of Mexico [28]. Approximately 20% of the country's population is in a situation of food poverty, and 5% are classified as malnourished [25]. This situation is even more critical in the rural environment, where agriculture represents 50% of the income of the family [29,30]. According to the 2018 report on the evolution of the Sustainable Development Goals (SDGs), 58.2% of the Mexican rural population lived in a situation of poverty [31]. This figure was as high as 71.9% among the indigenous population (a total of twelve million people) [31]. It is estimated that the children living in the rural areas have a growth delay of 43.4%, more than double that of the national average of Mexico, with negative effects on motor and cognitive development [32,33].

Mexico is a paradigmatic example of a country which has experienced a successful process of regional development based on the evolution of traditional agricultural models towards modern agricultural systems [28]. However, this development has occurred in an unplanned way and displays shortcomings in terms of sustainability [23,31]. Due to its location and climate conditions, Mexican agriculture is particularly sensitive to the problem of water. Some of the principal agricultural regions suffer from serious deficit problems in their water bodies. Furthermore, this country is located in an area particularly vulnerable to the impacts of global climate change, most of all in terms of water resources and agricultural management [21,25]. In addition, this development has been based on the use of poor environmental management practices, fundamentally with respect to the management of water resources and the unequal distribution of land and infrastructures [23,29]. As a result, this country is a perfect laboratory for studying the agricultural development experienced by developing countries. Therefore, the objective of this study is to analyse one of the Mexican regions (Costa de Hermosillo) that has experienced an agricultural modernisation process more intensely, based on the overexploitation of its groundwater resources. Furthermore, it seeks to identify the principal barriers and facilitators for obtaining sustainability in the management of water resources in this region. Finally, it attempts to find a series of measures that will contribute to the adoption of sustainable practices in water management in the agricultural region studied.

The state of Sonora holds the third position in terms of the value of national agricultural production, with more than 15327 million pesos (748 million US$), accounting for 13.7% of the national total and a cultivated area of 411,090 hectares. The Costa de Hermosillo represents 12% of the total surface area with 49524 hectares and 23.2% of the total value of production with 3556 million pesos (173 million US$) [34]. The agriculture of the Costa de Hermosillo has evolved from traditional production systems based on corn, wheat and cotton crops to an intensive agricultural model based on the use of new technologies and innovation processes in production, storage and distribution [35,36]. This transformation began with the coming into force of the North American Free Trade Agreement (NAFTA) in 1994 [27]. Currently, the predominant crops are tomatoes, pumpkins, asparagus, green chili, melon, citrus fruits, cucumber, watermelon, grapes and walnuts, which are mainly exported. The state has a vast hydraulic infrastructure made up of a system of dams and pipelines for irrigation, which is carried out principally through gravity and flooding [37].

The Colonisation Decree of 1949 establishes three forms of land ownership; small owners, settlers and ejido members [38]. This ownership structure gave rise to a concentration of water as a result of the prior concentration of land [38]. The small owners have the private ownership of a farm for which the volume of groundwater used for irrigation cannot exceed 100 ha based on Clause XV of Article 27 of the constitution. In practice, this condition is not fulfilled [39]. The small owners have farms of between 200 and 400 ha. Furthermore, different members of the same family own farms resulting in the formation of large family farms with thousands of hectares [40]. The settler sector is formed by 66 settler associations, fruit of the migration from other regions. These associations were granted the right to collectively farm the low-quality land close to the coast, which were affected by the salinisation due to the seawater intrusion into the aquifer [41]. These lands have now been abandoned and the settlers work as day labourers for the small owners or have emigrated, mostly to the United States [41]. Finally, the ejido sector is made up of 28 scattered rural villages which were established from 1964 [42]. The crop area of the ejidos is of a low quality and is leased to the small owners or used for subsistence production in small farms by the ejido members, who sell their produce directly to consumers [39].

The use and exploitation of the groundwater is regulated through the National Water Act of 1992 (LAN). Article 3 of this Act allows the exploitation of the aquifers for the use of the resources through an individual license or concession granted to the private farmers by the National Water Commission (CONAGUA), which must be registered in the Public Registry of Water Rights (REPDA) [43]. Article 4 provides that the authority and administration of the aquifer correspond to the Federal Executive Body, which exercises these responsibilities directly or through the CONAGUA which, in turn, must be made up of a technical board and should have close ties with the Basin Councils of the respective water basin body responsible for monitoring, administrating or managing the use of the water resources [43].

Table 1 presents a selection of previous literature on the adoption of sustainable practices in Mexican agriculture. Among these works, the study of soil conservation and water resource management are highlighted as priority issues. Of particular relevance is the study of traditional knowledge in subsistence agricultural production, as a basis for the development of the most vulnerable rural populations. For more detailed information, see the work of Ochoa-Noriega [23], a bibliometric review of sustainable agriculture in Mexico.

Table 1. Previous literature on sustainable agricultural management in Mexico.

Title	Author and Year
Adoption of phytodesalination as a sustainable agricultural practice for improving the productivity of saline soils	Lastiri-Hernández et al. 2021 [44]
Analysis of energy consumption for tomato production in low technology greenhouses of Mexico	Ramırez-Arias et al. 2020 [45]
Temporal Dynamics of Rhizobacteria Found in Pequin Pepper, Soybean, and Orange Trees Growing in a Semi-arid Ecosystem	Diaz-Garza et al. 2020 [46]
The Use of Water in Agriculture in Mexico and Its Sustainable Management: A Bibliometric Review	Ochoa-Noriega, et al. 2020 [23]
Sustainability prospective for water resources in Northwestern Mexico: Use of recycled concrete for Agricultural purpose water supply	Gutiérrez-Moreno et al. 2020 [47]
Ecological, Cultural, and Geographical Implications of Brahea dulcis (Kunth) Mart. Insights for Sustainable Management in Mexico	Pérez-Valladares et al. 2020 [48]
The sustainable cultivation of Mexican nontoxic Jatropha curcas to produce biodiesel and food in marginal rural lands	Pérez et al. 2019 [49]
Sustainability and environmental management in the Mexican vegetable sector	Padilla-Bernal et al. 2019 [50]
Vulnerability, innovation and social resilience in the maize (Zea mays L.) production: The case of the conservation tillage club of chiapas, Mexico	Díaz-José et al. 2018 [51]
The myth behind sustainable African palm crop. Socio-environmental impacts of palm oil in Chiapas, Mexico	León et al. 2017 [52]
TEK and biodiversity management in agroforestry systems of different socio-ecological contexts of the Tehuacán Valley	Vallejo-Ramos et al. 2016 [53]
Degree of sustainability of rural development in subsistence, intermediate, and commercial farmers, under an autopoietic view point	García et al. 2009 [54]

2. Materials and Methods

This study seeks to analyse a complex agricultural system that incorporates different types of agents with conflicting objectives. Moreover, it aims to reach a consensus-based proposal for the sustainable management of the water resources available in the system. In order to fulfil this objective, a participatory qualitative methodology has been developed. This type of research provides a more in-depth understanding of the topic of study, the variables involved, the relationships established between them and identifies the critical points, which enables us to appreciate the interactions in complex systems, such as the case of water management systems [55,56]. Finally, even though the potential for generalisation of case studies may be limited, these types of studies can offer a range of possible alternatives to test in similar contexts and can constitute a model with which to reach consensus-based measures in other contexts [57].

2.1. Case Study

The study was conducted in the Costa de Hermosillo, in the northeastern region of Mexico, in the central coastal plain of the state of Sonora (Figure 1). The Hermosillo Coast stretches 100 km in a straight line between the city of Hermosillo and Bahía de Kino, on the shores of the Gulf of California. This area has a semi-arid climate, with an annual average rainfall of less than 100 mm, concentrated in the summer months, an annual average temperature of 24 °C which can fluctuate between a maximum of 47 °C and a minimum of −3 °C, and high solar radiation [55].

The Costa de Hermosillo corresponds to irrigation district 051 created in 1953 for the management of its agricultural water resources [58]. This district is supplied by the water basin of the Sonora and Bacoachi rivers, which have irregular flows, a low volume and high infiltration [59]. The principal source of water for irrigation is underground, being one of the largest pump irrigation districts in the country [41]. In 1980, a total of 498 wells were drilled exclusively for agricultural use, accounting for 90% of the available water for this sector [60,61]. The main aquifer of the system is identified as 2619. This aquifer has an average annual recharge of 250 hm^3/year and an average extraction of 346 hm^3/year [41]. As a result, there is an average annual deficit of 96 hm^3/year, which has translated into a reduction in the total volume of water, giving rise to a process of water intrusion, contaminating the available freshwater [58]. It has been declared as one of the 17 aquifers with saltwater intrusion and as one of the 115 overexploited aquifers on a national level, having the highest deficit of the 61 existing in the state of Sonora [62]. As a consequence of the water resource situation, farms that are unproductive due to the salinity of the soil have been abandoned. The concession of new farms is unfeasible and the rivalry between the different users of water for irrigation has increased.

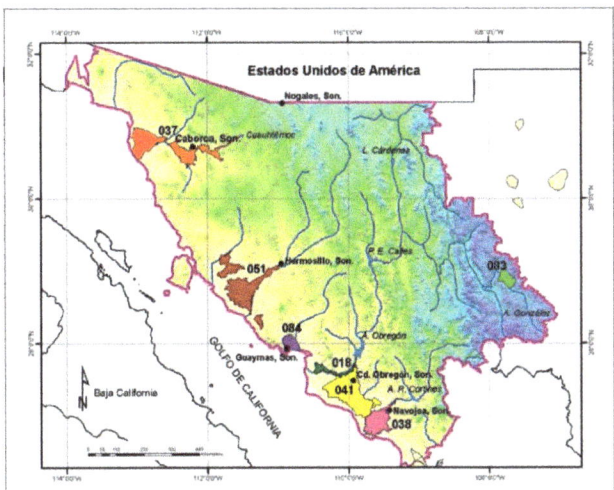

Figure 1. Location of irrigation districts of the basin organisation: Northwestern Mexico. District 051 Costa de Hermosillo, Sonora, Mexico [34]. Sonora borders the American states of Arizona and New Mexico.

The administration and management of the use of the groundwater in this area is carried out through a group license granted to the User's Association of District 051 (AUDR, 051) [63]. This has led to a greater concentration of resources, predominantly by the private farmers who have been able to afford the pumping costs and the investments necessary to meet the standards established in the destination countries for exported crops [40].

2.2. Methods

This study has used a series of methods aiming to compile both primary and secondary information, based on the different development stages of the research. First, a literature review was carried out in order to establish the conceptual framework to guide the rest of the process. Second, a series of interviews with experts was conducted on the topic in order to identify the possible management alternatives and principal barriers and facilitators for their adoption. Finally, a workshop was undertaken to assess the different points of view of the groups of stakeholders involved and to draw up a consensus-based proposal for the adoption of the measures to apply.

As a starting point, the literature review is considered as a necessary task in all research studies [63]. The objective of this methodological tool is to identify the most relevant contributions in which the concepts and theories that should be applied are defined and to structure the research problem [64]. In this way, the context is obtained and the theoretical and conceptual foundation is established based on previous studies on the topic and case studies [65]. Furthermore, the experience gained in other analyses enables us to identify the principal variables that intervene in the case study and to delimit their structure of relations, allowing us to establish starting hypotheses [66]. The literature review included both scientific and grey literature. In the first case, the main literature repositories were used, both in English and Spanish, such as Dialnet, Scielo and Scopus. The grey literature included documents published by official sources such as the National Water Commission, the Official Journal of Mexico, or the United Nations.

The interviews are more or less structured conversations which generate interaction between the parties involved with the objective of obtaining knowledge [67]. As an exploratory method of research, the interviews seek to find new aspects and develop research questions regarding topics that are not clearly defined [68]. In-depth interviews generate an exchange of ideas through interactive conversations with stakeholders with the objective of establishing a close relationship between the participants and the interviewer in order to obtain exhaustive and significant responses [69]. These interviews are not structured or semi-structured. They are based on a script with a series of open questions which are answered during the interview [70]. The method of sample selection was snowballing. This non-probability sampling technique is based on the fact that a small set of study subjects recruits future subjects from among their acquaintances. In this way, the statistical sample grows according to a snowball or domino effect [63]. There were two advantages to using this methodology. On the one hand, it made it possible to contact the right person for the purpose of the study. On the other hand, it allowed for a good predisposition on the part of the interviewee by having the recommendation of another person. A total of seven experts participated: two from academia, two from business, two from administration and one technical professional. The experts were selected from among persons of recognised prestige within the agricultural sector for their leadership position within a relevant organisation (public or private), number of scientific publications, and/or years of experience. A script for the open-ended interview is included in the supplementary material (Supplementary Material 1).

Finally, a workshop was conducted in order to fulfil the objective of designing a management proposal agreed by all of the parties. This methodology enables different stakeholders to collaborate in order to share their knowledge on the theme of study [71]. The workshop is a tool that allows the knowledge from different fields to be synthesised and assessed and conclusions to be drawn [72,73]. Furthermore, it can reinforce the connection between the researchers and policymakers, enabling the development of knowledge that can serve as a base with which to generate policies [74]. The use of this methodology seeks to present all of the knowledge obtained in the previous stages of the research, to incorporate the different points of view of the stakeholders and to reach a consensus-based proposal which allows the adoption of sustainable management practices. In the previous interview phase, farmers, policymakers and researchers were highlighted as the main stakeholder groups. In the case of farmers, it refers to private owners, as they are the main decision-makers in land management. The policymakers are responsible for setting policies and regulations, as well as incentives to encourage behaviour. Researchers are the main providers of knowledge. Through the snowballing procedure, an equal number of members were selected from each group, so that there would be a homogeneous representation of the different groups. In this way, all groups are in the same position to reach an unbiased consensus.

In order to establish a hierarchy with respect to the level of influence of the different factors identified regarding the adoption of each of the practices proposed, a workshop was carried out incorporating the most representative interested parties. The workshop was attended by representatives of farmers (private landowners), policy makers, and researchers. Farmers (private landowners) are the ones who are mainly affected by the proposed measures and who must carry out the practice. Policy makers need to regulate and set incentives to implement the practices. Finally, researchers are in charge of generating the necessary knowledge to guide the whole process. Each of these groups contributed with a total of three participants, so that the different interests and points of view were considered equally.

3. Results

The most urgent problem to be addressed, according to the perception of stakeholders, is the scarcity of water resources and the overexploitation of aquifer Costa de Hermosillo (designated in the National Water Law as aquifer 2619), caused by the development of agricultural activities on the Hermosillo Coast. Therefore, different practices have been identified to increase the supply of water for irrigation through diversification of sources. Of all the possible alternatives, two sustainable practices capable of contributing to the recovery of the aquifer through the reduction of abstractions have been selected:

- The harvesting and storage of rainwater (hereinafter P1—practice 1). Given the characteristics of the area of study, the majority of the rainwater is lost through evaporation or run-off. Rainwater can constitute a low-cost resource, requiring only the installation of a small infrastructure to enable its channeling and storage [75,76]. Another relevant aspect is the monitoring of rainfall in order to plan the water needs based on the harvesting of annual rainfall for the crops [75]. This rainfall monitoring should include the total duration of rainfall, the intensity (volume of rain per unit of time) and frequency (the number of precipitations in a given time and with certain characteristics). The compilation of these data enables the design of a climate prediction model for developing technical processes of infrastructures that control the harvesting and storage of rainwater for agricultural use.
- The desalination of seawater (hereinafter P2—practice 2). Desalination is a process in which the salts are eliminated from the water. Although there are different methods of desalination, the most commonly used is reverse osmosis. In this process, the water is conducted through semi-permeable membranes under pressure. The salts are retained in the membranes, while the water molecules circulate.

Five principal barriers and five facilitators were identified for adopting sustainable practices in the management of irrigation water in the area of study (Figure 2). These factors were classified into three different groups: institutional, technical and socio-economic. Barriers include (i) the lack of regulation and the high level of noncompliance with existing legislation; (ii) the current land ownership structure and the concentration of water use rights; (iii) the lack of technical knowledge regarding the proposed innovations; (iv) the low level of rainfall; and (v) the lack of environmental knowledge of farmers. The main facilitators are (i) the existence of institutional incentives for the adoption of sustainable practices; (ii) the continuous process of technological innovation in which the sector is immersed; (iii) the positive disposition of farmers towards technical change; (iv) the collaborative relationships between the different actors; (v) the sector's financing capacity.

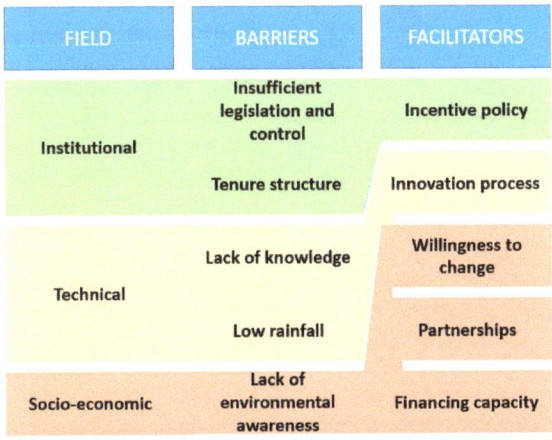

Figure 2. Main barriers and facilitators for the adoption of sustainable practices.

Figure 3 illustrates the results regarding the perception of each group of stakeholders in terms of the level of influence of each barrier to adopting the proposed practices. In this case, the farmers and policymakers show a higher level of agreement. The two groups coincide in considering that the principal barriers to adopting the rainwater harvesting systems are the lack of knowledge of the different aspects of the infrastructure, capacity and return on investment, and the erratic behaviour of the rainfall, which makes it difficult to forecast the water needs at any given time, particularly with the impact of climate change. Meanwhile, the researchers indicated a high degree of noncompliance with the applicable regulations, the low level of environmental awareness among the farmers and policymakers, and the power of the farmers to concentrate the water rights derived from the land ownership regime.

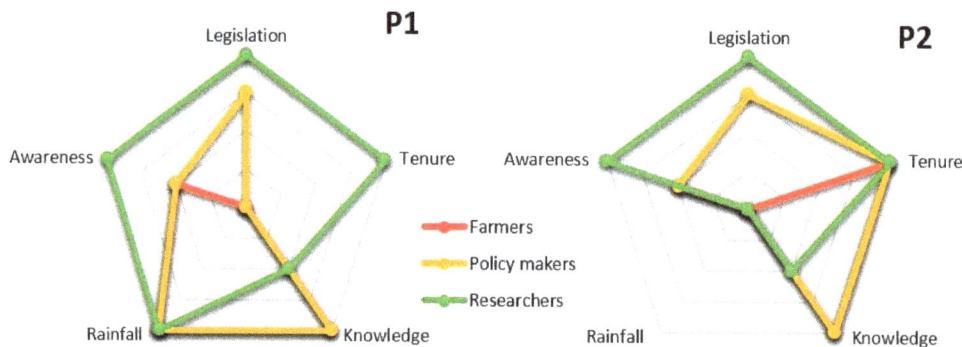

Figure 3. Main barriers to the adoption of sustainable practices by group. P1 (harvesting and storage of rainwater, practice 1); P2 (desalination of seawater, practice 2).

With respect to the installation of the desalination plant in the area of study, the three groups identify the ownership regime as the principal obstacle. However, there are different opinions with respect to the reason for this. The researchers and policymakers believe that the negotiating capacity of the farmers can impose objectives to increase the crop area instead of mitigating the overexploitation of the aquifers. On the other hand, the farmers highlight the need for finance from the administration, given that the group of farmers is very small and cannot undertake such a large investment which would be borne by a small number of entrepreneurs. In this case, the rainfall factor is not relevant, as the desalination of seawater does not depend on climate factors. The principal discrepancy regarding the different barriers resides in the fact that the researchers continue to denounce a lack of compliance with the regulations and environmental awareness. The farmers and policymakers claim that there is a gap in the knowledge on a local level regarding the impact of the use of desalinated water. In this respect, the researchers argue that there is sufficient research in favour of the use of this technology, although they acknowledge that more information on a local level is required even though previous studies have been carried out [77,78].

Concerning the factors acting as facilitators for the adoption of the proposed practices, the results are shown in Figure 4. In this case, the responses are more similar as they refer to both management alternatives. With regards to rainwater harvesting, the three groups indicate that the modernisation of agriculture experienced over the last few decades and the disposition of the farmers in following the continuous improvement process are the principal pillars for the adoption of these practices. The policymakers indicated that the administration has already made different proposals to encourage technological development in the region which should serve as an incentive to adopt these practices. Meanwhile, the researchers surveyed in this study support that the sector has sufficient financing capacity to cover the investment necessary for the installation of rainwater harvesting systems. However, with respect to the installation of the seawater desalination plant, the researchers surveyed in this study do not believe that the prior innovation process will be so positive, given that, to date, the sector has not made an investment of such a scale, so prior experience will not be useful for managing this new additional resource.

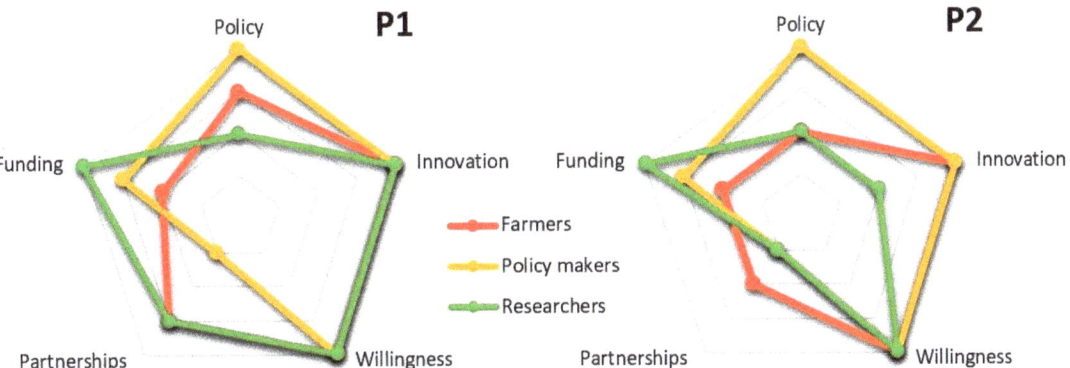

Figure 4. Main facilitators for adopting sustainable practices by group. P1 (harvesting and storage of rainwater, practice 1); P2 (desalination of seawater, practice 2).

As a result of the workshop and sharing the different points of view, the different groups represented established a series of common points that serve as a guide to design an action plan for the adoption of the different practices proposed. This action plan is based on three pillars. Firstly, commitment to the gradual reduction in groundwater extraction from the aquifer as new water sources become available. Secondly, closer collaboration and communication between the different groups to obtain and disseminate knowledge and to improve the level of environmental awareness. Finally, the design of a public-private financing strategy in order to undertake the investment necessary for the development of the proposed projects.

First, the three groups (farmers, policy makers, and researchers) agree that the situation of the aquifer is unsustainable and threatens the continuity of the agricultural activity in the area in the medium to long term, which is the case of other coastal territories of this region. To address this problem, the farmers consider as positive the reduction of the extraction of groundwater for irrigation until the aquifer has been recharged and other alternative water sources become available for crop irrigation. In this way, the crop areas will not be extended until the water supply is expanded and the possible surplus will be used for recharging the underground water bodies.

Second, to guarantee that this commitment is fulfilled, all of the groups agree that it is necessary to work together to design the best strategy for implementing the new processes. On the one hand, updated technical knowledge is required that will enable the optimisation of the investment. This knowledge should be obtained directly on the farms; so, close collaboration between researchers and farmers is required. Furthermore, the policymakers should provide coverage and get involved in all of the levels for the process to be successful. Therefore, there should be a three-way collaboration. On the other hand, the level of environmental awareness regarding the different processes related to agriculture should be improved, not only in terms of water management, but also those referring to the conservation of the soil and air pollution. These factors give rise to a better conservation of the water resources. Finally, a comprehensive management of the agricultural ecosystem is the only way to maintain the means of subsistence for future generations.

Third, undertaking the proposed investments, particularly the seawater desalination plant, requires a considerable volume of investment. According to the consensus reached, given the general interest in the conservation of the underground water bodies, while maintaining an agricultural production that supports a good part of the population in the area of study, the best option is the development of a joint financing project between the administration and the entrepreneurs. In this way, the negative externality (negative result of agricultural activity not incorporated in its costs) generated with the overexploitation of the aquifer by the farmers would be internalised and the farmers would be compensated by the positive externality (positive result from agricultural activity not incorporated in its profits) generated by the creation of jobs and the securing of the rural population.

4. Discussion

Among the different aspects that condition the sustainability of the agricultural productive model on the Hermosillo Coast, water supply is a pressing structural problem. This situation is aggravated by the impact of global climate change on temperature and rainfall patterns. Therefore, all actors involved agree that measures must be taken. This is mainly because producers are already aware of the decline in crop productivity due to high groundwater salinity, that is the only source of water for irrigation. Carrillo-Huerta and Gómez-Bretón [78] conducted a study on the evaluation of irrigation technology in a district with an overexploited aquifer in Puebla, Mexico. Among their results, they identified farmers and policy-makers as the key stakeholder groups in water resources management. These results are similar to those obtained in the present study. However, in their case, the path chosen to improve the management of water resources and the state of the aquifer was on the demand side, whereas in this study the focus was on increasing the supply of the resource. Dévora-Isiordia et al. [79] estimated an increase of one tonne per hectare in sorghum production by using desalinated water (0.1 dS/m) instead of brackish well water (10.3 dS/m) in the Yaki Valley, Sonora. In addition, these authors have tested the technical and economic feasibility of using desalinated water in the study area by comparing different technical developments [80]. Their results show that the price of desalinated water (obtained from brackish water) was 0.6 US$/1000L, while for seawater it was 1.2 US$/1000L. Other studies show that desalinated water can be more costly in monetary and energy terms when compared to other sources. A report comparing the cost of alternative water supply and efficiency options in California [81] cited that desalinated water was the most expensive source of water to groundwater at 2100$/acre foot for large projects and 2800–4000$/acre foot for smaller projects. The cheapest was captured stormwater (590$/acre foot). The second least expensive was desalinized brackish water (requires less energy than to desalinate seawater). It is shown that reused water could be a more economical alternative source than desalinated seawater. The reason why reused water was not initially included as an alternative in this study was that in the study area there are 44 water treatment plants for reuse for industrial purposes and irrigation of gardens and green areas. Thus, its availability for agriculture is lower. However, the authors of this work propose this as a future line of research due to the fact that some of these plants are underused, the price differential indicated, and the improvement in the circularity of urban water that reuse would entail.

Rainwater harvesting systems, adapted to different types of agricultural practices, are widely developed, and have demonstrated their viability to supplement irrigation in semi-arid environments [76]. Loera-Alvarado et al. [82] conducted a study to test the suitability of runoff water for agricultural use in the State of San Luis Potosí. From their results, they concluded that the runoff water stored in earthen dams is of excellent quality for agricultural use (even in soils with very low permeability) and to grow crops sensitive to salinity and sodium. However, they indicate that it is necessary to assess water suitability in conjunction with the soil-climatic characteristics of the site in order to establish an appropriate management system for each specific case. This would be especially necessary in the case of combining runoff water with desalinated seawater, based on the proposal of this study. López-Hernández et al. [83] compare a rainwater harvesting system with groundwater abstraction for domestic and agricultural use in a municipality in the State of Tlaxcala, Mexico. Their results show that rainwater can be more economically viable than ground-water abstraction when demand is low. A future line of research could compare the demand for the two types of water in the study area to establish the tipping point in the use of these two resources and influence demand control to minimise water use from the aquifer.

As already mentioned, in the area of study there is a regulation for the concession of the rights to extract groundwater for irrigation, establishing a maximum limit per farmer [39]. These rights are obtained through the User's Association of District 051 (AUDR, 051), up to the maximum allowed by law [63]. However, in practice this condition is not fulfilled. There is a lack of control in terms of the area that is irrigated with groundwater and the amount of water extracted by each concessionaire. On the other hand, the ownership structure of the land has enabled a small group of owners to control large areas [40]. This has given rise to the concentration, in parallel, of the water rights, and has relegated a large part of the population in the area to the role of day labourer. This, in turn, bestows a high degree of negotiating power to the private owners, with respect to the proletariat and the administration, given that their decisions have important repercussions on an environmental, economic and social level for the whole region. From a technical perspective, there is a lack of knowledge on the impact of the adoption of the practices proposed, due to their innovative nature in the area of study [84]. Furthermore, the scarcity of rainfall is a factor to consider, particularly in the case of the practice of rainfall harvesting. In the case of this specific practice, the development of scenarios to evaluate the technical and economic feasibility of the investment is much more relevant. Finally, from a social perspective, the short and medium-term economic criteria play a prominent role in decision-making. Conversely, the main social and environmental impacts are seen in the medium and long term [85]. Social impacts include inequality, job insecurity and deterioration of health, especially for the most vulnerable people [32]. Environmental impacts include the deterioration and depletion of water bodies, the transformation of the landscape, and infertility of the soil [6]. In this respect, there is a low level of awareness of the concept of sustainability among farmers. For their part, the labourers do not have the capacity to influence the decisions of the landowners. They receive low wages, which they supplement with subsistence farming on small, unproductive plots of land. In many cases, they are forced to migrate to improve their living conditions in the USA. Aznar-Sanchez et al. [86] studied the use of desalinated seawater as a measure to increase irrigation water supply and improve the sustainability of an overexploited aquifer in Spain. In their case, the main barriers on the farmers' side were the low level of knowledge about the impact of using this type of water, the increased costs (e.g. due to increased fertiliser use) and the price of water. These last two factors were not identified by the stakeholders in the present study.

Despite these barriers, the area of study has a series of facilitators for the adoption of sustainable practices in the management of water for irrigation. On the institutional level, there is a willingness to offer economic and technical consulting incentives for the adoption of technological innovation, leading to an improvement in exports, all under the umbrella of NAFTA. From a technical perspective, and also since the entry into force of NAFTA, the Costa de Hermosillo has experienced a process of innovation, on both a technological and organisational level [80]. The success of this process has generated great interest in continuous improvement among the farmers. Furthermore, during this period, ties have been established between the farmers through the official bodies and through professionals promoting common interests, such as water management or the marketing of products [39]. Carrillo-Huerta and Gómez-Bretón [78] identified technical assistance as the main contribution of public managers to the adoption of sustainable irrigation practices. On the farmers' side, these authors found that associationism around irrigation communities is the main facilitator towards sustainable management that allows aquifer recovery. In the same way, there is close contact between the agricultural business organisations and the Public Administration. These relationships constitute facilitators when designing legislative proposals and providing resources. Finally, as a result of the exporting activity and its attractiveness for investment, there is sufficient financial capacity to carry out the investments necessary to improve the agricultural production sector in the area of study, provided that a return on the investment can be gained. Carrillo-Huerta and Gómez-Bretón [78] identified water price as a determining factor in irrigation management in their study from the demand side. This factor has not been pointed out by the stakeholders in our study from the supply side perspective. Aznar-Sanchez et al. [86] identified the possibility of crop diversification and the lack of availability of other alternative sources as the main facilitators for the use of desalinated seawater for irrigation.

Carrillo-Huerta and Gómez-Bretón [78] point out that the lack of consensus in the design and planning of irrigation management measures is the main reason for the current state of deterioration of the aquifer, the result of overexploitation. Therefore, although the proposal made in this paper may be ambitious, having the agreement of the main stakeholders is a positive starting point. The project to build a seawater desalination plant implies the mobilisation of a large amount of resources, not only for the desalination facility, but also for the channelling and transport of water. On the other hand, in 2017, the governor of the state of Sonora, Claudia Pavlovich, led a proposal for the construction of a desalination plant with a capacity of 6,307,200 m^3 per year devoted to human consumption [87]. This project has not yet been implemented. However, it is proof that the proposal made in this paper has broad support and the backing of the political class.

Finally, it should be noted that the main limitation of this study is its exploratory nature and the qualitative information on which it is based. Therefore, the development of a broad stakeholder survey is proposed as a future line of research. The purpose of this survey would be to verify the real support of all stakeholders for the proposal, as well as to identify any possible conflicting points that may be detected.

5. Conclusions

The objective of this study is to elaborate a proposal to improve the management of the water resources of the Costa de Hermosillo which would be able to: (i) improve the situation of overexploitation of the underground water bodies, (ii) contribute to the sustainability of the agricultural activity in the area, and (iii) reach a consensus between the different parties involved in order to guarantee the success of its implementation.

The results show that the main concern for different stakeholders to ensure the sustainability of an agricultural system in a semi-arid environment is the availability of water. Technology offers a variety of alternatives to try to increase water supply through sources other than overexploited water bodies. In systems based on the use of groundwater with seawater intrusion problems, alternative water sources such as desalinized seawater, rainwater, brackish water, and reclaimed municipal water are potential alternative sources for groundwater and surface waters.

The results also show that the principal driving factors for adopting innovations in the management of agricultural irrigation are the existence of institutional incentives for adopting sustainable practices; the continual process of technological innovation in which the sector is immersed; the good disposition of the farmers towards technical change; the collaboration relationships between the different stakeholders; and the financing capacity of the sector. The principal elements that hinder the adoption of these practices are the lack of regulation and the high level of non-compliance with the legislation in force; the structure of the current land ownership and the concentration of the water use rights; the lack of technical knowledge pertinent to the innovations proposed; the low level of rainfall; and the lack of environmental knowledge of the farmers.

The principal contribution of this study is a proposal designed by the farmers, policymakers and researchers of the area to evaluate the implementation of rainwater harvesting systems and the construction of a seawater desalination plant. This proposal is based on three pillars of action: (i) the reduction of extractions, (ii) continuous cooperation and (iii) public-private financing. These pillars constitute the priority lines of work for stakeholders to carry out the plan designed to improve sustainability in the use of water resources for irrigation. Therefore, a strong commitment from all stakeholders in these three areas of action is essential. Furthermore, given that the concentration of land ownership in turn leads to a concentration of water use rights, it would be desirable to update the forms of water governance in a way that it decouples land use from water use.

Supplementary Materials: The following supporting information can be downloaded at: https://www.mdpi.com/article/10.3390/agronomy12010154/s1. Supplementary Material 1: Analysis of the acceptance of sustainable practices in water management for the intensive agriculture of the Costa de Hermosillo (Mexico).

Author Contributions: Conceptualization, C.O.-N., J.F.V.-M., J.A.A.-S. and B.L.-F.; Formal analysis, C.O.-N. and J.F.V.-M.; Investigation, C.O.-N.; Methodology, C.O.-N., J.F.V.-M., J.A.A.-S. and B.L.-F.; Project administration, J.A.A.-S.; Supervision, J.F.V.-M. and J.A.A.-S.; Validation, J.F.V.-M., J.A.A.-S. and B.L.-F.; Writing—original draft, C.O.-N.; Writing—review & editing, J.F.V.-M., J.A.A.-S. and B.L.-F. All authors have read and agreed to the published version of the manuscript.

Funding: This research received no external funding.

Acknowledgments: This research was partially supported by the Spanish Ministry of Economy and Competitiveness and the European Regional Development Fund by means of the research project ECO2017-82347-P, and from Junta de Andalucía and FEDER aid (project P18-RT-2327 and project UAL-2020-SEJ-D1931, Consejería de Transformación Económica, Industria, Conocimiento y Universidades). And by the FPU19/04549 Predoctoral Contract to Belén López-Felices.

Conflicts of Interest: The authors declare no conflict of interest.

References

1. Koren, O.; Bagozzi, B.E.; Benson, T.S. Food and water insecurity as causes of social unrest: Evidence from geolocated Twitter data. *J. Peace Res.* **2021**, *58*, 67–82. [CrossRef]
2. Oberle, B.; Bringezu, S.; Hatfeld-Dodds, S.; Hellweg, S.; Schandl, H.; Clement, J.; Cabernard, L.; Che, N.; Chen, D.; Droz-Georget, H.; et al. *Global Resources Outlook 2019: Natural Resources for the Future We Want. A Report of the International Resource Panel*; United Nations Environment Programme: Nairobi, Kenya, 2019. Available online: https://www.resourcepanel.org/file/1161/download?token=gnbLydMn (accessed on 12 July 2021).
3. United Nations Department of Economic and Social Affairs Population Division (UNDESA). Probabilistic population Projections Based on the World Population Prospects: The 2019 Revision. New York, USA. 2019. Available online: https://population.un.org/wpp (accessed on 12 July 2021).
4. Ceratti, M. *Dos Planetas Más Para Poder Vivir En Este*; World Bank: Washington, DC, USA, 2016. Available online: https://www.bancomundial.org/es/news/feature/2016/08/09/objetivo-desarrollo-sostenible-ods-12-consumo (accessed on 12 July 2021).
5. Hunter, M.C.; Smith, R.G.; Schipanski, M.E.; Atwood, L.W.; Mortensen, D.A. Agriculture in 2050: Recalibrating targets for sustainable intensification. *Bioscience* **2017**, *67*, 386–391. [CrossRef]
6. Lauretti, R.; Paço, A.; Mainardes, E.W. Sustainable Development in Agriculture and its Antecedents, Barriers and Consequences—An Exploratory Study. *Sustain. Prod. Consum.* **2021**, *27*, 298–311. [CrossRef]
7. Velasco-Muñoz, J.F.; Aznar-Sánchez, J.A. The economic valuation of ecosystem services in the agroecosystems in Spain: Conceptual framework and methodology. *Pecvnia* **2016**, *22*, 75–93. [CrossRef]
8. Velasco-Munoz, J.F.; Mendoza, J.M.F.; Aznar-Sanchez, J.A.; Gallego-Schmid, A. Circular economy implementation in the agricultural sector: Definition, strategies and indicators. *Resour. Conserv. Recycl.* **2021**, *170*, 105618. [CrossRef]
9. Pedro-Monzonís, M.; Solera, A.; Ferrer, J.; Estrela, T.; Paredes-Arquiola, J. A review of water scarcity and drought indexes in water resources planning and management. *J. Hydrol.* **2015**, *527*, 482–493. [CrossRef]
10. Mancosu, N.; Snyder, R.L.; Kyriakakis, G.; Spano, D. Water scarcity and future challenges for food production. *Water* **2015**, *7*, 975–992. [CrossRef]
11. Aznar-Sánchez, J.A.; Piquer-Rodríguez, M.; Velasco-Muñoz, J.F.; Manzano-Agugliaro, F. Worldwide research trends on sustainable land use in agriculture. *Land Use Pol.* **2019**, *67*, 104069. [CrossRef]
12. Maxwell, S.L.; Fuller, R.A.; Brooks, T.M.; Watson, J.E.M. Biodiversity: The ravages of guns, nets and bulldozers. *Nature* **2016**, *536*, 143–145. [CrossRef] [PubMed]
13. Kissinger, G.; Herold, M.; De Sy, V. *Drivers of Deforestation and Forest Degradation: A Synthesis Report for REDD+ Policymakers*; Lexeme Consulting: Vancouver, BC, Canada, August 2012. Available online: https://www.cifor.org/knowledge/publication/5167/ (accessed on 25 July 2021).
14. Forouzani, M.; Karami, E. Agricultural water poverty index and sustainability. *Agron. Sustain. Dev.* **2011**, *31*, 415–432. [CrossRef]
15. Velasco-Muñoz, J.F.; Aznar-Sánchez, J.A.; Batlles-delaFuente, A.; Fidelibus, M.D. Sustainable Irrigation in Agriculture: An Analysis of Global Research. *Water* **2019**, *11*, 1758. [CrossRef]
16. Aznar-Sánchez, J.A.; Velasco-Muñoz, J.F.; Belmonte-Ureña, L.J.; Manzano-Agugliaro, F. The worldwide research trends on water ecosystem services. *Ecol. Indic.* **2019**, *99*, 310–323. [CrossRef]

17. Aznar-Sánchez, J.A.; Belmonte-Ureña, L.J.; Velasco-Muñoz, J.F.; Manzano-Agugliaro, F. Economic analysis of sustainable water use: A review of worldwide research. *J. Clean. Prod.* **2018**, *198*, 1120–1132. [CrossRef]
18. Adeyemi, O.; Grove, I.; Peets, S.; Norton, T. Advanced monitoring and management systems for improving sustainability in precision irrigation. *Sustainability* **2017**, *9*, 353. [CrossRef]
19. Cunningham, S.A.; Attwood, S.J.; Bawa, K.S.; Benton, T.G.; Broadhurst, L.M.; Didham, R.K.; McIntyre, S.; Perfecto, I.; Sam-ways, M.J.; Tscharntke, T.; et al. To close the yield-gap while saving biodiversity will require multiple locally relevant strate-gies. *Agric. Ecosyst. Environ.* **2013**, *173*, 20–27. [CrossRef]
20. Fu, H.Z.; Wang, M.H.; Ho, Y.S. Mapping of drinking water research: A bibliometric analysis of research output during 1992–2011. *Sci. Total Environ.* **2013**, *443*, 757–765. [CrossRef]
21. UNESCO. UN-Water, 2020: United Nations World Water Development Report 2020: Water and Climate Change, Paris, UNESCO. Available online: https://unesdoc.unesco.org/ark:/48223/pf0000372985/PDF/372985eng.pdf.multi (accessed on 25 July 2021).
22. Ochoa-Noriega, C.A.; Velasco-Muñoz, J.F.; Aznar-Sánchez, J.A.; Mesa-Vázquez, E. Overview of Research on Sustainable Agricul-ture in Developing Countries. The Case of Mexico. *Sustainability* **2019**, *13*, 8563. [CrossRef]
23. Ochoa-Noriega, C.A.; Aznar-Sánchez, J.A.; Velasco-Muñoz, J.F.; Álvarez-Bejar, A. The Use of Water in Agriculture in Mexico and Its Sustainable Management: A Bibliometric Review. *Agronomy* **2020**, *10*, 1957. [CrossRef]
24. Food and Agricultural Organization of the United Nations. *El sistema Alimentario en México—Oportunidades Para el Campo Mexicano en la Agenda 2030 de Desarrollo Sostenible*; FAO: Ciudad de México, Mexico, 2019. Available online: http://www.fao.org/3/CA291 0ES/ca2910es.pdf (accessed on 25 July 2021).
25. Food, Agricultural and Fisheries Information Service. *2019 Food & Agriculture Overview*; SIAP: Mexico City, Mexico, 2019. Available online: https://nube.siap.gob.mx/gobmx_publicaciones_siap/pag/2019/Agricultural-Atlas-2019 (accessed on 21 July 2021).
26. World Trade Organization. World Trade Statistical Review. 2019. Available online: https://www.wto.org/english/res_e/statis_e/wts2019_e/wts19_toc_e.htm (accessed on 21 July 2021).
27. Dyer, G.A.; Hernández-Solano, A.; Meza-Pale, P.; Robles-Berlanga, H.; Yúnez-Naude, A. Mexican agriculture and policy under NAFTA. In *Serie Documentos de Trabajo del Centro de Estudios Económicos 2018–04*; El Colegio de México, Centro de Estudios Económicos: Mexico City, Mexico, 2018.
28. Garduño-Rivera, R. Regional Economic Development in Mexico: Past, Present, and Future. In *NAFTA's Impact on Mexico's Regional Development*; New Frontiers in Regional Science: Asian, Perspectives; De León-Arias, A., Aroca, P., Eds.; Springer: Singapore, 2021; Volume 51. [CrossRef]
29. Sosa-Baldivia, A.; Ruíz-Ibarra, G. Food availability in Mexico: An analysis of agricultural production over the last 35 years and its projection for 2050. *Pap. Poblac.* **2017**, *23*, 207–230. Available online: https://rppoblacion.uaemex.mx/article/view/9111 (accessed on 25 July 2021).
30. The World Bank. 2021. Available online: https://data.worldbank.org/indicator/SL.AGR.EMPL.ZS?end=2019&locations=MX&start=1991 (accessed on 21 July 2021).
31. Oficina de la Presidencia de la República y Secretario Ejecutivo del Consejo Nacional de la Agenda 2030 para el Desarrollo Sostenible. Informe Nacional Voluntario para el Foro Político de Alto Nivel sobre Desarrollo SOSTENIBLE 2018. Bases y Fundamentos en México Para Una Visión del Desarrollo Sostenible a Largo Plazo. Avance en el cumplimiento de la Agenda 2030 y los Objetivos de Desarrollo Sostenible. Available online: http://www.agenda2030.mx/docs/doctos/InfNalVol_FPAN_DS_2018 _es.pdf (accessed on 4 August 2021).
32. Carrasco-Quintero, M.R.; Ortiz-Hernández, L.; Roldán-Amaro, J.A.; Chávez-Villasana, A. Desnutrición y desarrollo cognitivo en infantes de zonas rurales marginadas de México. *Gac. Sanit.* **2016**, *30*, 304–307. [CrossRef]
33. Solovieva, Y.; Quintanar, R.; Lázaro, G. Efectos socioculturales sobre el desarrollo psicológico y neurológico en niños preescolares. *Cuad. Hispanoam. Psicol. México* **2006**, *6*, 9–20.
34. Comisión Nacional del Agua (CONAGUA). *Estadísticas Agrícolas de los Distritos de Riego, Año Agrícola 2017–2018*; CONAGUA: México D.F., Mexico, 2019. Available online: https://files.conagua.gob.mx/conagua/publicaciones/Publicaciones/EAUR_2017 -2018.pdf (accessed on 4 August 2021).
35. Bracamonte, A.; Valle, N.; Méndez, R. La nueva agricultura sonorense: Historia reciente de un viejo negocio. *Región Soc.* **2007**, *19*, 51–70. [CrossRef]
36. Secretaría de Agricultura, Ganadería, Recursos Hidráulicos, Pesca y Acuacultura (SAGHARPA). Programa Sectorial de Mediano Plazo de Agricultura, Ganadería, Recursos Hidráulicos, Pesca y Acuicultura 2016–2021. Hermosillo, Mexico. 2016. Available online: http://sagarhpa.sonora.gob.mx/portal_sagarhpa/images/archivos/PMP/PSMPAGRHPAPART1.pdf (accessed on 4 August 2021).
37. Camarena-Gómez, B.O.; Ochoa-Nogales, C.B.; Valenzuela-Quintanar, A.I. Comunicación y percepción del riego por compuestos orgánicos persistentes en jornaleros agrícolas de Sonora, México. *POLIS Rev. Latinoam.* **2014**, *13*, 275–300.
38. Bravo-Pérez, H.M.; Castro-Ramírez, J.C.; Magaña-Zamora, J.D.; Reyes-Martínez, A. Evaluación de políticas alternativas de suministro de agua en Hermosillo, Sonora, México. *Tecnol. Cien. Agua* **2013**, *4*, 163–169. Available online: http://www.scielo.org.mx/scielo.php?script=sci_arttext&pid=S2007-24222013000200011 (accessed on 25 July 2021).
39. Martínez-Peralta, C.M.; Moreno-Vázquez, L.M. Análisis de diseño institucional de las reglas génesis de la Asociación de Usuarios del DR 051-Costa de Hermosillo. *Estud. Soc.* **2016**, *47*, 41–69.

40. Moreno-Vázquez, J.L. Por Debajo del Agua. In *Sobreexplotación y Agotamiento del Acuífero de la Costa de Hermosillo, 1945–2005*; El Colegio de Sonora: Hermosillo, Mexico, 2005; p. 507. ISBN 9686755551.
41. Martínez-Peralta, C.M. El Dilema de los Comunes en la Gran Irrigación El Caso del Acuífero de la Costa de Hermosillo, Sonora, México, 1970−2010. Ph.D. Thesis, Colegio de Sonora (COLSON), Hermosillo, Mexico, 2014.
42. Pérez-López, E.P. Los Sobrevivientes del Desierto: Producción y Estrategias de Vida Entre los Ejidatarios de la Costa de Hermosillo, Sonora (1932–2010). Ph.D. Thesis, UAM-Xochimilco, México City, Mexico, 2011.
43. Diario Oficial. Ley de Aguas Nacionales. Secretaría de Agricultura y Recursos Hidráulicos. 1992. Available online: http://www.diputados.gob.mx/LeyesBiblio/ref/lan/LAN_orig_01dic92_ima.pdf (accessed on 4 August 2021).
44. Lastiri-Hernández, M.A.; Álvarez-Bernal, D.; Moncayo-Estrada, R.; Cruz-Cárdenas, G.; Silva García, J.T. Adoption of phytodesalination as a sustainable agricultural practice for improving the productivity of saline soils. *Environ. Dev. Sustain.* **2021**, *23*, 8798–8814. [CrossRef]
45. Ramırez-Arias, A.; Campos-Salazar, V.; Pineda-Pineda, J.; Fitz-Rodrıguez, E. Analysis of energy consumption for tomato production in low technology greenhouses of Mexico. *Acta Hortic.* **2020**, *1296*, 753–758. [CrossRef]
46. Diaz-Garza, A.M.; Fierro-Rivera, J.I.; Pacheco, A.; Schüßler, A.; Gradilla-Hernández, M.S.; Senés-Guerrero, C. Temporal Dynamics of Rhizobacteria Found in Pequin Pepper, Soybean, and Orange Trees Growing in a Semi-arid Ecosystem. *Front. Sustain. Food Syst.* **2020**, *419*, 602283. [CrossRef]
47. Gutiérrez-Moreno, M.; Sánchez-Atondo, A.; Mungaray-Moctezuma, A.; Salazar-Briones, C. Sustainability prospective for water resources in Northwestern Mexico: Use of recycled concrete for Agricultural purpose water supply. *Interciencia* **2020**, *45*, 370–377.
48. Pérez-Valladares, C.X.; Moreno-Calles, A.I.; Casas, A.; Rangel-Landa, S.; Blancas, J.; Caballero, J.; Velazquez, A. Ecological, cultural, and geographical implications of *Brahea dulcis* (Kunth) Mart. insights for sustainable management in Mexico. *Sustainability* **2020**, *12*, 412. [CrossRef]
49. Pérez, G.; Islas, J.; Guevara, M.; Suárez, R. The sustainable cultivation of Mexican nontoxic Jatropha curcas to produce biodiesel and food in marginal rural lands. *Sustainability* **2019**, *11*, 5823. [CrossRef]
50. Padilla-Bernal, L.E.; Lara-Herrera, A.; Vélez-Rodríguez, A.; Loureiro, M. Sustainability and environmental management in the Mexican vegetable sector. *Acta Hortic.* **2019**, *1258*, 163–170. [CrossRef]
51. Díaz-José, J.; Guevara-Hernandez, F.; Rodríguez-Larramendi, L.A.; Nahed-Toral, J.; Pinto-Ruiz, R.; Coss, A.L.; Aguirre-López, J.M. Vulnerability, innovation and social resilience in the maize (*Zea mays* L.) production: The case of the conservation tillage club of Chiapas, Mexico. *Trop. Subtrop. Agroecosyst.* **2018**, *21*, 399–408.
52. León, A.; Agustin, A.; Sulvaran, J. The myth behind sustainable african palm crop. Socio-environmental impacts of palm oil in Chiapas, Mexico. *Int. J. Ecol. Dev.* **2017**, *32*, 1–19.
53. Vallejo-Ramos, M.; Moreno-Calles, A.I.; Casas, A. TEK and biodiversity management in agroforestry systems of different socio-ecological contexts of the Tehuacán Valley. *J. Ethnobiol. Ethnomed.* **2016**, *12*, 31. [CrossRef]
54. García, L.B.; Dávila, J.P.; Acosta, F.O.; Lizán, S.S.; Acuña, I.J.; López, F.G. Degree of sustainability of rural development in subsistence, intermediate, and comercial farmers, under an autopoietic view point. *Rev. Cient.* **2009**, *19*, 650–658.
55. Wisser, D.; Frolking, S.; Douglas, E.M.; Fekete, B.M.; Schumann, A.H.; Vörösmarty, C.J. The significance of local water resources captured in small reservoirs for crop production—A global-scale analysis. *J. Hydrol.* **2010**, *384*, 264–275. [CrossRef]
56. Queirós, A.; Faria, D.; Almeida, F. Strengths and limitations of qualitative and quantitative research methods. *Eur. J. Educ. Stud.* **2017**, *3*, 369–387. [CrossRef]
57. Kuntosch, A.; König, B.; Bokelmann, W.A. Systemic Perspective to Horticultural Innovation—The Case of Energy Saving Innovations in German Horticulture Proc. II International Symposium on Horticulture in Europe ed J-C Mauget and S Godet. *Acta Hortic.* **2015**, *1099*, 503–510. [CrossRef]
58. Comisión Nacional del Agua (CONAGUA, 2020). Actualización de la Disponibilidad Media anual de Agua en el Acuífero Costa de Hermosillo (2619) Estado de Sonora 2020. Subdirección General Técnica, Gerencia de Aguas Subterráneas. Comisión Nacional del Agua: México D.F., Mexico. Available online: https://sigagis.conagua.gob.mx/gas1/Edos_Acuiferos_18/sonora/DR_2619.pdf (accessed on 6 August 2021).
59. Díaz-Caravantes, R.E.; Bravo-Peña, L.C.; Alatorre-Cejudo, L.C.; Sánchez-Flores, E. Presión antropogénica sobre el agua subterránea en México: Una aproximación geográfica. *Investig. Geográficas* **2013**, *82*, 93–103. Available online: http://www.scielo.org.mx/scielo.php?pid=S0188-46112013000300007&script=sci_abstract (accessed on 25 July 2021). [CrossRef]
60. Olavarrieta-Carmona, M.V. Beneficios de la cuota energética. Estudio de caso de la Costa de Hermosillo, Sonora, México, 2006–2007. *Región Soc.* **2010**, *22*, 146–164. Available online: http://www.scielo.org.mx/scielo.php?script=sci_arttext&pid=S1870-39252010000100007 (accessed on 25 July 2021). [CrossRef]
61. Hernández-Pérez, J.L. Los Cambios en el Patrón de Cultivos en Sonora a Partir del Proceso de Restauración Agrícola en México: El Caso de la Costa de Hermosillo. Master's Thesis, Centro de Investigación en Alimentación y Desarrollo (CIAD), Hermosillo, Mexico, 2012.
62. Manzanares-Rivera, J.L. Calidad de los recursos hídricos en el contexto de la actividad económica y patrones de salud en Sonora, México. *Salud Colect.* **2016**, *12*, 397–414. [CrossRef]
63. Flick, U. *Designing Qualitative Research*; SAGE Publications Ltd.: New York, NY, USA, 2007; ISBN 9780761949763. [CrossRef]
64. Grant, M.J.; Booth, A. A typology of reviews: An analysis of 14 review types and associated methodologies. *Health Inf. Libr. J.* **2009**, *26*, 91–108. [CrossRef]

65. Velten, S.; Leventon, J.; Jager, N.; Newig, J. What is sustainable agriculture? A systematic review. *Sustainability* **2015**, *7*, 7833–7865. [CrossRef]
66. Reiter, B. Theory and Methodology of Exploratory Social Science Research. *Int. J. Soc. Res. Methodol.* **2017**, *5*, 129–150. Available online: http://scholarcommons.usf.edu/gia_facpub/132 (accessed on 25 July 2021).
67. Qu, S.Q.; Dumay, J. The qualitative research interview. *Qual. Res. Acc. Manag.* **2011**, *8*, 238–264. [CrossRef]
68. Næss, P. Validating explanatory qualitative research: Enhancing the interpretation of interviews in urban planning and transportation research. *Appl. Mobilities* **2018**, *5*, 186–205. [CrossRef]
69. Rosenthal, M. Qualitative research methods: Why, when, and how to conduct interviews and focus groups in pharmacy research. *Curr. Pharm. Teach. Learn.* **2016**, *8*, 509–516. [CrossRef]
70. DiCicco-Bloom, B.; Crabtree, B.F. The qualitative research interview. *Med. Educ.* **2006**, *40*, 314–321. [CrossRef]
71. Ahmed, S.; Asraf, R.M. The workshop as a qualitative research approach: Lessons learnt from a "critical thinking through writing" workshop. *Turk. Online J. Des. Art Commun.* **2018**, *September 2018 Special Edition*, 1504–1510. [CrossRef]
72. MacMillan, D.C.; Marshall, K. The Delphi process—An expert-based approach to ecological modelling in data-poor environments. *Anim. Conserv.* **2006**, *9*, 11–19. [CrossRef]
73. Coleman, S.; Hurley, S.; Koliba, C.; Zia, A. Crowdsourced Delphis: Designing solutions to complex environmental problems with broad stakeholder participation. *Glob. Environ. Chang.* **2017**, *45*, 111–123. [CrossRef]
74. Oreszczyn, S.; Carr, S. Improving the link between policy research and practice: Using a scenario workshop as a qualitative research tool in the case of genetically modified crops. *Qual. Res.* **2008**, *8*, 473–497. [CrossRef]
75. Organización de las Naciones Unidas para la Agricultura y la Alimentación. *Captación y Almacenamiento de Agua de Lluvia—Opciones Técnicas Para la Agricultura Familiar en América Latina y el Caribe*; FAO: Santiago, Chile, 2013; ISBN 978-92-5-307580-5.
76. Velasco-Muñoz, J.F.; Aznar-Sánchez, J.A.; Batlles de la Fuente, A.; Fidelibus, M.D. Rainwater harvesting for agricultural irrigation: An analysis of global research. *Water* **2019**, *11*, 1320. [CrossRef]
77. Aznar-Sánchez, J.A.; Belmonte-Ureña, L.J.; Velasco-Muñoz, J.F.; Valera, D.L. Aquifer sustainability and the use of desalinated seawater for greenhouse irrigation in the Campo de Nijar, Southeast Spain. *Int. J. Environ. Res. Public Health* **2019**, *16*, 898. [CrossRef]
78. Carrillo-Huerta, M.M.; Gómez-Bretón, E. La tecnología en el uso sustentable del agua para riego en México. El caso del acuífero de Tecamachalco, Puebla, 2017. *Panor. Econ.* **2020**, *15*, 27–56. [CrossRef]
79. Dévora-Isiordia, G.E.; López, M.; Fimbres, G.; Álvarez, J.; Astorga, S. Desalación por ósmosis inversa y su aprovechamiento en la agricultura en el valle del Yaqui, Sonora, México. *Tecnol. Cien. Agua* **2016**, *3*, 155–169. Available online: http://www.scielo.org.mx/scielo.php?script=sci_arttext&pid=S2007-24222016000300155 (accessed on 25 July 2021).
80. Dévora-Isiordia, G.E.; González-Enríquez, R.; Ruiz-Cruz, S. Evaluación de procesos de desalinización y su desarrollo en México. *Tecnol. Cienc. Agua* **2013**, *4*, 27–46.
81. Cooley, H.; Phurisamban, R. The Cost of Alternative Water Supply and Efficiency Options in California. Pacific Institute: Oakland, CA, USA, 2016; ISBN 978-1-893790-75-9.
82. Loera-Alvarado, L.A.; Torres-Aquino, M.; Martínez-Montoya, J.F.; Cisneros-Almazán, R.; Martínez-Hernández, J.J. Calidad del agua de escorrentía para uso agrícola captada en bordos de almacenamiento. *Ecosistemas Recur. Agropecu.* **2019**, *6*, 283–295. [CrossRef]
83. López-Hernández, N.A.; Palacios-Vélez, O.L.; Anaya-Garduño, M.; Chávez-Morales, J.; Rubiños-Panta, J.E.; García-Carrillo, M. Diseño de sistemas de captación del agua de lluvia: Alternativa de abastecimiento hídrico. *Rev. Mex. Cienc. Agríc.* **2017**, *8*, 1433–1439. Available online: http://www.scielo.org.mx/scielo.php?script=sci_arttext&pid=S2007-09342017000601433&lng=es&tlng=es (accessed on 25 July 2021). [CrossRef]
84. Villa-Rodríguez, A.O.; Bracamonte-Sierra, A. Procesos de aprendizaje y modernización productiva en el agro noroeste de México: Los casos de la agricultura comercial de la Costa de Hermosillo, Sonora y la agricultura orgánica de la zona sur de Baja California Sur. *Estud. Front.* **2013**, *27*, 217–254. Available online: http://www.scielo.org.mx/scielo.php?script=sci_arttext&pid=S0187-6961 2013000100008 (accessed on 25 July 2021). [CrossRef]
85. Yáñez-Quijada, A.I.; Camarena-Gómez, B.O. Salud ambiental en localidades agrícolas expuestas en plaguicidas en Sonora. *Soc. Ambiente* **2019**, *7*, 55–82. [CrossRef]
86. Aznar-Sánchez, J.A.; Belmonte-Ureña, L.J.; Velasco-Muñoz, J.F.; Valera, D.L. Farmers' profiles and behaviours toward desalinated seawater for irrigation: Insights from South-east Spain. *J. Clean. Prod.* **2021**, *296*, 126568. [CrossRef]
87. Gobierno de Sonora. Desaladora Sonora. Available online: https://desaladora.sonora.gob.mx/ (accessed on 2 December 2021).

Article

Deficit Water Irrigation in an Almond Orchard Can Reduce Pest Damage

José Enrique González-Zamora *, Cristina Ruiz-Aranda, María Rebollo-Valera, Juan M. Rodríguez-Morales and Salvador Gutiérrez-Jiménez

Departamento de Agronomía, Universidad de Sevilla, 41013 Sevilla, Spain;
cristinaruizaranda@gmail.com (C.R.-A.); mariarv16@hotmail.com (M.R.-V.); islab_control@yahoo.es (J.M.R.-M.);
rockerodepara@hotmail.com (S.G.-J.)
* Correspondence: zamora@us.es

Abstract: Irrigated almond orchards in Spain are increasing in acreage, and it is pertinent to study the effect of deficit irrigation on the presence of pests, plant damage, and other arthropod communities. In an orchard examined from 2017 to 2020, arthropods and diseases were studied by visual sampling under two irrigation treatments (T1, control and T2, regulated deficit irrigation (RDI)). Univariate analysis showed no influence of irrigation on the aphid *Hyalopterus amygdali* (Blanchard) (Hemiptera: Aphididae) population and damage, but *Tetranychus urticae* Koch (Trombidiformes: Tetranychidae) damage on leaves was significantly less (50–60% reduction in damaged leaf area) in the T2 RDI treatment compared to the full irrigation T1 control in 2019 and 2020. Typhlocybinae (principal species *Asymmetrasca decedens* (Paoli) (Hemiptera: Cicadellidae)) population was also significantly lower under T2 RDI treatment. Chrysopidae and Phytoseiidae, important groups in the biological control of pests, were not affected by irrigation treatment. The most important diseases observed in the orchard were not, in general, affected by irrigation treatment. The multivariate principal response curves show significant differences between irrigation strategies in 2019 and 2020. In conclusion, irrigation schemes with restricted water use (such as T2 RDI) can help reduce the foliar damage of important pests and the abundance of other secondary pests in almond orchards.

Keywords: leaf damage; *Tetranychus urticae*; *Asymmetrasca decedens*; *Stigmina carpophila*; *Polystigma amygdalinum*

Citation: González-Zamora, J.E.; Ruiz-Aranda, C.; Rebollo-Valera, M.; Rodríguez-Morales, J.M.; Gutiérrez-Jiménez, S. Deficit Water Irrigation in an Almond Orchard Can Reduce Pest Damage. *Agronomy* **2021**, *11*, 2486. https://doi.org/10.3390/agronomy11122486

Academic Editors: Pantazis Georgiou and Dimitris Karpouzos

Received: 3 November 2021
Accepted: 3 December 2021
Published: 7 December 2021

1. Introduction

Water used for crop irrigation faces shortages in the near future due to lower rainfall in the Mediterranean basin and increased evapotranspiration, according to the latest study published by the European Environmental Agency on climate change, impacts, and vulnerability in Europe [1].

Almond is considered a drought-tolerant species and its response to water scarcity has been defined in many studies under deficit irrigation, which minimizes loss of production and increases fruit quality [2–5]. This, together with the good prices for fruit and the good future prospects [6], stimulated a steady growth in the total almond crop area in Spain by 36.3% between 2014 and 2020 [7,8]. This increase occurred mainly in new plantations that substituted other, less profitable, crops in areas with irrigation rights [9], which means an increase in irrigated almond acreage of 152.7% (118,202 ha in 2020, around 25% in Andalucía; MAPA 2015, 2020). Furthermore, studies of efficient water use were also extended to other typical Mediterranean crops such as olives [10–12].

Recommended deficit irrigation (DI) for Spanish and Portuguese almonds varies between 1300–1500 and 6500 $m^3 \cdot ha^{-1}$ of water [3,4,13,14], depending on the limitations imposed in the irrigation schedule and the objectives of the study. Reduced irrigation has an effect on kernel production per ha (with less irrigation, there is less kernel production), but the efficiency of water use (in terms of kernel production (kg) per water (m^3) used) is significantly improved [2–5,14]. Secondarily, deficit irrigation in almond has also led to studies on how it can affect the nutritional quality of the kernel, which generally find clear improvements in the nutritional and sensory qualities of kernels produced under such an irrigation regime [15–18].

Almond pests (and diseases) were little studied in Spain until this recent change in acreage and management, but new interest has led to studies on how they affect the crop in this new situation [19–23]. Briefly and focused in Andalucía, the main arthropod pests of almond orchards are aphids (mainly *Hyalopterus amygdali* (Blanchard)) and mites (especially *Tetranychus urticae* Koch, with other secondary species); other groups can have importance in some locations and moments, such as certain Hemiptera (*Asymmetrasca decedens* (Paoli), *Monosteira unicostata* (Mulsant and Rey), *Parlatoria oleae* (Colvée)), and sometimes Coleoptera (*Capnodis tenebrionis* (L.)) and Lepidoptera (*Anarsia lineatella* Zeller). The almond crop has a variety of pathogens that affect it, although the cultivated variety and management influence the severity of the damage. The most important diseases of the aerial part are fungi as *Colletotrichum acutatum* species complex, *Monilinia laxa* (Aderh. and Ruhland) Honey, *Stigmina carpophila* (Lév.) M.B. Ellis, *Taphrina deformans* (Berk.) Tul., and others, and in soil *Phytophthora* de Bary spp. and *Verticillium dahliae* Kleb. are the most important. In other areas where almonds are of particular importance, such as California (USA), studies on the main pests and diseases of the crop have been carried out for a long time [24–26].

Arthropod communities in plants (thus specific pests that affect crops) can be influenced by the water status of plants, which influences different physiological processes and nutritional quality, as many studies have revealed [27–33]. Therefore, a rational approach to sustainable use of water in different crops should include the effects on the most relevant biotic factors (pests and diseases) that affect the crops. The effects of deficit irrigation on arthropod populations and diseases are not usually considered in scientific production, but recent changes in crop management in Spain to more productive methods have promoted such studies in super-intensive olive [34,35] and irrigated almond [23,36] orchards.

The present study focused on how crop irrigation management can impact the presence and population of some arthropod pests and the damage they produce, as well as the effect on beneficial arthropods, in an almond orchard, and collaterally also the presence of the most important diseases observed during the study. Specifically, this study compared two irrigation regimes in an almond orchard over four years, providing a more complete view of their effects on the crop. The most important result obtained after this long-term study is that T2, with regulated deficit irrigation (RDI) treatment, produced a sensible reduction in damage inflicted by two-spotted spider mites and a smaller population of leafhoppers compared to T1, with more irrigation.

2. Materials and Methods

2.1. Experimental Design

The experiment was conducted in an orchard in Dos Hermanas (province of Sevilla, Spain), with coordinates 37°13.805′ N 5°54.823′ W. It has an area of 29,423 m², and the experiment was carried out on 7968 m². The orchard has 2 cultivated almond (*Prunus dulcis* (Mill) DA Webb) varieties, "Vairo" and "Guara", planted in paired lines, with a tree spacing of 6 m × 8 m, and the research was carried out with the cultivar "Vairo". The trees were 7 years old at the beginning of the experiment in 2017, which lasted until 2020 (a total of 4 years). The orchard was fertilized and controlled for pests, diseases, and weeds using the criteria of the owner and advisor technicians. The timing and products used in the 4 years are listed in Table A1 (Appendix A). Samplings were performed before the application schedule or several days after it to reduce contact with residues.

The statistical design used complete randomized blocks with 4 blocks and 2 irrigation treatments. Each experimental plot had 12 trees (4 rows with 3 trees in each row), with the 2 central trees in each plot used for sampling purposes (corresponding to the "Vairo" trees). A repetition of each irrigation treatment was randomly assigned within each block, making 4 repetitions of each irrigation treatment for the whole experiment.

This study focused on two irrigation strategies: T1, irrigation control, and T2, regulated deficit irrigation (RDI). The control plots received irrigation to avoid any water stress in the trees and to meet their evapotranspiration (ETc) needs. With RDI, the use of water was decreased during a specific growing state of the trees: water stress was applied during kernel filling (phase II), and the full irrigated conditions were maintained for the rest of the season (phase I, which ran from full bloom until the beginning of kernel filling, and phase III, the postharvest period). The RDI plots were irrigated according to this strategy, but limited to the water resources allowed by the Guadalquivir River Water Authority (Confederación Hidrográfica del Guadalquivir); this resulted in a 78% reduction in the total amount of water used for irrigation compared to the control treatment. The irrigation parameters for each treatment can be found in Tables A2 and A3 (Appendix B), and in more detail in Martín-Palomo et al. [14]. The average annual water irrigation provided in each treatment during the 4 years of this study was T1 = 594.0 ± 117.7 mm and T2 = 130.5 ± 18.1 mm.

2.2. Sampling Procedure

The sampling period was from March to September/October in each year of the study, except in 2020, when sampling started in mid-May when COVID-19 pandemic lockdown restrictions were relaxed. Sampling was performed biweekly, with 18 dates in 2017, 18 in 2018, 13 in 2019, and 14 in 2020.

The 2 central trees of each plot were scouted and 2 shoots (each around 6 cm, with 3–4 leaves) in each cardinal direction per tree (16 branches per plot were observed on each sampling date) were randomly selected in each sampling date; for statistical analysis, the mean of each cardinal direction was used, which means 4 values per plot, 16 per treatment, and 32 on each sampling date. The same procedure was followed for fruits when they were formed until harvest.

Visual sampling was carried out in different ways (Table 1): presence/absence of arthropods and diseases; presence/absence of symptoms of damage by feeding of some arthropods, and in some cases estimated leaf area damage (with an ordinal scale: 0, no damage; 1, 1–20% of surface damaged; 2, 21–50% of surface damaged; 3, >50% of surface damaged); and direct count of certain arthropods. Two diseases were easily detected on leaves and had an important presence: *Stigmina carpophila* (Lév.) M.B. Ellis (shothole blight, SB) and *Polystigma amygdalinum* P.F. Cannon (red leaf blotch, RLB).

Arthropod samples were taken to the laboratory to confirm or elucidate the species. The specimens were separated following different generic taxonomic guides [37,38] and specific works [20,39]. Several species were determined with the help of experts only in particular cases when they were important in relation to the crop. Samples of the most relevant specimens are kept in the laboratory collection.

2.3. Data Analysis

Repeated-measures ANOVA was used to analyze how the different observed variables were individually affected by irrigation treatment with the analysis of time-series abundance data. SPSS (v15.0 for Windows) was used to test whether irrigation treatment (between-subject effect, with two treatments), time (within-subject effect), and interaction of time and irrigation treatment were significant in the response variables for each year of the study. GLM analysis was also performed, pooling data from the 3 or 4 years for each response variable (using similar types of data) with treatment (fixed factor), year (random factor), and interaction treatment × year to test whether a general pattern was present. Data transformations [40] appear in Table 1.

A multivariate principal response curve (PRC) was used for synthesis and global observation of the possible effects of the treatments under study each year when multiple variables were concerned. This method was used in agricultural entomology [34,41–43] with the same objective of analyzing and interpreting the effect of treatment on a complex of observed variables (taxon or other).

Table 1. Most important parameters registered for almond sampling in 2017–2020, with type of measure and transformation used in statistical analysis.

	2017		2018 to 2020	
	Type of Measure	**Transformation**	**Type of Measure**	**Transformation**
Hyalopterus amygdali				
Population	Scale (0–4), then continous value	Log (x + 1)	Scale (0–4), then continous value	Log (x + 1)
Damage	Scale (0–3)		Scale (0–3)	
Tetranychus urticae				
Population	Proportion of organs occupied	Arcsin $\sqrt{p}$	Population count	Log (x + 1)
Damage	Scale (0–3), then proportion of leaf area with damage	Arcsin $\sqrt{p}$	Scale (0–3), then proportion of leaf area with damage	Arcsin $\sqrt{p}$
Asymmetrasca decedens				
Population	Proportion of organs occupied	Arcsin $\sqrt{p}$	Population count	Log (x + 1)
Damage	Proportion of organs with symptoms	Arcsin $\sqrt{p}$	Proportion of organs with symptoms	Arcsin $\sqrt{p}$
Phyllonorycter cerasicolella	Population count	$\sqrt{x}$	Population count	$\sqrt{x}$
Monosteira unicostata				
Population	Proportion of organs occupied	Arcsin $\sqrt{p}$	Population count	Log (x + 1)
Damage	Scale (0–3), then proportion of leaf surface with damage	Arcsin $\sqrt{p}$	Scale (0–3), then proportion of leaf surface with damage	Arcsin $\sqrt{p}$
Hemiberlesia rapax	Proportion of organs occupied	Arcsin $\sqrt{p}$	Proportion of organs occupied	Arcsin $\sqrt{p}$
Chrysopidae sp.	Population count	$\sqrt{x}$	Population count	$\sqrt{x}$
Euseius stipulatus	Population count	$\sqrt{x}$	Population count	$\sqrt{x}$
Other arthropods	Population count	$\sqrt{x}$	Population count	$\sqrt{x}$
Stigmina carpophila	Proportion of organs occupied	Arcsin $\sqrt{p}$	Proportion of organs occupied	Arcsin $\sqrt{p}$
Polystigma amygdalinum	Proportion of organs occupied	Arcsin $\sqrt{p}$	Proportion of organs occupied	Arcsin $\sqrt{p}$

In PRC, the community response under study is represented by a canonical coefficient, which measures the response to abundance by a designated control, expressed as deviations from a control community over time. The treatment designated as the control is represented by a horizontal line, which serves as a reference to assess its relationship with the other treatment [44]. PRC analysis generates a species weight (or weights of higher taxonomic groups and observations in our case), plotted on the right vertical axis; weights are used to indicate which ones follow the plotted community pattern, but only weights higher than |0.4–0.5| are considered significant [44]. A visual interpretation of the PRC graphs can be found in Auber et al. [45].

Quantitative tests to determine whether a PRC diagram displays significant variance due to treatment were performed in R (v3.6.3) with the package "vegan" (v2.5-2), which uses a Monte Carlo procedure to generate up to 999 permutations. Count data were transformed with log (x + 1), and presence/absence and leaf damage data were transformed with arcsin ($\sqrt{p}$) prior to the application of PRC.

3. Results and Discussion

Control of pests, diseases, and weeds was carried out following basically the criteria of the owner and technicians, and the authors did not interfere with them, although in some moments certain changes in the timing and products were suggested to coordinate our sampling schedule with the normal activity in the orchard. Although the effect of the pesticides could interfere with arthropods and diseases, they were applied throughout the orchard at the same time, and the only differential factor was the water used in the two irrigation treatments.

Four years of studying the effect of deficit irrigation in an almond orchard give a general idea (as shown in the PRC results) that the effect on arthropods and diseases was not clear at the beginning of the study, and, as happened in a similar study carried out in a super-intensive olive orchard [34], there was a progressive effect over subsequent years.

PRC showed a pattern in which a general effect of the irrigation treatment was not observed in the first two years of the study (2017 and 2018), either in population as density ($p = 0.323$ and $p = 0.413$, respectively, Figure 1a,b), population as proportion ($p = 0.893$ and $p = 0.348$, respectively, Figure 2a,b), or damage ($p = 0.457$ and $p = 0.106$, respectively, Figure 3a,b). In the third year (2019), PRC showed a significant effect of irrigation treatment on damage ($p = 0.011$; Figure 3c), especially due to the activity of *Tetranychus urticae* Koch (Trombidiformes: Tetranychidae), and population (as proportion) was almost significant ($p = 0.05$; Figure 2c). However, in the fourth year of the study (2020), the three PRCs showed a significant effect of irrigation treatment on population density ($p = 0.001$; Figure 1d), the population as a proportion ($p = 0.001$; Figure 2d), and damage ($p = 0.001$; Figure 3d), with particular importance of *T. urticae*, with higher population density and damage in treatment T1, as shown by its weight on the right vertical axis, which is always opposite to the canonical value of treatment T2. Other populations also had high weight values, especially in 2019 (*Stigmina*, whose weight is opposite to T2, indicating more presence in T1; Figure 2c) and 2020 (*Polystigma*, Typhlocibinae; Figure 2d), indicating an effect of irrigation treatment on their populations.

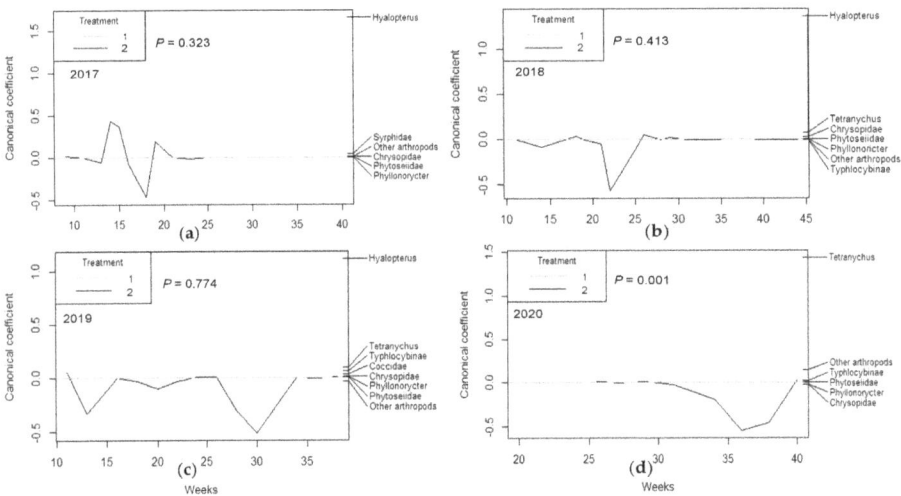

Figure 1. Principal response curves (PRCs) for the most important arthropod taxa whose population density was evaluated on almond shoots for (**a**) 2017, (**b**) 2018, (**c**) 2019, and (**d**) 2020. The *p* values denote the significance of treatment T2 (RDI), represented as a plotted line, relative to control treatment on all dates based on an F–type permutation test. Arthropod taxa are shown on the right vertical axis with their weights, which have the same scale as canonical coefficients on the left vertical axis.

The particular study of arthropods and diseases showed a differential response to irrigation treatment, hence the final perception of the effect of deficit irrigation. The most representative groups are described below.

The most important variables observed over four years are shown in Tables 2 and A4 (Appendix C). Aphid *Hyalopterus amygdali* (Blanchard) (Hemiptera: Aphididae) was one of the most important pests present in the orchard in the studied period. Its population was not statistically different between irrigation treatments in any year or taking all years together ($p = 0.707$; Table 2), and the damage observed was only significant in 2019 ($p < 0.01$; Table 2). The *H. amygdali* population and the damage it caused were not affected by the irrigation treatment in the three years during which this species was observed in this study. The irrigation treatment started in mid-March with the blossom stage, and the population normally had its peak in May, when deficit irrigation (in T2) was starting. The small difference in water used in both treatments (and thus the little stress produced) at the beginning of the season may explain the similar populations and damage observed in both treatments [30].

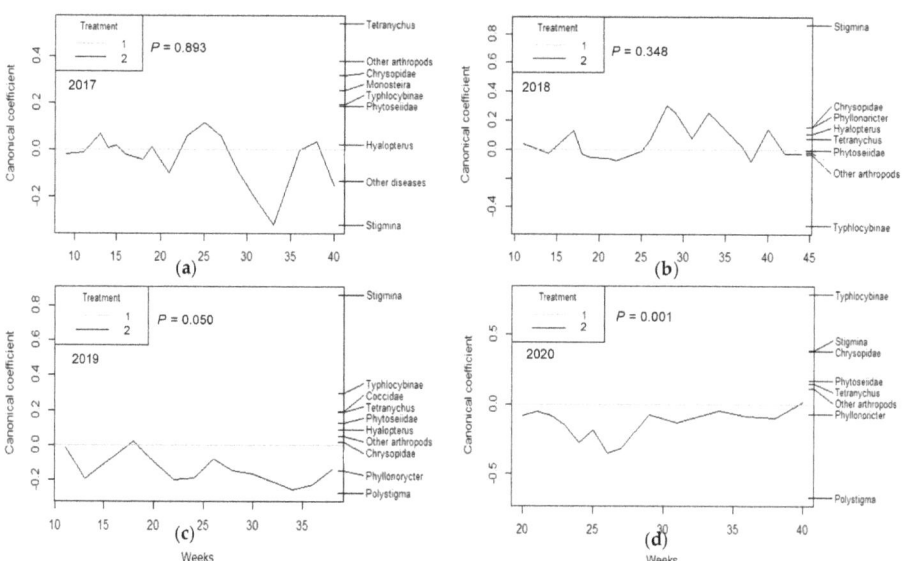

Figure 2. Principal response curves (PRCs) of the most important arthropod taxa and diseases, whose presence were evaluated as the proportion of almond shoots (or fruits) occupied for (**a**) 2017, (**b**) 2018, (**c**) 2019, and (**d**) 2020. The *p* values denote the significance of treatment T2 (RDI), represented as a plotted line, relative to control treatment on all dates based on an F−type permutation test. Arthropod taxa and diseases are shown on the right vertical axis with their weights, which have the same scale as canonical coefficients on the left vertical axis.

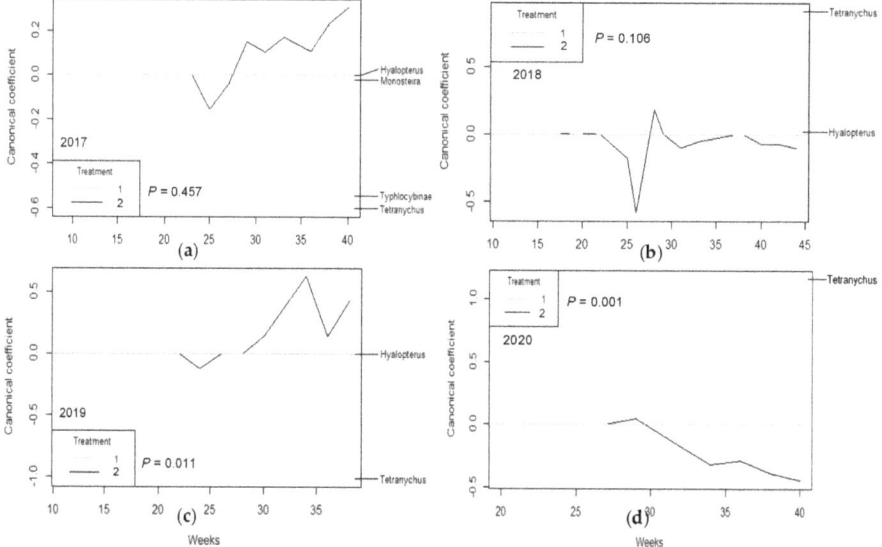

Figure 3. Principal response curves (PRCs) of most important arthropod taxa showing any type of damage on almond shoots for (**a**) 2017, (**b**) 2018, (**c**) 2019, and (**d**) 2020. The *p* values denote significance of treatment T2 (RDI), represented as a plotted line, relative to control treatment on all dates based on an F−type permutation test. Arthropod taxa are shown on the right vertical axis with their weights, which have the same scale as canonical coefficients on the left vertical axis.

Table 2. Significance (*p*) of treatment (Tr) and interaction treatment × sampling date (Tr × SD) for important parameters registered in almond sampling in 2017–2020.

	2017		2018		2019		2020		2017–2020		
	Tr	Tr × SD	Tr	Tr × SD	Tr	Tr × SD	Tr	Tr × SD	Tr	Year	Tr × Year
Hyalopterus amygdali											
Population	0.644	0.084	0.166	0.039 *	0.184	0.577	-[2]	-	0.707	0.089	0.181
Damage [1]	0.952	-	0.309	-	<0.01 **	-	-[2]	-	0.062	-	-
Tetranychus urticae											
Population	0.166	0.028 *	0.368	0.260	0.826	0.272	<0.01 **	<0.01 **	0.297	0.133	0.080
Damage	0.114	0.069	0.012 *	<0.01 **	<0.01 **	<0.01 **	<0.01 **	<0.01 **	0.075	0.180	<0.01 **
Asymmetrasca decedens	0.013 *	0.802	<0.01 **	0.034 *	<0.01 **	0.063	<0.01 **	<0.01 **	0.082	0.469	<0.01 **
Phyllonorycter cerasicolella	-[3]	-	0.714	0.269	0.814	0.042 *	0.312	0.624	0.621	0.562	0.575
Monosteira unicostata	0.509	0.771	-[3]	-	-[3]	-	-[3]	-			
Hemiberlesia rapax	-[3]	-	-[3]	-	0.108	0.760	-[3]	-			
Chrysopidae sp.	0.497	0.093	0.932	0.046 *	0.606	0.423	<0.01 **	0.030 *	0.175	0.030 *	0.584
Euseius stipulatus	0.781	0.115	0.269	0.100	0.072	0.382	0.186	0.097	0.316	<0.01 **	0.205
Other arthropods	0.117	0.214	0.417	0.700	0.403	0.302	0.070	0.313	0.319	0.057	0.231
Stigmina carpophila	0.795	0.187	0.096	0.052	0.014 *	0.308	0.232	0.107	0.420	0.025 *	<0.01 **
Polystigma amygdalinum	-[3]	-	-[3]	-	0.146	0.328	<0.01 **	0.341	0.128	0.113	0.326

Repeated-measures ANOVA was used to analyze most of the data in each individual year. GLM analysis was used to analyze the years together, although for some response variables (*Tetranychus urticae* population, *Asymmetrasca decedens*) the data included only three available years (2018 to 2020) to use similar population density data. [1] Damage produced by *Hyalopterus amygdali* was analyzed with the non-parametric Wilcoxon signed-rank test. [2] COVID-19 restrictions from 15 March to 15 May prevented adequate sampling of this insect. [3] Not present in the sampling period, or with such low presence that it was not included in the analysis. * $p < 0.05$; and ** $p < 0.01$.

Tetranychus urticae was present during the four years, but its population was not influenced much by irrigation treatment: only in 2020, the population was significantly higher in T1 than in T2 ($p < 0.01$; Table 2), and taking the density counts of the three years (2018–2020) together, there were no differences between treatments ($p = 0.297$; Table 2). However, the same was not the case for the damage observed in leaves: in three out of the four years (2018, 2019, and 2020; Table 2), there was a significant effect of the irrigation treatment, with a reduction in *T. urticae* damage in T2 compared with T1. Furthermore, although taking the four years together there was no significant effect of treatment ($p = 0.075$; Table 2), the treatment × year interaction was significant ($p < 0.01$; Table 2), indicating that in some years (that is, in 2019 and 2020) there were significant differences between irrigation treatments. More statistical results are presented in Table A4 (Appendix C).

The population of the two-spotted spider mite *T. urticae* was not (in general) affected by irrigation treatment, except in 2020 (Table 2, Figure 4), but the damage on leaves was different (Table 2, Figure 5): In 2019 there was low leaf surface damage, corresponding to low mite populations, but in 2018 and 2020 the damage was much more evident and substantial, reflecting that *T. urticae* is one of the most important pests in almond crops in Spain [20,46]. In some ways, this agrees with Hodson and Lampinen [47], who found that the *Tetranychus pacificus* McGregor population or damage increased with high water availability on leaves and decreased with intermediate water stress in different almond cultivars in California. Prgomet et al. [13] also observed that almond leaves with RDI treatment had less water availability compared to full irrigation treatment, although no mite interaction was studied.

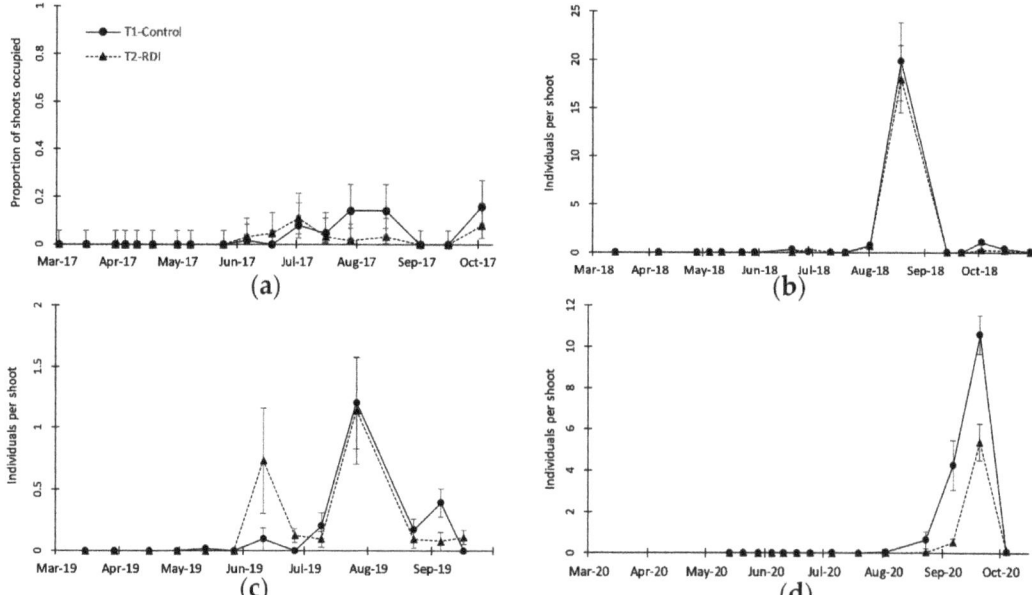

Figure 4. Seasonal patterns of *Tetranychus urticae* as (**a**) proportion of shoots occupied (2017) and (**b–d**) population density in shoots (2018, 2019, and 2020, respectively). Note the different y-scales. Solid line represents treatment T1 (Control) and dotted line represents treatment T2 (RDI). Vertical bars represent exact confidence interval of proportion (**a**) and standard error of the mean (**b–d**).

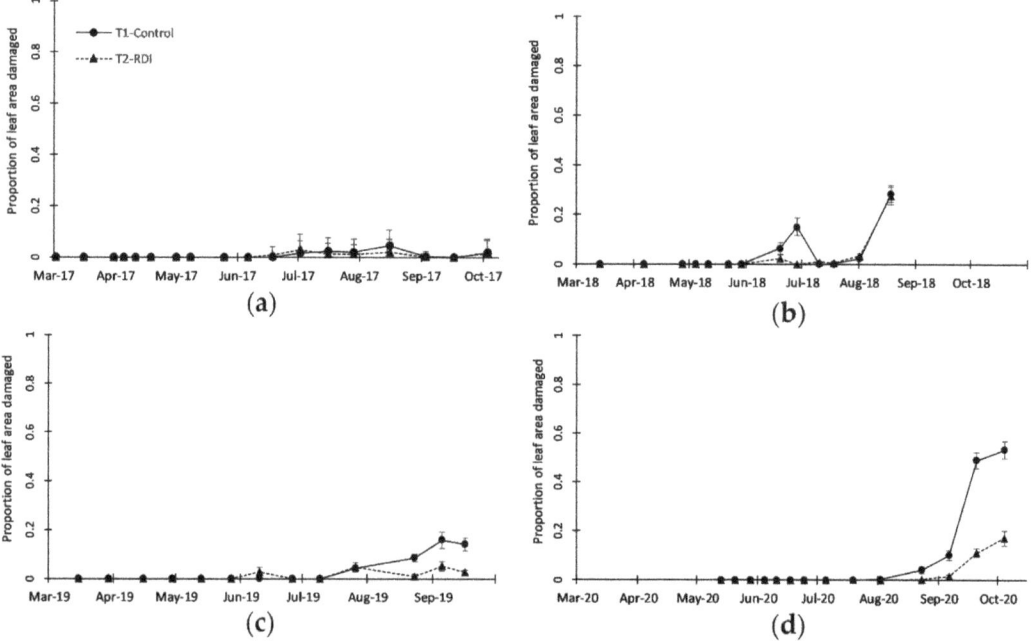

Figure 5. Seasonal patterns of *Tetranychus urticae* damage on leaves, measured as the proportion of leaf area with feeding symptoms for (**a**) 2017, (**b**) 2018, (**c**) 2019, and (**d**) 2020. Solid line represents treatment T1 (Control) and dotted line represents treatment T2 (RDI). Vertical bars represent standard error of the mean.

The *Tetranychus urticae* population was not clearly different between irrigation treatments from 2017 to 2019 (Figure 4a–c and Table 2) as mentioned above, and only in 2020, there were clear (and significant) differences between the treatments, with a higher population peak in T1 than in T2 (Figure 4d and Table 2). There was great variability in the densities reached in 2018–2020 (Figure 4b–d), from peaks of around 1 individual per shoot in 2019 (Figure 4c) to nearly 20 individuals per shoot in 2018 (Figure 4b). In 2017 and 2020 (Figure 4b,d) there was a significant effect of the sampling date (Table 2), showing that *T. urticae* increased its population from mid-June or later. Signs of damage on leaves produced by *T. urticae* also varied between years, from very low levels in 2017 (Figure 5a) to rather high levels in 2020 (Figure 5d), when nearly 55% of the leaf area was affected in irrigation treatment T1. Leaf damage was clearly observed in 2018, 2019, and 2020 (Figure 5b–d), showing significantly higher damage in T1 than in T2 over the three years (Table 2). Damage was more evident at the end of the sampling period (September), especially in 2019 and 2020 (Figure 5c,d), coinciding with the beginning of leaf abscission, reaching, in general, a 50–60% reduction in leaf damage in T2 compared with T1.

Other authors have studied the effect of irrigation on almond mite populations, such as Goldhammer et al. [2], who found no differences in *T. urticae* populations using different water doses (from 580 to 860 mm) and timing. However, it must be noted that they evaluated *T. urticae* in April–June (soon to develop mite populations) and their lowest irrigation treatment used an annual average of 580 mm (a 33% reduction compared with their highest treatment), similar to the highest irrigation treatment in this study. Youngman and Barnes [48] reported a more severe attack of spider mites on water-stressed almond trees, but it was not repeated in the second year. Using herbaceous plants, English-Loeb [27,49] observed an effect of irrigation treatment on a *T. urticae* population: the mites were more abundant in well-watered and severely stressed plants, and least abundant in slightly to moderately stressed plants, with a non-monotonic effect of water stress on their population, an effect about which Hodson and Lampinen [47] discussed. Studies conducted in soybean [50] showed no significant differences in spider mite populations in moisture-unstressed and stressed plants, but reductions in photosynthetic rate by spider mites were greater in the former. Studies have indicated that a moderate water stress level can save protein production, increasing other components that can play a role in the defense against phytophagous [33]. Other authors [28] have reported that during water-deficit stress, foliar nitrogen concentrations can increase in stressed plants, providing a valuable increase in nutritional quality for herbivores, but this can be counteracted by a reduction in water potential and water content, which can reduce herbivore feeding, especially those with piercing-sucking mouthparts.

The lower damage observed by *T. urticae* in the T2-RDI treatment may have consequences in the next season, because in the post-harvest period (see Tables A2 and A3), irrigation resumed to almost normal levels, and trees with less damage (especially in the leaves by spider mites) could store more nutrients to use for better blooming and sprouting in the next season [51].

Other secondary pests were present in the orchard, such as Typhlocybinae (with principal species *Asymmetrasca decedens* (Paoli) (Hemiptera, Cicadellidae)). In the four years, there were significant differences between irrigation treatments, with a greater presence (2017) or population (2018–2020) in T1, the more irrigated treatment, than in T2 ($p = 0.013$ and $p < 0.01$, respectively; Table 2). Taking the three years 2018 to 2020 together, there was no significant evidence of the effect of irrigation treatment ($p = 0.082$; Table 2), but the treatment × year interaction was significant ($p < 0.01$; Table 2) in 2018 and 2020. Only in 2017, this group caused damage, with significantly more damage in treatment T1 than in treatment T2 (not shown).

The presence of Typhlocybinae was constant during the four years of sampling, although always with low populations (no more than 0.35 individuals per shoot; Figure 6), and almost no damage was observed on leaves and shoots most of the time. The presence of this group was more noticeable from the end of May to July, and the differences in population between the two irrigation treatments were more important in 2018 and 2020 (Figure 6b,d), with a significantly higher population in T1 than in T2. Leafhoppers (Typhlocybinae) presented a small population in the orchard, but under T2, with less irrigation, had a significantly smaller population in each of the four years of study. This may be related to better resource availability in T1, and equally to the less suitable environment in less irrigated treatment, as suggested by Sconiers and Eubanks [28] about arthropods with piercing-sucking mouthparts. Several leafhoppers are known for their ability to transmit diseases in different crops [52], and if their population can be maintained at lower levels, then this can be considered a positive effect.

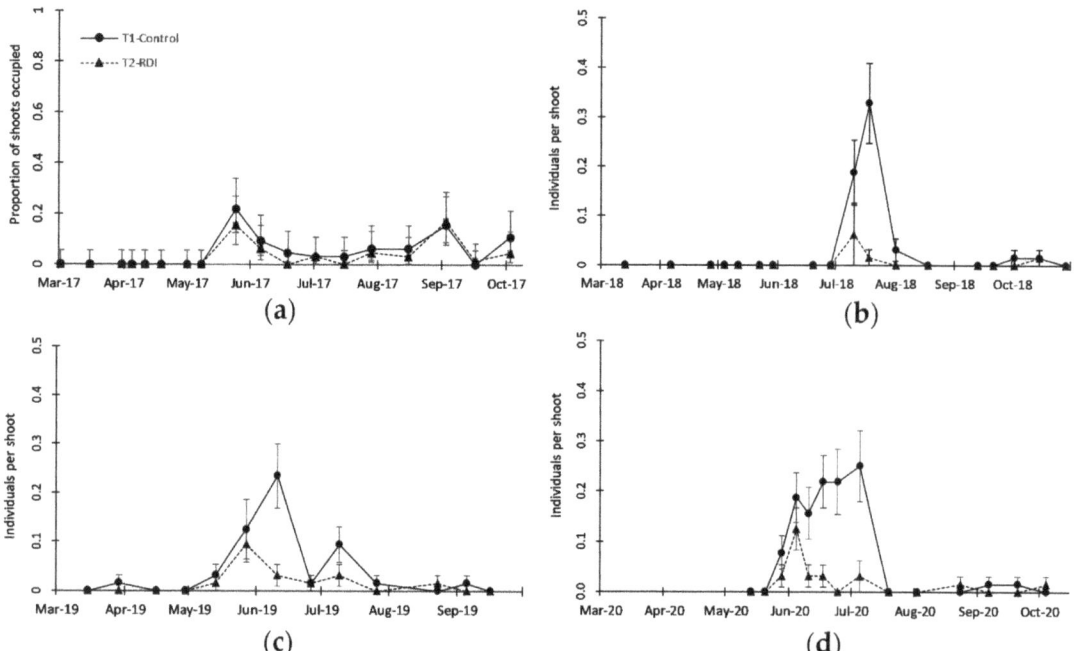

Figure 6. Seasonal patterns of leafhoppers (Typhlocybinae) in terms of (**a**) proportion of shoots occupied in 2017 and (**b–d**) population density on shoots in 2018, 2019 and 2020, respectively. Solid line represents treatment T1 (Control) and dotted line represents treatment T2 (RDI). Vertical bars represent exact confidence interval of (**a**) proportion and (**b–d**) standard error of the mean.

Regarding Lepidoptera (principal species was *Phyllonorycter cerasicolella* (Herrich-Schäffer) (Lepidoptera: Gracillariidae)), there were no differences in the three years individually or together ($p = 0.621$; Table 2). The same was the case with several species or groups relevant only in one year, such as *Monosteira unicostata* (Mulsant and Rey) (Hemiptera: Tingidae) in 2017 ($p = 0.509$; Table 2) and Coccidae (principal species *Hemiberlesia rapax* (Comstock) (Hemiptera: Diaspididae)) in 2019 ($p = 0.108$; Table 2). More statistical results are presented in Table A4 (Appendix C).

Two groups of natural enemies were consistently observed during the sampling period. Chrysopidae (order Neuroptera, with no species identified) was the most regularly observed in the orchard during the four years of the study; the population was similar in both irrigation treatments in three years, and taken together there was no effect of the irrigation treatment ($p = 0.175$; Table 2). Chrysopidae are among the most important predator groups in many crops and are also relevant in nut crops such as almonds [20,53], and in olives, no effect of irrigation treatment was observed on their population [34]. Phytoseiidae (most frequent species was *Euseius stipulatus* (Athias-Henriot) (Mesostigmata: Phytoseiidae)) was the second predator group to appear, periodically observed on leaves (mainly in May to June), and there was no effect of irrigation treatment on its population in the four years or when the years were analyzed together ($p = 0.316$; Table 2). More statistical results are presented in Table A4 (Appendix C).

Chrysopidae are generalist predators that can prey on a wide range of arthropods (small larvae/nymphs, their eggs, etc.), but were not observed preying on *T. urticae*, probably because of the heavy web produced by the mite, nor in general on any other pest observed in the study. Regarding phytoseids, *E. stipulatus* does not prey on mites that produce a lot of webs, as happens with *T. urticae*. Additionally, this phytoseid appears at the end of spring, for a short period of time, earlier than the mite, and feeds on small arthropods and secretion of the leaves. Although the presence of predators could influence the population of some arthropods, we do not think that their presence altered the effect that differential irrigation exerted on pest population or damage, as in *T. urticae* in 2020 (Figure 4d). These predators were present in both irrigation treatments, but the different water applied in each irrigation treatment was the definite factor that could affect arthropods.

The category "other arthropods" includes arthropod groups (mainly Coleoptera, Thysanoptera, Hymenoptera, Heteroptera and Araneae) of little quantitative or qualitative importance in the orchard, and in the four years there was no effect of irrigation treatment on their population, or when the four years were analyzed together ($p = 0.319$; Table 2). More statistical results are presented in Table A4 (Appendix C).

Symptoms of several diseases were observed in the orchard during the sampling period. The most frequent in the four years was *S. carpophila* (SB), but in three years there was no effect of irrigation treatment on its presence in the leaves, and no effect when the four years were taken together ($p = 0.420$; Table 2), although the treatment × year interaction was significant ($p < 0.01$; Table 2), namely in 2019. *Polystigma amygdalinum* (RLB) was less frequent, only clearly detected in 2019 (Table 2), without an effect of irrigation treatment on its presence, and 2020 ($p < 0.01$; Table 2), when T2 had more effect than T1, but taking both years together, the effect of treatment was not significant ($p = 0.128$; Table 2). More statistical results are presented in Table A4 (Appendix C).

Stigmina carpophila was the most remarkable disease in the orchard, with a constant presence during the four-year sampling period, reaching a large presence in some years (such as in 2018, with a peak of 80% of shoots with symptoms; Figure 7b). Only in 2019, there was a significant difference between irrigation treatments (Figure 7c and Table 2), with more symptoms in T1, with more irrigation, than in T2.

The two diseases most frequently observed in the orchard, *S. carpophila* (SB) and *P. amygdalinum* (RLB), did not present clear evidence of the effect of the irrigation treatment on their occurrence. *S. carpophila*, *P. amygdalinum*, and other diseases are common in almond crops [20,46], especially when the crop is managed in an intensive way, but the cultivar "Vairo" is not particularly affected by these two diseases, especially RLB [23,54].

The interest of deficit irrigation and its interaction with tree physiology and pests and diseases impact is also present in other parts of the world: Smith et al. [55] have studied the positive effect that a combination of RDI and early harvest has on several pests and diseases in California almonds, thus improving the long-term sustainability of the crop and IPM programs.

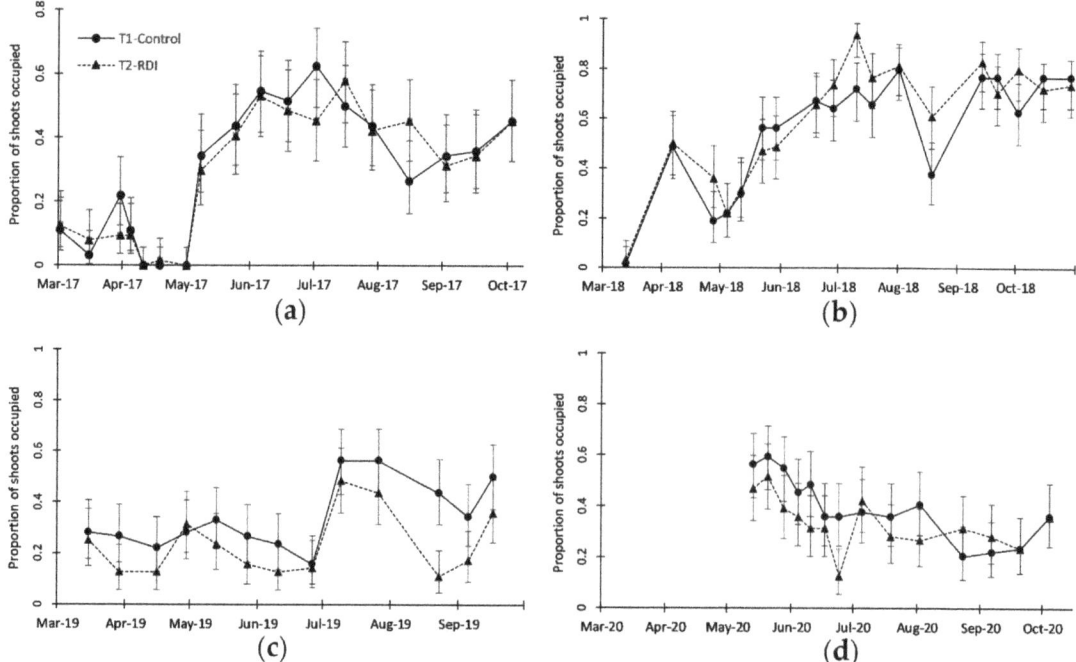

Figure 7. Seasonal patterns of *Stigmina carpophila* (shothole blight, SB) presence on leaves, measured as proportion of shoots with symptoms in (**a**) 2017, (**b**) 2018, (**c**) 2019 and (**d**) 2020. Solid line represents treatment T1 (Control), and dotted line represents treatment T2 (RDI). Vertical bars represent exact confidence intervals of proportions.

4. Conclusions

The conclusions obtained in this research are based on one orchard, but for a long period of four seasons, from 2017 to 2020. The main interest of this work is that the regulated deficit irrigation (RDI) treatment, T2, as applied in this study, produced a sensible reduction in damage to trees inflicted by *T. urticae* than the more irrigated treatment, T1. The positive effect of T2-RDI was also observed in the form of a smaller leafhopper population compared to T1. Furthermore, the irrigation treatments applied in this study did not have a differential effect on two groups of natural enemies (Chrysopidae, Phytoseiidae) or on the diseases observed in the orchard. Regulated deficit irrigation strategies help to reduce water use in crops, which in the case of almonds, implies better efficiency of the water used to obtain the harvest, and in addition to having a positive effect on reducing the presence and damage of several pests, also helps to regulate the vigor of the plants and, therefore, helps in different cultivation operations, such as pruning.

Author Contributions: Conceptualization and methodology, J.E.G.-Z.; formal analysis, J.E.G.-Z.; investigation, J.E.G.-Z., C.R.-A., M.R.-V., J.M.R.-M., S.G.-J.; resources, J.E.G.-Z.; data curation, J.E.G.-Z., C.R.-A., M.R.-V, J.M.R.-M., S.G.-J.; writing—original draft preparation, J.E.G.-Z.; writing—review and editing, J.E.G.-Z., C.R.-A., M.R.-V., J.M.R.-M., S.G.-J.; project administration, J.E.G.-Z. All authors have read and agreed to the published version of the manuscript.

Funding: This research was funded by the Ministerio de Economia y Competitividad (Ministry of Economy and Competitivity) within the project titled "Optimizacion de recursos hídricos limitados en cultivos leñosos mediterráneos principales: olivo y almendro" with code AGL2016-75794-C4-4-R.

Institutional Review Board Statement: Not applicable.

Informed Consent Statement: Not applicable.

Data Availability Statement: Not applicable.

Acknowledgments: The authors thank J.M. Durán and A. Serrano from the Entomology Laboratory of the Consejeria de Agricultura de la Junta de Andalucia in Montequinto (Sevilla, Spain) for their help in identifying some of the species cited in this work. Authors thank also the owners of the orchard "La Florida" (Dos Hermanas, Sevilla) for their collaboration in this work, specially to Carlos Angulo.

Conflicts of Interest: The authors declare no conflict of interest. The funders had no role in the design of the study; in the collection, analyses, or interpretation of data; in the writing of the manuscript, or in the decision to publish the results.

Appendix A

Table A1. Treatments against pests and diseases in the four years of the study.

Date	Product Used	Used against
31 January 2017	Copper oxichloride 52%	Diseases
	Paraffinic oil 83%	Eggs and immature arthropods
16–17 February 2017	Thiophanate-methyl 70%	Diseases
1–2 March 2017	Boscalid 26.7% + Pyraclostrobin 6.7%	Diseases
18–21 March 2017	Boscalid 26.7% + Pyraclostrobin 6.7%	Diseases
	Metconazol 9%	
	Mancozeb 75%	Diseases
	Deltamethrin 2.5%	Aphids
26 April 2017	Azoxystrobin 25%	Diseases
16–17 May 2017	Fluopyram 20% + Tebuconazole 20%	Diseases
	Tau-fluvalinate 24%	Aphids, leafhoppers
	Copper oxichloride 52%	Diseases
9–11 June 2017	Tau-fluvalinate 24%	
	Hexythiazox 10%	Two-spotted spider mite
	Abamectin 1.8 %	
19 August 2017	Thiram 50%	Diseases
9 September 2017	Imidacloprid 20%	Two-spotted spider mite, *Monosteira*,
	Dimethoate 40%	*Capnodis* (beetles)
7 March 2018	Fenbuconazole 2.5 %	Diseases
20 March 2018	Tebuconazole 50 % + Trifloxystrobin 25 %	Diseases
5 April 2018	Fluxapyroxad 7.5 % + Pyraclostrobin 15 %	Diseases
	Deltamethrin 2.5%	Aphids
	Fluopyram 20% + Tebuconazole 20%	Diseases
12 May 2018	Imidacloprid 20%	Aphids, leafhoppers, *Monosteira*,
		Capnodis (beetles)
9 July 2018	Deltamethrin 2.5%	Lepidoptera, leafhoppers
21 July 2018	Imidacloprid 20%	Aphids, leafhoppers, *Monosteira*,
		Capnodis (beetles)
	Thiacloprid 48 %	Lepidoptera
	Deltamethrin 2.5%	Lepidoptera, leafhoppers
	Fenpyroximate 5.12 %	Two-spotted spider mite
	Mancozeb 75%	Diseases
5 October 2018	Acetamiprid 20 %	Leafhoppers
14 February 2019	Thiophanate-methyl 70%	Diseases
	Copper oxichloride 52%	
15–17 March 2019	Boscalid 26.7% + Pyraclostrobin 6.7%	Diseases
	Mancozeb 75%	Diseases
	Trifloxystrobin 50%	
	Deltamethrin 2.5%	Aphids
12–13 April 2019	Folpet 40 %	Diseases
	Thiophanate-methyl 70%	
27 April 2019	Boscalid 26.7% + Pyraclostrobin 6.7%)	Diseases
	Copper oxichloride 52%	Diseases
20–21 June 2019	Tau-fluvalinate 24%	Two-spotted spider mite, *Monosteira*, leafhoppers

Table A1. *Cont.*

Date	Product Used	Used against
3–5 March 2020	Metconazole 9 %	
	Boscalid 26.7% + Pyraclostrobin 6.7%	Diseases
17–18 March 2020	Tebuconazole 25 %	Diseases
	Trifloxystrobin 50%	
	Deltamethrin 2.5%	
	Acetamiprid 20 %	Aphids
7 April 2020	Thiophanate-methyl 70%	Diseases
22 April 2020	Difenoconazole 4% + Isopyrazam 10%	Diseases
	Difenoconazole 25 %	
	Azoxystrobin 20 % + Cyproconazole 8 %	
	Copper oxichloride 52%	
	Mancozeb 75%	
20 May 2020	Dodine 40 %	Diseases
	Mancozeb 75%	
	Deltamethrin 2.5%	Aphids, leafhoppers
1 June 2020	Captan 47.5 %	Diseases
	Deltamethrin 2.5%	Leafhoppers
	Mancozeb 75%	Diseases
17 June 2020	Tau-fluvalinate 24%	Mites, leafhoppers, lepidoptera
	Fenpyroximate 5.12 %	
	Copper oxichloride 52%	Diseases
	Deltamethrin 2.5%	Leafhoppers, lepidoptera
	Acetamiprid 20 %	

Appendix B

The trees were irrigated with a line of drip emitters ($3.8 \text{ L} \cdot \text{h}^{-1}$) separated by 0.4 m. Irrigation scheduling was performed daily with a remote programming device (Ciclon, C-146 v 3.53, Maher, Almeria, Spain). This device controls each plot in the experimental orchard. Data from the previous day were used to change the current scheduling. Then, irrigation was changed daily and the water applied in RDI treatments was estimated according to the maximum daily shrinkage (MDS) of the trunk and stem water potential (SWP). Daily irrigation was based on the estimated maximum daily ETc (3 mm) when a difference of more than 30% of the threshold was measured and was reduced to 1.5 mm and 0.75 mm when the differences were between 20–30% and 10–20%, respectively. If the differences were less than 10% or the measured value indicated a better-than-expected water status, the trees were not irrigated. Irrigation was measured in each plot with a water meter at the beginning of the measured tree line.

The crop–water status was measured weekly using the midday stem water potential (ψ) and a pressure chamber (PMS Instrument Company, Albany, OR, USA) for two trees per plot in each irrigation treatment during the experiment (March to October in the four years of the study). A full description of the methodology is given in [14]. The irrigation plan followed in the treatments is provided in Table A2.

Table A2. Irrigation doses and stress levels in irrigation treatments.

Treatment	Phase	Irrigation [1]	Threshold I
1. Control		100% Et_c	
2. RDI	I (full bloom to kernel filling)	600 $\text{m}^3 \cdot \text{ha}^{-1}$	$\psi = -1.2$ Mpa; signal MDS = 1
	II (kernel filling to harvest)	100 $\text{m}^3 \cdot \text{ha}^{-1}$	$\psi = -2.0$ Mpa; signal MDS = 2.75
	III (post-harvest hydration)	300 $\text{m}^3 \cdot \text{ha}^{-1}$	$\psi = -1.2$ Mpa; signal MDS = 1

[1] Theorical amount of irrigation water to be received in treatment 2 (RDI) was around 1000 $\text{m}^3 \cdot \text{ha}^{-1}$ per year, but the final annual average of water used was 1305 $\text{m}^3 \cdot \text{ha}^{-1}$.

The water stress integral (Table A3) was calculated by Equation (A1) to describe the cumulative effect of deficit irrigation strategies in the irrigation period:

$$\text{SI} = |\Sigma(\psi - (-0.2)) \times n| \tag{A1}$$

where SI is the stress integral, ψ is the average midday stem water potential for any interval, and n is the number of days in the interval.

Table A3. Average water stress integral (SI, MPa $\times$ day) ($\pm$ SE) in irrigation treatments in the four years of the study.

Treatment	Year	Total	Phase I (until Fruit Filling)	Phase II (until Harvest)	Phase III (Post-Harvest Hydration)
1. Control	2017	128.6 $\pm$ 6.3	25.4 $\pm$ 1.6	54.2 $\pm$ 3.5	53.3 $\pm$ 3.9
	2018	99.4 $\pm$ 7.4	6.1 $\pm$ 0.8	53.4 $\pm$ 5.2	39.9 $\pm$ 2.7
	2019	67.2 $\pm$ 6.9	7.5 $\pm$ 1.4	30.6 $\pm$ 2.8	29.1 $\pm$ 3.0
	2020	84.1 $\pm$ 14.3	8.8 $\pm$ 2.6	40.0 $\pm$ 5.9	35.2 $\pm$ 6.6
2. RDI	2017	207.8 $\pm$ 12.1	31.7 $\pm$ 3.0	94.9 $\pm$ 4.9	86.3 $\pm$ 5.7
	2018	148.7 $\pm$ 21.9	7.9 $\pm$ 1.3	83.8 $\pm$ 10.3	57.0 $\pm$ 10.7
	2019	206.1 $\pm$ 12.4	14.8 $\pm$ 1.7	110.7 $\pm$ 10.4	80.6 $\pm$ 3.1
	2020	174.5 $\pm$ 14.5	18.6 $\pm$ 2.7	90.8 $\pm$ 7.1	65.1 $\pm$ 5.9

Phase II (from kernel filling to harvest) occurred in the following periods: day 151 to 221 (2017), day 168 to 245 (2018), day 135 to 225 (2019), day 149 to 224 (2020).

Appendix C

The between-subjects analysis of the repeated-measures ANOVA used the following factors: treatment (two levels of irrigation), block (four levels), and treatment $\times$ block. Tables 2 and A3 only show the results of the treatment factor.

There were four sets of observations for each plot (one for each cardinal direction), so the total number of observations is 32 for each sampling date. They are distributed in the ANOVA as treatment, 1 d.f.; block, 3 d.f.; treatment $\times$ block, 3 d.f.; error, 24 d.f.; and intersection, 1 d.f.

The within-subjects analysis in the repeated-measures ANOVA was performed with the time factor (sampling date) and its interaction with the other between-subject factors (treatment, block, and treatment $\times$ block), but only the results of treatment $\times$ sampling date are presented in Tables 2 and A3. First, we tested whether Mauchly's sphericity test was significant. In most cases, the test was significant and the Greenhouse–Geisser degree of freedom correction was applied.

A GLM analysis of the years taken together is also in the table, although for some response variables (*Tetranychus urticae* population, *Asymmetrasca decedens*) the data used included only three of the four available years (2018 to 2020), to use similar population density data. The factors were treatment (fixed factor), year (random factor), and interaction treatment $\times$ year. Degrees of freedom are explained in a footnote to the table.

Table A4. Statistics and degree of freedom of repeated-measures ANOVA in the four years of study showing F statistics of irrigation treatment (Tr) and interaction treatment × sampling date (Tr × SD) for the most important parameters registered.

	2017		2018		2019		2020		2017–2019		
	Tr	Tr × SD	Tr	Tr × SD	Tr	Tr × SD	Tr	Tr × SD	Tr	Year	Tr × Year
Hyalopterus amygdali											
Population	0.219	2.224 (3.4, 81.8)	2.038	2.630 (4.0, 94.8)	1.869	0.658 (2.9, 70.5)	–[2]	–	0.187	10.20	1.713
Damage[1]	Z = −0.061	–	Z = −1.016	–	Z = −2.690	–	–[2]	–	Z = −1.870		
Tetranychus urticae											
Population	2.045	2.441 (6, 144)	0.842	1.379 (2.4, 58.2)	0.049	1.333 (2.7, 65.7)	51.4	21.03 (2.7, 65.0)	1.953	6.520	2.528
Damage	2.690	2.235 (4.1, 98.5)	7.370	10.710 (3.5, 83.9)	27.77	9.152 (3.5, 83.5)	170.3	41.23 (3.3, 78.7)	7.147	3.249	4.642
Asymmetrasca decedens	7.12	0.469 (5.1, 122.2)	28.58	4.105 (1.5, 37.2)	8.084	2.443 (3.4, 81.1)	26.7	4.60 (4.6, 109.4)	9.880	1.133	5.606
Phyllonorycter cerasicolella	–[3]	–	0.138	1.298 (5, 120)	0.056	1.988 (9, 216)	1.07	0.738 (6.2, 148)	0.316	0.779	0.553
Monosteira unicostata	0.449	0.406 (3.4, 80.7)	–[3]	–	–[3]	–	–[3]	–			
Hemiberlesia rapax	–[3]	–	–[3]	–	2.793	0.494 (4.5, 107.1)	8.01	2.25 (7.3, 175.6)	2.996	13.593	0.649
Chrysopidae sp.	0.476	1.805 (6.7, 161.2)	0.007	1.776 (13, 299)	0.272	0.992 (4.7, 113.5)	1.86	1.87 (5.5, 132.4)	1.420	39.793	1.533
Euseius stipulatus	0.079	2.137 (2.5, 60.4)	1.279	2.026 (3.9, 92.6)	3.548	1.053 (3.7, 89.2)			1.417	8.385	1.435
Other arthropods	2.643	1.366 (8, 192)	0.683	0.618 (5.5, 131.2)	0.725	1.226 (4.9, 116.7)	3.59	1.21 (2.8, 66.7)	0.868	15.470	5.765
Stigmina carpophila	0.069	1.434 (7.8, 186.7)	2.996	1.886 (9.4, 226.5)	7.076	1.165 (12, 288)	1.50	1.52 (13, 312)	23.958	31.041	0.965
Polystigma amygdalinum	–[3]	–	–[3]	–	2.229	1.161 (7, 168)	21.3	1.14 (7.5, 181.1)			

Degrees of freedom for F statistics of treatment (Tr) in separate years: 1, 24. Degrees of freedom for F statistics of treatment × sampling date (Tr × SD) are in brackets after applying the Greenhouse–Geisser's degree of freedom correction. Degree of freedom for F statistics in the four years together: Tr (1, n° years-1), Y (n° years-1), Tr x Y (n° years-1, >700). [1] Damage produced by *Hyalopterus amygdali* was analyzed with non-parametric Wilcoxon signed-rank test. [2] COVID-19 restrictions from 15 March to 15 May prevented adequate sampling of this insect. [3] Not present in the sampling period, or with such low presence that it was not included in the analysis.

References

1. EEA Climate Change, Impacts and Vulnerability in Europe 2016—European Environment Agency. Available online: https://www.eea.europa.eu/publications/climate-change-impacts-and-vulnerability-2016 (accessed on 5 October 2021).
2. Goldhamer, D.A.; Viveros, M.; Salinas, M. Regulated deficit irrigation in almonds: Effects of variations in applied water and stress timing on yield and yield components. *Irrig. Sci.* **2006**, *24*, 101–114. [CrossRef]
3. Egea, G.; Nortes, P.A.; Domingo, R.; Baille, A.; Pérez-Pastor, A.; González-Real, M.M. Almond agronomic response to long-term deficit irrigation applied since orchard establishment. *Irrig. Sci.* **2013**, *31*, 445–454. [CrossRef]
4. Gutiérrez-Gordillo, S.; Durán-Zuazo, V.H.; García-Tejero, I.F. Response of three almond cultivars subjected to different irrigation regimes in Guadalquivir river basin. *Agric. Water Manag.* **2019**, *222*, 72–81. [CrossRef]
5. Gutiérrez-Gordillo, S.; Lipan, L.; Zuazo, V.H.D.; Sendra, E.; Hernández, F.; Hernández-Zazueta, M.S.; Carbonell-Barrachina, Á.A.; García-Tejero, I.F. Deficit irrigation as a suitable strategy to enhance the nutritional composition of hydrosos almonds. *Water* **2020**, *12*, 3336. [CrossRef]
6. MAPA Serie Histórica Almendro 2014–2018. Available online: https://www.mapa.gob.es/estadistica/pags/anuario/2019/CAPITULOSPDF/CAPITULO07/pdfc07_10.1.1.pdf (accessed on 21 October 2021).
7. MAPA Superficie y Producción de Almendro 2014. Available online: https://www.mapa.gob.es/estadistica/pags/anuario/2015/TABLAS%20PDF/CAPITULO%2013/pdfc13_10.1.2.pdf (accessed on 21 October 2021).
8. MAPA Avance de Superficie y Producción de Almendro 2020. Available online: https://www.mapa.gob.es/es/estadistica/temas/estadisticas-agrarias/agricultura/superficies-producciones-anuales-cultivos/ (accessed on 21 October 2021).
9. Junta de Andalucia Caracterización del Sector de la Almendra en Andalucía. Available online: https://www.juntadeandalucia.es/export/drupaljda/estudios_informes/16/12/Caracterizaci%C3%B3n%20del%20sector%20de%20la%20almendra_0.pdf (accessed on 21 October 2021).
10. Fernández, J.E.; Perez-Martin, A.; Torres-Ruiz, J.M.; Cuevas, M.V.; Rodriguez-Dominguez, C.M.; Elsayed-Farag, S.; Morales-Sillero, A.; García, J.M.; Hernandez-Santana, V.; Diaz-Espejo, A. A regulated deficit irrigation strategy for hedgerow olive orchards with high plant density. *Plant Soil* **2013**, *372*, 279–295. [CrossRef]
11. Gómez del Campo, M.; García, J.M. Summer Deficit-Irrigation Strategies in a Hedgerow Olive cv. Arbequina Orchard: Effect on Oil Quality. *J. Agric. Food Chem.* **2013**, *61*, 8899–8905. [CrossRef] [PubMed]
12. Moriana, A.; Pérez-López, D.; Prieto, M.H.; Ramírez-Santa-Pau, M.; Pérez-Rodriguez, J.M. Midday stem water potential as a useful tool for estimating irrigation requirements in olive trees. *Agric. Water Manag.* **2012**, *112*, 43–54. [CrossRef]
13. Prgomet, I.; Pascual-Seva, N.; Morais, M.C.; Aires, A.; Barreales, D.; Castro Ribeiro, A.; Silva, A.P.; Barros, A.I.; Gonçalves, B. Physiological and biochemical performance of almond trees under deficit irrigation. *Sci. Hortic.* **2020**, *261*, 108990. [CrossRef]
14. Martín-Palomo, M.J.; Corell, M.; Girón, I.; Andreu, L.; Trigo, E.; López-Moreno, Y.E.; Torrecillas, A.; Centeno, A.; Pérez-López, D.; Moriana, A. Limitations of using trunk diameter fluctuations for deficit irrigation scheduling in almond orchards. *Agric. Water Manag.* **2019**, *218*, 115–123. [CrossRef]
15. García-Tejero, I.F.; Lipan, L.; Gutiérrez-Gordillo, S.; Durán Zuazo, V.H.; Jančo, I.; Hernández, F.; Cárceles Rodríguez, B.; Carbonell-Barrachina, Á.A. Deficit Irrigation and Its Implications for HydroSOStainable Almond Production. *Agronomy* **2020**, *10*, 1632. [CrossRef]
16. Lipan, L.; Cano-Lamadrid, M.; Hernández, F.; Sendra, E.; Corell, M.; Vázquez-Araújo, L.; Moriana, A.; Carbonell-Barrachina, Á.A. Long-term correlation between water deficit and quality markers in hydrosostainable almonds. *Agronomy* **2020**, *10*, 1470. [CrossRef]
17. Lipan, L.; Moriana, A.; López-Lluch, D.; Cano-Lamadrid, M.; Sendra, E.; Hernández, F.; Vázquez-Araújo, L.; Corell, M.; Carbonell-Barrachina, A. Nutrition Quality Parameters of Almonds as Affected by Deficit Irrigation Strategies. *Molecules* **2019**, *24*, 2646. [CrossRef]
18. Lipan, L.; Collado-González, J.; Wojdyło, A.; Domínguez-Perles, R.; Gil-Izquierdo, Á.; Corell, M.; Moriana, A.; Cano-Lamadrid, M.; Carbonell-Barrachina, Á. How does water stress affect the low molecular weight phenolics of hydroSOStainable almonds? *Food Chem.* **2021**, *339*, 127756. [CrossRef] [PubMed]
19. Torguet Pomar, L.; Batlle Caravaca, I.; Alegre, S.; Miarnau i Prim, X. Nuevas plagas y enfermedades emergentes, una amenaza para el cultivo del almendro en España. *Rev. Frutic.* **2016**, *49*, 152–165.
20. Durán Alvaro, J.M.; Cabello Yuste, J.; Fernández Gonzalez, M.I.; Flores González, R.; Morera Oliveros, B.; Páez Sánchez, J.I.; Sánchez Megías, A.; Serrano Caballos, A.; Vega Guillén, J.M. *Plagas y Enfermedades del Almendro*; Junta de Andalucía, Consejeria de Agricultura, Pesca y Desarrollo Sostenible, Eds.; Secretaria General Técnica. Servicio de Publicaciones y Divulgación: Sevilla, España, 2017; Available online: https://www.juntadeandalucia.es/export/cdn-micrositios/documents/71753/17493429/Plagas+y+enfermedades+del+almendro/47775ba0-7ef6-44ab-8478-56a15c82007f (accessed on 5 October 2021).
21. Sánchez-Ramos, I.; Pascual, S.; Fernández, C.E.; Marcotegui, A.; González-Núñez, M. Effect of temperature on the survival and development of the immature stages of Monosteira unicostata (Hemiptera: Tingidae). *Eur. J. Entomol.* **2015**, *112*, 664–675. [CrossRef]
22. Ollero-Lara, A.; López-Moral, A.; Lovera Manzanares, M.; Raya Ortega, M.C.; Roca Castillo, L.F.; Arquero Quilez, O.; Trapero, A. Las enfermedades del almendro en Andalucía. *Rev. Frutic.* **2016**, 166–183.

23. Ollero-Lara, A.; Agustí-Brisach, C.; Lovera, M.; Roca, L.F.; Arquero, O.; Trapero, A. Field susceptibility of almond cultivars to the four most common aerial fungal diseases in southern Spain. *Crop Prot.* **2019**, *121*, 18–27. [CrossRef]
24. Welter, S.C.; Barnes, M.M.; Ting, I.P.; Hayashi, J.T. Impact of Various Levels of Late-Season Spider Mite (Acari: Tetranychidae) Feeding Damage on Almond Growth and Yield. *Environ. Entomol.* **1984**, *13*, 52–55. [CrossRef]
25. Shorey, H.H.; Gerber, R.G. Use of Puffers for Disruption of Sex Pheromone Communication Among Navel Orangeworm Moths (Lepidoptera: Pyralidae) in Ahnonds, Pistachios, and Walnuts. *Environ. Entomol.* **1996**, *25*, 1154–1157. [CrossRef]
26. Cabrera-La Rosa, J.C.; Johnson, M.W.; Civerolo, E.L.; Chen, J.; Groves, R.L. Seasonal population dynamics of Draeculacephala minerva (Hemiptera: Cicadellidae) and transmission of Xylella fastidiosa. *J. Econ. Entomol.* **2008**, *101*, 1105–1113. [CrossRef] [PubMed]
27. English-Loeb, G.M. Plant Drought Stress and Outbreaks of Spider Mites: A Field Test. *Ecology* **1990**, *71*, 1401–1411. [CrossRef]
28. Sconiers, W.B.; Eubanks, M.D. Not all droughts are created equal? The effects of stress severity on insect herbivore abundance. *Arthropod. Plant. Interact.* **2017**, *11*, 45–60. [CrossRef]
29. Frampton, G.K.; Van Den Brink, P.J.; Gould, P.J.L.L. Effects of spring drought and irrigation on farmland arthropods in southern Britain. *J. Appl. Ecol.* **2000**, *37*, 865–883. [CrossRef]
30. Mody, K.; Eichenberger, D.; Dorn, S. Stress magnitude matters: Different intensities of pulsed water stress produce non-monotonic resistance responses of host plants to insect herbivores. *Ecol. Entomol.* **2009**, *34*, 133–143. [CrossRef]
31. Weldegergis, B.T.; Zhu, F.; Poelman, E.H.; Dicke, M. Drought stress affects plant metabolites and herbivore preference but not host location by its parasitoids. *Oecologia* **2015**, *177*, 701–713. [CrossRef] [PubMed]
32. Tariq, M.; Wright, D.J.; Rossiter, J.T.; Staley, J.T. Aphids in a changing world: Testing the plant stress, plant vigour and pulsed stress hypotheses. *Agric. For. Entomol.* **2012**, *14*, 177–185. [CrossRef]
33. Gely, C.; Laurance, S.G.W.; Stork, N.E. How do herbivorous insects respond to drought stress in trees? *Biol. Rev.* **2020**, *95*, 434–448. [CrossRef]
34. González-Zamora, J.E.; Alonso-López, M.T.; Gómez-Regife, Y.; Ruiz-Muñoz, S. Decreased water use in a super-intensive olive orchard mediates arthropod populations and pest damage. *Agronomy* **2021**, *11*, 1337. [CrossRef]
35. Agustí-Brisach, C.; Jiménez-Urbano, J.P.; del Carmen Raya, M.; López-Moral, A.; Trapero, A. Vascular fungi associated with branch dieback of olive in super-high-density systems in Southern Spain. *Plant Dis.* **2021**, *105*, 797–818. [CrossRef]
36. Agustí-Brisach, C.; Moldero, D.; Raya, M.D.C.; Lorite, I.J.; Orgaz, F.; Trapero, A. Water stress enhances the progression of branch dieback and almond decline under field conditions. *Plants* **2020**, *9*, 1213. [CrossRef]
37. Chinery, M. *Guía de Campo de los Insectos de España y de Europa*; Omega: Barcelona, Spain, 2005; ISBN 84-282-0469-1.
38. Barrientos, J.A. *Bases Para un Curso Práctico de Entomología*; Asociación Española de Entomología: Salamanca, Spain, 1988; ISBN 84-404-2417-5.
39. Ferragut, F.; Pérez-Moreno, I.; Iraola, V.; Escudero, A. *Ácaros Depredadores de la Familia Phytoseiidae en las Plantas Cultivadas*; Ediciones Agrotécnicas S.L.: Madrid, Spain, 2010; ISBN 9788487480539.
40. Perry, J.N. Statistical aspects of field experiments. In *Methods in Ecological and Agricultural Entomology*; Dent, D.R., Walton, M.P., Eds.; CAB International: Oxford, UK, 1997; pp. 171–202. ISBN 0851991327.
41. Prasifka, J.R.; Hellmich, R.L.; Dively, G.P.; Lewis, L.C. Assessing the Effects of Pest Management on Nontarget Arthropods: The Influence of Plot Size and Isolation. *Environ. Entomol.* **2005**, *34*, 1181–1192. [CrossRef]
42. Whitehouse, M.E.A.; Wilson, L.J.; Fitt, G.P. A comparison of arthropod communities in transgenic Bt and conventional cotton in Australia. *Environ. Entomol.* **2005**, *34*, 1224–1241. [CrossRef]
43. Naranjo, S.E. Long-term assesment of the effects of transgenic Bt cotton on the abundance of nontarget arthropod natural enemies. *Environ. Entomol.* **2005**, *34*, 1193–1210. [CrossRef]
44. Van Den Brink, P.J.; Ter Braak, C.J.F. Principal response curves: Analysis of time-dependent multivariate responses of biological community to stress. *Environ. Toxicol. Chem.* **1999**, *18*, 138–148. [CrossRef]
45. Auber, A.; Travers-Trolet, M.; Villanueva, M.C.; Ernande, B. A new application of principal response curves for summarizing abrupt and cyclic shifts of communities over space. *Ecosphere* **2017**, *8*, e02023. [CrossRef]
46. Gil Martin, A.; Arrivas Carrasco, G.; Barrios Sanroma, G. *Guía de Gestión Integrada de Plagas: Almendro*; Ministerio de Agricultura Alimentación y Medio Ambiente, Ed.; Secretaría General Técnica. Centro de Publicaciones: Madrid, Spain, 2015; ISBN 9788449114434.
47. Hodson, A.K.; Lampinen, B.D. Effects of cultivar and leaf traits on the abundance of Pacific spider mites in almond orchards. *Arthropod. Plant. Interact.* **2018**, *13*, 453–463. [CrossRef]
48. Youngman, R.R.; Barnes, M.M. Interaction of Spider Mites (Acari: Tetranychidae) and Water Stress on Gas-exchange Rates and Water Potential of Almond Leaves. *Environ. Entomol.* **1986**, *15*, 594–600. [CrossRef]
49. English-Loeb, G.M. Nonlinear responses of spider mites to drought-stressed host plants. *Ecol. Entomol.* **1989**, *14*, 45–55. [CrossRef]
50. Haile, F.J.; Higley, L.G. Changes in Soybean Gas-Exchange After Moisture Stress and Spider Mite Injury. *Environ. Entomol.* **2003**, *32*, 433–440. [CrossRef]
51. Goldhamer, D.A.; Viveros, M. Effects of preharvest irrigation cutoff durations and postharvest water deprivation on almond tree performance. *Irrig. Sci.* **2000**, *19*, 125–131. [CrossRef]
52. Daane, K.M.; Wistrom, C.M.; Shapland, E.B.; Sisterson, M.S. Seasonal Abundance of Draeculacephala minerva and Other Xylella fastidiosa Vectors in California Almond Orchards and Vineyards. *J. Econ. Entomol.* **2011**, *104*, 367–374. [CrossRef] [PubMed]

53. Szentkirályi, F. Lacewings in fruit and nut crops. In *Lacewings in the Crop Environment*; McEwen, P., New, T., Whittington, A., Eds.; Cambridge University Press: Cambridge, UK, 2001; pp. 172–238.
54. Marimon, N.; Luque, J.; Vargas, F.J.; Alegre, S.; Miarnau, X. Susceptibilidad varietal a la 'mancha ocre' (Polystigma ochraceum (Whalenb.) Sacc.) en el cultivo del almendro. (Poster) Generalitat de Catalunya, IRTA-Investigación y Tecnología Agroalimentaria: Barcelona, Spain. 2011. Available online: https://www.researchgate.net/profile/Jordi-Luque/publication/234297104_Susceptibilidad_varietal_a_la_\T1\textquoterightmancha_ocre\T1\textquoteright_Polystigma_ochraceum_Whalenb_Sacc_en_el_cultivo_del_almendro/links/5704e75608ae13eb88b812d8/Susceptibilidad-varietal-a-la-mancha-ocre-Polyst (accessed on 5 October 2021).
55. Smith, E.E.; Brown, P.H.; Andrews, E.M.; Shackel, K.A.; Holtz, B.A.; Rivers, D.J.; Haviland, D.R.; Khalsa, S.D.S. Early almond harvest as a strategy for sustainable irrigation, pest and disease management. *Sci. Hortic.* **2022**, *293*, 110651. [CrossRef]

MDPI AG
Grosspeteranlage 5
4052 Basel
Switzerland
Tel.: +41 61 683 77 34

Agronomy Editorial Office
E-mail: agronomy@mdpi.com
www.mdpi.com/journal/agronomy